UNDERSTANDING
PURE AND APPLIED
MATHS II
FOR ADVANCED LEVEL

A Dawson and R Parsons

Stanley Thornes (Publishers) Ltd

First published 1990

© A Dawson and R Parsons 1990

© Original line illustrations
Stanley Thornes (Publishers) Ltd

Phototypesetting by
Cotswold Typesetting Ltd, Gloucester

Printed and bound in Great Britain at
The Bath Press, Avon.

British Library Cataloguing in Publication Data
Dawson, A. (Anthony)
  Understanding pure and applied mathematics II.
  1. Mathematics
  I. Title   II. Parsons, R. A. (Roderick A.)
  510
  ISBN 0-7487-0215-6

Acknowledgements

The publishers would like to thank the following for
permission to reproduce questions from past examination
papers:

  Joint Matriculation Board
  Northern Ireland Schools Examination Council
  Oxford and Cambridge Schools Examination Board
  University of London School Examinations Board
  Welsh Joint Education Committee

and the following for providing photographs:

  All-Sport
  Aston Martin
  British Coal
  Casio Electronics
  National Aeronautics and Space Administration, USA
  North Scotland Hydro-Electric Board
  Popperfoto
  Pyramids Centre, Southsea
  The Science Museum, London
  Clare Starkey
  Watchmaker, Jeweller and Silversmith

The authors wish to thank Mr D Crocker, now Vice Principal
of Queen Mary Sixth Form College, Basingstoke, for his
advice and for reading the manuscripts of the two books.

Contents

Contents

Introduction

This book is the second in the series designed for students studying Advanced Level Pure and Applied Mathematics. It aims to:

(a) extend some of the concepts, ideas and techniques introduced in Book 1;

(b) cover the remaining topics required for the Advanced Level syllabus;

(c) offer students the opportunity to revise and practise some of the work covered in Book 1 through revision exercises;

(d) provide some help on examination techniques and give some past examination questions for students to attempt;

(e) be a possible text for the Advanced Supplementary examination.

It is written in a clear, straightforward style that will be helpful to students using it as a class textbook or to those working largely on their own. Many diagrams, graphs and drawings have been used to illustrate the mathematical ideas developed in the written text.

The Worked Examples show solutions written out step by step so that students can see how to present mathematical answers on paper in a clear and logical way.

Many Investigations have been included and those form a vital part of the text since they introduce or develop particular points but they do require the reader to do more of the work to discover some of the important results and to see some of the finer aspects of mathematics. In most cases the conclusions are used in the following text and some of the answers are given to enable students to check their results.

The open, clear presentation also allows the book to be used as a suitable text for students preparing for the Advanced Supplementary examination by selecting the appropriate material.

Mathematical competence and insight are gained through experience and a lot of hard work. Students using this book to help them reach such standards are encouraged to do more than just read the text. It is vital to have a pen and paper to hand so that the working of the questions and investigations can be written down and understood.

The reader will find some results and formulae highlighted for ease of reference but this also serves as a reminder that certain facts must be learnt by heart before they can be applied to problem solving. All students should make a point of learning these basic results thoroughly.

We have used an integrated approach to the topics of Pure and Applied Mathematics as in Book 1, in that students will develop both aspects side by side and use the relevant techniques of Pure Mathematics in the Applied Mathematics sections. Vectors have been used extensively throughout the book, where appropriate.

The Revision Exercises are intended to enable students to revise the work covered in Book 1 before moving on to the next section in Book 2 where the topics will be developed further. Successful completion of these examples will indicate to the student that the work of Book 1 is actually known thoroughly. The questions are based entirely on the content of Book 1.

Often, different examination boards use different notations. We have incorporated most of the usual forms but students should ensure that they know which notation has been adopted by their examination board. A list of symbols is provided on page 7.

Finally, we trust that you, the reader, will find these two books helpful and stimulating in developing your understanding of mathematics and that they will allow you to prepare thoroughly for the Advanced Level or Advanced Supplementary examinations.

Notation

$=$	is equal to
$\neq$	is not equal to
$\equiv$	is identically equal to
$\simeq$	is approximately equal to
$>$	is greater than
$\geqslant$	is greater than or equal to
$<$	is less than
$\leqslant$	is less than or equal to
∞	infinitely large
$\Rightarrow$	implies
$\Leftarrow$	is implied by
$\Leftrightarrow$	implies and is implied by
$\rightarrow$	tends to
$\parallel$	parallel to
$\perp$	perpendicular to
$+\mathrm{ve}$	positive
$-\mathrm{ve}$	negative
$\triangle$	triangle
$\hat{A}$	moments about A

m	metres
s	seconds
kg	kilograms
N	newtons
J	joules
$\mathrm{m\,s^{-1}}$	metres per second
$\mathrm{m\,s^{-2}}$	metres per second per second
M, L, T	dimensions of mass, length and time
a	acceleration
s	displacement
v	velocity
F	force
g	acceleration of gravity, $g \simeq 9.8 \mathrm{\ m\ s^{-2}}$
μ	coefficient of friction
λ	angle of friction or modulus of elasticity
ω	angular velocity
$i =$	$\sqrt{-1}$

e exponential $e \simeq 2.71828$
or coefficient of restitution
or eccentricity

$\in$	is a member of
$\notin$	is not a member of
$\subset$	A is a subset of B
$\supset$	A contains B
ϕ	empty set
ξ	universal set
A'	complement of set A
$n(A)$	number of elements in set A

$\cup$	union
$\cap$	intersection
$\mathbb{N}$	the set of natural numbers
$\mathbb{Z}$	the set of integers
$\mathbb{Q}$	the set of rational numbers
$\mathbb{R}$	the set of real numbers
$\mathbb{C}$	the set of complex numbers
$\mathbb{F}$	the set of irrational numbers

Σ	the sum of
Δ	determinant
$\mathbf{PQ} = \vec{PQ}$	vector represented by PQ
$|\mathbf{PQ}| = PQ$	magnitude of vector $\mathbf{PQ}$
$|\mathbf{a}| = a$	magnitude of vector $\mathbf{a}$
$\mathbf{i, j, k}$	unit vectors along the x, y and z axes
$\mathbf{r} = x\mathbf{i} + y\mathbf{j} + z\mathbf{k}$	position vector
$\arg z$	argument of z
$\bar{z}$ or z^*	complex conjugate

$n!$ n factorial

^nP_r number of permutations of r objects chosen from n different items $= \dfrac{n!}{(n-r)!}$

$\dbinom{n}{r}$ the binomial coefficient; $\dfrac{n!}{r!(n-r)!} = {}^nC_r$ the number of combinations of r objects chosen from n

$x \rightarrow f(x)$ x is mapped onto $f(x)$

$f: x \rightarrow x^2$ the function f mapping x onto x^2

$|x|$ the modulus of x

f^{-1} inverse of function f

gf function f followed by function g

$\sqrt{x}$ non-negative square root of x

$[x]$ integral part of x

$\lim\limits_{x \to a} f(x)$ limit of $f(x)$ as x tends to a

δx small increment in x

$\dfrac{dy}{dx} = f'(x)$ first derivative of $y = f(x)$

$\dfrac{d^2y}{dx^2} = f''(x)$ second derivative of $y = f(x)$

$\dot{x} = \dfrac{dx}{dt}; \quad \ddot{x} = \dfrac{d^2x}{dt^2}$

$\displaystyle\int f(x)\,dx$ indefinite integral of $f(x)$ with respect to x

$\displaystyle\int_a^b f(x)\,dx$ definite integral of $f(x)$ with respect to x

between the limits of a and b

Revision Exercises A

These Exercises should enable you to revise work covered in Book 1.

Algebra

1 Find the number which when added to its square gives 30.

2 Find the dimensions of **(a)** a square **(b)** a rectangle, whose area is equal to its perimeter.

3 Make x the subject of the formulae,

  (a) $y = \dfrac{3x+4}{5x+7}$ **(b)** $y = \dfrac{7x-4}{3-5x}$ **(c)** $y = x^2 + 2x + 3$

4 Solve **(a)** $2^x = 64$ **(b)** $2^x = 0.125$ **(c)** $2^x = 6$ **(d)** $2^x = 1$ **(e)** $2^x = 0$ **(f)** $2^x = 8$

5 Evaluate **(a)** $16^{\frac{1}{4}}$ **(b)** $16^{\frac{3}{2}}$ **(c)** $16^{\frac{1}{2}}$ **(d)** $16^{1\frac{1}{4}}$ **(e)** $16^{\frac{3}{4}}$ **(f)** $16^{-\frac{1}{2}}$

6 Solve **(a)** $2^{2x} = 64$ **(b)** $(2^x)^2 = 64$ **(c)** $2^{x^2} = 64$ **(d)** $2^{2^x} = 64$

7 Solve **(a)** $\log_2 8 = x$ **(b)** $\log_2 x = 8$ **(c)** $\log_8 2 = x$ **(d)** $\log_8 x = 2$

8 Evaluate **(a)** $(\sqrt{8})^{\frac{2}{3}}$ **(b)** $\sqrt{8^{\frac{2}{3}}}$ **(c)** $\sqrt{8} \times \sqrt{2}$ **(d)** $\sqrt{8} \div \sqrt{2}$

9 Solve **(a)** $3^{x+1} = 9^x$ **(b)** $3^{x+1} + 9^x = 0$ **(c)** $9^x = 3^{x+1} + 4$

10 Solve **(a)** $\log_2 x = 4$ **(b)** $\log_x 2 = 4$ **(c)** $3 \log_x 2 + \log_2 x = 4$

11 Solve **(a)** $\log_3 x + \log_3(x+8) = 2$ **(b)** $\log_x 27 + \log_3 x = 3.5$

12 Separate into partial fractions,

  (a) $\dfrac{2}{x^2-1}$ **(b)** $\dfrac{2x}{x^2-1}$ **(c)** $\dfrac{2x^2}{x^2-1}$ **(d)** $\dfrac{2}{x^2+x}$ **(e)** $\dfrac{2}{x^2-x}$ **(f)** $\dfrac{2}{x^3-x}$

13 Plot the graphs of **(a)** $y = 2^x$ and **(b)** $y = 2^{-x}$. Specify the domain and range.

Functions

1 Use your calculator to work out $\frac{1}{13}, \frac{2}{13}, \frac{3}{13}, \frac{4}{13}, \ldots, \ldots$, What do you discover?
   From your calculator value for $\frac{1}{13}$, subtract $0.076\,923$ to get $0.000\,000\,076\,923$.

$$\Rightarrow \quad \tfrac{1}{13} = 0.076\,923\,076\,923 \ldots = 0.\dot{0}76\,92\dot{3}$$

  Check with $999\,999 \div 076\,923 = 13$.

  Can you do the same with $\frac{1}{17}, \frac{2}{17}, \frac{3}{17}$ etc?

2 Evaluate $(1+t)^{1/t}$ for $t=1$, $t=0.1$, $t=0.01$, $t=0.001$, etc. What is the limit of $(1+t)^{1/t}$ as $t\to 0$?

3 Evaluate $a_2 = 2 + \dfrac{1}{2!}$ $\quad a_3 = 2 + \dfrac{1}{2!} + \dfrac{1}{3!}$ $\quad a_4 = 2 + \dfrac{1}{2!} + \dfrac{1}{3!} + \dfrac{1}{4!}$ $\quad a_5 = 2 + \dfrac{1}{2!} + \dfrac{1}{3!} + \dfrac{1}{4!} + \dfrac{1}{5!}$ $\quad a_6,\ a_7,$

etc. What is a_{10} equal to?

4 $1\frac{1}{2}^2 = 1^2 + 1 + \frac{1}{4} = 2\frac{1}{4}$; $\quad 2\frac{1}{2}^2 = 2^2 + 2 + \frac{1}{4} = 6\frac{1}{4}$; $\quad 3\frac{1}{2}^2 = 3^2 + 3 + \frac{1}{4} = 12\frac{1}{4}$.

State the result for the number x and prove it in general.

5 Find functions which map

	(a)	(b)	(c)	(d)	(e)
	$1\to 1$	$1\to 3$	$1\to 1$	$1\to 2$	$1\to 2$
	$2\to 3$	$2\to 2$	$2\to 2$	$2\to 5$	$2\to 3$
	$3\to 5$	$3\to 1$	$3\to 4$	$3\to 10$	$3\to 5$

6 Find inverse functions for question **5**.

7 If the function in **5(a)** is $A(x)$, in **5(b)** is $B(x)$ etc., find the composite functions

(a) $AB(x)$, (b) $BA(x)$, (c) $AC(x)$, (d) $CA(x)$, (e) $AD(x)$ (f) $DA(x)$ (g) $CD(x)$.

8 For $f(x)=2x+1$ and $g(x)=3x-2$, find in terms of x in its simplest form

(a) $fg(x)$ (b) $gf(x)$ (c) $ff(x)$ (d) $gg(x)$ (e) $f^{-1}(x)$ (f) $g^{-1}(x)$ (g) $f^{-1}g^{-1}(x)$
(h) $g^{-1}f^{-1}(x)$.

9 Complete the table for the composite functions

$f(x)=\dfrac{1}{x}$ and $g(x)=1-x$

e.g. $ff(x)=x=I(x)$ (the identity)

$fg(x)=\dfrac{1}{1-x}$

$gf(x)=1-\dfrac{1}{x}$

Define $h=fgf$ and simplify gfg.

Remember $ff=gg=I$.

	I	f	g	fg	gf	h
I	I	f	g	fg		
f	f	I	fg			
g	g	gf				
fg						
gf	gf					
h						

10 With reference to question **9**, find out what is meant by the set of six functions being **closed** under the operation of combining functions.

11 Using $f(x)=2x-1$, $g(x)=3x-2$ and $h(x)=x^2$ simplify

(a) $f(gh)$ and $(fg)h$ (b) $f(hg)$ and $(fh)g$

What is the **associative** law?

12 Which pairs of functions in question **9** are **commutative**?

13 Find the roots p and q of $x^2-5x+6=0$ and the equations with roots

(a) $1/p$ and $1/q$ (b) p^2 and q^2

14 The roots of $x^2+5x+3=0$ are p and q. Find the equation with roots (a) $1/p$ and $1/q$ (b) p^2 and q^2.

Co-ordinates

1 Using Greenwich as the origin, work out the **(a)** Cartesian co-ordinates **(b)** Polar co-ordinates **(c)** Latitude and Longitude of your present position.

2 Our hotel room was number 321 as we were on floor 3 and our friends on floor 5 had room number 512, but there were only 200 rooms (!!)

 Does your school (or college) have a system for numbering rooms?

3 Draw the straight line through A (1, 3) and B (2, 5) and find its equation.

 (a) Reflect A and B in $x = 0$ to A_1 and B_1 and find the equation of $A_1 B_1$.
 (b) Reflect A and B in $y = 0$ to A_2 and B_2 and find the equation of $A_2 B_2$.
 (c) Reflect A and B in $y = x$ to A_3 and B_3 and find the equation of $A_3 B_3$.
 (d) Reflect A and B in $y = -x$ to A_4 and B_4 and find the equation of $A_4 B_4$.
 (e) Rotate A and B through $-90°$ to A_5 and B_5 and find the equation of $A_5 B_5$.
 (f) Rotate A and B through $180°$ to A_6 and B_6 and find the equation of $A_6 B_6$.
 (g) Rotate A and B through $+90°$ to A_7 and B_7 and find the equation of $A_7 B_7$.

4 O is (0, 0), A is (2, 1), B is (1, 3).

 (a) Find D where $OADB$ is a parallelogram.
 (b) Find C where $OABC$ is a parallelogram.
 (c) Find E where $OBAE$ is a parallelogram.
 (d) Find F where $DCFE$ is a parallelogram.
 (e) Find angle OAB.
 (f) Find angle BAC.
 (g) Find the area of $\triangle OAB$.
 (h) Find the equation of CD.

5 O is (0, 0), A is (6, 0), B is (2, 6).

 (a) Find the equation of OM where M is the mid-point of AB.
 (b) Find the equation of AN where N is the mid-point of OB.
 (c) Find G, the intersection of AN and OM.
 (d) If P is (3, 0), show that BP passes through G.
 (e) Find Q, Q on OA where $OQ:QA = 1:2$.
 (f) Find the point R which divides BA in the ratio $1:2$.
 (g) Show that G lies on QR.
 (h) What is the ratio $NG:GA$?

6 Write down the three inequalities, using the figure shown,

 (a) which specify triangle T;
 (b) which specify triangle S.
 (c) Find the area of T.
 (d) Find the area of S.
 (e) Find the distance of the line AB from the origin.
 (f) Find $\sin O\widehat{B}A$.

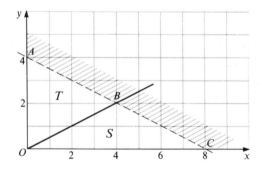

7 Find, using the figure,

 (a) the equation of the mediator (perpendicular bisector) of AB;
 (b) the mediator of OB.
 (c) Show that the mediators of $\triangle OAB$ are concurrent at W, giving the co-ordinates of W;
 (d) check that $WO = WA = WB$, describing W in relation to $\triangle OAB$;
 (e) find the intersection of the mediators of $\triangle OBC$;
 (f) where is the intersection of the mediators of $\triangle OAC$?

Differentiation

1 Differentiate the following functions of x with respect to x.

   (a) $f(x) = 5x^2 - 4x + 1$ (b) $f(x) = x^4 - \dfrac{3}{x^2}$

   (c) $f(x) = (x-2)(x+5)$ (d) $f(x) = x(x+1)(x+3)$

2 Differentiate the following functions with respect to x.

   (a) $y = (x-1)^2(x+4)^3$ (b) $y = \dfrac{(2x+1)}{3-x}$ (c) $y = \dfrac{\sin x}{1+x}$

   (d) $y = x \tan x$ (e) $y = (4x^2 - x)^5$ (f) $y = \cos^4 x$

3 Find the equation of the tangent to the curve $y = 4x^2 - \dfrac{1}{x}$, (a) at the point where $\dfrac{dy}{dx} = 0$ and (b) when it is parallel to the line $7x + y = 1$.

4 If $y = \sqrt{x^2 - 1}$, show that $y\dfrac{d^2y}{dx^2} + \left(\dfrac{dy}{dx}\right)^2 = 1$.

5 Find the co-ordinates and the nature of the turning points of the curve whose equation is $y = \dfrac{x^2 - x + 1}{x^2 + x + 1}$.

6 (a) If $y = \sin^{-1}(\cos x)$, show that $\dfrac{dy}{dx} = -1$.

   (b) If $x^2 \sin y + y \cos x = 2$, find $\dfrac{dy}{dx}$ in terms of x and y.

7 Find the equation of the tangent to the curve $xy^2 - 2xy + x^3 = 1$ at the point $(1, 0)$.

8 Find the gradient of the curve given parametrically by the equations $x = t^3$, $y = t^2 - t$. Hence, find the co-ordinates of any turning points.

9 The parametric equations of a curve are $x = \theta - \cos\theta$, $y = \sin\theta$. Show that the gradient of the curve is given by $-\sqrt{\dfrac{1-y}{1+y}}$.

10 A spherical ball is being pumped up so that its volume increases at a rate of 3 cm^3 s^{-1}. Find the rate of increase of the radius when the volume is 36π cm^3.

11 (a) If $y = x\tan^{-1} x$, show that $(1+x^2)^2 \dfrac{d^2y}{dx^2} = 2$.

   (b) If $y = (1-x^2)^{\frac{1}{2}} \sin^{-1} x$, show that $(1-x^2)\dfrac{dy}{dx} + xy = 1 - x^2$.

   Show also that $x(1-x^2)\dfrac{d^2y}{dx^2} - \dfrac{dy}{dx} + (1+x^2) = 0$.

12 Find the equations of the normals to the curve $x = 4t - t^2$, $y = 2 + t^3$, which are parallel to the line $2x - y = 1$.

13 A capsule is formed from a cylinder of radius r and height h, with a hemisphere of radius r attached with its plane surface in contact with one end of the cylinder. If the total volume of the capsule is 360π cm³, find h in terms of r and show that the total surface area is given by $A = \dfrac{720\pi}{r} + \dfrac{5\pi r^2}{3}$. If this area is to be a minimum, find the radius of the cylinder. (The area of the curved surface of a hemisphere $= 2\pi r^2$.)

14 If the height of a cylinder is three times its radius, find the percentage change in the volume if the radius is increased by 3%.

Integration

1 Integrate the following functions with respect to x.

    (a) $x^2 + 2x + 3$ **(b)** $2x^{-4}$ **(c)** $(x^2 - 3)^2$ **(d)** $\dfrac{x^3 + 4x^2 - 2}{x^2}$

2 Find $f(x)$, if $f'(x)$ equals,

    (a) $x - 5$ **(b)** $(x-2)(x+5)$ **(c)** $\sqrt{x}$ **(d)** $\sqrt{x}\left(\sqrt{x} - \dfrac{1}{x}\right)$

3 Integrate with respect to x,

    (a) $x^3 - \dfrac{1}{x^3}$ **(b)** $x^{\frac{1}{3}} + x^{-\frac{1}{3}}$ **(c)** $\left(x - \dfrac{1}{x}\right)^2$ **(d)** $\sqrt{x^3} + \dfrac{2}{\sqrt{x}}$ **(e)** $\dfrac{x^3 - x}{\sqrt{x}}$.

4 Find the equation of the curve which passes through the point $(-1, 7)$ and whose gradient is $3x^2 - 5$.

5 A curve has a gradient function $\sec^2 x + \cos x$ and passes through the point $(0, 3)$. Find the equation of the curve.

6 A particle has an acceleration at time t given by $\dfrac{dv}{dt} = 4 \sin t$. Find an expression for the velocity of the particle, v m s⁻¹, at time t seconds, if $v = 4$ when $t = \frac{1}{3}\pi$.

7 Evaluate the definite integrals.

    (a) $\displaystyle\int_0^1 (1 + x + x^2)\, dx$ **(b)** $\displaystyle\int_1^4 \sqrt{x}(x - 1)\, dx$ **(c)** $\displaystyle\int_2^3 (x - 2)^2\, dx$ **(d)** $\displaystyle\int_{-2}^{-1} x(1 - x^2)\, dx$

8 Find the area bounded by the curve $y = x^2 - 3x$, the x-axis and the lines $x = 3$ and $x = 4$.

9 Find the area contained between the curve $y^2 = x$, the y-axis and the line $y = 2$.

10 Find the area bounded by the curve $y = 3 + 2x - x^2$ and the line $y = x + 1$.

11 Find the area contained between the curves $y = x^2 - 3x$ and $y = 2 - x^2$.

12 Find the volume generated when the area bounded by the curve $y = x^2 - 4x$ and the x-axis, is rotated about the x-axis.

13 Find the volume generated when the area bounded by the curve $y = \dfrac{2}{x}$, the lines $x = 2$, $y = 2$ and the axes, is rotated about the y-axis.

14 Find the volume formed when the area bounded by the curve $y = x^2 - x + 3$ and the line $y = 3$ is rotated completely about the line $y = 3$.

15 Find **(a)** $\displaystyle\int_0^{1/2} \frac{1}{\sqrt{1-x^2}}\, dx$ **(b)** $\displaystyle\int_0^1 \frac{2}{1+x^2}\, dx$ **(c)** $\displaystyle\int \frac{3}{\sqrt{4-x^2}}\, dx.$

Trigonometry

1 **(a)** Find AC in terms of t. For $t = 2, 3 \ldots$ work out AB, BC and AC.

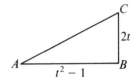

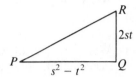

(b) Find PR in terms of s and t. For combinations of $s = 2, 3, \ldots$ and $t = 1, 2, 3 \ldots$ work out PQ, QR and RP.

2 A firm manufacturing a product packed in cartons of hexagonal cross-section is considering two methods of packing the cartons into boxes.

(a) How many cartons are packed into each box?
(b) What are the dimensions of each box?
(c) How much space is wasted?
(d) What are the advantages and disadvantages of each?

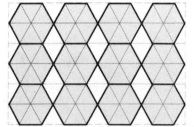

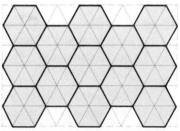

(e) Consider the two methods applied to a box which would just pack 4×4 in the first style. Consider larger sizes.
(f) Could these methods be applied to bottles or milk cartons?

3 Angles A and B are acute. If $\sin A = 3/5$ and $\cos B = 12/13$, find the values of

(a) $\cos A$ **(b)** $\tan A$ **(c)** $\sin B$ **(d)** $\tan B$ **(e)** $\sin(A+B)$ **(f)** $\cos(A+B)$
(g) $\sin(A-B)$ **(h)** $\cos(A-B)$ **(i)** $\sin 2A$ **(j)** $\cos 2A$ **(k)** $\tan 2A$ **(l)** $\sin 2B$
(m) $\cos 2B$ **(n)** $\tan 2B$; **(o)** Estimate $2B$ **(p)** Estimate B.

4 Find the angles of the triangle with sides 2, 3 and 4 cm.

5 Find the angles of the triangle with sides 8, 15 and 17 cm.

6 Solve the triangle ABC defined by $AB = 5$ cm, $AC = 6$ cm and $\angle B = 40°$.

7 Simplify **(a)** $\dfrac{\sin x}{\sqrt{1-\sin^2 x}}$ **(b)** $\dfrac{\tan x}{\sqrt{1-\cos^2 x}}$ **(c)** $\dfrac{\tan x}{\sqrt{1+\tan^2 x}}.$

8 Solve for $0° \leqslant x \leqslant 360°$ **(a)** $3\cos^2 x + 5\sin x = 1$ **(b)** $3\cos^2 x + 5\sin x = 5$ **(c)** $3\sin^2 x + 5\cos x = 1.$

9 For $0° \leqslant x \leqslant 360°$, draw an accurate graph of $y = \sin x$ using 1 cm $= 30°$ on the x-axis and 2 cm $= 1$ unit on the y-axis. On the same axes draw the graphs of $y = \sin 2x$ and $y = \sin 3x$.

10 Using the same scales as in question **9**, draw the graph of $y = \sin x$ and square the values to draw $y = \sin^2 x$ on the same axes.

11 Repeat question **10** for $y = \cos x$ and $y = \cos^2 x$.

12 With the same scale as in question **10**, draw the graph $y = \cos 2x$ for $0° \leqslant x \leqslant 360°$. Add 1 to each value to draw $y = 1 + \cos 2x$.
Halve each value to draw $y = \frac{1}{2}(1 + \cos 2x)$ and compare it with $y = \cos^2 x$.

13 An indoor running track of 200 metres per lap has straights of 50 m and curves (ABC) of length 50 m.

(a) Find the radius of the inside curves OA.
(b) If the track has 4 lanes of width 1 m, find the length of one 'lap' in the outside lane.
(c) What is the length of stagger for a one lap race?
(d) What is the length of stagger for a two lap race (400 m)?
(e) What are the advantages of running in the inside or outside lane?
(f) What are the dimensions of the rectangle which will enclose the track?
(g) Why not make the rectangle 'squarer', i.e. the straights shorter and the bends wider?
(h) Why do aircraft when stacking (waiting to land) fly on a racetrack (as above) course and not a circular path?

14 (a) A 50 pence coin has 'radius' of 16 mm and a 'diameter' of 30 mm (not 32 mm). Explain why.

The 'radius' of the 20 p coin is 11 mm and the 'diameter' is 21 mm. Explain.

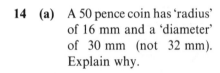
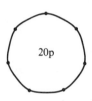
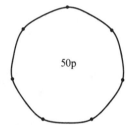

(b) My designs for a new 10p and 5p coin have 'radii' 8 mm and 5 mm. Find their diameters. Each 'side' (edge of coin) is an arc of a circle with centre at the opposite vertex.

(c) Find the areas of each of the four coins. Are the areas in proportion to their value? Are the volumes in proportion to their value? Do five 20p coins weigh the same as two 50p coins if they are made of the same material?

(d) Why are the coins this shape and not circular?

1 Exponential and logarithmic functions

Definitions

$$2^3 = 8 \quad \Leftrightarrow \quad \log_2 8 = 3$$
$$a^x = N \quad \Leftrightarrow \quad \log_a N = x$$

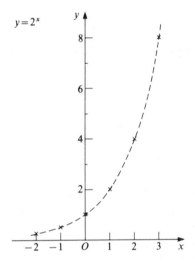

$y = 2^x$

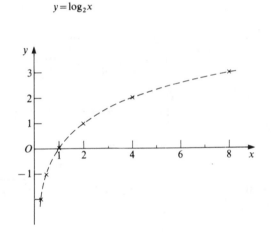

$y = \log_2 x$

Investigation 1

To differentiate a^x

(a) Draw accurately the graph of $y = 2^x$ for $-2 \leqslant x \leqslant 3$.

What is $2^{0.5}$ and $2^{1.5}$?

Measure the gradients and complete the table.

What do you notice?

x	-2	-1	0	1	2	3
$y = 2^x$	0.25	0.5	1	2	4	8
$\dfrac{dy}{dx}$						
$\dfrac{dy}{dx} \Big/ 2^x$						

(b) Repeat for $y = 3^x$.

In what ways are the graphs of $y = 2^x$ and $y = 3^x$ similar and different?

What can you say about the graph of $y = a^x$ (for $a > 1$)?

(cont.)

Investigation 1—continued

(c) You should find that $\dfrac{d}{dx}(2^x) = k_2 2^x$ and $\dfrac{d}{dx}(3^x) = k_3 3^x$ and that k_2 is the value of the gradient of 2^x at $x = 0$ and k_3 is the gradient of 3^x at $x = 0$.

Can you prove this for $y = a^x$? i.e. that $\dfrac{d}{dx}(a^x) = Ka^x$ where K is the gradient of a^x at $x = 0$.

Hint   Let $y = a^x \Rightarrow y + \delta y = a^{x + \delta x} \Rightarrow \delta y = \ldots$ etc.

Solution: $\delta y = a^{x + \delta x} - a^x = a^x(a^{\delta x} - 1)$

$$\frac{dy}{dx} = \underset{\delta x \to 0}{\text{limit}}\ \frac{\delta y}{\delta x} = \lim_{\delta x \to 0} \frac{a^x(a^{\delta x} - 1)}{\delta x} = a^x \lim_{\delta x \to 0} \frac{(a^{\delta x} - 1)}{\delta x} = a^x K$$

$x = 0 \Rightarrow a^x = a^0 = 1 \Rightarrow K = \lim\limits_{\delta x \to 0} \dfrac{(a^{\delta x} - 1)}{\delta x}$ which is the gradient of a^x when $x = 0$.

(d) For $y = 2^x$, $k_2 \simeq 0.7$ and for 3^x, $k_3 \simeq 1.1$ but is there a $y = a^x$ where $k_a = 1$?

For what value of a is the constant equal to 1? Obviously between 2 and 3. So there is a function $y = e^x$ where $\dfrac{dy}{dx} = 1 \times e^x$ i.e. $\dfrac{dy}{dx} = y$ (!!!)

(We shall find out later that $e = 2.7183 \ldots$)

Investigation 2

To differentiate $y = \log_2 x$, $y = \log_{10} x$, $y = \log_e x$.

(a) For $y = \log_2 x$ complete the table.

Deduce the form of $\dfrac{dy}{dx}$.

(b) Repeat (a) for $y = \log_3 x$
and $y = \log_{10} x$.

x	0.5	1	2	4	8
$y = \log_2 x$	-1	0	1	2	3
$\dfrac{dy}{dx}$					
$x\dfrac{dy}{dx}$					

(c) Differentiate $y = \log_{10} x$ from first principles.

Hint $y = \log_{10} x \Rightarrow y + \delta y = \log_{10}(x + \delta x)$

$$\Rightarrow \quad \delta y = \log_{10}(x + \delta x) - \log_{10} x = \log_{10}\left(\frac{x + \delta x}{x}\right)$$

Solution

$$\frac{dy}{dx} = \lim_{\delta x \to 0} \frac{\delta y}{\delta x} = \lim_{\delta x \to 0}\left[\frac{1}{\delta x}\log_{10}\frac{(x + \delta x)}{x}\right] = \lim_{\delta x \to 0}\left[\frac{1}{\delta x}\log_{10}\left(1 + \frac{\delta x}{x}\right)\right]$$

(cont.)

Investigation 2—continued

To make the limit easier, put $t = \dfrac{\delta x}{x}$ so $\dfrac{1}{\delta x} = \dfrac{1}{tx}$ and $t \to 0$ as $\delta x \to 0$.

$$\frac{dy}{dx} = \lim_{t \to 0} \frac{1}{tx} \log_{10}(1+t) = \lim_{t \to 0} \frac{1}{x} \log_{10}(1+t)^{1/t} = \frac{1}{x} \lim_{t \to 0} \log_{10}(1+t)^{1/t}$$

Find the limit of $(1+t)^{1/t}$ as $t \to 0$ by taking values of t from $t = 0.1, 0.01, 0.001$ etc. by using your calculator.

It would appear that the limit is close to 2.7183 which we denote by e.

t	$(1+t)^{1/t}$	
0.1	1.1^{10}	$= 2.5937$
0.01	1.01^{100}	$= 2.7048$
0.001	1.001^{1000}	$= 2.7169$
0.0001	$1.0001^{10\,000}$	$= 2.7181$
0.000\,01	$1.000\,01^{100\,000}$	$= 2.7183$

$e \simeq 2.718\,281\,828\ldots$ and is an irrational number (like π)

So we have $\dfrac{d}{dx}(\log_{10}x) = \dfrac{1}{x}\log_{10}e = \dfrac{1}{x} \times 0.434$.

$\dfrac{d}{dx}(\log_2 x) = \dfrac{1}{x}\log_2 e = \dfrac{1}{x} \times 1.44$ and $\dfrac{d}{dx}(\log_3 x) = \dfrac{1}{x}\log_3 e = \dfrac{1}{x} \times 0.91$

and
$$\boxed{\frac{d}{dx}(\log_e x) = \frac{1}{x}\log_e e = \frac{1}{x}}$$

Check this result by using Investigation 1, i.e. $y = \log_e x \iff x = e^y$

$$\Rightarrow \frac{dx}{dy} = e^y \quad \Rightarrow \quad \frac{dy}{dx} = \frac{1}{e^y} = \frac{1}{x}$$

Investigation 3

To integrate $\dfrac{1}{x}$. $\displaystyle\int \frac{1}{x}\,dx = \int x^{-1}\,dx = \frac{x^0}{0}$ by the rule for integration. This is not finite, although the area under the curve is defined.

Define $L(t) = \displaystyle\int_1^t \frac{1}{x}\,dx.$ Show that $L(t)$ satisfies the rules of logarithms i.e.

(a) $L(1) = 0$ **(b)** $L(a) + L(b) = L(ab)$ **(c)** $L(a^n) = nL(a)$;

and **(d)** find the base of the logarithmic function, L.

Solution

(a) $L(1) = \displaystyle\int_1^1 \frac{1}{x}\,dx = 0$ since the limits are the same.

(b) $L(ab) = \displaystyle\int_1^{ab} \frac{1}{x}\,dx = \int_1^a \frac{1}{x}\,dx + \int_a^{ab} \frac{1}{x}\,dx = L(a) + \int_a^{ab} \frac{1}{x}\,dx$

(cont.)

Investigation 3—continued

To simplify $\displaystyle\int_a^{ab} \frac{1}{x}\,dx$ substitute $x = au$

so that the limits $x = a \;\Rightarrow\; u = 1$ and $x = ab \;\Rightarrow\; u = b$

$$\int_{x=a}^{x=ab} \frac{1}{x}\,dx = \int_{u=1}^{u=b} \frac{1}{au}\,a\,du = \int_{u=1}^{u=b} \frac{1}{u}\,du = L\,(b) \;\Rightarrow\; L\,(ab) = L\,(a) + L\,(b)$$

(c) To show $L\,(a^n) = \displaystyle\int_1^{a^n} \frac{1}{x}\,dx.$

Let $x = u^n \;\Rightarrow\; dx = nu^{n-1}\,du$ and $x = 1 \;\Rightarrow\; u = 1 \quad x = a^n \;\Rightarrow\; u = a$

So $L\,(a^n) = \displaystyle\int_{x=1}^{x=a^n} \frac{1}{x}\,dx = \int_{u=1}^{u=a} n\,\frac{u^{n-1}}{u^n}\,du = \int_1^a n\,\frac{1}{u}\,du = n\,L\,(a)$

$L\,(x)$ is therefore a logarithmic function.

(d) The base e of a logarithm satisfies $\log_e e = 1$ so $\displaystyle\int_1^e \frac{1}{x}\,dx = 1.$

Consider $L\,(3)$ which lies between the shaded area of rectangles (lower bound) and the larger rectangles (upper bound).

Check that this gives $0.95 < L\,(3) < 1.28.$

The average of these gives the area using the trapezium rule to give $L\,(3) < 1.115.$

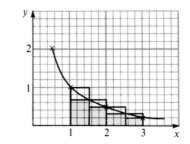

So e is just less than 3. More accuracy is obtained by using narrower rectangles.

Work out a value for $L\,(2.8)$ using trapezia of width 0.1.

$e \simeq 2.718$ and Chapter 12 on Series will give ways of calculating e to any accuracy.

The function $L\,(x)$ gives logarithms in base $e\,(\simeq 2.718)$ called **natural** or **napierian** logarithms after John Napier, and are denoted by

$$\log_e x \quad \text{or more usually} \quad \ln x$$

My calculator gives $e = 2.718\,281\,828\ldots$ but the recurring bank of figures is only momentary. Remember when you study series to work out the value of e correct to 10 decimal places to see if this is so.

Important results

$$\frac{d}{dx}(e^x) = e^x \quad \ldots\ldots\ldots\ldots \text{ (1)}$$

$$\frac{d}{dx}(\ln x) = \frac{1}{x} \quad \ldots\ldots\ldots\ldots \text{ (2)}$$

$$\int \frac{1}{x}\,dx = \ln x + c$$

$$= \ln Ax \quad \ldots\ldots\ldots\ldots \text{ (3)}$$

$$(\text{where } c = \ln A)$$

$$\frac{d}{dx}(a^x) = a^x \ln a$$

$$\frac{d}{dx}(\log x) = \frac{1}{x}\log e$$

e^x and $\ln x$ are inverse functions, so $e^{\ln x} = x$ and $\ln e^x = x.$

(1) $\Rightarrow$ **(2)** since

$$y = e^x \quad \Rightarrow \quad x = \ln y \quad \Rightarrow \quad \frac{dx}{dy} = \frac{1}{dy/dx} = \frac{1}{e^x} = \frac{1}{y} \quad \Rightarrow \quad \frac{d}{dy}(\ln y) = \frac{1}{y}$$

(2) $\Rightarrow$ **(1)** since

$$y = \ln x \quad \Rightarrow \quad x = e^y \quad \text{and} \quad \frac{dy}{dx} = \frac{1}{x} \quad \Rightarrow \quad \frac{dx}{dy} = x \quad \Rightarrow \quad \frac{d}{dy}(e^y) = e^y$$

Differentiating exponential functions

WORKED EXAMPLES

Differentiate **(a)** e^{2x} **(b)** e^{x^2} **(c)** $e^{f(x)}$ **(d)** e^{x^2+x}

(a) $y = e^{2x} \quad \Rightarrow \quad \ln y = 2x \quad \Rightarrow \quad \frac{1}{y}\frac{dy}{dx} = 2 \quad \Rightarrow \quad \frac{dy}{dx} = 2y = 2e^{2x}$

(b) $y = e^{x^2} \quad \Rightarrow \quad \ln y = x^2 \quad \Rightarrow \quad \frac{1}{y}\frac{dy}{dx} = 2x \quad \Rightarrow \quad \frac{dy}{dx} = 2xy = 2xe^{x^2}$

(c) $y = e^{f(x)} \quad \Rightarrow \quad \ln y = f(x) \quad \Rightarrow \quad \frac{1}{y}\frac{dy}{dx} = \frac{df}{dx} = f'(x) \quad \Rightarrow \quad \frac{dy}{dx} = yf'(x) = f'(x)e^{f(x)}$

(d) $y = e^{x^2+x} \quad \Rightarrow \quad \ln y = x^2+x \quad \Rightarrow \quad \frac{1}{y}\frac{dy}{dx} = 2x+1 \quad \Rightarrow \quad \frac{dy}{dx} = (2x+1)e^{x^2+x}$

Or differentiate $e^{x^2+x} = e^{x^2} \times e^x$ as a product of two functions e^{x^2} and e^x.

$$y = e^{x^2} \times e^x \quad \Rightarrow \quad \frac{dy}{dx} = 2xe^{x^2}e^x + e^{x^2}e^x = (2x+1)e^{x^2}e^x = (2x+1)e^{x^2+x}$$

Exercise 1.1

Differentiate the functions in **1** to **5** with respect to x.

1 **(a)** 2^x **(b)** 3^x **(c)** 4^x **(d)** e^x **(e)** $(e^x)^2$

2 **(a)** 2^{x^2} **(b)** 3^{x^3} **(c)** e^{x^4} **(d)** $(e^x)^4$ **(e)** e^{4x}

3 **(a)** $5e^{2x}$ **(b)** $e^x e^{2x}$ **(c)** e^{3x+2} **(d)** $e^{\sqrt{x}}$ **(e)** $e^{1/x}$

4 **(a)** e^{-x} **(b)** $-e^{-x}$ **(c)** $e^x + e^{-x}$ **(d)** $e^x - e^{-x}$

5 **(a)** $e^{\sin x}$ **(b)** $e^{\cos x}$ **(c)** $e^{\tan x}$ **(d)** $e^{\sec x}$

6 Sketch on the same axes the graphs of e^x, e^{-x}, $\dfrac{(e^x + e^{-x})}{2}$.

7 Sketch on the same axes the graphs of e^x, e^{-x}, $\dfrac{(e^x - e^{-x})}{2}$.

8 Sketch the following graphs and find their maximum points.

(a) $y = xe^{-x}$ (b) $y = x^2 e^{-x}$ (c) $y = x^n e^{-x}$

9 For $y = xe^x$ find:

(a) $\dfrac{dy}{dx}$ (b) $\dfrac{d^2 y}{dx^2}$ (c) $\dfrac{d^n y}{dx^n}$ (d) $\left(\dfrac{dy}{dx}\right)^2$

10 For the functions $c(x) = \dfrac{(e^x + e^{-x})}{2}$ and $s(x) = \dfrac{(e^x - e^{-x})}{2}$

(a) find their derivatives   (b) $c^2 - s^2$
(c) plot their graphs   (d) simplify $c(-x)$ and $s(-x)$
(e) simplify $c(a)c(b) - s(a)s(b)$ (f) simplify $s(a)c(b) + c(a)s(b)$

Differentiating logarithmic functions

WORKED EXAMPLES

Differentiate (a) $\log_{10} x$ (b) $\ln x$ (c) $\ln f(x)$

(a) **Either** remember $\dfrac{d}{dx}(\log_a x) = \dfrac{1}{x}\log_a e \;\Rightarrow\; \dfrac{d}{dx}(\log_{10} x) = \dfrac{1}{x}\log_{10} e$

   Or change to base e,

   so $y = \log_{10} x = \dfrac{\ln x}{\ln 10} \;\Rightarrow\; \dfrac{dy}{dx} = \dfrac{1}{x \ln 10} = \dfrac{1}{x}\log_{10} e$

   Or $y = \log_{10} x \;\Rightarrow\; x = 10^y \;\Rightarrow\; \dfrac{dx}{dy} = 10^y \ln 10 \;\Rightarrow\; \dfrac{dy}{dx} = \dfrac{1}{10^y \ln 10} = \dfrac{1}{x}\log_{10} e$

(b) $y = \ln x \;\Rightarrow\; \dfrac{dy}{dx} = \dfrac{1}{x}$

   Or $y = \ln x \;\Rightarrow\; x = e^y \;\Rightarrow\; \dfrac{dx}{dy} = e^y \;\Rightarrow\; \dfrac{dy}{dx} = \dfrac{1}{e^y} = \dfrac{1}{x}$

   Remember this result and use the change of base rule for $\log_{10} x$ to avoid confusion over remembering $\log_{10} e$ or $\log_e 10$ in the answer.

(c) $y = \ln f(x) \;\Rightarrow\; \dfrac{dy}{dx} = \dfrac{1}{f(x)} f'(x)$ using the composite function rule.

   Or $y = \ln f(x) \;\Rightarrow\; f(x) = e^y \;\Rightarrow\; \dfrac{df}{dx} = e^y \dfrac{dy}{dx} \;\Rightarrow\; \dfrac{dy}{dx} = \dfrac{1}{e^y} f'(x) = \dfrac{f'(x)}{f(x)}$

WORKED EXAMPLES

Differentiate   **(a)** $\ln(x^2+x)$ **(b)** $\ln 7x^6$ **(c)** $\ln(\tan x)$

(a)   Using **(c)** above,

$$y=\ln(x^2+x) \quad\Rightarrow\quad \frac{dy}{dx}=\frac{2x+1}{x^2+x} \quad \left(\frac{f'(x)}{f(x)} \text{ where } f(x)=x^2+x\right)$$

(b) $y=\ln(7x^6) \quad\Rightarrow\dfrac{dy}{dx}=\dfrac{7\times 6x^5}{7x^6}=\dfrac{6}{x}$

Or $y=\ln(7x^6)=\ln 7+\ln x^6=\ln 7+6\ln x \quad\Rightarrow\quad \dfrac{dy}{dx}=0+\dfrac{6\times 1}{x}=\dfrac{6}{x}.$

Try to simplify the function using rules of logarithms **before** differentiating.

Product rule $y=uv \quad\Rightarrow\quad \ln y=\ln u+\ln v \quad\Rightarrow\quad \dfrac{1}{y}\dfrac{dy}{dx}=\dfrac{1}{u}\dfrac{du}{dx}+\dfrac{1}{v}\dfrac{dv}{dx}$

$$\frac{dy}{dx}=\frac{y}{u}\frac{du}{dx}+\frac{y}{v}\frac{dv}{dx}=v\frac{du}{dx}+u\frac{dv}{dx}$$

which is a nice simple proof.

Exercise 1.2

1   Differentiate   **(a)** $\ln(x+1)$ **(b)** $5\ln(x+4)$ **(c)** $\ln\sec x$ **(d)** $(\ln x)^2$

2   Differentiate and simplify   **(a)** $\ln 3x$ **(b)** $\ln x^3$ **(c)** $\ln(x^2-1)$ **(d)** $\ln\cot x$

3   Simplify and differentiate   **(a)** $\ln 3x$ **(b)** $\ln x^3$ **(c)** $\ln(x^2-1)$ **(d)** $\ln\cot x$

Differentiate the following.

4   **(a)** $\ln x^2$ **(b)** $\ln(x^2+1)$ **(c)** $\ln(x+1)^2$ **(d)** $\ln(x+1)^3$ **(e)** $\ln(x+1)^n$

5   **(a)** $\ln x^3$ **(b)** $\ln x^4$ **(c)** $\ln(x^3+x^4)$ **(d)** $\ln(x^3\times x^4)$ **(e)** $\ln\dfrac{x^3}{x^4}$ **(f)** $\dfrac{\ln x^3}{\ln x^4}$

6   **(a)** $\ln\sqrt{x}$ **(b)** $\ln\sqrt{x+1}$ **(c)** $\ln\sqrt{x^2+1}$ **(d)** $\ln\dfrac{x+1}{x+2}$

7   **(a)** $\ln\dfrac{1}{x}$ **(b)** $\ln\dfrac{1}{x+1}$ **(c)** $\ln\dfrac{1}{x^2}$ **(d)** $\ln\dfrac{2}{(x+1)^3}$

8   **(a)** $\ln\sin x$ **(b)** $\ln\cos x$ **(c)** $\ln\sec x$ **(d)** $\ln\operatorname{cosec} x$

9   **(a)** $\ln\sin 2x$ **(b)** $\ln\cos^3 x$ **(c)** $\ln\sin x^3$ **(d)** $\ln(\sin x+\cos x)$

10   **(a)** $x\ln x$ **(b)** $x^2\ln x$ **(c)** $x^3\ln x$ **(d)** $x^n\ln x$

11   **(a)** $\ln(\tan x+\sec x)$ **(b)** $\ln(\operatorname{cosec} x+\cot x)$

12   **(a)** $e^{\ln x}$ **(b)** $\ln e^x$ **(c)** $e^x\ln x$ **(d)** $\ln(xe^x)$

Integration with exponential and logarithmic functions

Remember

$$\frac{d}{dx}\left(e^{f(x)}\right) = f'(x)\cdot e^{f(x)} \quad\Rightarrow\quad \int f'(x)e^{f(x)}dx = e^{f(x)} + K \qquad \dotsb \quad (1)$$

$$\frac{d}{dx}\ln f(x) = \frac{f'(x)}{f(x)} \quad\Rightarrow\quad \int \frac{f'(x)}{f(x)}dx = \ln|f(x)| + c \qquad \dotsb \quad (2)$$

The reason for the modulus sign in **(2)** is explained in Worked Example **2(b)**, which follows.

WORKED EXAMPLES

1 Integrate with respect to x, **(a)** e^{2x} **(b)** $2xe^{x^2}$ **(c)** $\dfrac{1}{x+1}$ **(d)** $\dfrac{2x}{x^2-1}$.

(a) $\dfrac{d}{dx}(e^{2x}) = 2e^{2x} \quad\Rightarrow\quad \displaystyle\int e^{2x}\,dx = \tfrac{1}{2}e^{2x} + c$

(b) Using **(1)** $\displaystyle\int 2x\,e^{x^2}dx = e^{x^2} + k$ by recognition or inspection.

Or by substitution:

Let $e^{x^2} = u \quad\Rightarrow\quad \dfrac{du}{dx} = 2x\,e^{x^2} \quad\Rightarrow\quad \displaystyle\int 2x\,e^{x^2}dx = \int du = u + k = e^{x^2} + k$

(c) $\displaystyle\int \frac{1}{x+1}\,dx = \ln|x+1| + c$ **but** $\displaystyle\int \frac{1}{x^2+1}\,dx \neq \ln(x^2+1)$ (check by differentiating $\ln(x^2+1)$)

(d) Using **(2)** $\displaystyle\int \frac{2x}{x^2-1}\,dx = \ln(x^2-1) + k$ being of the form $\displaystyle\int \frac{f'(x)}{f(x)}$

Or by Partial Fractions $\displaystyle\int \frac{2x}{x^2-1}\,dx = \int \frac{1}{x-1} + \frac{1}{x+1}\,dx = \ln(x-1) + \ln(x+1) = \ln(x^2-1) + c$

2 (a) $\displaystyle\int_{-1}^{1} e^{x+1}dx$ **(b)** $\displaystyle\int_{-3}^{-2} \frac{1}{x}\,dx$ **(c)** $\displaystyle\int_{-1}^{2} \frac{1}{x}\,dx$ **(d)** $\displaystyle\int x\,e^x\,dx$

(a) $\displaystyle\int_{-1}^{1} e^{x+1}\,dx = \left[e^{x+1}\right]_{-1}^{1} = e^2 - e^0 = e^2 - 1$

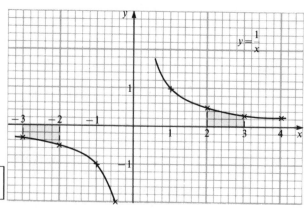

(b) $\displaystyle\int_{-3}^{-2} \frac{1}{x}\,dx = \left[\ln x\right]_{-3}^{-2} = \ln(-2) - \ln(-3)$

which is not defined, but from the graph the areas

for $\displaystyle\int_{-3}^{-2} \frac{1}{x}\,dx$ and $\displaystyle\int_{2}^{3} \frac{1}{x}\,dx$ are the same except

that the first is the negative of the second, so

$$\left[\int_{-3}^{-2} \frac{1}{x}\,dx = -\int_{2}^{3} \frac{1}{x}\,dx = -(\ln 3 - \ln 2) = \ln 2 - \ln 3\right]$$

We obtain a correct answer by saying

$$\int_{-3}^{-2} \frac{1}{x}\, dx = \left[\ln|x|\right]_{-3}^{-2} = \ln 2 - \ln 3 = \ln \frac{2}{3}$$

However in **(c)** our limits of integration include a discontinuity of the function at $x = 0$ so we cannot perform this integral.

It would be tempting to say
$$\int_{-1}^{2} \frac{1}{x}\, dx = \int_{-1}^{0} \frac{1}{x}\, dx + \int_{0}^{1} \frac{1}{x}\, dx + \int_{1}^{2} \frac{1}{x}\, dx$$
$$= \quad A \quad + \quad B \quad + \quad C$$

and $A = -B$ by symmetrical areas so

$$\int_{-1}^{2} \frac{1}{x}\, dx = C = \int_{1}^{2} \frac{1}{x}\, dx = \ln 2$$

But $\ln 0$ is undefined, being infinite (or $-$ infinity (!)).

(d) $\dfrac{d}{dx}(x\, e^x) = x\, e^x + e^x \quad \Rightarrow \quad \displaystyle\int (x\, e^x + e^x) = x\, e^x + c$

$$\Rightarrow \quad \int x\, e^x\, dx + \int e^x\, dx = x\, e^x + c$$

$$\Rightarrow \quad \int x\, e^x\, dx = x\, e^x - \int e^x\, dx + c$$

$$= x\, e^x - e^x + c$$

But it is more usual to integrate $x\, e^x$ by parts (see the next chapter).

Exercise 1.3

Integrate   **(a)** as a function of x **(b)** between the limits 1 and 2,

1 e^{x+2} **2** $e^x + 2$ **3** $e^x \cdot e^x$ **4** e^{x-2} **5** e^{-2x}

Integrate   **(a)** as a function of x **(b)** between the limits 1 and 3,

6 e^{3x} **7** $e^x + e^{2x}$ **8** $(e^x)^3$ **9** $x^2 e^{x^3}$ **10** e^{nx}

Integrate   **(a)** as a function of x **(b)** between the limits 0 and $\pi/4$,

11 $\cos x\, e^{\sin x}$ **12** $\sin x\, e^{\cos x}$ **13** $(1 + \tan^2 x)\, e^{\tan x}$

Integrate   **(a)** as a function of x **(b)** between the limits 1 and 4,

14 $x^{-2} e^{1/x}$ **15** $\dfrac{e^{\sqrt{x}}}{\sqrt{x}}$ **16** $6\, e^{x/7}$ **17** $(1 + e^x)^2$

18 $e^{\ln x}$ **19** $\ln e^x$ **20** $\ln x$ (**Hint**   differentiate $x \ln x$)

21 $\dfrac{1}{x+2}$ **22** $\dfrac{x}{x+1}$ **23** $\dfrac{6}{x}$ **24** $\dfrac{1}{2x-1}$ **25** $\dfrac{x}{x^2+2}$

Integrate **(a)** as a function of x **(b)** between the limits 2 and 3 or the limits indicated,

26 $\dfrac{1}{1-2x}$ **27** $\dfrac{x}{x^2-1}$ **28** $\dfrac{x^2}{x+1}$ **29** $\dfrac{1}{x^2-1}$

30 $\displaystyle\int_0^{\pi/4} \tan x\,dx$ **31** $\displaystyle\int_{\pi/6}^{\pi/3} \cot x\,dx$ **32** $\displaystyle\int_{\pi/4}^{\pi/3} \sec x\,dx$

Exercise 1.4   Miscellaneous

1 Solve **(a)** $\log_2 x+\log_x 2=2.5$ **(b)** $\log_3 x+\log_x 3=2.5$ **(c)** $\log_2 x+\log_x 2=2$

2 Solve **(a)** $2^x=3$ **(b)** $3^x=4$ **(c)** $2^x\times 3^{x+1}=36$ **(d)** $2^x\times 4^{x+1}=32$

3 Solve **(a)** $9^x-3^{x+1}+2=0$ **(b)** $3^{2x}-5\times 3^x+6=0$

4 Given that $2\log_4 x+\log_2 y=4$, show that $xy=16$. Hence solve the simultaneous equations $2\log_4 x+\log_2 y=4$ and $\log_{10}(x+y)=1$.

5 Differentiate with respect to x,

   (a) $\ln\tan x$ **(b)** a^x **(c)** $\log_{10}x$ **(d)** $\ln(x^3)$ **(e)** $(\ln x)^3$

6 Solve **(a)** $e^{\ln x}=3$ **(b)** $\ln e^x=4$ **(c)** $e^{\ln x}+\ln e^x=5$

7 Find the minimum point on the curve $y=x^x$ and sketch the curve for $0<x<2$.

8 Make x the subject of **(a)** $y=3e^{2x}$ **(b)** $y=6+e^{x-1}$ **(c)** $\ln y=3+2\ln x$

9 If $y=e^{3x}\cos 2x$, show that $\dfrac{d^2y}{dx^2}-6\dfrac{dy}{dx}+13y=0$.

10 Solve **(a)** $e^x=6$ **(b)** $e^{-x}=3$ **(c)** $e^x-e^{-x}=3$

2 Integration 2

Standard integrals

In Book 1 we considered the meaning of integration and by thinking of the process as the reverse of differentiation we found several **standard results** which are listed below. These need to be learnt thoroughly.

$$\frac{d}{dx}(ax^n) = nax^{n-1} \qquad \int ax^n\, dx = \frac{ax^{n+1}}{n+1} + C \qquad (n \neq -1)$$

$$\frac{d}{dx}(\sin x) = \cos x \qquad \int \cos x\, dx = \sin x + C$$

$$\frac{d}{dx}(\cos x) = -\sin x \qquad \int \sin x\, dx = -\cos x + C$$

$$\frac{d}{dx}(\tan x) = \sec^2 x \qquad \int \sec^2 x\, dx = \tan x + C$$

$$\frac{d}{dx}(\cot x) = -\mathrm{cosec}^2 x \qquad \int \mathrm{cosec}^2 x\, dx = -\cot x + C$$

$$\frac{d}{dx}(\sec x) = \sec x \tan x \qquad \int \sec x \tan x\, dx = \sec x + C$$

$$\frac{d}{dx}(\mathrm{cosec}\, x) = -\mathrm{cosec}\, x \cot x \qquad \int \mathrm{cosec}\, x \cot x\, dx = -\mathrm{cosec}\, x + C$$

This list of integrals allows us to integrate any function expressed in one of these forms although often the functions are more complicated and other methods are required. In this chapter we shall consider various special techniques for dealing with specific types of functions.

Inspection

It is possible that the expression to be integrated can be recognized as the result of a particular differential. In such cases the integral can be written down straight away.

Hence since $\dfrac{d}{dx}(x^3 + 4x^2 + 7x) = 3x^2 + 8x + 7$

$$\Rightarrow \quad \int (3x^2 + 8x + 7)\, dx = x^3 + 4x^2 + 7x + C$$

Investigation 1

Consider $\int x^2(2x^3+5)^3 dx$ by finding the differential of $(2x^3+5)^4$.

Use the chain rule to show that $\dfrac{d}{dx}(2x^3+5)^4 = 24x^2(2x^3+5)^3$.

Deduce that $\int 24x^2(2x^3+5)^3 dx = (2x^3+5)^4 + A$.

Since we can divide by a constant, deduce that $\int x^2(2x^3+5)^3 dx = \dfrac{1}{24}(2x^3+5)^4 + C$.

This process can be seen more clearly if we write out the differential fully.

Consider $\dfrac{d}{dx}(5x^2+7)^6$.

Let $y = (5x^2+7)^6$ and $u = 5x^2+7 \Rightarrow y = u^6$.

Hence, $\dfrac{dy}{du} = 6u^5$ and $\dfrac{du}{dx} = 10x$.

Using the chain rule for differentiation, $\dfrac{dy}{dx} = \dfrac{dy}{du} \times \dfrac{du}{dx}$

$$\therefore \quad \frac{dy}{dx} = 10x \times 6u^5 = 60x(5x^2+7)^5.$$

Notice that the result is formed by differentiating $(5x^2+7)^6$ as though it is a function of the form x^n and then multiplying by the differential of the expression inside the bracket (i.e. $5x^2+7) = 10x$.

Thus, $\dfrac{d}{dx}(5x^2+7)^6 = 6(5x^2+7)^5 \times 10x = 60x(5x^2+7)^5$

and hence, $\int x(5x^2+7)^5 dx = \dfrac{1}{60}(5x^2+7)^6 + C$.

It is vital to recognize the term to be integrated (the **integrand**) as a product of a power of an expression and its differential.

In general,
$$\int \left[f(x) \right]^n f'(x)\, dx = \frac{1}{n+1}\left[f(x) \right]^{n+1} + C$$

Investigation 2

Find the integrals of the following by differentiating the given expressions with respect to x.

(a) $\int x(4x^2-1)^5 dx$ consider $\dfrac{d}{dx}(4x^2-1)^6$

(b) $\int x^3(3x^4+2)^7 dx$ consider $\dfrac{d}{dx}(3x^4+2)^8$

(c) $\int (2x-3)(x^2-3x+1)^3 dx$ consider $\dfrac{d}{dx}(x^2-3x+1)^4$

(cont.)

Investigation 2—continued

(d) $\displaystyle\int \frac{x+1}{\sqrt{x^2+2x-3}}\,dx$ consider $\dfrac{d}{dx}(x^2+2x-3)^{\frac{1}{2}}$

(e) $\displaystyle\int \cos x \sin^3 x\,dx$ consider $\dfrac{d}{dx}(\sin^4 x)$

(Remember that $\sin^3 x$ means $(\sin x)^3$)

▷ **WORKED EXAMPLES**

1 Find **(a)** $\displaystyle\int x(2x^2+1)^3 dx$ and   **(b)** $\displaystyle\int \frac{3x-2}{\sqrt{3x^2-4x+1}}\,dx.$

(a) Since $\dfrac{d}{dx}(2x^2+1)=4x$ the integrand consists of a product of a power of $(2x^2+1)$ and its differential (apart from the constant multiple of 4).

Consider $\dfrac{d}{dx}(2x^2+1)^4 = 4(2x^2+1)^3 \times 4x = 16x(2x^2+1)^3$

Hence $\displaystyle\int x(2x^2+1)^3 dx = \frac{1}{16}(2x^2+1)^4 + C$

(b) In this case we rearrange the integrand.

$$\int \frac{3x-2}{\sqrt{3x^2-4x+1}}\,dx = \int (3x-2)(3x^2-4x+1)^{-\frac{1}{2}}\,dx$$

Now $\dfrac{d}{dx}(3x^2-4x+1)^{\frac{1}{2}} = \frac{1}{2}(3x^2-4x+1)^{-\frac{1}{2}}(6x-4)$

$$= (3x-2)(3x^2-4x+1)^{-\frac{1}{2}}$$

Hence $\displaystyle\int (3x-2)(3x^2-4x+1)^{-\frac{1}{2}}\,dx = (3x^2-4x+1)^{\frac{1}{2}} + C$

or $\displaystyle\int \frac{3x-2}{\sqrt{3x^2-4x+1}}\,dx = \sqrt{3x^2-4x+1} + C.$

2 Find **(a)** $\displaystyle\int \cos^4 x \sin x\,dx$ **(b)** $\displaystyle\int \sec^4 x \tan x\,dx.$

(a) Since $\cos^4 x \sin x$ actually means $(\cos x)^4 \sin x$ and $\dfrac{d}{dx}(\cos x) = -\sin x$, the integrand consists of a product of a power of $\cos x$ and its differential.

So consider $\dfrac{d}{dx}(\cos^5 x) = 5\cos^4 x(-\sin x)$

Hence, it follows that $\displaystyle\int \cos^4 x \sin x\,dx = -\frac{1}{5}\cos^5 x + C.$

(b) In this case we must recognize that $\dfrac{d}{dx}(\sec x) = \sec x \tan x$ and so the integrand is a product of $\sec^3 x$ and the differential of $\sec x$ (i.e. $\sec x \tan x$).

So $\qquad \displaystyle\int \sec^4 x \tan x \, dx = \int \sec^3 x \, (\sec x \tan x) \, dx$

Now consider $\dfrac{d}{dx}(\sec^4 x) = 4 \sec^3 x \, (\sec x \tan x) = 4 \sec^4 x \tan x$

Hence, $\qquad \displaystyle\int \sec^4 x \tan x \, dx = \frac{1}{4} \sec^4 x + C$

Exercise 2.1

Find the following integrals with respect to x.

1 $\displaystyle\int \cos 2x \, dx$
2 $\displaystyle\int (x-3)^2 \, dx$
3 $\displaystyle\int \operatorname{cosec}^2 x \, dx$

4 $\displaystyle\int \sin(5x-2) \, dx$
5 $\displaystyle\int x(2x+1)^2 \, dx$
6 $\displaystyle\int x(x^2-1)^3 dx$

7 $\displaystyle\int \cos x \sin^3 x \, dx$
8 $\displaystyle\int \sin x \sqrt{\cos x} \, dx$
9 $\displaystyle\int x(x^2-5)^{-2} dx$

10 $\displaystyle\int \frac{2x+1}{(x^2+x-3)^2} \, dx$
11 $\displaystyle\int \sin 2x \, dx$
12 $\displaystyle\int 2 \sin x \cos x \, dx$

13 $\displaystyle\int \frac{x^3-1}{\sqrt{x^4-4x}} \, dx$
14 $\displaystyle\int (x^2+1)^3 dx$
15 $\displaystyle\int \sec^2 x \tan x \, dx$

16 $\displaystyle\int x \sin(x^2) \, dx$
17 $\displaystyle\int \frac{\cos \sqrt{x}}{\sqrt{x}} \, dx$
18 $\displaystyle\int \operatorname{cosec}^2 x \cot^3 x \, dx$

19 $\displaystyle\int (x-3)(x^2-6x+1)^4 dx$
20 $\displaystyle\int \sec^3 x \tan x \, dx$
21 $\displaystyle\int \frac{1}{(2x+3)^2} \, dx$

Evaluate the definite integrals.

22 $\displaystyle\int_0^{\pi/4} \cos^2 x \sin x \, dx$
23 $\displaystyle\int_1^2 2\sqrt{5x-1} \, dx$
24 $\displaystyle\int_0^{1/2} \frac{x}{(1-x^2)^2} \, dx$

25 $\displaystyle\int_{-1}^4 \frac{1}{\sqrt{2x+3}} \, dx$
26 $\displaystyle\int_0^{\pi/4} \sin^2 x \cos x \, dx$
27 $\displaystyle\int_1^2 x\sqrt{4-x^2} \, dx$

Powers of sin *x* and cos *x*

Whilst it is possible to integrate the sine or cosine of a multiple angle, it is not necessarily possible to integrate powers or products of trigonometrical functions directly.

Thus, as
$$\frac{d}{dx}(\cos 4x) = -4 \sin 4x \qquad \text{(using the \textbf{chain rule} with } u = 4x)$$

$$\Rightarrow \quad \int \sin 4x \, dx = -\tfrac{1}{4} \cos 4x + C$$

and in general
$$\int \sin ax \, dx = -\frac{1}{a} \cos ax + C$$

and
$$\int \cos ax \, dx = \frac{1}{a} \sin ax + C$$

However, we **cannot** find $\displaystyle\int \sin^2 x \, dx$ or $\displaystyle\int \cos^3 x \, dx$ as they stand.

When integrating the trigonometrical functions it is possible to change the integrand into an equivalent form before integrating. The problem is to find a suitable alternative by using the trigonometrical identities. Some special cases occur as follows.

(i) Odd powers of sin *x* or cos *x*

In these examples we change the integrand by using the identities

$$\cos^2 x \equiv 1 - \sin^2 x \quad \text{or} \quad \sin^2 x \equiv 1 - \cos^2 x$$

consider
$$\int \sin^5 x \, dx = \int \sin^4 x \, (\sin x) \, dx$$

$$= \int (\sin^2 x)^2 \sin x \, dx$$

$$= \int (1 - \cos^2 x)^2 \sin x \, dx$$

$$= \int (1 - 2 \cos^2 x + \cos^4 x) \sin x \, dx$$

$$= \int (\sin x - 2 \cos^2 x \sin x + \cos^4 x \sin x) \, dx$$

Each of these terms can be integrated by inspection.

Hence $\displaystyle\int \sin^5 x \, dx = -\cos x + \tfrac{2}{3} \cos^3 x - \tfrac{1}{5} \cos^5 x + C$

Investigation 3

Use the technique above to show that

$$\int \cos^5 x \, dx = \int \cos^4 x \, (\cos x) \, dx = \int (1 - \sin^2 x)^2 \cos x \, dx$$

$$= \sin x - \tfrac{2}{3} \sin^3 x + \tfrac{1}{5} \sin^5 x + C$$

(ii) Even powers of sin x or cos x

In these cases we again change the integrand but this time we choose the appropriate identity from

$$\cos^2 x \equiv \tfrac{1}{2}(1+\cos 2x) \quad \text{.............} \textbf{(1)} \quad \text{or} \quad \sin^2 x \equiv \tfrac{1}{2}(1-\cos 2x) \quad \text{.............} \textbf{(2)}$$

To evaluate $\displaystyle\int \cos^4 x \, dx$ we write it as $\displaystyle\int (\cos^2 x)^2 \, dx$ and use identity **(1)**.

Hence, $\displaystyle\int_0^{\pi/4} \cos^4 x \, dx = \int_0^{\pi/4} (\cos^2 x)^2 \, dx = \int_0^{\pi/4} [\tfrac{1}{2}(1+\cos 2x)]^2 \, dx$

$$= \frac{1}{4} \int_0^{\pi/4} (1+2\cos 2x + \cos^2 2x) \, dx$$

Since the term $\cos^2 2x$ still cannot be evaluated, we repeat the application of the equivalent identity to **(1)**, $\cos^2 2x = \tfrac{1}{2}(1+\cos 4x)$.

$$\therefore \int_0^{\pi/4} \cos^4 x \, dx = \frac{1}{4} \int_0^{\pi/4} \left[1+2\cos 2x + \frac{1}{2}(1+\cos 4x) \right] dx$$

$$= \frac{1}{4} \int_0^{\pi/4} \left(\frac{3}{2} + 2\cos 2x + \frac{1}{2}\cos 4x \right) dx$$

$$= \frac{1}{4} \left[\frac{3x}{2} + \frac{2\sin 2x}{2} + \frac{1}{2}\left(\frac{\sin 4x}{4} \right) \right]_0^{\pi/4}$$

$$= \left[\frac{3x}{8} + \frac{1}{4}\sin 2x + \frac{1}{32}\sin 4x \right]_0^{\pi/4}$$

$$= \left(\frac{3\pi}{32} + \frac{1}{4} + 0 \right) - 0 = \frac{1}{32}(3\pi+8)$$

Investigation 4

Use the technique above to show that

(a) $\displaystyle\int \sin^4 x \, dx = \frac{3x}{8} - \frac{1}{4}\sin 2x + \frac{1}{32}\sin 4x + C$

(b) $\displaystyle\int \cos^6 x \, dx = \frac{5x}{16} + \frac{1}{4}\sin 2x + \frac{3}{64}\sin 4x - \frac{1}{48}\sin^3 2x + C$

(iii) Products of sin x and cos x

It may be possible to integrate certain combinations of $\sin x$ and $\cos x$ directly.

For example, $\displaystyle\int \sin^3 x \cos x \, dx = \tfrac{1}{4}\sin^4 x + C$ by inspection.

However, for products in the form $\sin ax \cos bx$, $\sin ax \sin bx$ and $\cos ax \cos bx$ we need to use the **factor formulae** to rearrange the integrand. These were derived on page 174 of Book 1 and are listed below.

$$\sin P + \sin Q \equiv 2\sin\tfrac{1}{2}(P+Q)\cos\tfrac{1}{2}(P-Q) \quad \text{.............} \textbf{(1)}$$

$$\sin P - \sin Q \equiv 2\cos\tfrac{1}{2}(P+Q)\sin\tfrac{1}{2}(P-Q) \quad \text{.............} \textbf{(2)}$$

$$\cos P + \cos Q \equiv 2\cos\tfrac{1}{2}(P+Q)\cos\tfrac{1}{2}(P-Q) \quad \text{.............} \textbf{(3)}$$

$$\cos P - \cos Q \equiv -2\sin\tfrac{1}{2}(P+Q)\sin\tfrac{1}{2}(P-Q) \quad \text{.............} \textbf{(4)}$$

Consider $\displaystyle\int \sin 6x \cos 2x \, dx$

Using the identity **(1)** above, we have $\frac{1}{2}(P+Q) = 6x \Rightarrow P+Q = 12x$

$$\text{and} \quad \tfrac{1}{2}(P-Q) = 2x \Rightarrow P-Q = 4x$$

Adding and subtracting gives $\quad 2P = 16x \Rightarrow P = 8x$

$$\text{and} \quad 2Q = 8x \Rightarrow Q = 4x$$

Hence, $\displaystyle\int \sin 6x \cos 2x \, dx = \int \tfrac{1}{2}(\sin 8x + \sin 4x) \, dx$

$$= -\frac{\cos 8x}{16} - \frac{\cos 4x}{8} + C$$

Investigation 5

Use the identities above to show that

(a) $\displaystyle\int \cos 3x \sin x \, dx = -\frac{\cos 4x}{8} + \frac{\cos 2x}{4} + C$

(b) $\displaystyle\int_0^{\pi/6} \sin 5x \sin x \, dx = \left[-\frac{\sin 6x}{12} + \frac{\sin 4x}{8} \right]_0^{\pi/6} = \frac{\sqrt{3}}{16}$

(c) $\displaystyle\int \cos \tfrac{1}{2}x \cos 2x \, dx = \frac{1}{5}\sin\left(\frac{5x}{2}\right) + \frac{1}{3}\sin\left(\frac{3x}{2}\right) + C$

Exercise 2.2

1 By using the identity $\cos^2\theta + \sin^2\theta \equiv 1$, evaluate

(a) $\displaystyle\int_0^{\pi/4} \sin^3\theta \, d\theta$ **(b)** $\displaystyle\int_0^{\pi/4} \cos^3\theta \, d\theta$

2 By using the identity $\cos 2x \equiv 2\cos^2 x - 1 \equiv 1 - 2\sin^2 x$ evaluate

(a) $\displaystyle\int_0^{\pi/2} \sin^2 x \, dx$ **(b)** $\displaystyle\int_0^{\pi/2} \cos^2 x \, dx$

3 Find   **(a)** $\displaystyle\int \sin\frac{x}{2}\cos\frac{3x}{2} \, dx$ **(b)** $\displaystyle\int \cos 3x \cos 5x \, dx$

4 Find   **(a)** $\displaystyle\int \cos^6 2x \, dx$ **(b)** $\displaystyle\int \sin^3\left(\frac{x}{4}\right) dx$

5 Show that $\frac{1}{8}(\cos 4\theta + 4\cos 2\theta + 3) = \cos^4\theta$ and find $\displaystyle\int_0^{\pi/2} \cos^4\theta \, d\theta$

Integration 2

Substitution

In the previous section we changed the integrand to a different form before integrating by using the trigonometrical identities. We can also change the integrand by making a **substitution** which changes the variable.

This can be done in two ways.

(i)   To integrate $\int x\sqrt{x+2}\,dx$ we could use a substitution $u = x+2$.

(ii)   To integrate $\int \dfrac{1}{9+x^2}\,dx$ we could use a substitution $x = 3\tan\theta$.

The purpose of this section is to see whether it is appropriate to use a substitution with u as a function of x or whether it is necessary to let x be a function of θ.

Method 1 – where u is defined as a function of x

Consider $y = \displaystyle\int x(x^2-4)^5\,dx$.

In this case we could choose the substitution $u = x^2 - 4$.

Hence $\dfrac{du}{dx} = 2x$ and since $y = \displaystyle\int x(x^2-4)^5\,dx$ we have $\dfrac{dy}{dx} = x(x^2-4)^5$

Using the chain rule $\dfrac{dy}{du} = \dfrac{dy}{dx} \times \dfrac{dx}{du}$

$$= x(x^2-4)^5 \frac{dx}{du}$$

Integrating both sides with respect to u gives $y = \displaystyle\int x(x^2-4)^5 \frac{dx}{du}\,du$

Using the values $u = (x^2-4)$ and $x\dfrac{dx}{du} = \dfrac{1}{2}$ we can write the integral in terms of u only.

Hence, $y = \displaystyle\int (u^5)\left(\frac{1}{2}\right) du = \int \frac{1}{2}u^5\,du$

$\therefore\quad y = \dfrac{u^6}{12} + C = \dfrac{1}{12}(x^2-4)^6 + C$

This integral could have been found by inspection. Check that you obtain the same result. However, the **method of substitution** can be applied to many other integrals which cannot be evaluated by inspection.

In general, to find $y = \displaystyle\int f(x)\,dx$ .... **(1)**

Let u be an appropriate function of x.   From **(1)**, $\dfrac{dy}{dx} = f(x)$.

By using the chain rule $\dfrac{dy}{du} = \dfrac{dy}{dx} \times \dfrac{dx}{du}$

$$= f(x)\frac{dx}{du}$$

Integrating with respect to u, gives $\quad y = \int f(x)\left(\dfrac{dx}{du}\right) du$

$$\boxed{\int f(x)\, dx = \int f(x)\left(\frac{dx}{du}\right) du}$$

WORKED EXAMPLE

Find $\quad \displaystyle\int x\sqrt{5x+1}\, dx.$

The appropriate substitution may not be given and must be chosen to suit the function to be integrated. In this case we could use $u = \sqrt{5x+1}$ or $u = 5x+1$.

Let $\quad u = \sqrt{5x+1} \quad \Rightarrow \quad u^2 = 5x+1$

Differentiating implicitly with respect to x gives

$$2u\frac{du}{dx} = 5 \quad \Rightarrow \quad \frac{dx}{du} = \frac{2u}{5}$$

Now, using the method of substitution $\quad \displaystyle\int f(x)\, dx = \int f(x)\left(\frac{dx}{du}\right) du$

$$\int x\sqrt{5x+1}\, dx = \int x\sqrt{5x+1}\left(\frac{dx}{du}\right) du$$

Substituting for $f(x)$ and $\dfrac{dx}{du}$ in terms of u gives

$$\int x\sqrt{5x+1}\, dx = \int \left(\frac{u^2-1}{5}\right)(u)\left(\frac{2u}{5}\right) du$$

$$= \frac{2}{25}\int (u^2-1)u^2\, du = \frac{2}{25}\int (u^4 - u^2)\, du$$

$$= \frac{2}{25}\left(\frac{u^5}{5} - \frac{u^3}{3}\right) + C$$

This needs to be returned to an expression in terms of x by using the substitution $u^2 = 5x+1$ but it is often easier to simplify in terms of u first.

$$\therefore \quad \int x\sqrt{5x+1}\, dx = \frac{2u^3}{25}\left(\frac{u^2}{5} - \frac{1}{3}\right) + C$$

$$= \frac{2u^3}{375}(3u^2 - 5) + C$$

$$= \frac{2(5x+1)^{3/2}}{375}[3(5x+1) - 5] + C$$

$$= \frac{2(5x+1)^{3/2}}{375}(15x - 2) + C$$

Now try to repeat this problem by using the substitution $u = 5x+1$.

Investigation 6

Use the substitution $u = 2x - 1$ to find $\int x(2x-1)^8 \, dx$.

Show that $\dfrac{dx}{du} = \dfrac{1}{2}$ and that $\int x(2x-1)^8 \, dx = \int \dfrac{u^8}{4}(u+1) \, du$.

Integrate with respect to u and show that the result after simplification is $\dfrac{1}{360}(18x+1)(2x-1)^9 + C$.

Investigation 7

The choice of substitution is not unique but a particular option may make the working longer or more complicated.

Use the substitutions **(i)** $u = \sqrt{x-1}$ and **(ii)** $u = x - 1$ to evaluate $\displaystyle\int \dfrac{x}{\sqrt{x-1}} \, dx$.

In case **(i)** show that $x = u^2 + 1$ and $\dfrac{dx}{du} = 2u$ which gives

$$\int \frac{x}{\sqrt{x-1}} \, dx = \int 2(u^2+1) \, du = \frac{2}{3}(x+2)(x-1)^{\frac{1}{2}} + C$$

In case **(ii)**, show that $x = u + 1$ and $\dfrac{dx}{du} = 1$ which gives

$$\int \frac{x}{\sqrt{x-1}} \, dx = \int \frac{u+1}{u^{\frac{1}{2}}} \, du = \int (u^{\frac{1}{2}} + u^{-\frac{1}{2}}) \, du$$

$$= \frac{2}{3}(x+2)(x-1)^{\frac{1}{2}} + C$$

> **WORKED EXAMPLE**

Find $\displaystyle\int \cos^2 x \sin^3 x \, dx$.

In this case the substitution must involve a trigonometrical function, say, $u = \cos x$.

Let $u = \cos x \Rightarrow \dfrac{du}{dx} = -\sin x \Rightarrow \dfrac{dx}{du} = -\dfrac{1}{\sin x}$

$$\therefore \quad \int \cos^2 x \sin^3 x \, dx = \int \cos^2 x \sin^3 x \left(\frac{dx}{du}\right) du$$

$$= \int \cos^2 x \sin^3 x \left(\frac{-1}{\sin x}\right) du$$

$$= -\int \cos^2 x \sin^2 x \, du = -\int \cos^2 x (1 - \cos^2 x) \, du$$

$$= -\int u^2(1-u^2)\,du = -\int (u^2-u^4)\,du$$

$$= -\left(\frac{u^3}{3}-\frac{u^5}{5}\right)+C = \frac{u^3}{15}(3u^2-5)+C$$

Replacing u by $\cos x$ we have

$$\int \cos^2 x \sin^3 x \, dx = \frac{1}{15}\cos^3 x \,(3\cos^2 x-5)+C$$

Investigation 8

Repeat the above Worked Example by using the identity $\cos^2 x \equiv 1-\sin^2 x$ to show that

$$\int \cos^2 x \sin^3 x \, dx = \int (\sin^3 x - \sin^5 x)\,dx$$

Use the method for odd powers of $\sin x$ to show that the same result is obtained.

Exercise 2.3

Find the following integrals using the given substitutions.

1 $\displaystyle\int (x-3)^6\,dx; \quad u=x-3$

2 $\displaystyle\int (3x+1)^4\,dx; \quad u=3x+1$

3 $\displaystyle\int (1-4x)^3\,dx; \quad u=1-4x$

4 $\displaystyle\int x(2+x)^5\,dx; \quad u=2+x$

5 $\displaystyle\int x(3-x^2)^3\,dx; \quad u=3-x^2$

6 $\displaystyle\int x\sqrt{x+1}\,dx; \quad u=x+1$

7 $\displaystyle\int x\sqrt{2x-3}\,dx; \quad u=2x-3$

8 $\displaystyle\int 2x\sqrt{x^2+9}\,dx; \quad u^2=x^2+9$

9 $\displaystyle\int \cos x \sin^4 x \, dx; \quad u=\sin x$

10 $\displaystyle\int \tan^4 x \sec^2 x \, dx; \quad u=\tan x$

11 $\displaystyle\int \sin x \cos^7 x \, dx; \quad u=\cos x$

12 $\displaystyle\int (x-2)^5\,(x+1)^2\,dx; \quad u=x-2$

13 $\displaystyle\int \frac{x+2}{\sqrt{2x-1}}\,dx; \quad u=\sqrt{2x-1}$

14 $\displaystyle\int (x-3)\sqrt{x-4}\,dx; \quad u=\sqrt{x-4}$

Find the following integrals.

15 $\displaystyle\int x(1-x)^4\,dx$

16 $\displaystyle\int x(2-x^2)^5\,dx$

17 $\displaystyle\int \sqrt{3+4x}\,dx$

18 $\displaystyle\int \frac{2x}{\sqrt{x+2}}\,dx$

19 $\displaystyle\int \frac{x+5}{(x-3)^2}\,dx$

20 $\displaystyle\int 2x^2\sqrt{x^3+1}\,dx$

21 $\displaystyle\int \tan x\sqrt{\sec x}\,dx$ **22** $\displaystyle\int \cos x\sqrt{\sin x}\,dx$ **23** $\displaystyle\int \frac{\cos\sqrt{x}}{\sqrt{x}}\,dx$

24 Use the substitution **(i)** $u=\sqrt{\cos x}$ and **(ii)** $u=\cos x$ to find $\displaystyle\int \frac{\sin^3 x}{\sqrt{\cos x}}\,dx$ showing the results to be the same.

Method 2 – where x is defined as a function of θ

Consider $y=\displaystyle\int g(x)\,dx$ and let $x=f(\theta)$.

Using the chain rule $\dfrac{dy}{d\theta}=\dfrac{dy}{dx}\times\dfrac{dx}{d\theta}$ $\Rightarrow$ $\dfrac{dy}{d\theta}=g(x)\dfrac{dx}{d\theta}$.

Integrating with respect to θ gives a similar result to that obtained in Method 1:

$$\int g(x)\,dx = \int g(x)\left(\frac{dx}{d\theta}\right)d\theta$$

WORKED EXAMPLE

Find $\displaystyle\int \frac{1}{\sqrt{1-x^2}}\,dx.$

In this case the substitution $u=\sqrt{1-x^2}$ gives $\dfrac{du}{dx}=-x(1-x^2)^{-\frac{1}{2}}$ by the approach of Method 1.

Hence, $\dfrac{dx}{du}=-\dfrac{\sqrt{1-x^2}}{x}$ and since $x=\sqrt{1-u^2}$ we have

$$\int \frac{1}{\sqrt{1-x^2}}\,dx = \int \frac{1}{\sqrt{1-x^2}}\left(\frac{dx}{du}\right)du$$

$$= -\int \frac{u}{u\sqrt{1-u^2}}\,du$$

$$= -\int \frac{1}{\sqrt{1-u^2}}\,du$$

Now this is basically the same integral as the original and hence it cannot be integrated by the chosen substitution.

We try the substitution $x=\sin\theta$ instead, since $\sqrt{1-x^2}$ can be written as $\sqrt{1-\sin^2\theta}=\cos\theta$.

So, let $x=\sin\theta$ $\Rightarrow$ $\dfrac{dx}{d\theta}=\cos\theta.$

Hence,

$$\int \frac{1}{\sqrt{1-x^2}}\,dx = \int \frac{1}{\sqrt{1-x^2}} \left(\frac{dx}{d\theta}\right) d\theta$$

$$= \int \frac{1}{\sqrt{1-\sin^2\theta}} (\cos\theta)\,d\theta = \int \frac{\cos\theta}{\cos\theta}\,d\theta$$

$$= \int 1\,d\theta = \theta + C$$

But as $x = \sin\theta$, $\theta = \sin^{-1}x$. Remember that $\sin^{-1}x$ is the inverse sine function and is sometimes written as arcsin x.

$$\int \frac{1}{\sqrt{1-x^2}}\,dx = \sin^{-1}x + C$$

Investigation 9

Use the substitution $x = \tan\theta$ to find $\displaystyle\int \frac{1}{1+x^2}\,dx$.

Show that $\dfrac{dx}{d\theta} = \sec^2\theta$ and hence $\displaystyle\int \frac{1}{1+x^2}\,dx = \int 1\,d\theta$

Deduce that

$$\int \frac{1}{1+x^2}\,dx = \tan^{-1}x + C$$

Investigation 10

Find $\displaystyle\int \frac{1}{a^2+x^2}\,dx$ by using the substitution $x = a\tan\theta$.

(a) Show that $\dfrac{dx}{d\theta} = a\sec^2\theta$ and that $\displaystyle\int \frac{1}{a^2+x^2}\,dx = \int \frac{1}{a}\,d\theta$

Deduce that

$$\int \frac{1}{a^2+x^2}\,dx = \frac{1}{a}\tan^{-1}\frac{x}{a} + C$$

(b) Check your result by differentiating $\dfrac{1}{a}\tan^{-1}\dfrac{x}{a}$.

Let $y = \dfrac{1}{a}\tan^{-1}\dfrac{x}{a}$ which gives $\tan ay = \dfrac{x}{a}$.

Differentiate implicitly with respect to x and show that

$$\frac{dy}{dx} = \frac{1}{a^2\sec^2 ay} = \frac{1}{a^2(1+\tan^2 ay)} = \frac{1}{a^2+x^2}$$

Investigation 11

(a) Find $\displaystyle\int \frac{1}{\sqrt{a^2-x^2}}\,dx$ by using the substitution $x = a\sin\theta$.

You should find that

$$\int \frac{1}{\sqrt{a^2-x^2}}\,dx = \sin^{-1}\frac{x}{a} + C$$

(b) Check your result by differentiating implicitly with respect to x.

Thus if $y = \sin^{-1}\dfrac{x}{a}$ then $\sin y = \dfrac{x}{a}$.

Deduce that $\dfrac{dy}{dx} = \dfrac{1}{a\cos y} = \dfrac{1}{\sqrt{a^2-x^2}}$.

(c) Use the substitution $x = a\cos\theta$ to show that this result can also be obtained in the form,

$$\int \frac{1}{\sqrt{a^2-x^2}}\,dx = -\cos^{-1}\frac{x}{a} + C$$

WORKED EXAMPLES

1 Find $\displaystyle\int \frac{1}{49+4x^2}\,dx.$

This integral can be performed by making the substitution $x = \frac{7}{2}\tan\theta$ but it is permissible to use the standard result found in Investigation 10.

The integrand needs a slight modification so that it is in the form $\dfrac{1}{a^2+x^2}$.

So, $\displaystyle\int \frac{1}{49+4x^2}\,dx = \frac{1}{4}\int \frac{1}{(\frac{49}{4}+x^2)}\,dx = \frac{1}{4}\int \frac{1}{(\frac{7}{2})^2+x^2}\,dx$

Hence, in this case $a = \frac{7}{2}$

Thus $\displaystyle\int \frac{1}{49+4x^2}\,dx = \frac{1}{4}\left(\frac{2}{7}\right)\tan^{-1}\left(\frac{2x}{7}\right) + C = \frac{1}{14}\tan^{-1}\left(\frac{2x}{7}\right) + C$

2 Find $\displaystyle\int \sqrt{9-x^2}\,dx$

Let $x = 3\sin\theta \;\Rightarrow\; \dfrac{dx}{d\theta} = 3\cos\theta$

So $\displaystyle\int \sqrt{9-x^2}\,dx = \int \sqrt{9-x^2}\left(\frac{dx}{d\theta}\right)d\theta = \int \sqrt{9-9\sin^2\theta}\,(3\cos\theta)\,d\theta = \int 3\cos\theta\,(3\cos\theta)\,d\theta = \int 9\cos^2\theta\,d\theta$

Since $\cos 2\theta = 2\cos^2\theta - 1 \Rightarrow \cos^2\theta = \frac{1}{2}(1+\cos 2\theta)$

$$\int \sqrt{9-x^2}\, dx = \int \frac{9}{2}(1+\cos 2\theta)\, d\theta$$

$$= \frac{9}{2}\theta + \frac{9}{4}\sin 2\theta + C$$

$$= \frac{9}{2}\sin^{-1}\left(\frac{x}{3}\right) + \frac{9}{4}\sin 2\theta + C$$

$$= \frac{9}{2}\sin^{-1}\left(\frac{x}{3}\right) + \frac{9}{4}(2\cos\theta\sin\theta) + C$$

$$= \frac{9}{2}\sin^{-1}\left(\frac{x}{3}\right) + \frac{x}{2}\sqrt{9-x^2} + C$$

3 Find $\displaystyle\int \frac{1}{x^2-2x+5}\, dx.$

Again the integrand can be rearranged by completing the square in the denominator.

So, $x^2-2x+5 \equiv (x^2-2x+1)+4 \equiv (x-1)^2+4$

and $\displaystyle\int \frac{1}{x^2-2x+5}\, dx = \int \frac{1}{4+(x-1)^2}\, dx$

Now, let $u=x-1 \Rightarrow \dfrac{du}{dx}=1$ and $\dfrac{dx}{du}=1$

Hence, $\displaystyle\int \frac{1}{x^2-2x+5}\, dx = \int \frac{1}{4+u^2}\, du = \frac{1}{2}\tan^{-1}\left(\frac{u}{2}\right) + C$

$$= \frac{1}{2}\tan^{-1}\left(\frac{x-1}{2}\right) + C$$

Exercise 2.4

Find the following integrals using the substitutions given.

1 $\displaystyle\int \frac{1}{4+x^2}\, dx; \quad x=2\tan\theta$ **2** $\displaystyle\int \frac{1}{\sqrt{4-x^2}}\, dx; \quad x=2\sin\theta$

3 $\displaystyle\int \sqrt{16-x^2}\, dx; \quad x=4\sin\theta$ **4** $\displaystyle\int \frac{1}{\sqrt{4-9x^2}}\, dx; \quad x=\frac{2}{3}\sin\theta$

Find the following integrals.

5 $\displaystyle\int \frac{3}{\sqrt{1-x^2}}\, dx$ **6** $\displaystyle\int \sqrt{25-x^2}\, dx$ **7** $\displaystyle\int \frac{1}{16+9x^2}\, dx$

Find the following integrals by completing the square in the integrand.

8 $\displaystyle\int \frac{1}{x^2-4x+5}\, dx$ **9** $\displaystyle\int \frac{1}{x^2-6x+18}\, dx$ **10** $\displaystyle\int \frac{1}{3+2x+x^2}\, dx$

11 Use the substitution $2x = 5\tan\theta$ to find $\displaystyle\int \frac{2}{25 + 4x^2}\, dx$

12 By using the substitution $x = \sin\theta$ show that

$$\int \sqrt{1 - x^2}\, dx = \tfrac{1}{2}(\sin^{-1} x + x\sqrt{1 - x^2}) + C$$

13 Find **(a)** $\displaystyle\int \frac{1}{17 + 4x + 4x^2}\, dx$ **(b)** $\displaystyle\int \frac{2x^2}{\sqrt{1 - x^2}}\, dx$

14 Use the substitution $x = a\sin\theta$ to show that

$$\int \sqrt{a^2 - x^2}\, dx = \tfrac{1}{2}x\sqrt{a^2 - x^2} + \tfrac{1}{2}a^2 \sin^{-1}\left(\frac{x}{a}\right) + C$$

Definite integrals and changing the limits

When using the method of substitution to evaluate **definite integrals** it is important to remember that the given limits are values of x and not values of the new variable say, u.

The limits could be inserted at the end of the problem after replacing u in terms of x, but it is more convenient to change the limits as well as the integrand into the new variable.

WORKED EXAMPLES

1 Calculate the area under the curve $y = \sin^3 x$ from $x = 0$ to $x = \pi/2$.

Since the area under a curve is given by

$$A = \int_a^b y\, dx$$

it follows that, in this case,

$$A = \int_0^{\pi/2} \sin^3 x\, dx$$

Use the substitution $u = \cos x$.

Hence $\dfrac{du}{dx} = -\sin x$

$\Rightarrow \quad \dfrac{dx}{du} = -\dfrac{1}{\sin x}$

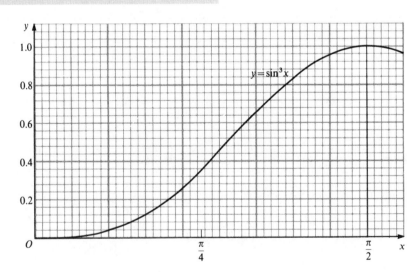

Now

$$\int_0^{\pi/2} \sin^3 x\, dx = \int_{x=0}^{x=\pi/2} \sin^3 x \left(\frac{dx}{du}\right) du = -\int_{x=0}^{x=\pi/2} \sin^2 x\, du$$

Since the integral is now with respect to u, we change the limits.

If $u = \cos x$ then when $x = 0$, $u = 1$

and when $x = \pi/2$, $u = 0$

Hence $\displaystyle\int_0^{\pi/2} \sin^3 x \, dx = -\int_1^0 (1 - \cos^2 x) \, du$

$\displaystyle = \int_1^0 (u^2 - 1) \, du$

$\displaystyle = \left[\frac{u^3}{3} - u \right]_1^0$

$= 0 - (\tfrac{1}{3} - 1)$

Hence the area under the curve is $\frac{2}{3}$ unit^2.

---

2 Find $\displaystyle\int_0^2 \frac{1}{\sqrt{16 - x^2}} \, dx$ by changing the variable and the limits using the substitution $x = 4 \sin \theta$.

Since $x = 4 \sin \theta$, it follows that $\dfrac{dx}{d\theta} = 4 \cos \theta$.

Also, when $x = 0$, $\theta = 0$

and when $x = 2$, $\theta = \dfrac{\pi}{6}$

$\therefore \displaystyle\int_0^2 \frac{1}{\sqrt{16 - x^2}} \, dx = \int_{x=0}^{x=2} \frac{1}{\sqrt{16 - x^2}} \frac{dx}{d\theta} \, d\theta = \int_0^{\pi/6} \frac{1}{\sqrt{16(1 - \sin^2 \theta)}} (4 \cos \theta) \, d\theta$

$\displaystyle = \int_0^{\pi/6} \frac{4 \cos \theta}{4 \cos \theta} \, d\theta = \int_0^{\pi/6} 1 \, d\theta$

$\displaystyle = \Big[\theta \Big]_0^{\pi/6} = \left(\frac{\pi}{6} - 0 \right)$

$\displaystyle = \frac{\pi}{6}$

---

3 Evaluate $\displaystyle\int_1^5 \frac{4x + 1}{\sqrt{2x - 1}} \, dx$.

Let $u = \sqrt{2x - 1} \;\Rightarrow\; u^2 = 2x - 1$.

Differentiating implicitly with respect to x gives

$$2u \frac{du}{dx} = 2 \;\Rightarrow\; \frac{du}{dx} = \frac{1}{u} \;\Rightarrow\; \frac{dx}{du} = u$$

Also, when $x = 1$, $u = 1$

and when $x = 5$, $u = 3$

$$\text{So} \quad \int_1^5 \frac{4x+1}{\sqrt{2x-1}} \, dx = \int_{x=1}^{x=5} \frac{4x+1}{\sqrt{2x-1}} \left(\frac{dx}{du}\right) du$$

$$= \int_1^3 \left(\frac{2u^2+3}{u}\right) u \, du = \int_1^3 (2u^2+3) \, du$$

$$= \left[\frac{2u^3}{3} + 3u\right]_1^3 = (18+9) - (\tfrac{2}{3}+3)$$

$$= 24 - \tfrac{2}{3} = 23\tfrac{1}{3}$$

4 Sketch the curve whose parametric equations are $x = t^2 - 1$ and $y = 2t^3 - 1$. Find the area enclosed by the x-axis, the part of the curve for which $1 \leqslant t \leqslant 2$ and the lines $x = 0$, $x = 3$.

The shape of the curve can be found by constructing a table of values of x and y for specific values of t and plotting these on a pair of axes.

t	-3	-2	-1	0	1	2	3
x	8	3	0	-1	0	3	8
y	-55	-17	-3	-1	1	15	53

The sketch of the curve is shown below with the area to be found shaded.

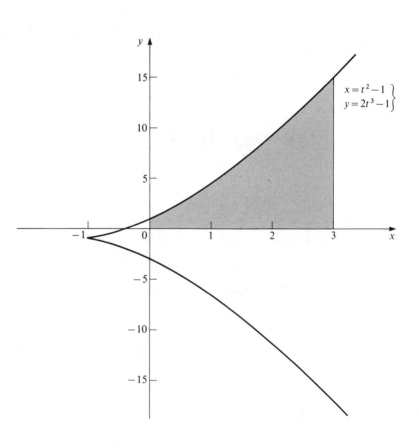

Now, Area $= \int y \, dx$, which we can write as $A = \int y \left(\frac{dx}{dt}\right) dt$.

Since $x = t^2 - 1 \quad \Rightarrow \quad \frac{dx}{dt} = 2t$

The limits of the parameter t are already given as $t=1$ and $t=2$.

Thus,
$$\text{Area} = \int_1^2 y\left(\frac{dx}{dt}\right) dt = \int_1^2 (2t^3-1)(2t)\, dt$$
$$= \int_1^2 (4t^4-2t)\, dt$$
$$= \left[\frac{4t^5}{5}-t^2\right]_1^2$$
$$= \left(\frac{128}{5}-4\right)-\left(\frac{4}{5}-1\right)$$
$$= 21.8$$

The area of the shaded portion is 21.8 units^2.

Exercise 2.5

Evaluate the following integrals.

1 $\displaystyle\int_1^{\sqrt{3}} \frac{2}{1+x^2}\, dx$

2 $\displaystyle\int_{\frac{1}{2}}^1 \frac{5}{\sqrt{1-x^2}}\, dx$

3 $\displaystyle\int_0^2 \sqrt{4-x^2}\, dx$

4 $\displaystyle\int_0^{1/4} \frac{1}{\sqrt{1-4x^2}}\, dx$

5 $\displaystyle\int_0^{4/3} \frac{1}{16+9x^2}\, dx$

6 $\displaystyle\int_0^{\sqrt{3}/2} \frac{1}{\sqrt{3-x^2}}\, dx$

Evaluate the following definite integrals by changing the variable and the limits.

7 $\displaystyle\int_0^1 x(x+1)^4\, dx$

8 $\displaystyle\int_1^5 9x\sqrt{3x+1}\, dx$

9 $\displaystyle\int_0^{\pi/6} \sec^4 x \tan x\, dx$

10 $\displaystyle\int_0^{\frac{1}{2}} \frac{x}{\sqrt{1-x^2}}\, dx$

11 $\displaystyle\int_1^2 (x-1)(x-2)^3\, dx$

12 $\displaystyle\int_{-2}^2 x^2\sqrt{(2-x)}\, dx$

13 Use the substitution $u=x^2$ to show that $\displaystyle\int_0^1 \frac{x}{1+x^4}\, dx = \frac{\pi}{8}$.

14 Sketch the curve whose parametric equations are $x=t^2$ and $y=t^3$.
Find the area enclosed by the x-axis, the line $x=4$, and the part of the curve for which $0\leqslant t\leqslant 2$.

15 Find the area bounded by the x-axis, the line $x=\frac{\pi}{2}-1$, and the curve whose parametric equations are $x=\theta-\sin\theta$ and $y=1-\cos\theta$ for $0\leqslant\theta\leqslant\frac{1}{2}\pi$.

16 Find the area enclosed by the x-axis and the curve $y=2x(1-x)^5$.

17 The parametric equations of an ellipse are given by $x=4\cos\theta$ and $y=3\sin\theta$. Find the area of the ellipse.

18 Use the substitution $x=4\sin^2\theta$ to show that $\displaystyle\int_0^2 \sqrt{x(4-x)}\, dx = \pi$.

Integration by parts

We have seen that it is possible to integrate a product of two functions in certain circumstances. For example,

$$\left.\begin{array}{c} \displaystyle\int 2x(x^2+3)^4\,dx = \tfrac{1}{5}(x^2+3)^5 + C \\[2mm] \displaystyle\int \cos x \sin^3 x\,dx = \tfrac{1}{4}\sin^4 x + C \\[2mm] \displaystyle\int 2x\,e^{x^2}\,dx = e^{x^2} + C \end{array}\right\} \text{ By inspection}$$

However, if the integrand is a product of two functions of x but not in a form where it can be integrated by inspection, it may be possible to apply a technique based upon the **product rule for differentiation**.

If $y = uv$ where u and v are functions of x, then the product rule gives

$$\frac{dy}{dx} = u\frac{dv}{dx} + v\frac{du}{dx} \quad \text{or} \quad \frac{d}{dx}(uv) = u\frac{dv}{dx} + v\frac{du}{dx}$$

Integrating with respect to x

$$\int \left[\frac{d}{dx}(uv)\right]dx = \int u\frac{dv}{dx}\,dx + \int v\frac{du}{dx}\,dx$$

$$uv = \int u\frac{dv}{dx}\,dx + \int v\frac{du}{dx}\,dx$$

Rearranging gives

$$\boxed{\int u\frac{dv}{dx}\,dx = uv - \int v\frac{du}{dx}\,dx}$$

This rule provides a method for integrating a product of two functions of x. It does not produce the final result immediately but gives a second integral which, hopefully, is easier to integrate.

Notice that it must be possible to identify in the original product two terms denoted by u (which can be differentiated with respect to x) and $\dfrac{dv}{dx}$ (which can be integrated with respect to x). These differential and integral functions make up the new integral to be evaluated.

This technique is called **integration by parts**.

WORKED EXAMPLE

Find $\displaystyle\int x \cos x\,dx$.

It is necessary to decide which part of the integrand will be represented by u and which part by $\dfrac{dv}{dx}$.

Let $\quad u = x \quad \Rightarrow \quad \dfrac{du}{dx} = 1$

and $\quad \dfrac{dv}{dx} = \cos x \quad \Rightarrow \quad v = \sin x$

At this stage we do not need a constant of integration when integrating $\dfrac{dv}{dx}$ since any particular integral will do (i.e. the case when the constant of integration is zero).

Using the result for integration by parts

$$\int u \frac{dv}{dx} dx = uv - \int v \frac{du}{dx} dx$$

$$\therefore \quad \int x \cos x \, dx = x \sin x - \int (\sin x \times 1) \, dx$$

$$\Rightarrow \quad \int x \cos x \, dx = x \sin x + \cos x + C$$

For definite integrals evaluated using integration by parts, we use the standard result with appropriate limits.

$$\int_a^b u \frac{dv}{dx} dx = \left[uv \right]_a^b - \int_a^b v \frac{du}{dx} dx$$

WORKED EXAMPLES

1 Evaluate $\displaystyle\int_0^1 xe^{3x} \, dx$.

Let $\quad u = x \quad \Rightarrow \quad \dfrac{du}{dx} = 1$

and $\quad \dfrac{dv}{dx} = e^{3x} \quad \Rightarrow \quad v = \tfrac{1}{3}e^{3x}$

Integrating by parts

$$\int_a^b u \frac{dv}{dx} dx = \left[uv \right]_a^b - \int_a^b v \frac{du}{dx} dx$$

$$\therefore \quad \int_0^1 xe^{3x} \, dx = \left[\tfrac{1}{3}xe^{3x} \right]_0^1 - \int_0^1 (\tfrac{1}{3}e^{3x} \times 1) \, dx$$

$$= \tfrac{1}{3}e^3 - \int_0^1 \tfrac{1}{3}e^{3x} \, dx$$

$$= \tfrac{1}{3}e^3 - \left[\tfrac{1}{9}e^{3x} \right]_0^1$$

$$= \tfrac{1}{3}e^3 - \tfrac{1}{9}e^3 + \tfrac{1}{9}$$

$$= \tfrac{1}{9}(2e^3 + 1)$$

2 Find $\displaystyle\int_0^{\pi/2} x^2 \sin 2x \, dx$.

In some examples it is possible that the new integral still cannot be evaluated although it might be in a form where we can apply the method of integration by parts again.

Let $\quad u = x^2 \quad \Rightarrow \quad \dfrac{du}{dx} = 2x$

and $\quad \dfrac{dv}{dx} = \sin 2x \quad \Rightarrow \quad v = -\tfrac{1}{2}\cos 2x$

Integrating by parts

$$\int_a^b u\frac{dv}{dx}\,dx = \left[uv\right]_a^b - \int_a^b v\frac{du}{dx}\,dx \qquad \ldots\ldots\ldots\ (1)$$

$$\therefore \quad \int_0^{\pi/2} x^2 \sin 2x\,dx = \left[-\tfrac{1}{2}x^2\cos 2x\right]_0^{\pi/2} - \int_0^{\pi/2} 2x(-\tfrac{1}{2}\cos 2x)\,dx$$

$$\Rightarrow \quad \int_0^{\pi/2} x^2 \sin 2x\,dx = \left(\frac{\pi^2}{8} - 0\right) + \int_0^{\pi/2} x\cos 2x\,dx$$

$$= \frac{\pi^2}{8} + \int_0^{\pi/2} x\cos 2x\,dx \qquad \ldots\ldots\ldots\ (2)$$

This new integral cannot be evaluated as it stands but we can apply the technique of integration by parts to it.

Consider $\quad \displaystyle\int_0^{\pi/2} x\cos 2x\,dx$

Let $\quad u = x \quad \Rightarrow \quad \dfrac{du}{dx} = 1$

and $\quad \dfrac{dv}{dx} = \cos 2x \quad \Rightarrow \quad v = \tfrac{1}{2}\sin 2x$

Integrating by parts using Equation **(1)** above

$$\int_0^{\pi/2} x\cos 2x\,dx = \left[\tfrac{1}{2}x\sin 2x\right]_0^{\pi/2} - \int_0^{\pi/2} \tfrac{1}{2}\sin 2x\,dx$$

$$= 0 - \frac{1}{2}\int_0^{\pi/2}\sin 2x\,dx$$

$$= \left[\tfrac{1}{4}\cos 2x\right]_0^{\pi/2} = (-\tfrac{1}{4} - \tfrac{1}{4}) = -\tfrac{1}{2}$$

Substituting into Equation **(2)** we have

$$\int_0^{\pi/2} x^2 \sin 2x\,dx = \frac{\pi^2}{8} - \frac{1}{2}$$

Investigation 12

The choice of which terms should be represented by u and $\dfrac{dv}{dx}$ is important since, if they are selected unwisely, it may mean that the new integral created cannot be evaluated or even that $\dfrac{dv}{dx}$ itself cannot be integrated.

(a) Let $\quad u = \ln x \quad$ and $\quad \dfrac{dv}{dx} = x \quad$ and show that

$$\int x\ln x\,dx = \tfrac{1}{2}x^2\ln x - \tfrac{1}{4}x^2 + C$$

(cont.)

Investigation 12—continued

(b) What happens if we choose $u = x$ and $\dfrac{dv}{dx} = \ln x$?

(c) By choosing $u = \cos x$ and $\dfrac{dv}{dx} = x$ show that

$$\int x \cos x \, dx = \tfrac{1}{2} x^2 \cos x - \int \tfrac{1}{2} x^2 \sin x \, dx$$

This is clearly not a useful choice since the new integral is more difficult to evaluate than the original.

Compare the working and the result obtained in the Worked Example on page 44.

WORKED EXAMPLE

Find $\displaystyle\int \tan^{-1} x \, dx$

Sometimes it is possible to apply the technique to a function which is not a product. In such cases we take $\dfrac{dv}{dx} = 1$.

Let $\quad u = \tan^{-1} x \quad \Rightarrow \quad \dfrac{du}{dx} = \dfrac{1}{1+x^2}$

and $\dfrac{dv}{dx} = 1 \quad \Rightarrow \quad v = x$

Integrating by parts

$$\int u \frac{dv}{dx} \, dx = uv - \int v \frac{du}{dx} \, dx$$

$$\therefore \quad \int [(\tan^{-1} x) \times 1] \, dx = x \tan^{-1} x - \int \frac{x}{1+x^2} \, dx$$

We can integrate the new integral since it is almost in the form $\displaystyle\int \frac{f'(x)}{f(x)} \, dx$ which was discussed on page 22.

$$\therefore \quad \int \tan^{-1} x \, dx = x \tan^{-1} x - \tfrac{1}{2} \ln (1+x^2) + C$$

Investigation 13

(a) Use the method of the previous Worked Example to show that

$$\int \ln x \, dx = x \ln x - x \quad \text{by choosing} \quad u = \ln x \quad \text{and} \quad \frac{dv}{dx} = 1$$

(b) By choosing $u = x$ and $\dfrac{dv}{dx} = \ln x$ and using the result obtained in part **(a)**, show that

$$\int x \ln x \, dx = (x^2 \ln x - x^2) - \int x \ln x \, dx + \int x \, dx$$

(cont.)

Investigation 13—continued

Since the term $\int x \ln x \, dx$ occurs also on the right hand side of this equation it can be combined with the same term on the left hand side.

Show that

$$2 \int x \ln x \, dx = x^2 \ln x - x^2 + \tfrac{1}{2}x^2 + A$$

Deduce that

$$\int x \ln x \, dx = \tfrac{1}{2}x^2 \ln x - \tfrac{1}{4}x^2 + C$$

This was the same as the result obtained in Investigation 12(**a**), although that method was rather more direct.

The technique of producing the original integral in the working can be used to advantage in a number of situations.

WORKED EXAMPLES ⟩

Find $\displaystyle\int e^{2x} \cos x \, dx$

Let $\quad u = e^{2x} \quad \Rightarrow \quad \dfrac{du}{dx} = 2e^{2x}$

and $\dfrac{dv}{dx} = \cos x \quad \Rightarrow \quad v = \sin x$

Integrating by parts

$$\int u \frac{dv}{dx} \, dx = uv - \int v \frac{du}{dx} \, dx$$

$$\therefore \quad \int e^{2x} \cos x \, dx = e^{2x} \sin x - 2 \int e^{2x} \sin x \, dx \quad \text{.............. (1)}$$

The new integral still cannot be evaluated as it stands but we can apply integration by parts again.

Consider $\displaystyle\int e^{2x} \sin x \, dx$

Let $\quad u = e^{2x} \quad \Rightarrow \quad \dfrac{du}{dx} = 2e^{2x}$

and $\dfrac{dv}{dx} = \sin x \quad \Rightarrow \quad v = -\cos x$

Integrating by parts gives

$$\int e^{2x} \sin x \, dx = -e^{2x} \cos x - \int 2e^{2x}(-\cos x) \, dx$$

$$= -e^{2x} \cos x + \int 2e^{2x} \cos dx$$

Substituting into (1) gives

$$\int e^{2x} \cos x \, dx = e^{2x} \sin x - 2\left\{ -e^{2x} \cos x + \int 2e^{2x} \cos x \, dx \right\}$$

$$= e^{2x}(\sin x + 2 \cos x) - 4 \int e^{2x} \cos x \, dx$$

Hence,
$$5 \int e^{2x} \cos x \, dx = e^{2x}(\sin x + 2 \cos x) + A$$

$$\int e^{2x} \cos x \, dx = \tfrac{1}{5}e^{2x}(\sin x + 2 \cos x) + C \qquad (C = \tfrac{1}{5}A)$$

Exercise 2.6

Find the following integrals.

1 $\displaystyle\int x \sin x \, dx$ **2** $\displaystyle\int xe^{2x} \, dx$ **3** $\displaystyle\int x \cos 3x \, dx$

4 $\displaystyle\int xe^{-x} \, dx$ **5** $\displaystyle\int \sin^{-1} x \, dx$ **6** $\displaystyle\int x \tan^{-1} x \, dx$

7 $\displaystyle\int x^2 \cos x \, dx$ **8** $\displaystyle\int x^3 e^x \, dx$ **9** $\displaystyle\int x^2 \ln x \, dx$

10 $\displaystyle\int x \, (\ln x)^2 \, dx$ **11** $\displaystyle\int x \tan^2 x \, dx$ **12** $\displaystyle\int \ln 2x \, dx$

13 $\displaystyle\int e^x \sin x \, dx$ **14** $\displaystyle\int e^{-x} \cos x \, dx$ **15** $\displaystyle\int e^{2x} \sin 2x \, dx$

16 $\displaystyle\int x^{-2} \ln x \, dx$ **17** $\displaystyle\int x \sec^2 x \, dx$ **18** $\displaystyle\int x \cos nx \, dx$

Find the following definite integrals.

19 $\displaystyle\int_1^e \ln x \, dx$ **20** $\displaystyle\int_0^2 x^2 e^x \, dx$ **21** $\displaystyle\int_0^{\pi/4} x \cos 2x \, dx$

22 $\displaystyle\int_0^1 x^2 e^{3x} \, dx$ **23** $\displaystyle\int_1^2 x^n \ln x \, dx$ **24** $\displaystyle\int_1^2 \ln (x^2) \, dx$

25 $\displaystyle\int_0^{\pi/2} x \cos^2 x \, dx$ **26** $\displaystyle\int_0^{\pi/2} e^x \cos 2x \, dx$ **27** $\displaystyle\int_0^{\frac{1}{2}} \cos^{-1} x \, dx$

28 By writing $x^5 e^{x^3}$ as $x^3(x^2 e^{x^3})$ find $\displaystyle\int x^5 e^{x^3} \, dx$.

29 Find the area contained by the curve $y = x \ln x$, the x-axis and the line $x = 2$.

30 Find the area contained by the curve $y = e^{2x} \sin x$ and the x-axis between $x = 0$ and $x = \pi$.

31 Find $\displaystyle\int_0^N e^{-2x} \cos 3x \, dx$ and deduce that $\displaystyle\int_0^\infty e^{-2x} \cos 3x \, dx = \tfrac{2}{13}$.

32 Use the substitution $u = \sin x$ to show that $\displaystyle\int_0^{\pi/6} \sec x \, dx = \tfrac{1}{2} \ln 3$.

Hence use integration by parts to show that $\displaystyle\int_0^{\pi/6} \sec^3 x \, dx = \tfrac{1}{3} + \tfrac{1}{4} \ln 3$.

Further substitutions

There are many special techniques which we can use for integration. However, these are not always included in every advanced level syllabus. This section illustrates a useful method when the integrand contains a fraction whose denominator consists of the sum of sin x and cos x. Readers should check their syllabuses to see if these methods are included.

The substitution $t = \tan \frac{1}{2}x$ can be used, for example, in the following cases:

$$\int \frac{1}{5 + 4\cos x}\, dx \qquad \int \frac{\sin x}{1 - \cos x}\, dx \qquad \int \frac{1}{\sin x}\, dx$$

Investigation 14

(a) Use the identity $\tan A \equiv \dfrac{2\tan\frac{1}{2}A}{1 - \tan^2\frac{1}{2}A}$ to show that if $t = \tan\frac{1}{2}x$ then $\tan x = \dfrac{2t}{1 - t^2}$.

Using a right-angled triangle with sides $2t$ and $(1 - t^2)$ such that $\tan x = \dfrac{2t}{1 - t^2}$ show that the hypotenuse is $(1 + t^2)$.

Hence deduce that $\sin x = \dfrac{2t}{1 + t^2}$ and $\cos x = \dfrac{1 - t^2}{1 + t^2}$.

(b) Differentiate $t = \tan\frac{1}{2}x$ with respect to x and show that

$$\frac{dx}{dt} = \frac{2}{1 + \tan^2\frac{1}{2}x} = \frac{2}{1 + t^2}$$

The results obtained in parts (a) and (b) of Investigation 14 form the basic substitutions when integrating functions of this type.

If $t = \tan\frac{1}{2}x$

$$\frac{dx}{dt} = \frac{2}{1 + t^2} \qquad \sin x = \frac{2t}{1 + t^2} \qquad \cos x = \frac{1 - t^2}{1 + t^2}$$

WORKED EXAMPLES ⟩

Find $\displaystyle\int_0^{\pi/2} \frac{1}{5 + 4\cos x}\, dx$.

Using the substitution $t = \tan\frac{1}{2}x$ we have $\cos x = \dfrac{1 - t^2}{1 + t^2}$ and $\dfrac{dx}{dt} = \dfrac{2}{1 + t^2}$.

Also when $x = 0$, $t = 0$; and when $x = \frac{1}{2}\pi$; $t = 1$.

Hence,
$$\int_0^{\pi/2} \frac{1}{5+4\cos x}\,dx = \int_0^1 \frac{1}{5+4\cos x}\left(\frac{dx}{dt}\right)dt$$

$$= \int_0^1 \frac{1}{[5+4(1-t^2)/(1+t^2)]}\left(\frac{2}{1+t^2}\right)dt$$

$$= \int_0^1 \frac{2}{[5(1+t^2)+4(1-t^2)]}\,dt$$

$$= \int_0^1 \frac{2}{9+t^2}\,dt = \left[\tfrac{2}{3}\tan^{-1}\left(\frac{t}{3}\right)\right]_0^1$$

$$= \tfrac{2}{3}\tan^{-1}(\tfrac{1}{3})-\tfrac{2}{3}\tan^{-1}(0)$$

$$= \tfrac{2}{3}\tan^{-1}(\tfrac{1}{3})$$

Investigation 15

This method often produces a final integral which needs to be split using **partial fractions**, as discussed on page 21 of Book 1. Consider an integral similar to the one in the Worked Example but where the numerical values have been changed.

Show by using the substitution $t=\tan\tfrac{1}{2}x$, that
$$\int_0^{\pi/2} \frac{1}{4+5\cos x}\,dx = \int_0^1 \frac{2}{9-t^2}\,dt$$

By using partial fractions, show that this can be written as
$$\int_0^1 \frac{2}{9-t^2}\,dt = \int_0^1 \left[\frac{1}{3(3-t)}+\frac{1}{3(3+t)}\right]dt.$$

Hence, show that $\displaystyle\int_0^{\pi/4} \frac{1}{4+5\cos x}\,dx = \tfrac{1}{3}\ln 2.$

Investigation 16

If the integrand contains sines or cosines of **multiple angles**, then the substitution can be modified to $t=\tan$ (half the given angle).

So for $\displaystyle\int \frac{1}{3+5\sin 4x}\,dx$ choose $t=\tan 2x.$

Using this substitution, show that $\dfrac{dx}{dt} = \dfrac{1}{2(1+t^2)}$

and that $\sin 4x = \dfrac{2t}{1+t^2};\quad \cos 4x = \dfrac{1-t^2}{1+t^2}.$

Hence, show that $\displaystyle\int_0^{\pi/8} \frac{1}{3+5\sin 4x}\,dx = \frac{1}{2}\int_0^1 \frac{1}{(t+3)(3t+1)}\,dt$

Use partial fractions to show that this can be written as $\displaystyle\frac{1}{16}\int_0^1 \left(\frac{3}{(3t+1)}-\frac{1}{(t+3)}\right)dt = \tfrac{1}{16}\ln 4$

Investigation 17

If the integrand contains **even powers** of $\sin x$ or $\cos x$, then try $t = \tan x$ as a substitution.

Given that $t = \tan x$, show that $\dfrac{dx}{dt} = \dfrac{1}{1+t^2}$

and that $\sin^2 x = \dfrac{t^2}{1+t^2}; \quad \cos^2 x = \dfrac{1}{1+t^2}.$

Show that $\displaystyle\int \frac{1}{1+\sin^2 x}\, dx = \int \frac{1}{1+2t^2}\, dt$

$$= \frac{1}{\sqrt{2}} \tan^{-1}\left(\sqrt{2}t\right) + C$$

$$= \frac{1}{\sqrt{2}} \tan^{-1}\left(\sqrt{2}\tan x\right) + C$$

Exercise 2.7

Find the following integrals using the substitution $t = \tan \frac{1}{2}x$.

1 $\displaystyle\int_{0}^{\pi/2} \frac{1}{13 + 12\cos x}\, dx$ **2** $\displaystyle\int \frac{1}{3 + 5\cos x}\, dx$ **3** $\displaystyle\int \frac{1}{5 + 3\cos x}\, dx$

4 $\displaystyle\int \frac{1}{1 + \sin x}\, dx$ **5** $\displaystyle\int \frac{1}{3\sin x - 4\cos x}\, dx$ **6** $\displaystyle\int_{\pi/3}^{\pi/2} \frac{dx}{1 + \sin x + \cos x}$

7 Use the substitution $t = \tan \frac{1}{2}x$ to show that $\displaystyle\int \operatorname{cosec} x\, dx = \ln(\tan \frac{1}{2}x) + C.$

8 Use the substitution $t = \tan x$ to show that $\displaystyle\int_{0}^{\pi/4} \frac{dx}{1 + \sin 2x} = \frac{1}{2}.$

9 Using the substitution given find

    (a) $\displaystyle\int \frac{1}{\cos 4x}\, dx; \quad t = \tan 2x$ **(b)** $\displaystyle\int \frac{1}{5 + 3\cos \frac{1}{2}x}\, dx; \quad t = \tan \frac{1}{4}x.$

10 Use the substitution $t = \tan \frac{1}{2}x$ to show that

$$\int \sec x\, dx = \ln\left(\frac{1 + \tan \frac{1}{2}x}{1 - \tan \frac{1}{2}x}\right) + C$$

Deduce that $\displaystyle\int \sec x\, dx = \ln(\sec x + \tan x) + C.$

By writing $\sec x = \operatorname{cosec}(x + \frac{1}{2}\pi)$ and using the result of Question **7**, show that

$$\int \sec x\, dx = \ln \tan\left(\frac{x}{2} + \frac{\pi}{4}\right) + C$$

11 Find, using the substitution $t = \tan x$

    (a) $\displaystyle\int \frac{dx}{1 + 3\sin^2 x}$ **(b)** $\displaystyle\int \frac{dx}{1 + \cos^2 x}$

12 Use the substitution $t = \tan x$ to show that

$$\int \frac{1}{1 - 2\sin^2 x}\, dx = \tfrac{1}{2}\ln\left[\tan\left(x + \frac{\pi}{4}\right)\right] + C$$

13 Evaluate the following integrals

(a) $\displaystyle\int_0^{\pi/3} \frac{1}{5\cos^2\theta - 1}\, d\theta$ (b) $\displaystyle\int_0^{\pi/4} \frac{4}{4\cos^2\theta - 9\sin^2\theta}\, d\theta$ (c) $\displaystyle\int_0^{\pi/4} \frac{\sin^2 x}{1 + \cos^2 x}\, dx$

Exercise 2.8   Miscellaneous

1 Integrate with respect to x,   (a) $x\sqrt{x^2 + 3}$ (b) $\dfrac{x(x^2 - 2)}{\sqrt{x^4 - 4x^2 - 2}}$.

2 Evaluate (a) $\displaystyle\int_0^{\pi} (\cos x + \sin x)^2\, dx$ (b) $\displaystyle\int_0^{\pi/3} \sin 3x \cos x\, dx$.

3 Evaluate (a) $\displaystyle\int_1^3 \frac{x - 2}{x + 1}\, dx$ (b) $\displaystyle\int_0^{\pi/2} \frac{\sin\theta}{\sqrt{\cos\theta}}\, d\theta$.

4 Express $\dfrac{3}{9 - x^2}$ in partial fractions and hence find $\displaystyle\int \frac{3}{9 - x^2}\, dx$.

5 Evaluate (a) $\displaystyle\int_1^2 \frac{2 - x - 3x^2}{x^3}\, dx$ (b) $\displaystyle\int_0^2 (1 + x^2)^2\, dx$.

6 Find (a) $\displaystyle\int \mathrm{cosec}^2\, 2x\, dx$ (b) $\displaystyle\int \sec x \tan x\, dx$.

7 Find (a) $\displaystyle\int \cos^4 x \sin x\, dx$ (b) $\displaystyle\int \sec^5 x \tan x\, dx$.

8 Find the following integrals by changing the variable,

(a) $\displaystyle\int 4x(3x - 2)^5\, dx$ (b) $\displaystyle\int \frac{1}{2x\sqrt{1 - 4x^2}}\, dx$

9 Use the substitution $t = \tan\tfrac{1}{2}\theta$ to evaluate

(a) $\displaystyle\int_0^{\pi/2} \frac{1}{1 + \sin\theta}\, d\theta$ (b) $\displaystyle\int_0^{\pi/2} \frac{\sin\theta}{2 - \cos\theta}\, d\theta$

10 Use the substitution $t = \tan x$ to evaluate

(a) $\displaystyle\int_0^{\pi/3} \frac{dx}{9 - 8\sin^2 x}$ (b) $\displaystyle\int_0^{\pi/4} \frac{dx}{\cos^2 x + 3\sin^2 x}$

11 Use the substitution $u = \pi - x$ to show that

$$\int_0^{\pi} \frac{x\, dx}{1 + \sin x} = \pi \int_0^{\pi/2} \frac{dx}{1 + \sin x}$$

12 Use the substitution $x = \pi - u$ to show that

$$\int_0^\pi \frac{x \sin x}{1 + \cos^2 x}\, dx = \frac{\pi}{2} \int_0^\pi \frac{\sin x}{1 + \cos^2 x}\, dx$$

Hence by letting $\cos x = \tan \theta$ show that $\displaystyle\int_0^\pi \frac{\sin x}{1 + \cos^2 x}\, dx = \frac{\pi}{2}$.

Deduce that $\displaystyle\int_0^\pi \frac{x \sin x}{1 + \cos^2 x}\, dx = \frac{\pi^2}{4}$.

13 Show that, if $a \neq b$ and a and b are positive integers, $\displaystyle\int_0^\pi \cos ax \cos bx \, dx = 0$.

14 Find the volume of the solid of revolution formed when the area contained by the curve $y = \cos^2 x$ between $x = \dfrac{\pi}{2}$ and $x = \dfrac{3\pi}{2}$ and the x-axis is rotated completely about the x-axis.

15 Find the area between the curve $y = xe^{-2x}$, the axes and the line $x = 2$.

16 Find **(a)** $\displaystyle\int e^{2x} \cos 3x \, dx$ **(b)** $\displaystyle\int e^x \sin^2 x \, dx$.

3 Partial fractions

Introduction

Investigation 1

Fill in numbers for the right-hand sides of the identities, e.g. $\dfrac{7}{8} = \dfrac{1}{2} + \dfrac{1}{4} + \dfrac{1}{8}$ $\dfrac{1}{6} = \dfrac{1}{2} - \dfrac{1}{3}$

(a) $\dfrac{3}{8} = \dfrac{?}{2} + \dfrac{?}{4} + \dfrac{?}{8}$

(b) $\dfrac{5}{6} = \dfrac{?}{2} + \dfrac{?}{3}$

(c) $\dfrac{5}{8} = \dfrac{?}{2} + \dfrac{?}{4} + \dfrac{?}{8}$

(d) $\dfrac{11}{12} = \dfrac{?}{4} + \dfrac{?}{3} + \dfrac{?}{2}$

(e) $\dfrac{7}{12} = \dfrac{?}{6} + \dfrac{?}{4} + \dfrac{?}{3} + \dfrac{?}{2}$

(f) $\dfrac{11}{15} = \dfrac{?}{3} + \dfrac{?}{5}$

(g) $\dfrac{13}{8} = \dfrac{?}{8} + \dfrac{?}{?} + \dfrac{?}{2}$

(h) $\dfrac{15}{16} = \dfrac{1}{?} + \dfrac{1}{?} + \dfrac{1}{?} + \dfrac{1}{?} + \dfrac{1}{?}$

(i) $\dfrac{5}{18} = \dfrac{?}{?} + \dfrac{?}{9} + \dfrac{?}{3}$

(j) $\dfrac{19}{17} = \dfrac{?}{?} + \dfrac{?}{?}$

In some cases there is more than one solution.

Can you establish any rules?

Can **every** fraction be expressed as a combination of simpler fractions?

Investigation 2

We considered simple partial fractions in Book 1.

Try to separate the following expressions into the partial fraction format suggested on the right-hand side.

(a) $\dfrac{2x+1}{(x^2+1)(x+2)} \equiv \dfrac{p}{(x^2+1)} + \dfrac{q}{(x+2)}$

(b) $\dfrac{2x+1}{(x^2+1)(x+2)} \equiv \dfrac{px+q}{(x^2+1)} + \dfrac{r}{(x+2)}$

(c) $\dfrac{2x+1}{(x+1)^2(x+2)} \equiv \dfrac{p}{(x+1)^2} + \dfrac{q}{(x+2)}$

(d) $\dfrac{2x+1}{(x+1)^2(x+2)} \equiv \dfrac{px+q}{(x+1)^2} + \dfrac{r}{(x+2)}$

(cont.)

(e) $$\frac{3x+2}{(x+1)^2} \equiv \frac{p}{(x+1)^2} + \frac{q}{(x+1)}$$

(f) $$\frac{2x+1}{(x+1)^2(x+2)} \equiv \frac{p}{(x+1)^2} + \frac{q}{(x+1)} + \frac{r}{(x+2)}$$

Check which of your answers are correct by reversing the process, i.e. by putting the separate fractions together again. Now see the following sections.

Quadratic factors

Example   (a) $$\frac{2x+1}{(x^2+1)(x+2)} \equiv \frac{p}{(x^2+1)} + \frac{q}{(x+2)}$$

$$\Rightarrow \quad \frac{2x+1}{(x^2+1)(x+2)} \equiv \frac{p(x+2)}{(x^2+1)(x+2)} + \frac{q(x^2+1)}{(x^2+1)(x+2)}$$

$$\Rightarrow \quad 2x+1 \equiv p(x+2) + q(x^2+1)$$

Comparing coefficients of x^2 $\Rightarrow$ $0 = q$ which leads to no value for p since $2x+1$ cannot equal $p(x+2)$. So we must try the format suggested in **(b)**.

Example   (b) $$\frac{2x+1}{(x^2+1)(x+2)} \equiv \frac{px+q}{(x^2+1)} + \frac{r}{(x+2)}$$

$$\Rightarrow \quad 2x+1 \equiv (px+q)(x+2) + r(x^2+1)$$

$$x = -2 \Rightarrow \quad -3 = 0 + 5r \quad \Rightarrow \quad r = -3/5$$

$$x = 0 \Rightarrow \quad 1 = 2q + r \quad \Rightarrow \quad q = 4/5$$

Coefficients of x^2 $\Rightarrow \quad 0 = p + r \quad \Rightarrow \quad p = 3/5$

So, $$\frac{2x+1}{(x^2+1)(x+2)} \equiv \frac{3x+4}{5(x^2+1)} - \frac{3}{5(x+2)}$$

Repeated factors

Example   (c) $$\frac{2x+1}{(x+1)^2(x+2)} \equiv \frac{p}{(x+1)^2} + \frac{q}{(x+2)} \quad \text{............ (A)}$$

$$\Rightarrow \quad 2x+1 \equiv p(x+2) + q(x+1)^2$$

$x = -1 \Rightarrow -1 = p + 0 \Rightarrow p = -1$ and $x = -2 \Rightarrow -3 = q$

$$\Rightarrow \quad \frac{2x+1}{(x+1)^2(x+2)} = \frac{-1}{(x+1)^2} + \frac{-3}{(x+2)}$$

But $$\frac{-1}{(x+1)^2} + \frac{-3}{(x+2)} = \frac{-1(x+2) - 3(x+1)^2}{(x+1)^2(x+2)} = \frac{-3x^2 - 7x - 5}{(x+1)^2(x+2)} \neq \frac{2x+1}{(x+1)^2(x+2)}$$

So the identity **(A)** is **not** correct and the format **(d)** must be tried.

Example **(d)** $$\frac{2x+1}{(x+1)^2(x+2)} \equiv \frac{px+q}{(x+1)^2} + \frac{r}{(x+2)}$$

$$\Rightarrow \qquad 2x+1 \equiv (px+q)(x+2)+r(x+1)^2$$

$x=-2 \Rightarrow -3=r$ and $x=-1 \Rightarrow -p+q=-1$

Coefficients of x^2 give $p+r=0 \Rightarrow p=3 \Rightarrow q=+2$

So $$\frac{2x+1}{(x+1)^2(x+2)} \equiv \frac{3x+2}{(x+1)^2} + \frac{-3}{(x+2)} = \frac{3x+2}{(x+1)^2} - \frac{3}{(x+2)}$$

Example **(e)** shows that $\dfrac{3x+2}{(x+1)^2}$ can be simplified to $\dfrac{a}{(x+1)^2} + \dfrac{b}{(x+1)}$.

Method 1 $$\frac{3x+2}{(x+1)^2} \equiv \frac{a}{(x+1)^2} + \frac{b}{(x+1)} \quad \Rightarrow \quad 3x+2 \equiv a+b(x+1)$$

$x=-1 \Rightarrow -5=a$ and comparing coefficients of $x \Rightarrow b=3$

So $$\frac{3x+2}{(x+1)^2} \equiv \frac{-1}{(x+1)^2} + \frac{3}{(x+1)}$$

The **covering-up rule** (substitute $x=-1$ in LHS) gives the value of $a=-1$ (the numerator of $(x+1)^2$), but in this type of example it is safer to use Method 1.

Method 2 Express the numerator $3x+2$ in terms of the denominator $x+1$ i.e.

$$\frac{3x+2}{(x+1)^2} = \frac{3x+3-1}{(x+1)^2} = \frac{3(x+1)}{(x+1)^2} - \frac{1}{(x+1)^2} = \frac{3}{(x+1)} - \frac{1}{(x+1)^2}$$

Example **(d)** can now be **completely separated** into **3 partial fractions**

$$\frac{2x+1}{(x+1)^2(x+2)} \equiv \frac{3}{(x+1)} - \frac{1}{(x+1)^2} - \frac{3}{(x+2)}$$

Summary

> *Quadratic factors*
> Use the linear numerator $px+q$.
>
> For example $\dfrac{2x+1}{(x^2+1)(x+2)} \equiv \dfrac{px+q}{(x^2+1)} + \dfrac{r}{(x+2)}$

> *Repeated factors*
>
> $$\frac{2x+1}{(x+1)^2(x+2)} \equiv \frac{p}{(x+1)^2} + \frac{q}{(x+1)} + \frac{r}{(x+2)}$$

Partial fractions

1 Separate $\dfrac{3x}{(x+1)(x^2+2)}$ into partial fractions.

$$\frac{3x}{(x+1)(x^2+2)} \equiv \frac{a}{(x+1)} + \frac{bx+c}{(x^2+2)} \quad \Rightarrow \quad 3x \equiv a(x^2+2)+(bx+c)(x+1)$$

$x=-1 \quad \Rightarrow \quad -3=3a \quad \Rightarrow \quad a=-1$

and coefficients of $x^2 \quad \Rightarrow \quad 0=a+b \quad \Rightarrow \quad b=1$

Comparing coefficients of $x^2 \quad \Rightarrow \quad 3=b+c \quad \Rightarrow \quad c=2$

So
$$\frac{3x}{(x+1)(x^2+2)} \equiv \frac{-1}{(x+1)} + \frac{x+2}{(x^2+2)}$$

2 Separate $\dfrac{4x-8}{(x+1)(x-1)^2}$ into partial fractions.

$$\frac{4x-8}{(x+1)(x-1)^2} \equiv \frac{a}{x+1} + \frac{b}{(x-1)^2} + \frac{c}{x-1} \quad \Rightarrow \quad 4x-8 \equiv a(x-1)^2+b(x+1)+c(x+1)(x-1)$$

$x=1 \quad \Rightarrow \quad -4=2b \quad \Rightarrow \quad b=-2; \quad x=-1 \quad \Rightarrow \quad -12=4a \quad \Rightarrow \quad a=-3$

Comparing coefficients of $x^2 \quad \Rightarrow \quad 0=a+c \quad \Rightarrow \quad c=3$

So
$$\frac{4x-8}{(x+1)(x-1)} \equiv -\frac{3}{x+1} - \frac{2}{(x-1)^2} + \frac{3}{x-1}$$

Exercise 3.1

Separate the following into partial fractions.

1 (a) $\dfrac{2}{(x-1)(x+1)}$ **(b)** $\dfrac{2x}{(x+1)(x-1)}$ **(c)** $\dfrac{2x^2}{(x+1)(x-1)}$ **(d)** $\dfrac{2x^3}{(x+1)(x-1)}$

2 (a) $\dfrac{1}{(x-2)(x^2+2)}$ **(b)** $\dfrac{x}{(x-1)(x^2+2)}$ **(c)** $\dfrac{x^2}{(x-3)(x^2+1)}$ **(d)** $\dfrac{x^3}{x^3-x^2+x-1}$

3 (a) $\dfrac{4}{(x-2)(x+1)^2}$ **(b)** $\dfrac{4x+1}{(x-1)(x+2)^2}$ **(c)** $\dfrac{4x^2+2}{(x-3)(x+1)^2}$ **(d)** $\dfrac{4x^3+2x}{(x-1)(x+2)^2}$

4 (a) $\dfrac{2x+1}{(x+2)^3}$ **(b)** $\dfrac{2x+1}{(x^2+3)(x-2)}$ **(c)** $\dfrac{2x-3}{(x^2-2)(x-1)}$ **(d)** $\dfrac{3x-2}{(x^2-3)(x+2)}$

5 (a) $\dfrac{3}{x^2(x+2)}$ **(b)** $\dfrac{3x}{x^2(x-1)}$ **(c)** $\dfrac{3x^2+1}{x^2(x-2)}$ **(d)** $\dfrac{3x^3+1}{x^2(x+1)}$

Integration using partial fractions

Investigation 3

Integrate the following and check by differentiating your answer.

(a) $\displaystyle\int \frac{1}{1+x}\,dx$ **(b)** $\displaystyle\int \frac{x}{1+x^2}\,dx$ **(c)** $\displaystyle\int \frac{1}{1+x^2}\,dx$ **(d)** $\displaystyle\int \frac{2x}{1+x^2}\,dx$

(e) $\displaystyle\int \frac{2x}{x^2-1}\,dx$ **(f)** $\displaystyle\int \frac{2}{x^2-1}\,dx$ **(g)** $\displaystyle\int \frac{1}{(x+1)^2}\,dx$ **(h)** $\displaystyle\int \frac{1}{(x+1)(x+2)}\,dx$

(i) $\displaystyle\int \frac{2x}{(x+1)(x+2)}\,dx$ **(j)** $\displaystyle\int \frac{2x+3}{(x+1)(x+2)}\,dx$ **(k)** $\displaystyle\int_0^1 \frac{2x}{(x^2+1)(x+2)}\,dx$

(l) $\displaystyle\int_2^3 \frac{2x}{(x+1)^2}\,dx$ **(m)** $\displaystyle\int_1^2 \frac{2x}{(x+1)^2(x+2)}\,dx$

Answers to Investigation 3

(a) $\displaystyle\int \frac{1}{1+x}\,dx = \ln|1+x| + C$

(b) $\displaystyle\int \frac{x}{1+x}\,dx = \int \frac{x+1-1}{1+x}\,dx = \int \left(1 - \frac{1}{1+x}\right)dx = x - \ln|1+x| + C$

(c) $\displaystyle\int \frac{1}{1+x^2}\,dx$ is **not** equal to $\ln|1+x^2|$ which differentiates to $\dfrac{2x}{1+x^2}$.

We need to substitute $x = \tan u \ \Rightarrow\ \dfrac{dx}{du} = \sec^2 u$.

$$\int \frac{1}{1+x^2}\,dx = \int \frac{\sec^2 u}{1+\tan^2 u}\,du = \int \frac{\sec^2 u}{\sec^2 u}\,du = \int 1\,du = u + k = \tan^{-1}x + k$$

(d) $\displaystyle\int \frac{2x}{1+x^2}\,dx = \ln|1+x^2| + k$ from above by inspection,

                                     or by substituting $u = 1 + x^2$

(e) This is similar to **(d)** and can be recognized (the top being the derivative of the bottom).

$$\int \frac{2x}{x^2-1}\,dx = \ln|x^2-1| + k \quad \textbf{or by substituting } u = x - 1$$

                                   or by partial fractions, as for **(f)**

(f) This needs **partial fractions**

$$\int \frac{2}{x^2-1}\,dx = \int \frac{2}{(x-1)(x+1)}\,dx = \int \left(\frac{1}{x-1} - \frac{1}{x+1}\right)dx = \ln|x-1| - \ln|x+1| + k = \ln\left|\frac{x-1}{x+1}\right| + k$$

(g) $\displaystyle\int \frac{dx}{(x+1)^2} = \int (x+1)^{-2}\,dx = [-(x+1)^{-1}] = -\frac{1}{x+1} + k$

(h) $\displaystyle\int \frac{1}{(x+1)(x+2)}\,dx = \int \left(\frac{1}{x+1} - \frac{1}{x+2}\right)dx = \ln|x+1| - \ln|x+2| = \ln\left|\frac{x+1}{x+2}\right| + k$

(i) $\displaystyle\int \frac{2x}{(x+1)(x+2)}\,dx = \int\left(\frac{-2}{(x+1)} + \frac{4}{(x+2)}\right)dx = -2\ln|x+1| + 4\ln|x+2| = \ln\left(\frac{(x+2)^4}{(x+1)^2}\right) + k$

(j) $\displaystyle\int \frac{2x+3}{(x+1)(x+2)}\,dx = \int \frac{1}{x+1} + \frac{1}{x+2}\,dx = \ln|x+1| + \ln|x+2| = \ln|x+1|\,|x+2| + k$

or, if you recognize that the top is the derivative of the bottom,

$$\int \frac{2x+3}{(x+1)(x+2)}\,dx = \int \frac{2x+3}{x^2+3x+2}\,dx = \ln|x^2+3x+2| + k$$

(k) $\displaystyle\int_0^1 \frac{2x}{(x^2+1)(x+2)}\,dx$ needs partial fractions of the form $\displaystyle\frac{ax+b}{(x^2+1)} + \frac{c}{x+2}$.

$$\frac{2x}{(x^2+1)(x+2)} \equiv \frac{ax+b}{x^2+1} + \frac{c}{x+2} \quad\Rightarrow\quad 2x \equiv (ax+b)(x+2) + c(x^2+1)$$

$x=-2 \;\Rightarrow\; -4 = 5c \;\Rightarrow\; c = -4/5; \qquad x=0 \;\Rightarrow\; 0 = 2b+c \;\Rightarrow\; b = 2/5$

Coefficients of $x^2 \;\Rightarrow\; 0 = a+c \;\Rightarrow\; a = 4/5$

$$\int_0^1 \frac{2x}{(x^2+1)(x+2)}\,dx = \int_0^1\left(\frac{4x+2}{5(x^2+1)} - \frac{4}{5(x+2)}\right)dx = \int_0^1\left(\frac{4x}{5(x^2+1)} + \frac{2}{5(x^2+1)} - \frac{4}{5(x+2)}\right)dx$$

$$= \frac{2}{5}\int_0^1 \frac{2x}{x^2+1}\,dx + \frac{2}{5}\int_0^1 \frac{1}{x^2+1}\,dx - \frac{4}{5}\int_0^1 \frac{1}{x+2}\,dx$$

$$= \left[\frac{2}{5}\ln(x^2+1) + \frac{2}{5}\tan^{-1}x - \frac{4}{5}\ln|x+2|\right]_0^1 = \left[\frac{2}{5}\ln\left(\frac{(x^2+1)}{(x+2)^2}\right) + \frac{2}{5}\tan^{-1}x\right]_0^1$$

$$= \frac{2}{5}\ln\frac{2}{9} + \frac{2}{5}\tan^{-1}1 - \frac{2}{5}\ln\frac{1}{4} - \frac{2}{5}\tan^{-1}0$$

$$= \frac{2}{5}\ln\frac{2/9}{1/4} + \frac{2}{5}\tan^{-1}1 = \frac{2}{5}\ln\frac{8}{9} + \frac{\pi}{10} = 0.267$$

(l) $\displaystyle\int_2^3 \frac{2x}{(x+1)^2}\,dx = \int_2^3 \frac{(2x+2-2)}{(x+1)^2}\,dx = \int_2^3\left(\frac{2(x+1)}{(x+1)^2} - \frac{2}{(x+1)^2}\right)dx = \int_2^3\left(\frac{2}{x+1} - \frac{2}{(x+1)^2}\right)dx$

$$= \left[2\ln|x+1| + \frac{2}{x+1}\right]_2^3 = 2\ln 4 + \frac{1}{2} - 2\ln 3 - \frac{2}{3} = 2\ln\frac{4}{3} - \frac{1}{6} = 0.409$$

This example can also be done by a simple substitution $u = x+1$ which really leads to the same working.

(m) This needs partial fractions of the p, q, r form (**repeated factors**).

$$\int \frac{2x}{(x+1)^2(x+2)}\,dx \equiv \frac{p}{(x+1)^2} + \frac{q}{x+1} + \frac{r}{x+2} \quad\Rightarrow\quad 2x \equiv p(x+2) + q(x+1)(x+2) + r(x+1)^2$$

$x=-1 \;\Rightarrow\; -2 = p; \qquad x=-2 \;\Rightarrow\; -4 = r; \qquad 0 = qx^2 + rx^2 \;\Rightarrow\; q = -r = 4$

$$\int_1^2 \frac{2x}{(x+1)^2(x+2)}\,dx = \int_1^2\left(\frac{-2}{(x+1)^2} + \frac{4}{x+1} - \frac{4}{x+2}\right)dx = \left[\frac{2}{x+1} + 4\ln|x+1| - 4\ln|x+2|\right]_1^2$$

$$= \tfrac{2}{3} + 4\ln 3 - 4\ln 4 - 1 - 4\ln 2 + 4\ln 3 = 4\ln\tfrac{9}{8} - \tfrac{1}{3} = 0.138$$

As you will have seen in Chapter 1, you will need different methods for different questions, but the methods of partial fractions do **simplify** the functions to be integrated.

Exercise 3.2

This Exercise asks you to integrate the functions in Exercise 3.1, for which you have found partial fractions, so refer back for your answers.

In the questions which have limits, answers are also to be given in terms of x.

1 (a) $\int \dfrac{2}{(x-1)(x+1)}\,dx$ **(b)** $\int \dfrac{2x}{(x+1)(x-1)}\,dx$ **(c)** $\int \dfrac{2x^2}{(x+1)(x-1)}\,dx$

(d) $\int \dfrac{2x^3}{(x+1)(x-1)}\,dx$

2 (a) $\int \dfrac{1}{(x-2)(x^2+2)}\,dx$ **(b)** $\int_2^3 \dfrac{x}{(x-1)(x^2+2)}\,dx$ **(c)** $\int \dfrac{x^2}{(x-3)(x^2+1)}\,dx$

(d) $\int \dfrac{x^3}{x^3-x^2+x-1}\,dx$

3 (a) $\int \dfrac{4}{(x-2)(x+1)^2}\,dx$ **(b)** $\int_2^3 \dfrac{(4x+1)}{(x-1)(x+2)^2}\,dx$ **(c)** $\int_4^5 \dfrac{(4x^2+2)}{(x-3)(x+1)^2}\,dx$

(d) $\int_2^3 \dfrac{(4x^3+2x)}{(x-1)(x+2)^2}\,dx$

4 · (a) $\int_0^1 \dfrac{(2x+1)}{(x+2)^3}\,dx$ **(b)** $\int \dfrac{(2x+1)}{(x^2+3)(x-2)}\,dx$ **(c)** $\int_2^3 \dfrac{(2x-3)}{(x^2-2)(x-1)}\,dx$

(d) $\int \dfrac{(3x-2)}{(x^2-3)(x+2)}\,dx$

5 (a) $\int_1^2 \dfrac{3}{x^2(x+2)}\,dx$ **(b)** $\int_2^3 \dfrac{3x}{x^2(x-1)}\,dx$ **(c)** $\int_3^4 \dfrac{(3x^2+1)}{x^2(x-2)}\,dx$

(d) $\int_1^2 \dfrac{(3x^3+1)}{x^2(x+1)}\,dx$

Summation of series

Investigation 4

Find the sums of the following series.

(a) $\displaystyle\sum_{i=1}^{i=n} \dfrac{1}{2^i} = \dfrac{1}{2}+\dfrac{1}{4}+\cdots+\dfrac{1}{2^n}$ **(b)** $\displaystyle\sum_{i=1}^{\infty} \dfrac{1}{2^i} = \dfrac{1}{2}+\dfrac{1}{4}+\dfrac{1}{8}+\cdots$

(c) $\displaystyle\sum_{i=1}^{i=n} \dfrac{1}{i(i+1)} = \dfrac{1}{2}+\dfrac{1}{6}+\dfrac{1}{12}+\cdots+\dfrac{1}{n(n+1)}$ **(d)** $\dfrac{1}{1\times3}+\dfrac{1}{3\times5}+\cdots$ to n terms.

(e) $\dfrac{1}{1\times2\times3}+\dfrac{1}{2\times3\times4}+\cdots$ to n terms. **(f)** $\dfrac{1}{3}+\dfrac{1}{15}+\dfrac{1}{35}+\cdots+\dfrac{1}{4n^2-1}$

(cont.)

Investigation 4—continued

The series in **(a)** is a Geometric Progression with first term $a = \frac{1}{2}$ and common ratio $r = \frac{1}{2}$.

The sum to n terms is given by $S_n = \dfrac{a(1-r^n)}{1-r} = \dfrac{\frac{1}{2}[1-(\frac{1}{2})^n]}{1-\frac{1}{2}} = 1 - (\tfrac{1}{2})^n = \dfrac{2^n - 1}{2^n}$

(b) This is the sum to infinity of **(a)**, so $S_\infty = \dfrac{a}{1-r} = \dfrac{\frac{1}{2}}{1-\frac{1}{2}} = 1$

(c) This is summed using partial fractions noticing that $\dfrac{1}{i(i+1)} = \dfrac{1}{i} - \dfrac{1}{i+1}$. (See Book 1, page 276.)

So $\displaystyle\sum_1^n \frac{1}{i(i+1)} = \sum_1^n \left(\frac{1}{i} - \frac{1}{i+1}\right) = \left[1 - \frac{1}{2}\right] + \left[\frac{1}{2} - \frac{1}{3}\right] + \left[\frac{1}{3} - \frac{1}{4}\right] + \cdots + \left[\frac{1}{n} - \frac{1}{n+1}\right]$

Fractions at the end of each bracket cancel with the fraction at the beginning of the next bracket leaving

$$S_n = \left(1 - \frac{1}{n+1}\right) = \frac{n+1-1}{n+1} = \frac{n}{n+1}$$

Examples **(d)** and **(f)** are included in Exercise 3.3 while **(e)** has a variation.

$\dfrac{1}{1 \times 2 \times 3} + \dfrac{1}{2 \times 3 \times 4} + \ldots = \displaystyle\sum_{i=1}^{i=n} \frac{1}{i(i+1)(i+2)} = \sum_1^n \left(\frac{\frac{1}{2}}{i} + \frac{-1}{i+1} + \frac{\frac{1}{2}}{i+2}\right)$ using the **covering-up rule**

$2S_n = \displaystyle\sum_1^n \left(\frac{1}{i} - \frac{2}{i+1} + \frac{1}{i+2}\right) = \left[\frac{1}{1} - \frac{2}{2} + \frac{1}{3}\right]$

$\qquad\qquad + \left[\frac{1}{2} - \frac{2}{3} + \frac{1}{4}\right]$

$\qquad\qquad + \left[\frac{1}{3} - \frac{2}{4} + \frac{1}{5}\right]$

$\qquad\qquad + \ldots\ldots\ldots$

$\qquad\qquad + \left[\frac{1}{n-1} - \frac{2}{n} + \frac{1}{n+1}\right]$

$\qquad\qquad + \left[\frac{1}{n} - \frac{2}{n+1} + \frac{1}{n+2}\right]$

This time the fractions from 3 successive brackets cancel out leaving 3 terms

$$\frac{1}{1} - \frac{2}{2} + \frac{1}{2}$$

from the first 2 brackets and 3 terms

$$\frac{1}{n+1} - \frac{2}{n+1} + \frac{1}{n+2}$$

from the last 2 brackets.

$2S_n = 1 - \dfrac{2}{2} + \dfrac{1}{2} + \dfrac{1}{n+1} - \dfrac{2}{n+1} + \dfrac{1}{n+2} = \dfrac{1}{2} - \dfrac{1}{n+1} + \dfrac{1}{n+2} = \dfrac{1}{2} - \dfrac{[n+2-(n+1)]}{(n+1)(n+2)}$

$\qquad = \dfrac{1}{2} - \dfrac{1}{(n+1)(n+2)} = \dfrac{(n+1)(n+2) - 2}{(n+1)(n+2)} = \dfrac{n^2 + 3n}{2(n+1)(n+2)} = \dfrac{n(n+3)}{2(n+1)(n+2)}$

$$\Rightarrow S_n = \frac{n(n+3)}{4(n+1)(n+2)}$$

(cont.)

Investigation 4—continued

We can check by putting $n=2$; our formula gives $S_2 = \dfrac{2(5)}{4(3)(4)} = \dfrac{5}{24}$

The first two terms are $\dfrac{1}{6} + \dfrac{1}{24} = \dfrac{5}{24}$

Can you find the **sum** of all fractional series of this form using partial fractions or will it depend on the nature of the partial fractions?

Exercise 3.3

Find the sums of the following series.

1 (a) $\displaystyle\sum_1^n \left(\frac{2}{3}\right)^i$ (b) $\displaystyle\sum_1^\infty \left(\frac{2}{3}\right)^i$ (c) $\displaystyle\sum_2^{10} \frac{1}{i(i+1)}$ (d) $\displaystyle\sum_0^8 \frac{1}{(i+2)(i+3)}$

2 (a) $\dfrac{1}{1\times3} + \dfrac{1}{3\times5} + \dfrac{1}{5\times7} \ldots$ to n terms (b) $\displaystyle\sum_1^n \frac{1}{i(i+2)}$ (c) $\displaystyle\sum_1^\infty \frac{1}{i(i+2)}$

3 (a) $\dfrac{1}{2\times4} + \dfrac{1}{4\times6} + \dfrac{1}{6\times8} \ldots$ to n terms (b) $\displaystyle\sum_1^n \frac{1}{2i(2i+2)}$ (c) $\displaystyle\sum_1^\infty \frac{1}{4i(i+1)}$

4 (a) $\dfrac{1}{1\times3} + \dfrac{1}{2\times4} + \dfrac{1}{3\times5} \ldots$ to $2n$ terms (b) $\displaystyle\sum_1^{2n} \frac{1}{i(i+2)}$ (c) $\displaystyle\sum_0^\infty \frac{1}{(i+1)(i+3)}$

5 (a) $\dfrac{1}{3} + \dfrac{1}{15} + \dfrac{1}{35} + \ldots + \dfrac{1}{4n^2-1}$ (b) $\displaystyle\sum_1^n \left(\frac{1}{2i-1} - \frac{1}{2i+1}\right)$

6 (a) $\dfrac{1}{3} + \dfrac{1}{8} + \dfrac{1}{15} + \dfrac{1}{24} + \ldots$ to n terms (b) $\displaystyle\sum_2^n \frac{1}{i^2-1}$

7 (a) $(1\times2\times3) + (3\times4\times5) + \ldots$ to n terms (b) $\dfrac{1}{1\times2\times3} + \dfrac{1}{2\times3\times4} + \ldots$ to n terms

Investigation 5

Find the first five terms in each expansion.

(a) $\dfrac{1}{1+x} = (1+x)^{-1} =$

(b) $\dfrac{1}{1-x} = (1-x)^{-1} =$

(c) $\dfrac{1}{1+x} + \dfrac{1}{1-x} =$

(d) $\dfrac{1}{1-x^2} = (1-x^2)^{-1} =$

(e) $\dfrac{1}{(1+x)(1+2x)} = (1+x)^{-1}(1+2x)^{-1} =$

(f) $\dfrac{2}{1+2x} - \dfrac{1}{1+x} = 2(1+2x)^{-1} - (1+x)^{-1} =$

(g) $\dfrac{1}{1+3x+2x^2} = [1+(3x+2x^2)]^{-1} =$

Expanding series

The series in Investigation 5 may provide an alternative useful approach to series.

(a) $\dfrac{1}{1+x} \simeq 1-x+x^2-x^3+ \ldots$ valid for $|x|<1$

(b) $\dfrac{1}{1-x} \simeq 1+x+x^2+x^3+ \ldots$ valid for $|x|<1$

(c) $\dfrac{1}{1+x}+\dfrac{1}{1-x} \simeq 2+2x^2+2x^4+ \ldots$ valid for $|x|<1$

(d) $\dfrac{1}{1-x^2} \simeq 1+x^2+x^4+ \ldots$ valid for $|x|<1$

(e) $\dfrac{1}{(1+x)(1+2x)} = (1+x)^{-1}(1+2x)^{-1} = (1-x+x^2-x^3+ \ldots)(1-2x+4x^2-8x^3+ \ldots)$

$$= 1-3x+7x^2-15x^3 \ldots \text{ valid for } |x|<0.5$$

(f) $\dfrac{2}{1+2x}-\dfrac{1}{1+x} = 2(1+2x)^{-1}-(1+x)^{-1} = 2(1-2x+4x^2-8x^3+ \ldots)-(1-x+x^2-x^3+ \ldots)$

$$= 1-3x+7x^2-15x^3 \ldots$$

(g) $\dfrac{1}{1+3x+2x^2} = [1+(3x+2x^2)]^{-1} = 1-(3x+2x^2)+(3x+2x^2)^2-(3x+2x^2)^3+ \ldots = 1-3x+7x^2-15x^3$

(e), **(f)** and **(g)** are different forms of the same expression and **(f)** would appear to involve the least work and therefore possibly the less chance of making mistakes.

Exercise 3.4

Expand the series in the three ways of investigations **(e)**, **(f)** and **(g)** to get the first four terms.

1 (a) $\dfrac{2}{(1+x)(1-x)} = 2(1+x)^{-1}(1-x)^{-1} =$

(b) $\dfrac{1}{1+x}+\dfrac{1}{1-x} =$

(c) $\dfrac{2}{1-x^2} = 2(1-x^2)^{-1} =$

2 (a) $\dfrac{3}{(1+2x)(1-x)} = 3(1+2x)^{-1}(1-x)^{-1} =$

(b) $\dfrac{2}{1+2x}+\dfrac{1}{1-x} =$

(c) $\dfrac{3}{1+x-2x^2} = 3[1+(x-2x^2)]^{-1} =$

3 (a) $\dfrac{1+x}{1-x} = (1+x)(1-x)^{-1} =$

(b) $-1+\dfrac{2}{1-x} = -1+2(1-x)^{-1} =$

4 (a) $\dfrac{2+x}{(1+x)(1-x)} = (2+x)(1+x)^{-1}(1-x)^{-1} =$

(b) $\dfrac{\frac{1}{2}}{1+x}+\dfrac{\frac{3}{2}}{1-x} =$

(c) $(2+x)(1-x^2)^{-1} =$

5 Expand $\dfrac{5x}{(1-2x)(2+x)}$ in ascending powers of x as far as the term in x^3.

Revision Exercise B

This Exercise should enable you to revise work covered in Book 1.

Vectors

1   In a quadrilateral $PQRS$, the sides **PQ**, **QR** and **PS** are represented by the vectors **p**, **q** and **r** respectively. Express the vectors **PR**, **SR** and **QS** in terms of **p**, **q** and **r**.

2   $ABCDE$ is a regular pentagon in which **AB** = **a**, **BC** = **b** and **DC** = **c**. If O is the mid-point of AD, express **AD**, **OD** and **OB** in terms of **a**, **b** and **c**.

3   Find, by calculation, the resultant of two displacements given by 4.6 m, N 30° E and 3.5 m, S 21° W.

4   Find the resultant of the forces F_1, F_2 and F_3 which have magnitudes and directions given by 4 N along a line AB, 6 N at 60° to AB and $3\sqrt{2}$ N at 45° to AB, respectively, if F_2 and F_3 are on opposite sides of AB.

5   **OA**, **OB** and **OC** are three coplanar vectors represented by **a**, **b** and **c** respectively. If X divides the line segment AB in the ratio 1:2, and Y divides the line segment BC in the ratio 2:3, find **XY** in terms of **a** and **b**. If L and M are the mid-points of OX and OY respectively, show that **LM** $= \frac{1}{2}$**XY**.

6   Find the magnitude of the following vectors.

   (a) $\mathbf{i} + \mathbf{j}$ **(b)** $2\mathbf{i} - 3\mathbf{j} + \mathbf{k}$ **(c)** $-2\mathbf{i}$ **(d)** $\mathbf{i} - 3\mathbf{k}$

7   If **i**, **j** and **k** are the unit vectors along a set of mutually perpendicular axes Ox, Oy and Oz, respectively, find the magnitude and the direction vector of the resultant of the forces $F_1 = \mathbf{i} + \mathbf{j} + \mathbf{k}$, $F_2 = -2\mathbf{i} - \mathbf{j} + 4\mathbf{k}$, $F_3 = 2\mathbf{i} - \mathbf{j} - \mathbf{k}$ and $F_4 = \mathbf{i} - 3\mathbf{j} + \mathbf{k}$.

8   If $\mathbf{a} = \mathbf{i} + 3\mathbf{j} + 2\mathbf{k}$ and $\mathbf{b} = 2\mathbf{i} - \mathbf{j} + 3\mathbf{k}$, find $|\mathbf{a}|$ and $|2\mathbf{a} - \mathbf{b}|$. Find also the unit vector in the direction of $\mathbf{a} + 2\mathbf{b}$.

9   The line joining the points $A(1, 4)$ and $B(-7, 2)$ is divided internally in the ratio 3:1 by a point C. Find the position vector of C relative to **(a)** the origin and **(b)** a point $X(-2, -2)$.

10   The points A, B and C have position vectors $\mathbf{i} - 3\mathbf{k}$, $2\mathbf{i} - \mathbf{j} + 2\mathbf{k}$ and $3\mathbf{i} - 2\mathbf{j} - 3\mathbf{k}$ respectively. Find the vectors **AB**, **BC** and **AC** and deduce that the triangle ABC is isosceles.

11   The position vector of a particle at time t seconds is $\mathbf{r} = t^2\mathbf{i} + 2t\mathbf{j}$. Find its position vector at $t = 0, 1, 2, 3$ and 4. Find an expression for the velocity of the particle after t seconds. Deduce that the particle has a constant acceleration of 2 m s^{-2}.

12   A particle moves such that its velocity v at time t is given by $\mathbf{v} = 2t\mathbf{i} - 4\mathbf{j} + 3t^2\mathbf{k}$. After one second it is at the point $(0, 0, 4)$. Find its position vector after 2 seconds. Calculate its acceleration at this instant.

13   A particle is projected from a point O with an initial velocity $3\mathbf{i} + 4\mathbf{j}$ where **i** and **j** are the unit vectors in a horizontal and vertical direction. Find the velocity vector and position vector of the particle after t seconds.

14 A particle of mass 2 kg is acted upon by forces $\mathbf{F}_1 = 2\mathbf{i} + \mathbf{j}$, $\mathbf{F}_2 = 3\mathbf{i} + 2\mathbf{j} - \mathbf{k}$ and $\mathbf{F}_3 = \mathbf{i} - 5\mathbf{j} + 3\mathbf{k}$. Find the acceleration vector of the particle and its velocity vector after 4 seconds if it starts from rest. Deduce its position vector after 4 seconds if it is at the point with position vector $\mathbf{i} - \mathbf{j} + \mathbf{k}$ after 2 seconds.

15 A particle has a position vector given by $\mathbf{r} = \cos 2t\,\mathbf{i} + \sin 2t\,\mathbf{j} + t\mathbf{k}$. Find its velocity vector $\mathbf{v}$ and acceleration vector $\mathbf{a}$. Deduce that $|\mathbf{a}|$ is constant.

16 Two particles A and B move with velocity vectors $\mathbf{v}_1 = 2\mathbf{i} + \mathbf{j}$ and $\mathbf{v}_2 = -\mathbf{i} + 4\mathbf{j}$ respectively. Find the velocity of A relative to B in vector form.

17 To an observer on a ship A sailing with velocity $2\mathbf{i} - \mathbf{j}$, a ship B appears to be sailing with velocity $-\mathbf{i} + 3\mathbf{j}$. Find the actual velocity of the ship B if $\mathbf{i}$ and $\mathbf{j}$ are the unit vectors due east and due north, respectively.

18 To a runner moving at 16 km h^{-1} due north, the wind appears to blow from the east, but to a cyclist travelling at 24 km h^{-1} south-west the wind appears to blow from the south. If $a\mathbf{i} + b\mathbf{j}$ represents the actual velocity of the wind, where $\mathbf{i}$ and $\mathbf{j}$ are unit vectors due east and due north, respectively, write down the velocity vectors of the runner and the cyclist, and the apparent velocity vectors of the wind in each case. Hence find the true velocity vector of the wind.

19 A particle A starts from a point with position vector $9\mathbf{i} - \mathbf{j}$ and moves with a constant velocity $-\mathbf{i} + 2\mathbf{j}$. A second particle B starts from a point with position vector $2\mathbf{i} + 7\mathbf{j}$ one second later and moves with a constant velocity $2\mathbf{i} - \mathbf{j}$. Show that the particles collide and find the time that elapses before the impact.

4 Vectors

Introduction

In Book 1 we considered vector quantities (e.g. displacements, velocities, accelerations) which possess both magnitude and direction. In particular, we were able to specify the position of a point in space by defining its position vector in component form.

So, using $\mathbf{i}, \mathbf{j}$ and $\mathbf{k}$ as unit vectors along the co-ordinate axes Ox, Oy and Oz respectively, the position vector of the point $P(3, 2, 1)$ is given by

$$\mathbf{OP} = \mathbf{r} = 3\mathbf{i} + 2\mathbf{j} + \mathbf{k}$$

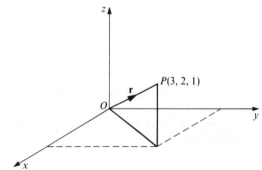

We also applied this idea to geometrical situations. In this chapter we shall develop our study of vectors to consider the geometry of straight lines and planes in three dimensions.

The equation of a straight line

In two dimensions

What do you understand by a linear equation? Consider, for example, the equation $2x + 3y = 9$.

The graph is shown here as a straight line passing through the points $(4\frac{1}{2}, 0)$ and $(0, 3)$.

The equation is really defining a relationship between two variables, x and y, such that the value of y is equal to $3 - \frac{2}{3}x$.

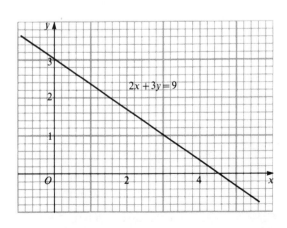

Now, if x and y define the co-ordinates of a point (x, y) on the graph, it follows that given values of x will produce corresponding values of y such that the point (x, y) lies on the straight line.

In other words, the equation defines a method for finding all the points lying on a particular straight line.

In three dimensions

Now you might assume that the linear equation in two dimensions, namely $ax+by=c$, can be extended to $ax+by+cz=d$ and that this will give the equation of a straight line in three dimensions. However, this is not the case although an equation of this form has an important meaning which will become clear in the following Investigation.

Investigation 1

The figure shows a cube of side 1 unit with one vertex at the origin and its edges parallel to the co-ordinate axes.

Consider the equation $x+y+z=2$ and choose values of x, y and z which satisfy the equation. Write your answers as a set of co-ordinates, e.g.

$$(1, 1, 0), (\tfrac{1}{2}, 1, \tfrac{1}{2}), (0, 0, 2)\ldots$$

Find enough solutions for which

$$0 \leqslant x \leqslant 1, \quad 0 \leqslant y \leqslant 1, \quad \text{and} \quad 0 \leqslant z \leqslant 1$$

and show these on a copy of the cube in the figure.

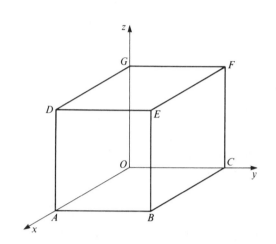

It should be clear that the points do not lie on a straight line. You can discover what is represented by the equation by marking the points $D(1, 0, 1)$, $B(1, 1, 0)$, $H(0, 0, 2)$, $X(1, \tfrac{1}{2}, \tfrac{1}{2})$ and $Y(\tfrac{1}{2}, 0, 1\tfrac{1}{2})$, drawing the lines BD, BH and DH and shading the triangle formed. The points X and Y should lie on BD and DH respectively.

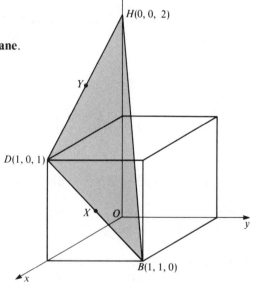

The triangle BDH defines a flat surface which is called a **plane**.

So the linear equation in three variables

$$ax+by+cz=d$$

defines a **plane** and not a straight line.

The work in Investigation 1 has not yielded the equation of a straight line in three dimensions and hence we must reconsider the basic concept i.e. an equation must define a means of specifying every point of a line.

In two dimensions we can find the equation of a line if we know

    (a) its gradient and one point through which it passes,

or **(b)** two points on the line (since the gradient can be found).

These methods were discussed on page 54 of Book 1.

We can use this basis to find the equation of a line in three dimensions. Such a line, *PQ*, passing through the point $(2, -1, 4)$ is shown in the figure.

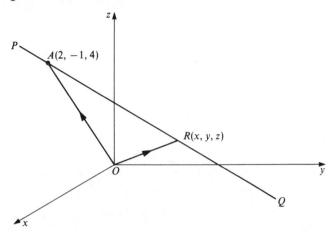

Remember that in three dimensions the direction of the line is specified by a **direction vector**. This will be the equivalent of the gradient in two dimensions.

Let the direction vector of the line *PQ* be $\mathbf{i}+3\mathbf{j}-2\mathbf{k}$ and let $R(x, y, z)$ be any other point of the line. Clearly, the position vector of *R* is given by **OR** and is denoted by **r**.

$$\mathbf{OR} = \mathbf{r} = \mathbf{OA} + \mathbf{AR} \qquad \text{............ (1)}$$

Now the position vector of $A(2, -1, 4)$ is $\mathbf{OA} = 2\mathbf{i}-\mathbf{j}+4\mathbf{k}$ and **AR** is a multiple of the direction vector of the line *PQ*.

$$\mathbf{AR} = \lambda(\mathbf{i}+3\mathbf{j}-2\mathbf{k}) \quad \text{where } \lambda \text{ is a scalar multiple}$$

Hence, substituting an Equation **(1)**

$$\mathbf{OR} = \mathbf{r} = (2\mathbf{i}-\mathbf{j}+4\mathbf{k}) + \lambda(\mathbf{i}+3\mathbf{j}-2\mathbf{k})$$
$$\mathbf{r} = (2+\lambda)\mathbf{i} + (3\lambda-1)\mathbf{j} + (4-2\lambda)\mathbf{k}$$

Since this result gives the position vector of a general point $R(x, y, z)$ relative to the origin, we can find specific positions of *R* by substituting values for λ. This means that *R* will take different positions along the line *PQ* as λ varies and hence the equation defines every point of the line.

We say that the **vector equation of the line** *PQ* is

$$\mathbf{r} = (2+\lambda)\mathbf{i} + (3\lambda-1)\mathbf{j} + (4-2\lambda)\mathbf{k}$$

Investigation 2

Choose different values of λ and substitute them into the equation above to find particular points of the line.

For example, if $\lambda = 2$ then $\mathbf{r} = 4\mathbf{i}+5\mathbf{j}$
           if $\lambda = -1$ then $\mathbf{r} = \mathbf{i}-4\mathbf{j}+6\mathbf{k}$

What value of λ would give point *A*?

When $\lambda = 2$ we obtained $\mathbf{r} = 4\mathbf{i}+5\mathbf{j}$ (i.e. the **k** component was zero). Hence the point $(4, 5, 0)$ lies on the line and is the point where the line cuts the *xy*-plane.
Show that the points where the line cuts the *xz*-plane and the *yz*-plane are $(2\frac{1}{3}, 0, 3\frac{1}{3})$ and $(0, -7, 8)$ respectively. Can you draw a sketch of the line on a set of mutually perpendicular axes?

General equation

It is possible to form **a general result for the vector equation of a straight line**.

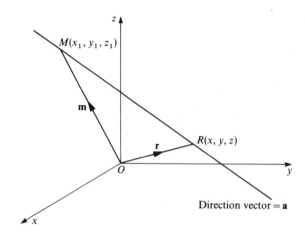

Consider a line in three dimensions with a direction vector given by

$$\mathbf{a} = a\mathbf{i} + b\mathbf{j} + c\mathbf{k}$$

which passes through the point $M(x_1, y_1, z_1)$ whose position vector relative to the origin is **m**.

Let the position vector of a general point of the line $R(x, y, z)$ be **r**.

Hence, $\mathbf{OR} = \mathbf{r} = \mathbf{OM} + \mathbf{MR} = \mathbf{m} + \lambda\mathbf{a}$

The vector equation of a straight line is

$$\mathbf{r} = \mathbf{m} + \lambda\mathbf{a}$$

where **a** is the direction vector of the line and **m** is the position vector of a fixed point of the line.

This is an unfamiliar format but it is really defining the position vector of a general point of the line. This is an important point to be remembered. For example,

$$\mathbf{r} = (\mathbf{i} + 3\mathbf{j} + 2\mathbf{k}) + \lambda(2\mathbf{i} - \mathbf{j} + \mathbf{k})$$

is the **vector equation of a line** through the point $(1, 3, 2)$ with a direction vector of $2\mathbf{i} - \mathbf{j} + \mathbf{k}$. It also defines the **position vector of a general point of the line**, i.e.

$$\mathbf{r} = (1 + 2\lambda)\mathbf{i} + (3 - \lambda)\mathbf{j} + (2 + \lambda)\mathbf{k}$$

which gives the point with co-ordinates $(1 + 2\lambda, 3 - \lambda, 2 + \lambda)$.

Investigation 3

Write down the vector equation of a line passing through the point $(0, 2, -3)$ with a direction vector $\mathbf{i} + \mathbf{j} - 2\mathbf{k}$.

By choosing values for λ, find the co-ordinates of six points on the line.

Show that the co-ordinates of a general point of the line are given by $(\lambda, 2 + \lambda, -3 - 2\lambda)$.

WORKED EXAMPLE

Find the vector equation of the line passing through the points $A(1, 2, 3)$ and $B(2, -1, 1)$.

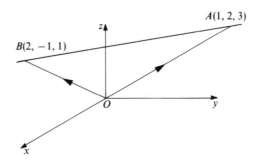

The position vector of A is $\mathbf{OA} = \mathbf{i} + 2\mathbf{j} + 3\mathbf{k}$.

The position vector of B is $\mathbf{OB} = 2\mathbf{i} - \mathbf{j} + \mathbf{k}$.

Thus the direction vector of the line AB is given by $\mathbf{AB}$.

Now,
$$\mathbf{AB} = \mathbf{AO} + \mathbf{OB} = \mathbf{OB} - \mathbf{OA}$$
$$= (2\mathbf{i} - \mathbf{j} + \mathbf{k}) - (\mathbf{i} + 2\mathbf{j} + 3\mathbf{k})$$
$$= \mathbf{i} - 3\mathbf{j} - 2\mathbf{k}$$

Since the line passes through the point A, the vector equation is
$$\mathbf{r} = \mathbf{OA} + \lambda(\mathbf{i} - 3\mathbf{j} - 2\mathbf{k})$$
$$\mathbf{r} = (\mathbf{i} + 2\mathbf{j} + 3\mathbf{k}) + \lambda(\mathbf{i} - 3\mathbf{j} - 2\mathbf{k})$$

This is often written in the form
$$\mathbf{r} = (1 + \lambda)\mathbf{i} + (2 - 3\lambda)\mathbf{j} + (3 - 2\lambda)\mathbf{k} \qquad \text{............ (1)}$$

General equation

In general, **the vector equation of a line passing through two points A and B** with position vectors $\mathbf{a}$ and $\mathbf{b}$ respectively, can be found as follows.

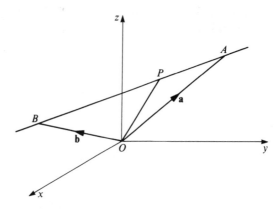

The direction of the line is given by $\mathbf{AB} = \mathbf{b} - \mathbf{a}$.

Hence, using point A as the fixed point, the position vector of a general point P is given by
$$\mathbf{r} = \mathbf{a} + \mathbf{AP}$$
$$\mathbf{r} = \mathbf{a} + \lambda(\mathbf{b} - \mathbf{a})$$

Thus, the vector equation of the line through two points with position vectors $\mathbf{a}$ and $\mathbf{b}$ is

$$\boxed{\mathbf{r} = (1 - \lambda)\mathbf{a} + \lambda\mathbf{b}}$$

So, for the points $A(1, 2, 3)$ and $B(2, -1, 1)$, the equation of the line will be

$$\mathbf{r} = (1-\lambda)(\mathbf{i}+2\mathbf{j}+3\mathbf{k}) + \lambda(2\mathbf{i}-\mathbf{j}+\mathbf{k})$$
$$= (\mathbf{i}+2\mathbf{j}+3\mathbf{k}) + \lambda(\mathbf{i}-3\mathbf{j}-2\mathbf{k})$$
$$= (1+\lambda)\mathbf{i} + (2-3\lambda)\mathbf{j} + (3-2\lambda)\mathbf{k}$$

This agrees with Equation **(1)** obtained in the Worked Example on page 71.

Investigation 4

Clearly, in the previous Worked Example, we could have chosen to use point B as the fixed point of the line. Show that the vector equation of the line in this case would be

$$\mathbf{r} = (2+\lambda)\mathbf{i} - (1+3\lambda)\mathbf{j} + (1-2\lambda)\mathbf{k} \qquad \text{............ (2)}$$

It is thus possible to obtain different equations which represent the same line. This means that different values of λ are needed in each of the two forms of the equation of the line to produce the same point.

i.e. in Equation **(1)**, $\lambda = 2$ gives the point $(3, -4, -1)$

in Equation **(2)**, $\lambda = 1$ would give the same point.

Find the values of λ needed in each equation to give the points $(1, 2, 3)$, $(-4, 17, 13)$ and $(4, -7, -3)$. Try some of your own examples.

For each of these points, write down the vector equation of the line with direction vector $\mathbf{i}-3\mathbf{j}-2\mathbf{k}$.

Investigation 5

How can we tell if **two different vector equations** represent the **same straight line**?

Consider the equations

$$\mathbf{r} = (2+\lambda)\mathbf{i} - (1+\lambda)\mathbf{j} + (4+2\lambda)\mathbf{k} \qquad \text{............ (1)}$$
$$\mathbf{r} = (-1+\lambda)\mathbf{i} + (2-\lambda)\mathbf{j} + (-2+2\lambda)\mathbf{k} \qquad \text{............ (2)}$$

Rearrange these two equations into the form $\mathbf{r} = \mathbf{m} + \lambda\mathbf{a}$ and show that they have the same direction vector, $\mathbf{a}$ (i.e. the lines are parallel).

Now find a point A on line **(1)** and a point B on line **(2)** by choosing values for λ in each equation. Find the direction of the line joining A and B.

If **AB** is a multiple of the direction vector $\mathbf{i}-\mathbf{j}+2\mathbf{k}$, then the line AB is parallel to both of the given lines.

This can only be true if they are the same line.

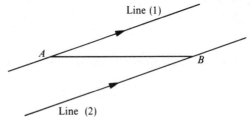

Exercise 4.1

1 Find the vector equation of the line which passes through a point M with a direction vector $\mathbf{a}$.

(a) $M(0, 1, 2)$; $\mathbf{a} = \mathbf{i} - 2\mathbf{j} + 2\mathbf{k}$ **(b)** $M(-2, 3, 1)$; $\mathbf{a} = -4\mathbf{i} + \mathbf{j} - 5\mathbf{k}$

(c) $M(3, \frac{1}{2}, 0)$; $\mathbf{a} = 3\mathbf{i} - \mathbf{k}$ **(d)** $M(0, 0, 0)$; $\mathbf{a} = 2\mathbf{i} + 3\mathbf{j} - \mathbf{k}$

2 Find the vector equation of the line which passes through the two points P and Q.

   (a) $P(1, -2, -3)$ and $Q(0, -2, 1)$ **(b)** $P(3, 1, 4)$ and $Q(0, 2, -1)$

   (c) $P(-2, -3, 1)$ and $Q(-1, 4, 1)$

3 For each of the following pairs of equations, decide whether they represent the same straight line.

   (a) $\mathbf{r} = (1+\lambda)\mathbf{i} + (3-\lambda)\mathbf{j} + (4+3\lambda)\mathbf{k}$ **(b)** $\mathbf{r} = (3+2\lambda)\mathbf{i} + 4\mathbf{j} + (1+4\lambda)\mathbf{k}$
         $\mathbf{r} = (2-\lambda)\mathbf{i} + (2+\lambda)\mathbf{j} + (7-3\lambda)\mathbf{k}$ $\mathbf{r} = (1+2\lambda)\mathbf{i} + 4\mathbf{j} - (3-4\lambda)\mathbf{k}$

   (c) $\mathbf{r} = (4-\lambda)\mathbf{i} + (-1+2\lambda)\mathbf{j} + (3+5\lambda)\mathbf{k}$ **(d)** $\mathbf{r} = (1+3\lambda)\mathbf{i} + (4-\lambda)\mathbf{j} + (2+\frac{1}{2}\lambda)\mathbf{k}$
         $\mathbf{r} = (5+2\lambda)\mathbf{i} - (3+4\lambda)\mathbf{j} - (2+10\lambda)\mathbf{k}$ $\mathbf{r} = (4-3\lambda)\mathbf{i} + (3+\lambda)\mathbf{j} + \frac{1}{2}(3-\lambda)\mathbf{k}$

4 Find the vector equation of the line which passes through the point with position vector $2\mathbf{i} - \mathbf{j} + \mathbf{k}$ and is parallel to the vector $3\mathbf{i} + \mathbf{j} - \mathbf{k}$.

5 The vector equation of a line is $\mathbf{r} = (1+\lambda)\mathbf{i} + (2-\lambda)\mathbf{j} + (3-2\lambda)\mathbf{k}$.

   Find **(a)** the direction vector of the line **(b)** the co-ordinates of the point where the line cuts the xy-plane and **(c)** the co-ordinates of the point where $x = 3$.

6 A vector equation is given by $\mathbf{r} = \mu\mathbf{a} + \lambda\mathbf{b}$ where μ and λ are constants and $\mathbf{a}$ and $\mathbf{b}$ are given vectors. Show that if $2\lambda + \mu = 1$, the equation represents a straight line through the points with position vectors $\mathbf{a}$ and $(\mathbf{b} - \mathbf{a})$.

7 Three points A, B, and C have position vectors $3\mathbf{a} + \mathbf{b}$, $2\mathbf{a} + 3\mathbf{b}$ and $5\mathbf{a} - 3\mathbf{b}$ respectively. Find the vector equation of the lines AB and AC. Deduce that A, B and C are collinear.

8 The vertices of a triangle ABC have position vectors $\mathbf{a}$, $\mathbf{b}$ and $\mathbf{c}$ relative to the origin O. If X is a point which divides BC in the ratio $2:1$ internally, find the position vector of X. Find also the vector $\mathbf{AX}$ and deduce the equation of the line passing through A and X.

9 A line AB has a vector equation given by $\mathbf{r} = (1-\lambda)\mathbf{i} + (3-2\lambda)\mathbf{j} + (4+\lambda)\mathbf{k}$. Find the vector equation of a parallel line PQ which passes through the point $(3, 0, -2)$.

The Cartesian equations of a straight line

The vector equation of a line is

$$\mathbf{r} = \mathbf{m} + \lambda\mathbf{a}$$

where the direction vector $\mathbf{a}$ is given by $a\mathbf{i} + b\mathbf{j} + c\mathbf{k}$ and the fixed point $M(x_1, y_1, z_1)$ has a position vector $\mathbf{m} = x_1\mathbf{i} + y_1\mathbf{j} + z_1\mathbf{k}$.

It is possible to find the equivalent Cartesian equations by choosing a general point $R(x, y, z)$ with position vector $\mathbf{r} = x\mathbf{i} + y\mathbf{j} + z\mathbf{k}$ and substituting into the vector equation.

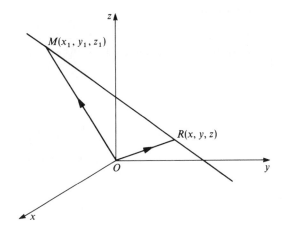

Consider the vector equation of the line in the figure, i.e. $\mathbf{r} = \mathbf{m} + \lambda\mathbf{a}$.

Substituting for $\mathbf{r}$, $\mathbf{m}$ and $\mathbf{a}$ gives

$$xi + yj + zk = (x_1\mathbf{i} + y_1\mathbf{j} + z_1\mathbf{k}) + \lambda(a\mathbf{i} + b\mathbf{j} + c\mathbf{k})$$

$$= (x_1 + \lambda a)\mathbf{i} + (y_1 + \lambda b)\mathbf{j} + (z_1 + \lambda c)\mathbf{k}$$

Vectors

Now, two vectors are equal **if** and **only if** their components are equal.

Hence,

$$x = x_1 + \lambda a \quad \ldots\ldots (1)$$
$$y = y_1 + \lambda b \quad \ldots\ldots (2)$$
$$z = z_1 + \lambda c \quad \ldots\ldots (3)$$

If $a \neq 0$, $b \neq 0$ and $c \neq 0$, these can be written as

$$\frac{x - x_1}{a} = \lambda; \qquad \frac{y - y_1}{b} = \lambda; \qquad \frac{z - z_1}{c} = \lambda$$

Hence the **equations of the line** can be written in the form

$$\boxed{\frac{x - x_1}{a} = \frac{y - y_1}{b} = \frac{z - z_1}{c}}$$

Since there is more than one equation here and as they relate to Cartesian co-ordinates, we call these the **Cartesian equations** (plural) **of a straight line.**

Notice that the values of a, b and c determine the direction ratios of the line. The equations are not unique (as the vector equation is not unique) since a different fixed point $M(x_1, y_1, z_1)$ can be selected.

There are special cases.

(a) If $a = 0$, then Equations **(2)** and **(3)** can still be written in the form

$$\frac{y - y_1}{b} = \frac{z - z_1}{c}$$

but Equation **(1)** becomes $x - x_1 = 0$.

(b) If $a = 0$ and $b = 0$, then Equations **(1)** and **(2)** will give $x - x_1 = 0$ and $y - y_1 = 0$ which will determine the line precisely.

Investigation 6

The vector equation of a straight line is $\mathbf{r} = (2 + 3\lambda)\mathbf{i} + (1 - \lambda)\mathbf{j} + (3 + 2\lambda)\mathbf{k}$. By writing $\mathbf{r} = x\mathbf{i} + y\mathbf{j} + z\mathbf{k}$ show that the Cartesian equations are given by

$$\frac{x - 2}{3} = \frac{y - 1}{-1} = \frac{z - 3}{2}$$

Notice the term in the centre, $\frac{y - 1}{-1}$. You probably obtained $\frac{-y + 1}{1}$ or $\frac{1 - y}{1}$, but it is usual to make the x, y or z term positive by multiplying the numerator and denominator by -1.

Compare the denominators with the direction vector obtained from the vector equation. What do you notice about the co-ordinates of the fixed point defined in the vector equation and the numerical values in the numerators of the Cartesian equations?

Clearly, $3, -1, 2$ are the direction ratios of the line and it passes through the point $(2, 1, 3)$.

The findings of Investigation 6 can be used to write down the Cartesian equations in one step.

For example, the Cartesian equations of the line through the point $(4, -2, -\frac{1}{2})$ with a direction vector $2\mathbf{i} - \frac{3}{4}\mathbf{j} + \mathbf{k}$ are

$$\frac{x-4}{2} = \frac{y-(-2)}{-\frac{3}{4}} = \frac{z-(-\frac{1}{2})}{1}$$

Usually written as

$$\frac{x-4}{2} = \frac{4y+8}{-3} = \frac{2z+1}{2}$$

Investigation 7

Show that the Cartesian equations of a line whose vector equation is $\mathbf{r} = (1-\lambda)\mathbf{i} + (1+2\lambda)\mathbf{j} + \mathbf{k}$ can be written as $y = 3 - 2x$ and $z = 1$.

This can be represented as shown in the figure by the plane $z = 1$, which is 1 unit above the xy-plane and the line $y = 3 - 2x$ in this plane, passing through the points $(1\frac{1}{2}, 0, 1)$ and $(0, 3, 1)$.

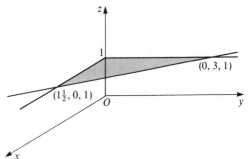

Repeat this for the following vector equations,

(a) $\mathbf{r} = \mathbf{i} + (2+\lambda)\mathbf{j} + (3+\lambda)\mathbf{k}$

(b) $\mathbf{r} = 2\lambda\mathbf{i} - \mathbf{j} + (2-\lambda)\mathbf{k}$

Investigation 8

Consider the line whose vector equation is $\mathbf{r} = (1-\lambda)\mathbf{i} + 2\mathbf{j} + \mathbf{k}$. Show that the Cartesian equations of the line are completely represented by $y = 2$ and $z = 1$.

Show this on a set of mutually perpendicular axes and compare your result with the first part of Investigation 7.

Repeat this for the lines whose vector equations are

(a) $\mathbf{r} = 3\mathbf{i} + \mathbf{j} + (2+\lambda)\mathbf{k}$ (b) $\mathbf{r} = -\mathbf{i} + (2-3\lambda)\mathbf{j} + 3\mathbf{k}$

WORKED EXAMPLE

Show that the following Cartesian equations represent the same straight line.

$$\frac{x-4}{5} = \frac{y+1}{-1} = \frac{z-3}{1} \qquad \text{............ (1)}$$

$$\frac{x+1}{1} = \frac{5y}{-1} = \frac{5z-10}{1} \qquad \text{............ (2)}$$

By writing Equation **(2)** in the form

$$\frac{x-x_1}{a} = \frac{y-y_1}{b} = \frac{z-z_1}{c}$$

we can see the direction ratios of the line and one point through which it passes.

$$\frac{x+1}{1} = \frac{y}{-\frac{1}{5}} = \frac{z-2}{\frac{1}{5}}$$

So the direction ratios of line **(1)** are $5, -1, 1$ which gives a direction vector of $5\mathbf{i}-\mathbf{j}+\mathbf{k}$.

The direction ratios of line **(2)** are $1, -\frac{1}{5}, \frac{1}{5}$, which gives a direction vector of $\mathbf{i}-\frac{1}{5}\mathbf{j}+\frac{1}{5}\mathbf{k}$.

Since $5\mathbf{i}-\mathbf{j}+\mathbf{k}$ is a multiple of $\mathbf{i}-\frac{1}{5}\mathbf{j}+\frac{1}{5}\mathbf{k}$, the lines are parallel.

Now line **(1)** passes through the point $(4, -1, 3)$, say A, and line **(2)** passes through the point $(-1, 0, 2)$, say B.

The direction of $\mathbf{AB} = -5\mathbf{i}+\mathbf{j}-\mathbf{k}$. Thus $\mathbf{AB}$ has the same direction as both line **(1)** and line **(2)** and this can only be true if the equations represent one and the same straight line.

Intersection of two lines

In two dimensions, two lines in a plane are either parallel or they intersect at one finite point. The graphs shown illustrate these two situations.

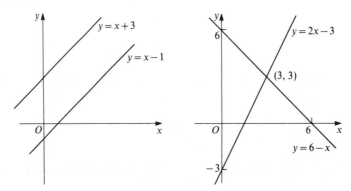

In three dimensions there is a further case. The lines may not be parallel and yet do not intersect. Such lines are called **skew lines**. The sketches illustrate the possibilities in three dimensions.

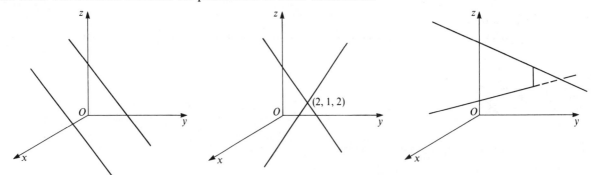

Consider a cuboid $ABCDEFGH$, as shown.

Examples of skew lines are: HB and CD; HB and AC; BG and CH; etc. Can you find some more? Which lines are parallel? Which lines intersect?

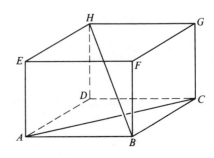

Clearly, if non-parallel lines in three dimensions **may** not intersect, then we must be able to discover whether they do or not and, if they do, find the point of intersection. The following Worked Example will illustrate how this can be achieved.

WORKED EXAMPLE

Show that the lines given by the following equations intersect and find the point of intersection:

$$\frac{x-2}{-1} = \frac{y-2}{1} = \frac{z-7}{2} \quad \text{and} \quad \frac{x-5}{1} = \frac{y-5}{2} = \frac{z-11}{3}$$

Write the equations in vector form.

Line 1 passes through $(2, 2, 7)$ and has a direction vector $-\mathbf{i}+\mathbf{j}+2\mathbf{k}$.
Line 2 passes through $(5, 5, 11)$ and has a direction vector $\mathbf{i}+2\mathbf{j}+3\mathbf{k}$.

Hence, the vector equations of the lines are

$$\mathbf{r} = (2\mathbf{i}+2\mathbf{j}+7\mathbf{k}) + \lambda(-\mathbf{i}+\mathbf{j}+2\mathbf{k})$$

and

$$\mathbf{r} = (5\mathbf{i}+5\mathbf{j}+11\mathbf{k}) + \mu(\mathbf{i}+2\mathbf{j}+3\mathbf{k})$$

Rewriting these gives

$$\mathbf{r} = (2-\lambda)\mathbf{i} + (2+\lambda)\mathbf{j} + (7+2\lambda)\mathbf{k} \qquad \text{............ (1)}$$

and

$$\mathbf{r} = (5+\mu)\mathbf{i} + (5+2\mu)\mathbf{j} + (11+3\mu)\mathbf{k} \qquad \text{............ (2)}$$

Now these equations also represent the position vectors of general points of the lines and so, if the lines intersect, there must be a value of λ and a value of μ for which these position vectors are the same.

Two vectors are equal if their components are equal:

$$2-\lambda = 5+\mu \quad \Rightarrow \quad \lambda+\mu = -3 \qquad \text{............ (3)}$$

$$2+\lambda = 5+2\mu \quad \Rightarrow \quad \lambda-2\mu = 3 \qquad \text{............ (4)}$$

$$7+2\lambda = 11+3\mu \quad \Rightarrow \quad 2\lambda-3\mu = 4 \qquad \text{............ (5)}$$

Subtracting Equation (4) from Equation (3) gives

$$3\mu = -6 \quad \Rightarrow \quad \mu = -2$$

Substituting into Equation (3) gives $\lambda = -1$

We must now check that these values also satisfy Equation (5). (Otherwise the $\mathbf{k}$ components may differ even though the $\mathbf{i}$ and $\mathbf{j}$ components match.)

In Equation (5), Left-hand side $= 2\lambda - 3\mu = 2(-1) - 3(-2)$

$$= 4 = \text{Right-hand side}$$

The values $\lambda = -1$ and $\mu = -2$ thus satisfy each of the Equations (3) to (5) and hence there is a common position vector and the lines intersect.

To find the point of intersection substitute either $\lambda = -1$ into Equation (1) or $\mu = -2$ into Equation (2).

Using Equation (1) with $\lambda = -1$ gives $\mathbf{r} = 3\mathbf{i}+\mathbf{j}+5\mathbf{k}$.

Hence the lines intersect at the point $(3, 1, 5)$.

Investigation 9

(a) Show that these lines do not intersect:

$$\frac{x+3}{-1} = \frac{y-5}{2} = \frac{z-1}{-1} \quad \text{and} \quad \frac{x-1}{1} = \frac{y-11}{5} = \frac{z-4}{1}$$

(b) Show that these lines intersect and find the point of intersection:

$$\frac{x-3}{1} = \frac{y-6}{3} = \frac{z+2}{-2} \quad \text{and} \quad \frac{x+2}{-3} = \frac{y-2}{2} = \frac{z-3}{1}$$

(c) Show that the unit vectors in the directions of the lines in part **(b)** are

$$\hat{\mathbf{a}} = \frac{1}{\sqrt{14}}(\mathbf{i}+3\mathbf{j}-2\mathbf{k}) \quad \text{and} \quad \hat{\mathbf{b}} = \frac{1}{\sqrt{14}}(-3\mathbf{i}+2\mathbf{j}+\mathbf{k})$$

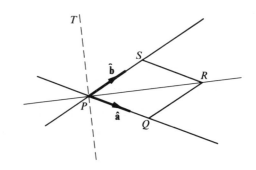

Hence, since $|\hat{\mathbf{a}}| = |\hat{\mathbf{b}}| = 1$, *PQRS* is a rhombus and its diagonal bisects the vertex at *P*. Show that the direction vector of the angle bisector *PR* is

$$\frac{1}{\sqrt{14}}(-2\mathbf{i}+5\mathbf{j}-\mathbf{k})$$

Use this result and the position vector of *P* to show that the equations of the angle bisector are

$$\frac{x-1}{-2} = \frac{y}{5} = \frac{z-2}{-1}$$

Show also that the other bisector, *PT*, has a direction vector $4\mathbf{i}+\mathbf{j}-3\mathbf{k}$ and find its Cartesian equations.

Exercise 4.2

1 Write the equations of the following lines in Cartesian form.

(a) $\mathbf{r} = (2+\lambda)\mathbf{i} - (1+2\lambda)\mathbf{j} + (3-\lambda)\mathbf{k}$

(b) $\mathbf{r} = (1-\lambda)\mathbf{i} + \mathbf{j} + (4+3\lambda)\mathbf{k}$

(c) $\mathbf{r} = (-1+2\lambda)\mathbf{i} + \frac{1}{2}(6-\lambda)\mathbf{j} + (2-\frac{3}{2}\lambda)\mathbf{k}$

(d) $\mathbf{r} = 2\mathbf{j} + \lambda(3\mathbf{i}+4\mathbf{k})$

(e) $\mathbf{r} = \lambda(\mathbf{i}+\mathbf{j}-2\mathbf{k})$

2 Write the equations of the following lines in vector form.

(a) $\dfrac{x-1}{4} = \dfrac{y-2}{3} = \dfrac{z-5}{2}$ **(b)** $\dfrac{x-3}{-1} = \dfrac{y}{5} = \dfrac{4-z}{2}$

(c) $\dfrac{2x-5}{3} = \dfrac{3y-1}{2} = \dfrac{z}{1}$ **(d)** $x = 4$ and $\dfrac{y}{3} = \dfrac{z+2}{1}$

3 Find the Cartesian equations of the lines through a point *A* with the given direction vector **d**.

(a) $A(1,2,3); \quad \mathbf{d} = \mathbf{i}-\mathbf{j}+\mathbf{k}$ **(b)** $A(0,0,6); \quad \mathbf{d} = \frac{1}{2}\mathbf{i}+\frac{1}{3}\mathbf{j}-\mathbf{k}$

(c) $A(-1,4,-3); \quad \mathbf{d} = 2\mathbf{i}-\frac{1}{4}\mathbf{j}+2\mathbf{k}$

4   Find the Cartesian equations of the lines joining the following points.

   (a) $(0, 0, 1)$ and $(1, -3, 2)$ **(b)** $(4, 6, -2)$ and $(3, 1, 2)$

   (c) $(3, -1, 4)$ and $(3, -1, -2)$

5   Find whether the following pairs of Cartesian equations represent the same straight line.

   (a) $\dfrac{x-4}{3} = \dfrac{y-2}{1} = \dfrac{z-5}{-2}$ and $\dfrac{x-1}{3} = \dfrac{y-1}{1} = \dfrac{z-7}{-2}$

   (b) $\dfrac{x-3}{1} = \dfrac{y}{-2} = \dfrac{z+2}{4}$ and $\dfrac{2(x-2)}{1} = \dfrac{y-2}{-1} = \dfrac{z+6}{2}$

   (c) $\dfrac{x+1}{-3} = \dfrac{y-2}{1} = \dfrac{z-4}{2}$ and $\dfrac{3x-6}{2} = \dfrac{2y-2}{1} = \dfrac{z-1}{1}$

   (d) $\dfrac{x-7}{-2} = \dfrac{y+2}{1} = \dfrac{z-3}{3}$ and $\dfrac{3(x-9)}{-2} = \dfrac{3(y+3)}{1} = \dfrac{z}{1}$

6   A, B and C are points with position vectors $\mathbf{i}+\mathbf{j}+\mathbf{k}$, $2\mathbf{i}-\mathbf{j}-3\mathbf{k}$ and $3\mathbf{i}-2\mathbf{j}-\mathbf{k}$ respectively. Find the vector and Cartesian equations of **(a)** the line through A and B and   **(b)** the line parallel to AB through the point C.

7   Three vertices of a parallelogram $ABCD$ are $A(1, 2, 3)$, $B(3, 5, 4)$ and $C(2, -4, 1)$.   Find the co-ordinates of D and the Cartesian equations of the lines AC and BD.

8   Show that the lines whose vector equations are

$$\mathbf{r} = (3+\lambda)\mathbf{i} + (-3-2\lambda)\mathbf{j} + (3+2\lambda)\mathbf{k}$$

$$\text{and} \quad \mathbf{r} = 2\mu\mathbf{i} + (-4+3\mu)\mathbf{j} + (-3+4\mu)\mathbf{k}$$

   intersect, and find the point of intersection.

9   Show that the lines whose Cartesian equations are

$$\frac{x-5}{2} = \frac{y-7}{1} = \frac{z+1}{-1} \quad \text{and} \quad \frac{x-9}{5} = \frac{y-8}{2} = \frac{z-4}{1}$$

   intersect, and find the point of intersection.

10   Given that the lines whose equations are $\mathbf{r} = 2\mathbf{i} + \lambda(\mathbf{i}+a\mathbf{j}+2\mathbf{k})$ and $\mathbf{r} = \mathbf{i}+2\mathbf{j}+\mathbf{k}+\mu(\mathbf{i}-2\mathbf{j}+\mathbf{k})$ intersect, find the value of a.

11   The points $A(1, 2, 3)$, $B(3, 4, 7)$ and $C(2, -3, 5)$ are three vertices of a parallelogram $ABCD$. Find the Cartesian equations of the diagonals AC and BD and hence find the co-ordinates of their point of intersection.

12   Find the Cartesian equations of the bisectors of the angle between the lines whose equations are

$$\frac{x-5}{4} = \frac{y+1}{-2} = \frac{z-3}{1} \quad \text{and} \quad \frac{x}{1} = \frac{y+1}{2} = \frac{z-6}{-4}$$

13   Find the position vector of the point of intersection of the lines given by $\mathbf{r} = \mathbf{a} + \lambda(\mathbf{a}-\mathbf{b})$ and $\mathbf{r} = 2\mathbf{b} + \mu(4\mathbf{a}-\mathbf{b})$.

14   Two particles start at the same instant from the points, $(1, -2, -7)$ and $(5, 0, -5)$ and move with velocities $\mathbf{i}+3\mathbf{j}+2\mathbf{k}$ and $-\mathbf{i}+2\mathbf{j}+\mathbf{k}$ respectively. Show that the Cartesian equations of their paths are given by

$$\frac{x-1}{1} = \frac{y+2}{3} = \frac{z+7}{2} \quad \text{and} \quad \frac{x-5}{-1} = \frac{y}{2} = \frac{z+5}{1}$$

   Show that the particles collide, find the time at which they meet and the co-ordinates of the point of intersection.

15 A particle starts from a point $A(2, 3, 0)$ and moves with a velocity $2\mathbf{i} + 3\mathbf{j} - \mathbf{k}$. A second particle starts from a point $B(4, 20, -7)$ one second earlier with a velocity $\mathbf{i} - 2\mathbf{j} + \mathbf{k}$. Show that the equations of their paths are

$$\mathbf{r} = 2(1+t)\mathbf{i} + 3(1+t)\mathbf{j} - t\mathbf{k} \quad \text{and}$$

$$\mathbf{r} = (5+t)\mathbf{i} + (18 - 2t)\mathbf{j} + (-6+t)\mathbf{k}$$

where t is the time after the first particle starts from A. Show that their paths intersect and find the point of intersection.

16 Two forces $\mathbf{F}_1 = 7\mathbf{i} + 2\mathbf{j} - 3\mathbf{k}$ and $\mathbf{F}_2 = 3\mathbf{i} - \mathbf{j} + 5\mathbf{k}$ act through the points $(8, 1, -1)$ and $(7, -3, 12)$ respectively. Find the vector equations of the lines of action of these forces. Show that these lines of action intersect and find the point of intersection. Also find the resultant force and deduce the Cartesian equations of its line of action.

17 Two forces $\mathbf{F}_1 = 4\mathbf{i} - 2\mathbf{j} + 8\mathbf{k}$ and $\mathbf{F}_2 = 3\mathbf{i} - 9\mathbf{j} + 6\mathbf{k}$ act through the points with position vectors $7\mathbf{i} - 2\mathbf{j} + \mathbf{k}$ and $2\mathbf{i} - 7\mathbf{j} - 9\mathbf{k}$ respectively. Find the resultant force and the Cartesian equations of its line of action.

Scalar product

Consider a ladder AB represented by a vector $\mathbf{l}$ leaning against a vertical wall at an angle θ to the horizontal.

If $|\mathbf{l}| = l$ then the distance of A from the foot of the wall is $l \cos \theta$.

$$\therefore \quad AN = l \cos \theta$$

AN is called the projection of AB in the direction AN.

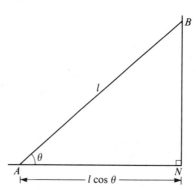

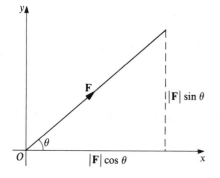

In statics, when we resolve a force into components we write $|\mathbf{F}| \cos \theta$ and $|\mathbf{F}| \sin \theta$ as the components in the directions Ox and Oy respectively.

Quantities of the form $|\mathbf{a}| \times \cos \theta$ occur frequently in mathematics and we now define an operation on vectors which incorporates this expression.

If $\mathbf{n}$ is a unit vector and $\mathbf{a}$ is a given vector, then the projection of $\mathbf{a}$ in the direction $\mathbf{n}$ is written $\mathbf{a} . \mathbf{n} = |\mathbf{a}| \cos \theta$ where θ is the angle between the vectors $\mathbf{a}$ and $\mathbf{n}$ and $|\mathbf{a}|$ is the modulus of $\mathbf{a}$.

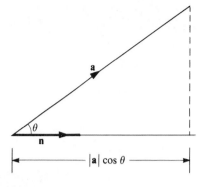

This product is known as the **scalar product** of the vectors **a** and **n** as it produces a scalar quantity for the result.

In general, the **scalar product of two vectors, a** and **b**, is denoted by **a . b** and is defined as $|\mathbf{a}||\mathbf{b}|\cos\theta$ where θ is the angle between **a** and **b**.

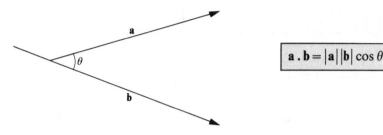

$$\mathbf{a} . \mathbf{b} = |\mathbf{a}||\mathbf{b}|\cos\theta$$

So **a . b** can be interpreted as

$\mathbf{a} . \mathbf{b} = |\mathbf{a}| \times (|\mathbf{b}|\cos\theta) = |\mathbf{a}| \times$ (projection of **b** in the direction of **a**)

$\quad\quad = |\mathbf{b}| \times (|\mathbf{a}|\cos\theta) = |\mathbf{b}| \times$ (projection of **a** in the direction of **b**)

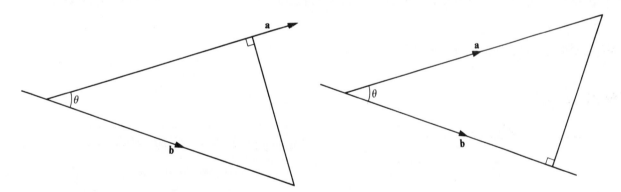

Note that the **.** is important and cannot be omitted. The scalar product **a . b** is read 'a dot b'.

It is important to pay particular attention to the directions of the vectors and the angle between them. For example,

if $|\mathbf{a}| = a$ and $|\mathbf{b}| = b$ then by definition $\mathbf{a} . \mathbf{b} = ab\cos\theta$

which may give a positive or negative result.

The following illustrations show some configurations.

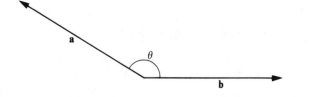

In this diagram, as θ is obtuse, $-1 < \cos\theta < 0$, and hence **a . b** will be negative.

81

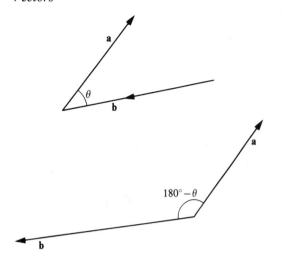

In this case, the vectors actually enclose an angle of $180° - \theta$, which is shown in the lower diagram.

$$\mathbf{a} \cdot \mathbf{b} = ab \cos (180° - \theta)$$

$$= -ab \cos \theta$$

Investigation 10

Consider the system of coplanar forces shown in the diagram.

Using the basic definition of scalar product find

(a) $\mathbf{F}_1 \cdot \mathbf{F}_2$ (b) $\mathbf{F}_1 \cdot \mathbf{F}_3$ (c) $\mathbf{F}_1 \cdot \mathbf{F}_4$

(d) $\mathbf{F}_1 \cdot \mathbf{F}_5$ (e) $\mathbf{F}_2 \cdot \mathbf{F}_5$

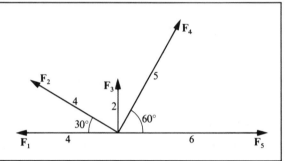

The work in Investigation 10 should have indicated some important basic results. Check that your answers were

(a) $8\sqrt{3}$ (b) 0 (c) -10 (d) -24 (e) $-12\sqrt{3}$

We can now identify several special results.

For parallel vectors

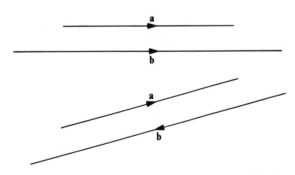

> If **a** and **b** are **like** parallel vectors then
>
> $\mathbf{a} \cdot \mathbf{b} = ab \cos 0° = ab$

> If **a** and **b** are **unlike** parallel vectors then
>
> $\mathbf{a} \cdot \mathbf{b} = ab \cos 180° = -ab$

When $\mathbf{a} = \mathbf{b}$ we write $\mathbf{a} \cdot \mathbf{a} = a^2$ (the square of the magnitude of **a**).

For perpendicular vectors

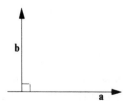

> If **a** and **b** are perpendicular vectors then
>
> $\mathbf{a} \cdot \mathbf{b} = ab \cos 90° = 0$

For the unit vectors **i**, **j** and **k**, we obtain the following special results.

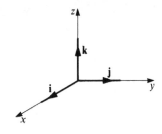

$$\mathbf{i}.\mathbf{i} = \mathbf{j}.\mathbf{j} = \mathbf{k}.\mathbf{k} = 1$$
$$\mathbf{i}.\mathbf{j} = \mathbf{j}.\mathbf{k} = \mathbf{k}.\mathbf{i} = 0$$
$$\mathbf{j}.\mathbf{i} = \mathbf{k}.\mathbf{j} = \mathbf{i}.\mathbf{k} = 0$$

Exercise 4.3

1 Find the scalar product of two vectors **a** and **b** whose directions enclose an angle θ.

(a) $|\mathbf{a}| = 5$; $|\mathbf{b}| = 2$; $\theta = 60°$ (b) $|\mathbf{a}| = 3$; $|\mathbf{b}| = 4$; $\theta = 120°$

(c) $\mathbf{a} = 7\mathbf{i} + 24\mathbf{j}$; $\mathbf{b} = \mathbf{i} + \mathbf{j}$; $\cos\theta = \dfrac{31\sqrt{2}}{50}$ (d) $\mathbf{a} = 3\mathbf{i} - 4\mathbf{j}$; $\mathbf{b} = \mathbf{i} + 2\mathbf{j}$; $\cos\theta = -\dfrac{1}{\sqrt{5}}$

2 If $\mathbf{p} = 3\mathbf{i} + \mathbf{j}$ and $\mathbf{q} = -4\mathbf{i} + 3\mathbf{j}$ are two vectors, draw a diagram to show these as displacements and find $\cos\theta$ where θ is the angle between **p** and **q**. Hence find **p**.**q**.

3 A rectangle $ABCD$ has sides $AB = 3$ cm and $AD = 2$ cm. Find the following scalar products.

(a) **AB**.**AD** (b) **AB**.**AC** (c) **AC**.**CB** (d) **AB**.**DC**

4 A regular hexagon $ABCDEF$ has sides of length 2 cm. Find (a) **AB**.**BC** (b) **AB**.**BD** (c) **AD**.**BC**
(d) **AF**.**BC**.

5 The figure shows the magnitudes and directions of five forces. Find (a) $\mathbf{F}_2.\mathbf{F}_4$ (b) $\mathbf{F}_3.\mathbf{F}_5$
(c) $\mathbf{F}_1.\mathbf{F}_5$ (d) $\mathbf{F}_2.\mathbf{F}_5$ (e) $\mathbf{F}_1.\mathbf{F}_2$.

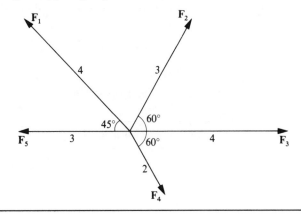

Properties of the scalar product

(1) Scalar product is commutative

From the definition $\mathbf{a}.\mathbf{b} = ab\cos\theta$

and $\mathbf{b}.\mathbf{a} = ba\cos\theta$

Since $ab = ba$,

$$\mathbf{a}.\mathbf{b} = \mathbf{b}.\mathbf{a}$$

(2) Multiplication by a scalar quantity

There is no equivalent of the associative rule for scalar products since **(a . b) . c** is not defined. However, if λ is a scalar quantity, then

$$\lambda(\mathbf{a} . \mathbf{b}) = (\lambda\mathbf{a}) . \mathbf{b} = \mathbf{a} . (\lambda\mathbf{b})$$

Again, using the basic definition, since

$$\lambda ab \cos \theta = (\lambda a)b \cos \theta = a(\lambda b) \cos \theta$$

$$\boxed{\lambda(\mathbf{a} . \mathbf{b}) = (\lambda\mathbf{a}) . \mathbf{b} = \mathbf{a} . (\lambda\mathbf{b})}$$

(3) Scalar product is distributive over addition

This means that, for three vectors **a**, **b** and **c**,

$$\mathbf{a} . (\mathbf{b} + \mathbf{c}) = \mathbf{a} . \mathbf{b} + \mathbf{a} . \mathbf{c}$$

Consider the figure which shows three vectors **a**, **b** and **c** together with **b** + **c**.

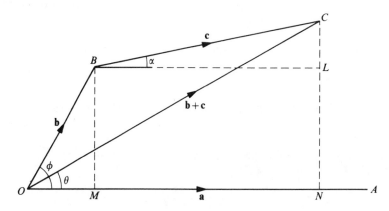

Now,

$$ON = OM + MN = OM + BL$$

$$|\mathbf{b} + \mathbf{c}| \cos \theta = |\mathbf{b}| \cos \phi + |\mathbf{c}| \cos \alpha$$

$$|\mathbf{a}||\mathbf{b} + \mathbf{c}| \cos \theta = |\mathbf{a}||\mathbf{b}| \cos \phi + |\mathbf{a}||\mathbf{c}| \cos \alpha$$

$$\boxed{\mathbf{a} . (\mathbf{b} + \mathbf{c}) = \mathbf{a} . \mathbf{b} + \mathbf{a} . \mathbf{c}}$$

WORKED EXAMPLE

Prove that $a^2 = b^2 + c^2 - 2bc \cos A$ by vector methods.

We can use the scalar product to prove geometrical results. In this case we have the cosine rule.

Consider a triangle ABC and let **BC**, **AC** and **AB** be represented by **a**, **b** and **c** respectively.

Also, let $|\mathbf{a}| = a$, $|\mathbf{b}| = b$ and $|\mathbf{c}| = c$.

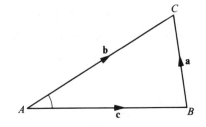

Now,
$$\mathbf{a} = \mathbf{b} - \mathbf{c}$$
$$\mathbf{a} \cdot \mathbf{a} = (\mathbf{b} - \mathbf{c}) \cdot (\mathbf{b} - \mathbf{c})$$
$$= \mathbf{b} \cdot (\mathbf{b} - \mathbf{c}) - \mathbf{c} \cdot (\mathbf{b} - \mathbf{c})$$
$$= \mathbf{b} \cdot \mathbf{b} - \mathbf{b} \cdot \mathbf{c} - \mathbf{c} \cdot \mathbf{b} + \mathbf{c} \cdot \mathbf{c}$$

Hence,
$$a^2 = b^2 + c^2 - 2(\mathbf{b} \cdot \mathbf{c})$$
$$a^2 = b^2 + c^2 - 2bc \cos A$$

Investigation 11

Consider the figure in which AB is the diameter of a circle, centre O, and $\mathbf{AO} = \mathbf{OB} = \mathbf{a}$.

P is any point on the circumference of the circle such that $\mathbf{OP} = \mathbf{b}$.

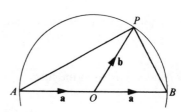

Find expressions for $\mathbf{AP}$, $\mathbf{PB}$ and $\mathbf{AP} \cdot \mathbf{PB}$. Deduce that $\mathbf{AP}$ is perpendicular to $\mathbf{PB}$ and that the angle in a semicircle is a right angle.

Investigation 12

Consider a triangle OAB and let $\mathbf{OA} = \mathbf{a}$ and $\mathbf{OB} = \mathbf{b}$.

If $|\mathbf{a}| = a$ and $|\mathbf{b}| = b$, find the unit vector, $\hat{\mathbf{b}}$, in the direction of $\mathbf{b}$. Use the fact that the projection of $\mathbf{a}$ in the direction of $\mathbf{b}$ is $\mathbf{a} \cdot \hat{\mathbf{b}}$ to show that

$$ON = \frac{(\mathbf{a} \cdot \mathbf{b})}{b}$$

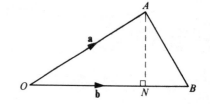

Use the theorem of Pythagoras in triangle OAN to show that

$$AN^2 = a^2 - \frac{(\mathbf{a} \cdot \mathbf{b})^2}{b^2}$$

Deduce that the area of triangle OAB is $\frac{1}{2}\sqrt{[a^2 b^2 - (\mathbf{a} \cdot \mathbf{b})^2]}$.

Scalar product of vectors in component form

Consider two vectors $\mathbf{a} = a_1 \mathbf{i} + a_2 \mathbf{j} + a_3 \mathbf{k}$ and $\mathbf{b} = b_1 \mathbf{i} + b_2 \mathbf{j} + b_3 \mathbf{k}$.

Hence,
$$\mathbf{a} \cdot \mathbf{b} = (a_1 \mathbf{i} + a_2 \mathbf{j} + a_3 \mathbf{k}) \cdot (b_1 \mathbf{i} + b_2 \mathbf{j} + b_3 \mathbf{k})$$

Using the properties of the scalar product we have

$$\mathbf{a} \cdot \mathbf{b} = a_1 b_1 (\mathbf{i} \cdot \mathbf{i}) + a_1 b_2 (\mathbf{i} \cdot \mathbf{j}) + a_1 b_3 (\mathbf{i} \cdot \mathbf{k}) + a_2 b_1 (\mathbf{j} \cdot \mathbf{i}) + a_2 b_2 (\mathbf{j} \cdot \mathbf{j}) + a_2 b_3 (\mathbf{j} \cdot \mathbf{k}) + a_3 b_1 (\mathbf{k} \cdot \mathbf{i}) + a_3 b_2 (\mathbf{k} \cdot \mathbf{j}) + a_3 b_3 (\mathbf{k} \cdot \mathbf{k})$$

Since $\mathbf{i} \cdot \mathbf{i} = \mathbf{j} \cdot \mathbf{j} = \mathbf{k} \cdot \mathbf{k} = 1$ and $\mathbf{i} \cdot \mathbf{j} = \mathbf{j} \cdot \mathbf{i} = \mathbf{i} \cdot \mathbf{k} = \mathbf{k} \cdot \mathbf{i} = \mathbf{j} \cdot \mathbf{k} = \mathbf{k} \cdot \mathbf{j} = 0$, then

$$\boxed{\mathbf{a} \cdot \mathbf{b} = a_1 b_1 + a_2 b_2 + a_3 b_3}$$

Vectors

WORKED EXAMPLE ▷

Calculate the scalar product of $\mathbf{a} = 3\mathbf{i} - 2\mathbf{j} - 4\mathbf{k}$ and $\mathbf{b} = \mathbf{i} + 3\mathbf{j} - 2\mathbf{k}$.

Now,

$$\mathbf{a} \cdot \mathbf{b} = (3\mathbf{i} - 2\mathbf{j} - 4\mathbf{k}) \cdot (\mathbf{i} + 3\mathbf{j} - 2\mathbf{k})$$

$$= (3 \times 1) + (-2 \times 3) + [-4 \times (-2)] = 3 - 6 + 8 = 5$$

The angle between two lines

If the angle between two vectors $\mathbf{a}$ and $\mathbf{b}$ is θ, then by definition,

$$\mathbf{a} \cdot \mathbf{b} = ab \cos \theta$$

This can be written in the form

$$\cos \theta = \frac{(\mathbf{a} \cdot \mathbf{b})}{ab}$$

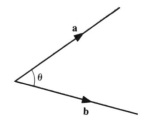

Remember that $\mathbf{a}$ and $\mathbf{b}$ will be the direction vectors of the two lines and if the scalar product $\mathbf{a} \cdot \mathbf{b} < 0$, then θ will be obtuse. In this case we usually say that the angle between the lines is the acute angle $180° - \theta$.

> If $\mathbf{a}$ and $\mathbf{b}$ are the direction vectors of two lines,
> then the angle between the lines is $\cos^{-1}\left(\dfrac{\mathbf{a} \cdot \mathbf{b}}{ab}\right)$.

WORKED EXAMPLES ▷

1   Find the angle between the lines

$$\frac{x-3}{-2} = \frac{y-2}{1} = \frac{z+9}{5} \quad \text{and} \quad \frac{x-2}{1} = \frac{y-4}{1} = \frac{z+2}{2}$$

These two lines have direction vectors $\mathbf{a} = -2\mathbf{i} + \mathbf{j} + 5\mathbf{k}$ and $\mathbf{b} = \mathbf{i} + \mathbf{j} + 2\mathbf{k}$ respectively.

Hence,

$$a = |\mathbf{a}| = \sqrt{(-2)^2 + 1^2 + 5^2} = \sqrt{30}$$

and

$$b = |\mathbf{b}| = \sqrt{1^2 + 1^2 + 2^2} = \sqrt{6}$$

Now, if θ is the angle between these directions then $\cos \theta = \dfrac{(\mathbf{a} \cdot \mathbf{b})}{ab}$.

$$\cos \theta = \frac{(-2\mathbf{i} + \mathbf{j} + 5\mathbf{k}) \cdot (\mathbf{i} + \mathbf{j} + 2\mathbf{k})}{\sqrt{30}\sqrt{6}} = \frac{-2 + 1 + 10}{\sqrt{30}\sqrt{6}} = \frac{9}{\sqrt{30}\sqrt{6}}$$

$$\Rightarrow \quad \theta = 47.87°$$

The angle between the given lines is $47.87°$.

2   Find the equation of the line through $A(-3, -2, -7)$ which meets the line

$$\frac{x-1}{2} = \frac{y-4}{3} = \frac{z-5}{-1} \quad \text{at right angles.}$$

Writing the given equation in vector form gives

$$\mathbf{r} = (1+2\lambda)\mathbf{i} + (4+3\lambda)\mathbf{j} + (5-\lambda)\mathbf{k}$$

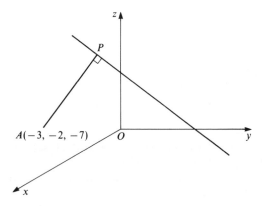

This is also the position vector of a general point P of the line.

Hence, $\quad\quad\quad\quad\quad\quad\quad\quad$ $\mathbf{OP} = (1+2\lambda)\mathbf{i} + (4+3\lambda)\mathbf{j} + (5-\lambda)\mathbf{k}$

Since $\quad\quad\quad\quad\quad\quad\quad\quad\quad$ $\mathbf{AP} = \mathbf{OP} - \mathbf{OA}$

We have, $\quad\quad\quad\quad\quad\quad\quad$ $\mathbf{AP} = (1+2\lambda)\mathbf{i} + (4+3\lambda)\mathbf{j} + (5-\lambda)\mathbf{k} - (-3\mathbf{i} - 2\mathbf{j} - 7\mathbf{k})$

$$\mathbf{AP} = (4+2\lambda)\mathbf{i} + (6+3\lambda)\mathbf{j} + (12-\lambda)\mathbf{k}$$

If $\mathbf{AP}$ is perpendicular to the line, it is perpendicular to the direction vector of the line and $\mathbf{AP} \cdot (2\mathbf{i} + 3\mathbf{j} - \mathbf{k}) = 0$.

So, $\quad\quad\quad\quad\quad\quad$ $[(4+2\lambda)\mathbf{i} + (6+3\lambda)\mathbf{j} + (12-\lambda)\mathbf{k}] \cdot (2\mathbf{i} + 3\mathbf{j} - \mathbf{k}) = 0$

$$8 + 4\lambda + 18 + 9\lambda - 12 + \lambda = 0$$

$$14\lambda = -14 \quad \Rightarrow \quad \lambda = -1$$

Hence, $\mathbf{OP} = -\mathbf{i} + \mathbf{j} + 6\mathbf{k}$ and P is the point $(-1, 1, 6)$

Also, $\mathbf{AP} = 2\mathbf{i} + 3\mathbf{j} + 13\mathbf{k}$

So the required line passes through the points $(-3, -2, -7)$ and $(-1, 1, 6)$ and has a direction vector $2\mathbf{i} + 3\mathbf{j} + 13\mathbf{k}$. Its Cartesian equations are

$$\frac{x+3}{2} = \frac{y+2}{3} = \frac{z+7}{13}$$

The shortest distance between two skew lines

As we have already seen on page 76, **skew lines** are lines in three dimensions which do not intersect and are not parallel.

Skew lines have a **common perpendicular** which is the least distance between the lines.

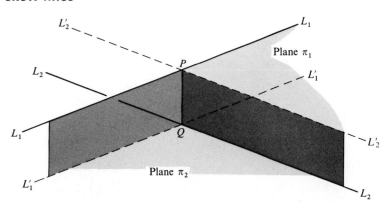

Vectors

Consider two skew lines L_1 and L_2. Draw a line L_2' parallel to L_2 to intersect L_1 at a point P, such that the plane defined by $L_2 L_2'$ is perpendicular to the plane $L_1 L_2'$.

Also draw a line L_1' parallel to L_1 to intersect L_2 at Q, such that the plane $L_1 L_1'$ is perpendicular to the plane $L_2 L_1'$.

Clearly, the planes $L_1 L_2'$ (say, π_1) and $L_2 L_1'$ (say, π_2) are parallel.

Thus the planes $L_1 L_1'$ and $L_2 L_2'$ are perpendicular to each of the planes π_1 and π_2.

Hence, PQ is perpendicular to the planes π_1 and π_2 and is thus perpendicular to the lines L_1 and L_2.

> The shortest distance between two skew lines is the length of the line which is perpendicular to each of the given lines.

WORKED EXAMPLE

Find the shortest distance between the two skew lines given by

$$\frac{x-6}{2} = \frac{y+4}{5} = \frac{z-2}{1} \quad \text{and} \quad \frac{x+1}{-4} = \frac{y-9}{5} = \frac{z-5}{7}$$

Find also the equation of the common perpendicular to these two lines.

Let the lines be denoted by L_1 and L_2 respectively.

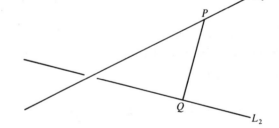

The direction vector of L_1 is

$$2\mathbf{i} + 5\mathbf{j} + \mathbf{k}$$

The direction vector of L_2 is

$$-4\mathbf{i} + 5\mathbf{j} + 7\mathbf{k}$$

The vector equation of L_1 is $\mathbf{r} = (6\mathbf{i} - 4\mathbf{j} + 2\mathbf{k}) + \lambda(2\mathbf{i} + 5\mathbf{j} + \mathbf{k})$.

The vector equation of L_2 is $\mathbf{r} = (-\mathbf{i} + 9\mathbf{j} + 5\mathbf{k}) + \mu(-4\mathbf{i} + 5\mathbf{j} + 7\mathbf{k})$.

So, if $\mathbf{r}_1$ is the position vector of a general point, P, on L_1 and $\mathbf{r}_2$ is the position vector of any point, Q, on L_2, then

$$\mathbf{r}_1 = (6 + 2\lambda)\mathbf{i} + (-4 + 5\lambda)\mathbf{j} + (2 + \lambda)\mathbf{k} \qquad \text{............ (1)}$$

$$\text{and} \quad \mathbf{r}_2 = (-1 - 4\mu)\mathbf{i} + (9 + 5\mu)\mathbf{j} + (5 + 7\mu)\mathbf{k} \qquad \text{............ (2)}$$

Hence, $\qquad \mathbf{PQ} = \mathbf{r}_2 - \mathbf{r}_1$

$$= (-7 - 4\mu - 2\lambda)\mathbf{i} + (13 + 5\mu - 5\lambda)\mathbf{j} + (3 + 7\mu - \lambda)\mathbf{k} \qquad \text{............ (3)}$$

Now $\mathbf{PQ}$ is perpendicular to L_1 and L_2 (i.e. perpendicular to the direction vectors of L_1 and L_2).

Thus, $\quad \mathbf{PQ} \cdot (2\mathbf{i} + 5\mathbf{j} + \mathbf{k}) = 0 \quad \text{and} \quad \mathbf{PQ} \cdot (-4\mathbf{i} + 5\mathbf{j} + 7\mathbf{k}) = 0$

$\therefore \quad [(-7 - 4\mu - 2\lambda)\mathbf{i} + (13 + 5\mu - 5\lambda)\mathbf{j} + (3 + 7\mu - \lambda)\mathbf{k}] \cdot (2\mathbf{i} + 5\mathbf{j} + \mathbf{k}) = 0$

and $\quad [(-7 - 4\mu - 2\lambda)\mathbf{i} + (13 + 5\mu - 5\lambda)\mathbf{j} + (3 + 7\mu - \lambda)\mathbf{k}] \cdot (-4\mathbf{i} + 5\mathbf{j} + 7\mathbf{k}) = 0$

Evaluating the scalar products gives

$$-14-8\mu-4\lambda+65+25\mu-25\lambda+3+7\mu-\lambda=0$$

and

$$28+16\mu+8\lambda+65+25\mu-25\lambda+21+49\mu-7\lambda=0$$

Simplifying gives $54+24\mu-30\lambda=0$ and $114+90\mu-24\lambda=0$

By inspection, the solutions are $\mu=-1$ and $\lambda=1$.

Substituting these values into Equations **(1)** and **(2)** gives the position vectors of P and Q, respectively.

Hence, $\mathbf{r}_1=8\mathbf{i}+\mathbf{j}+3\mathbf{k}$ and $\mathbf{r}_2=3\mathbf{i}+4\mathbf{j}-2\mathbf{k}$

Thus the co-ordinates of the feet of the perpendicular are $P(8,1,3)$ and $Q(3,4,-2)$.

The shortest distance between the lines is $|\mathbf{PQ}|=\sqrt{(5)^2+(-3)^2+(5)^2}$

$$=\sqrt{59}=7.68$$

Now the common perpendicular PQ passes through $P(8,1,3)$ and has a direction vector $5\mathbf{i}-3\mathbf{j}+5\mathbf{k}$.

The Cartesian equations are $\dfrac{x-8}{5}=\dfrac{y-1}{-3}=\dfrac{z-3}{5}$.

Exercise 4.4

1 Find the scalar product of $\mathbf{a}$ and $\mathbf{b}$ if

(a) $\mathbf{a}=\mathbf{i}+2\mathbf{j}+3\mathbf{k}$ and $\mathbf{b}=3\mathbf{i}+2\mathbf{j}+\mathbf{k}$

(b) $\mathbf{a}=-2\mathbf{i}+\mathbf{j}$ and $\mathbf{b}=3\mathbf{i}-2\mathbf{j}+4\mathbf{k}$

(c) $\mathbf{a}=\mathbf{i}-2\mathbf{j}+\mathbf{k}$ and $\mathbf{b}=2\mathbf{i}-5\mathbf{k}$

2 Find the angle between the vectors $\mathbf{a}$ and $\mathbf{b}$ if

(a) $\mathbf{a}=-\mathbf{i}+5\mathbf{j}+2\mathbf{k}$ and $\mathbf{b}=2\mathbf{i}-\mathbf{j}+4\mathbf{k}$

(b) $\mathbf{a}=5\mathbf{i}+3\mathbf{j}+2\mathbf{k}$ and $\mathbf{b}=2\mathbf{i}-4\mathbf{j}+\mathbf{k}$

(c) $\mathbf{a}=4\mathbf{i}-\mathbf{j}-8\mathbf{k}$ and $\mathbf{b}=\mathbf{i}+2\mathbf{j}-\mathbf{k}$

3 Evaluate (a) $(6\mathbf{i}-7\mathbf{j}+\mathbf{k})\cdot(-2\mathbf{i}-\mathbf{j}+4\mathbf{k})$ (b) $3\mathbf{i}\cdot(4\mathbf{i}-\mathbf{j}+2\mathbf{k})$

4 Use the properties of the scalar product to simplify

(a) $(\mathbf{a}+\mathbf{b})\cdot(\mathbf{a}+\mathbf{b})$ (b) $(\mathbf{p}+\mathbf{q})\cdot(\mathbf{p}-\mathbf{q})$

5 If $\mathbf{a}$, $\mathbf{b}$ and $\mathbf{c}$ are non-zero vectors such that $\mathbf{a}\cdot\mathbf{b}=\mathbf{a}\cdot\mathbf{c}$, show that either $\mathbf{b}=\mathbf{c}$, or $\mathbf{a}$ is perpendicular to $(\mathbf{b}-\mathbf{c})$.

6 If $\mathbf{a}$ and $\mathbf{b}$ are non-zero vectors such that $|\mathbf{a}|=|\mathbf{b}|$, show that $(\mathbf{a}+\mathbf{b})$ and $(\mathbf{a}-\mathbf{b})$ are perpendicular.

7 Simplify (a) $(\mathbf{a}+\mathbf{b})\cdot(\mathbf{a}-3\mathbf{b})$ (b) $(\mathbf{a}+\mathbf{b}+\mathbf{c})\cdot(\mathbf{a}-\mathbf{b}+\mathbf{c})$ if $\mathbf{a}$ is perpendicular to $\mathbf{c}$.

8 The vector $\mathbf{a}+t\mathbf{b}$ is perpendicular to a vector $\mathbf{c}$. If $\mathbf{a}\cdot\mathbf{c}=6$ and $\mathbf{b}\cdot\mathbf{c}=3$ find the value of t.

9 If $\mathbf{a}=2\mathbf{i}+\mathbf{j}-3\mathbf{k}$, $\mathbf{b}=\mathbf{i}-\mathbf{j}+\mathbf{k}$ and $\mathbf{c}=\mathbf{i}+\mathbf{j}-4\mathbf{k}$, find

   (a) $\mathbf{a}\cdot\mathbf{b}$ **(b)** $\mathbf{a}\cdot\mathbf{c}$ **(c)** $\mathbf{a}\cdot(\mathbf{b}+\mathbf{c})$ **(d)** $(2\mathbf{a}-\mathbf{b})\cdot\mathbf{c}$.

10 If $A(0,3,1)$, $B(2,-3,4)$ and $C(1,2,3)$ are the vertices of a triangle ABC, find the angles of the triangle.

11 Show that the vectors $\mathbf{u}=a\cos\theta\,\mathbf{i}+b\sin\theta\,\mathbf{j}$ and $\mathbf{v}=b\sin\theta\,\mathbf{i}-a\cos\theta\,\mathbf{j}$ are perpendicular.

12 Find the projection of the vector $\mathbf{i}+2\mathbf{j}-3\mathbf{k}$ in the direction of

   (a) $\mathbf{j}$ **(b)** $3\mathbf{i}+4\mathbf{j}$ **(c)** $-3\mathbf{i}+2\mathbf{j}+6\mathbf{k}$.

13 Find the angle between each of the following pairs of lines.

   (a) $\mathbf{r}=\mathbf{i}-\mathbf{j}+4\mathbf{k}+\lambda(2\mathbf{i}-3\mathbf{j}+\mathbf{k})$

   $\mathbf{r}=3\mathbf{i}+2\mathbf{j}-\mathbf{k}+\mu(\mathbf{i}-\mathbf{j}-\mathbf{k})$

   (b) $\dfrac{x-1}{2}=\dfrac{y+3}{-1}=\dfrac{z-5}{4}$ and $\dfrac{x-3}{-1}=\dfrac{y}{4}=\dfrac{z+1}{2}$

14 Show that $\mathbf{a}=\mathbf{i}+7\mathbf{j}+4\mathbf{k}$ is perpendicular to both $\mathbf{b}=3\mathbf{i}-\mathbf{j}+\mathbf{k}$ and $\mathbf{c}=\mathbf{i}+\mathbf{j}-2\mathbf{k}$.

15 The points $A(6,3,-1)$, $B(7,5,2)$ and $C(4,6,0)$ are the vertices of a triangle ABC. Find the angle ABC and hence find the area of the triangle.

16 Use vector methods to prove the theorem of Pythagoras.

17 Find the equations of the line through the point $(7,-4,4\tfrac{1}{2})$ which is perpendicular to the line
$\dfrac{x-2}{1}=\dfrac{y+3}{2}=\dfrac{z}{4}$, and intersects it.

18 Prove that the line drawn from the apex of an isosceles triangle to the mid-point of the base is perpendicular to the base.

19 Use vector methods to show that the diagonals of a rhombus bisect each other at right angles.

20 Prove that the altitudes of a triangle are concurrent.

   Hint   Let $\mathbf{a}$, $\mathbf{b}$ and $\mathbf{c}$ be the position vectors of A, B and C relative to H (the point of intersection of the altitudes from B and C).

   Consider $\mathbf{a}\cdot(\mathbf{b}-\mathbf{c})$.

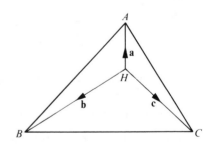

21 Show that, if AD is a median of a triangle, then
$$AB^2+AC^2=2AD^2+\tfrac{1}{2}BC^2$$

   This is the **theorem of Apollonius**.

   Hint   Consider $\mathbf{AD}\cdot\mathbf{AD}$ and $\mathbf{BC}\cdot\mathbf{BC}$.

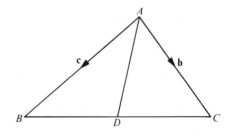

Equations of planes

We have already seen that a plane is a flat surface in three dimensions. However, we can define a plane in a variety of ways.

(a) Two intersecting lines define a plane.

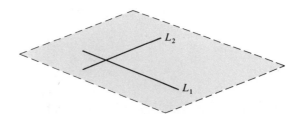

(b) Three non-collinear points in space define a plane.

(c) The normal vector from the origin to the plane and the distance of the plane from the origin, will define the plane.

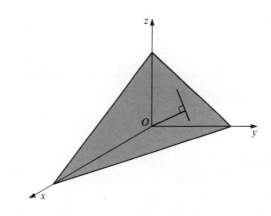

(d) The direction of any vector perpendicular to the plane and one point through which the plane passes, will define it uniquely.

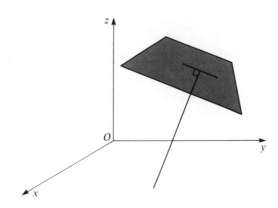

Remember that a plane is actually infinite in all directions but it is necessary to show a boundary when drawing it. The boundary can be shown as any shape, for example: a triangle; a square; a rectangle; a trapezium.

We can use each of the definitions **(a)** to **(d)** to investigate the various forms of the equation of a plane. Consider case **(c)** first, where the normal vector from the origin is known.

The vector equation of a plane

Consider a plane which is perpendicular to the unit vector $\hat{\mathbf{n}}$ from the origin and at a distance d from the origin, where d is a positive constant.

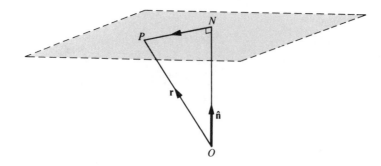

Let the vector from the origin perpendicular to the plane (called the normal vector) meet the plane at N where $\mathbf{ON} = d\hat{\mathbf{n}}$ and let P be any point of the plane with position vector $\mathbf{r}$ relative to the origin.

Since $\mathbf{NP}$ lies in the plane, it is perpendicular to $\mathbf{ON}$.

So, $\mathbf{NP} \cdot \mathbf{ON} = 0 \quad \Rightarrow \quad (\mathbf{r} - d\hat{\mathbf{n}}) \cdot (d\hat{\mathbf{n}}) = 0$

and $d(\mathbf{r} \cdot \hat{\mathbf{n}}) - d^2(\hat{\mathbf{n}} \cdot \hat{\mathbf{n}}) = 0 \quad \Rightarrow \quad \mathbf{r} \cdot \hat{\mathbf{n}} = d \quad$ (since $\hat{\mathbf{n}} \cdot \hat{\mathbf{n}} = 1$)

> The equation of a plane perpendicular to a unit vector $\hat{\mathbf{n}}$ at a distance d from the origin is
>
> $$\mathbf{r} \cdot \hat{\mathbf{n}} = d$$

Note that $\mathbf{r}$ is the position vector of any point of the plane, $\hat{\mathbf{n}}$ is the unit vector drawn from the origin perpendicular to the plane and d is the distance of the plane from the origin.

Clearly, if the equation is multiplied by a scalar quantity it would be in the form

$$\mathbf{r} \cdot \mathbf{n} = D$$

where $\mathbf{n}$ is a normal vector to the plane but **not** a unit vector. D is **not** the distance from the origin.

This still represents the equation of a plane perpendicular to $\mathbf{n}$ but at a distance $\dfrac{D}{|\mathbf{n}|}$ from the origin. That is,

$$\mathbf{r} \cdot \mathbf{n} = D \quad \Rightarrow \quad \mathbf{r} \cdot \frac{\mathbf{n}}{|\mathbf{n}|} = \frac{D}{|\mathbf{n}|} \quad \Rightarrow \quad \mathbf{r} \cdot \hat{\mathbf{n}} = \frac{D}{|\mathbf{n}|}$$

WORKED EXAMPLE

Find the vector equation of a plane whose normal vector from the origin is $6\mathbf{i} - 2\mathbf{j} + 3\mathbf{k}$.

In this case, $\mathbf{n} = 6\mathbf{i} - 2\mathbf{j} + 3\mathbf{k}$ which is not a unit vector.

Now, $d = |\mathbf{n}| = \sqrt{6^2 + (-2)^2 + 3^2} = \sqrt{49} = 7$

Hence, the unit vector in the direction of **n** is $\hat{\mathbf{n}} = \frac{1}{7}(6\mathbf{i} - 2\mathbf{j} + 3\mathbf{k})$.

The equation of the plane is $\mathbf{r} \cdot \hat{\mathbf{n}} = d$.

$$\mathbf{r} \cdot [\tfrac{1}{7}(6\mathbf{i} - 2\mathbf{j} + 3\mathbf{k})] = 7$$
$$\mathbf{r} \cdot (6\mathbf{i} - 2\mathbf{j} + 3\mathbf{k}) = 49$$

A plane perpendicular to a given vector passing through a fixed point

Now consider case (**d**) in which the normal vector to the plane, **a**, does not pass through the origin. Let P be a fixed point in the plane with position vector **p** and let R be any other point of the plane with position vector **r**.

Since **PR** lies in the plane,

$$\mathbf{PR} \cdot \mathbf{a} = 0$$
$$(\mathbf{r} - \mathbf{p}) \cdot \mathbf{a} = 0$$

or $\qquad\qquad \mathbf{r} \cdot \mathbf{a} = \mathbf{p} \cdot \mathbf{a}$

Thus, the vector equation of a plane through a point with position vector **p**, and normal to a vector **a** is

$$\boxed{\mathbf{r} \cdot \mathbf{a} = \mathbf{p} \cdot \mathbf{a}}$$

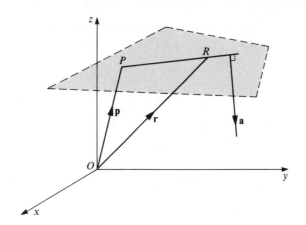

WORKED EXAMPLE

Find the equation of the plane normal to the vector $2\mathbf{i} - \mathbf{j} + 3\mathbf{k}$ which passes through the point $P(1, 0, 2)$ and find its distance from the origin.

The position vector of P is $\mathbf{i} + 2\mathbf{k}$ and the direction of the normal vector is $\mathbf{a} = 2\mathbf{i} - \mathbf{j} + 3\mathbf{k}$.

Hence, the vector equation of the plane is $\mathbf{r} \cdot \mathbf{a} = \mathbf{p} \cdot \mathbf{a}$.

$$\mathbf{r} \cdot (2\mathbf{i} - \mathbf{j} + 3\mathbf{k}) = (\mathbf{i} + 2\mathbf{k}) \cdot (2\mathbf{i} - \mathbf{j} + 3\mathbf{k})$$
$$\mathbf{r} \cdot (2\mathbf{i} - \mathbf{j} + 3\mathbf{k}) = 8$$

Now, $|\mathbf{a}| = \sqrt{2^2 + (-1)^2 + 3^2} = \sqrt{14}$ and thus the unit vector in the direction of **a** is $\dfrac{1}{\sqrt{14}}(2\mathbf{i} - \mathbf{j} + 3\mathbf{k})$.

We can write the equation as $\mathbf{r} \cdot \left(\dfrac{2\mathbf{i} - \mathbf{j} + 3\mathbf{k}}{\sqrt{14}}\right) = \dfrac{8}{\sqrt{14}}$ which is in the form $\mathbf{r} \cdot \hat{\mathbf{n}} = d$, and hence the distance from the origin is $\dfrac{8}{\sqrt{14}}$ units.

The Cartesian equation of a plane

Consider the vector equation of a plane in the form

$$\mathbf{r} \cdot \hat{\mathbf{n}} = d$$

Let $\hat{\mathbf{n}} = l\mathbf{i} + m\mathbf{j} + n\mathbf{k}$ and the position vector of a general point, P, be $\mathbf{r} = x\mathbf{i} + y\mathbf{j} + z\mathbf{k}$.

$$\therefore \quad (x\mathbf{i} + y\mathbf{j} + z\mathbf{k}) \cdot (l\mathbf{i} + m\mathbf{j} + n\mathbf{k}) = d$$

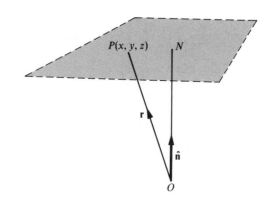

$$\boxed{lx + my + nz = d}$$

This is the **Cartesian equation of a plane.** The values of l, m and n are the **direction cosines of the normal** from the origin and d is the distance of the plane from the origin.

This result can also be derived from the vector equation $\mathbf{r} \cdot \mathbf{a} = \mathbf{p} \cdot \mathbf{a}$.

Let $\mathbf{a} = a\mathbf{i} + b\mathbf{j} + c\mathbf{k}$ and the position vector of a general point R be $x\mathbf{i} + y\mathbf{j} + z\mathbf{k}$. If P is a fixed point of the plane with position vector $\mathbf{p} = x_1\mathbf{i} + y_1\mathbf{j} + z_1\mathbf{k}$ then the equation becomes

$$(x\mathbf{i} + y\mathbf{j} + z\mathbf{k}) \cdot (a\mathbf{i} + b\mathbf{j} + c\mathbf{k}) = (x_1\mathbf{i} + y_1\mathbf{j} + z_1\mathbf{k}) \cdot (a\mathbf{i} + b\mathbf{j} + c\mathbf{k})$$
$$\Rightarrow \quad ax + by + cz = ax_1 + by_1 + cz_1$$

This can be written in the form $ax + by + cz = D$. In this case a, b and c are the **direction ratios of the normal** to the plane. The distance of the plane from the origin will be $\dfrac{D}{|\mathbf{a}|}$.

> WORKED EXAMPLE

Convert the vector equation of the plane $\mathbf{r} \cdot (2\mathbf{i} + \mathbf{j} - \mathbf{k}) = 5$ to Cartesian form.

Let $\mathbf{r} = x\mathbf{i} + y\mathbf{j} + z\mathbf{k}$, and the equation becomes

$$(x\mathbf{i} + y\mathbf{j} + z\mathbf{k}) \cdot (2\mathbf{i} + \mathbf{j} - \mathbf{k}) = 5$$
$$\Rightarrow \quad 2x + y - z = 5$$

Investigation 13

Consider the plane defined by the points $A(1, 2, -2)$, $B(2, 0, -1)$ and $C(3, -1, -3)$.

Show that $\mathbf{AB} = \mathbf{i} - 2\mathbf{j} + \mathbf{k}$ and $\mathbf{AC} = 2\mathbf{i} - 3\mathbf{j} - \mathbf{k}$.

If the normal vector to the plane is $\mathbf{a} = a\mathbf{i} + b\mathbf{j} + c\mathbf{k}$, show that

$$a - 2b + c = 0 \quad \text{and} \quad 2a - 3b - c = 0$$

by using the fact that $\mathbf{AB}$ and $\mathbf{AC}$ are perpendicular to $\mathbf{a}$.

Eliminate c to show that $b = \frac{3}{5}a$ and hence find c in terms of a.

Deduce that $\mathbf{a} = a\mathbf{i} + \frac{3}{5}a\mathbf{j} + \frac{1}{5}a\mathbf{k}$.

(cont.)

Investigation 13—continued

Thus, the direction vector of the normal to the plane can be represented by $5\mathbf{i}+3\mathbf{j}+\mathbf{k}$. Use this direction vector, the position of B, and the result $\mathbf{r}\cdot\mathbf{a}=\mathbf{p}\cdot\mathbf{a}$, to show that the vector equation of the plane is $\mathbf{r}\cdot(5\mathbf{i}+3\mathbf{j}+\mathbf{k})=9$.

Verify that the Cartesian equation of the plane is $5x+3y+z=9$.

This solution can be achieved in a different manner.

Let the equation of the plane be $ax+by+cz=d$.

Substitute the co-ordinates of the points A, B and C as the values of x, y and z into this equation and show that

$$a+2b-2c=d \qquad \text{............ (1)}$$

$$2a-c=d \qquad \text{............ (2)}$$

$$3a-b+3c=d \qquad \text{............ (3)}$$

Eliminate c from Equations **(1)** and **(2)** and from **(2)** and **(3)** and show that $d=\frac{9}{5}a$, $c=\frac{1}{5}a$ and $b=\frac{3}{5}a$.

Hence, deduce that the Cartesian equation is $5x+3y+z=9$, as before.

The parametric form for the vector equation of a plane

The work of Investigation 13 illustrated methods for dealing with the cases of a plane defined by two lines in the plane ($\mathbf{AB}$ and $\mathbf{AC}$) and by three points A, B and C. These are the cases **(a)** and **(b)** for defining a plane, as shown on page 91.

However, it is possible to approach these cases differently by expressing the position vector of a general point in terms of two parameters, λ and μ.

Consider a plane which contains the non-parallel vectors $\mathbf{b}$ and $\mathbf{c}$, and a given point A whose position vector is $\mathbf{a}$.

If P is a general point of the plane defined by the vectors $\mathbf{b}$ and $\mathbf{c}$, then the position vector of P relative to A will be the sum of multiples of $\mathbf{b}$ and $\mathbf{c}$.

$$\mathbf{AP}=\lambda\mathbf{b}+\mu\mathbf{c}$$

where λ and μ are independent parameters.

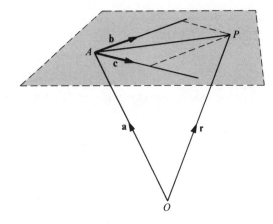

Thus, if $\mathbf{r}$ is the position vector of P relative to the origin, then

$$\mathbf{r}=\mathbf{a}+\mathbf{AP}$$

$$\mathbf{r}=\mathbf{a}+\lambda\mathbf{b}+\mu\mathbf{c}$$

Now since $\mathbf{r}$ defines the position vector of any point of the plane, it represents the equation of the plane.

> The parametric equation of a plane containing vectors $\mathbf{b}$ and $\mathbf{c}$ which passes through a point with position vector $\mathbf{a}$, is
>
> $$\mathbf{r}=\mathbf{a}+\lambda\mathbf{b}+\mu\mathbf{c}$$

Investigation 14

Consider a plane defined by three points, A, B and C whose position vectors relative to the origin are $\mathbf{a}$, $\mathbf{b}$ and $\mathbf{c}$, respectively.

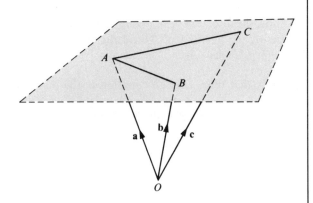

From the position vectors of A, B and C find expressions for the vectors $\mathbf{AB}$ and $\mathbf{AC}$.

Now $\mathbf{AB}$ and $\mathbf{AC}$ are two vectors in the plane which also passes through the fixed point A with position vector $\mathbf{a}$.

Show that the parametric equation of the plane can be written as

$$\mathbf{r} = (1 - \lambda - \mu)\mathbf{a} + \lambda\mathbf{b} + \mu\mathbf{c}$$

Writing $v = 1 - \lambda - \mu$ gives

$$\mathbf{r} = v\mathbf{a} + \lambda\mathbf{b} + \mu\mathbf{c}$$

> The parametric equation of a plane passing through the points with position vectors $\mathbf{a}$, $\mathbf{b}$ and $\mathbf{c}$ is
>
> $$\mathbf{r} = v\mathbf{a} + \lambda\mathbf{b} + \mu\mathbf{c} \quad \text{where} \quad v + \lambda + \mu = 1$$

WORKED EXAMPLES

---

1 Find the vector equation of the plane which passes through the points $A(1, 2, 0)$, $B(3, 1, -2)$ and $C(4, -5, 7)$ in **(a)** parametric form **(b)** Cartesian form and **(c)** scalar product form.

(a) The position vectors of the points A, B and C are given by

$$\mathbf{a} = \mathbf{i} + 2\mathbf{j}; \quad \mathbf{b} = 3\mathbf{i} + \mathbf{j} - 2\mathbf{k}; \quad \mathbf{c} = 4\mathbf{i} - 5\mathbf{j} + 7\mathbf{k}$$

The parametric equation of the plane is $\mathbf{r} = v\mathbf{a} + \lambda\mathbf{b} + \mu\mathbf{c}$

$\Rightarrow \quad \mathbf{r} = v(\mathbf{i} + 2\mathbf{j}) + \lambda(3\mathbf{i} + \mathbf{j} - 2\mathbf{k}) + \mu(4\mathbf{i} - 5\mathbf{j} + 7\mathbf{k})$

$$\text{(where } v + \lambda + \mu = 1)$$

$\Rightarrow \quad \mathbf{r} = (1 - \lambda - \mu)(\mathbf{i} + 2\mathbf{j}) + \lambda(3\mathbf{i} + \mathbf{j} - 2\mathbf{k}) + \mu(4\mathbf{i} - 5\mathbf{j} + 7\mathbf{k})$

$\Rightarrow \quad \mathbf{r} = (1 + 2\lambda + 3\mu)\mathbf{i} + (2 - \lambda - 7\mu)\mathbf{j} + (-2\lambda + 7\mu)\mathbf{k}$

Check that this is the same equation that you would obtain by finding the vectors $\mathbf{AB}$ and $\mathbf{AC}$ and substituting into $\mathbf{r} = \mathbf{a} + \lambda\mathbf{AB} + \mu\mathbf{AC}$.

What form do you obtain if you use the vectors $\mathbf{BA}$ and $\mathbf{BC}$, and substitute into $\mathbf{r} = \mathbf{b} + \lambda\mathbf{BA} + \mu\mathbf{BC}$?

(b) If $P(x, y, z)$ is a general point of the plane, then its position vector is given by $\mathbf{r} = x\mathbf{i} + y\mathbf{j} + z\mathbf{k}$.

Using the equation obtained in part **(a)** gives

$$x\mathbf{i} + y\mathbf{j} + z\mathbf{k} = (1 + 2\lambda + 3\mu)\mathbf{i} + (2 - \lambda - 7\mu)\mathbf{j} + (-2\lambda + 7\mu)\mathbf{k}$$

Hence,

$$x = 1 + 2\lambda + 3\mu \qquad \text{............ (1)}$$
$$y = 2 - \lambda - 7\mu \qquad \text{............ (2)}$$
$$z = -2\lambda + 7\mu \qquad \text{............ (3)}$$

Adding Equations **(1)** and **(3)**

$$x + z = 1 + 10\mu \qquad \text{............ (4)}$$

Adding Equation **(1)** and $2 \times$ Equation **(2)**

$$x + 2y = 5 - 11\mu \qquad \text{............ (5)}$$

Eliminating μ from Equations **(4)** and **(5)** gives

$$21x + 20y + 11z = 61$$

This is the Cartesian equation of the plane.

(c) We can find the scalar product form from the Cartesian equation since the direction vector of the normal to the plane is given by $21\mathbf{i} + 20\mathbf{j} + 11\mathbf{k}$ and the left-hand side of the Cartesian equation can be written as $\mathbf{r} \cdot (21\mathbf{i} + 20\mathbf{j} + 11\mathbf{k})$.

Hence, the scalar product form is $\mathbf{r} \cdot (21\mathbf{i} + 20\mathbf{j} + 11\mathbf{k}) = 61$.

Check that the answers to parts **(b)** and **(c)** could be obtained by using the methods of Investigation 13.

---

2 Find the point of intersection of the line $\dfrac{x-1}{1} = \dfrac{y+2}{-1} = \dfrac{z-4}{3}$ and the plane $x + y - 2z = 3$.

Let $P(x, y, z)$ be the point where the line meets the plane.

We can write the Cartesian equations in the form

$$\frac{x-1}{1} = \frac{y+2}{-1} = \frac{z-4}{3} = \lambda$$

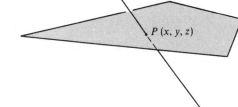

Hence, $x = 1 + \lambda$; $y = -2 - \lambda$; $z = 4 + 3\lambda$

Thus, since P is a point on the line, its position vector $\mathbf{r}$ is given by

$$\mathbf{r} = (1 + \lambda)\mathbf{i} + (-2 - \lambda)\mathbf{j} + (4 + 3\lambda)\mathbf{k}$$

and the co-ordinates of P are $(1 + \lambda, -2 - \lambda, 4 + 3\lambda)$.

Now, P is also a point of the plane and therefore these co-ordinates must satisfy the equation of the plane.

$$(1 + \lambda) + (-2 - \lambda) - 2(4 + 3\lambda) = 3$$
$$-9 - 6\lambda = 3 \quad \Rightarrow \quad 6\lambda = -12 \quad \Rightarrow \quad \lambda = -2$$

Using this value gives the position vector of P as $-\mathbf{i} - 2\mathbf{k}$ and hence the co-ordinates of the point of intersection are $(-1, 0, -2)$.

3 Find the Cartesian equation of the plane which passes through the point $A(1, -2, 5)$ and which contains the line

$$\frac{x+2}{-1} = \frac{y}{3} = \frac{z+1}{4}$$

Since the direction vector of the line is $-\mathbf{i} + 3\mathbf{j} + 4\mathbf{k}$, this represents a vector in the plane.

This line passes through the point $B(-2, 0, -1)$ and so $A(1, -2, 5)$ and $B(-2, 0, -1)$ are two points in the plane.

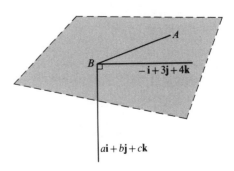

The vector **BA** lies in the plane.

$$\mathbf{BA} = 3\mathbf{i} - 2\mathbf{j} + 6\mathbf{k}$$

If the direction vector of the normal to the plane is $a\mathbf{i} + b\mathbf{j} + c\mathbf{k}$, then this vector is perpendicular to **BA** and to the vector $-\mathbf{i} + 3\mathbf{j} + 4\mathbf{k}$.

So, $\mathbf{BA} \cdot (a\mathbf{i} + b\mathbf{j} + c\mathbf{k}) = 0 \Rightarrow (3\mathbf{i} - 2\mathbf{j} + 6\mathbf{k}) \cdot (a\mathbf{i} + b\mathbf{j} + c\mathbf{k}) = 0$

$$\Rightarrow \quad 3a - 2b + 6c = 0 \quad \ldots\ldots\ldots \text{ (1)}$$

Also, $(-\mathbf{i} + 3\mathbf{j} + 4\mathbf{k}) \cdot (a\mathbf{i} + b\mathbf{j} + c\mathbf{k}) = 0$

$$\Rightarrow \quad -a + 3b + 4c = 0 \quad \ldots\ldots\ldots \text{ (2)}$$

Multiplying Equation **(1)** by 3 and Equation **(2)** by 2 gives

$$9a - 6b + 18c = 0$$

$$-2a + 6b + 8c = 0$$

Adding gives $\qquad\qquad\qquad\qquad 7a + 26c = 0 \Rightarrow c = -\dfrac{7a}{26}$

Substituting into Equation **(2)** gives

$$-a + 3b - \frac{28a}{26} = 0$$

$$\Rightarrow \quad -54a + 78b = 0 \Rightarrow b = \frac{9a}{13}$$

Hence, the direction vector of the normal is $a\mathbf{i} + \dfrac{9a}{13}\mathbf{j} - \dfrac{7a}{26}\mathbf{k}$.

This can be written in the form $26\mathbf{i} + 18\mathbf{j} - 7\mathbf{k}$.

The equation of a plane is $\mathbf{r} \cdot \mathbf{a} = \mathbf{p} \cdot \mathbf{a}$ where **a** is the normal vector and **p** is the position vector of a point in the plane.

Since A lies in the plane, $\mathbf{p} = \mathbf{i} - 2\mathbf{j} + 5\mathbf{k}$.

The equation of the plane is

$$\mathbf{r} \cdot (26\mathbf{i} + 18\mathbf{j} - 7\mathbf{k}) = (\mathbf{i} - 2\mathbf{j} + 5\mathbf{k}) \cdot (26\mathbf{i} + 18\mathbf{j} - 7\mathbf{k})$$

$$\Rightarrow \quad \mathbf{r} \cdot (26\mathbf{i} + 18\mathbf{j} - 7\mathbf{k}) = 26 - 36 - 35$$

The vector equation is $\mathbf{r} \cdot (26\mathbf{i} + 18\mathbf{j} - 7\mathbf{k}) = -45$

By letting $\mathbf{r} = x\mathbf{i} + y\mathbf{j} + z\mathbf{k}$ we can obtain the Cartesian equation

$$26x + 18y - 7z + 45 = 0$$

Exercise 4.5

1 Find the vector equation of the plane whose normal from the origin is given by

(a) $\mathbf{i} + 4\mathbf{j} - \mathbf{k}$ (b) $-2\mathbf{i} + 3\mathbf{j} + \mathbf{k}$

2 Find the distance of the following planes from the origin, given the normal vector, $\mathbf{n}$, and the position vector, $\mathbf{p}$, of a point in the plane.

(a) $\mathbf{n} = \mathbf{j} + 2\mathbf{k}$; $\mathbf{p} = 3\mathbf{i} - 2\mathbf{j} + \mathbf{k}$ (b) $\mathbf{n} = 3\mathbf{i} + \mathbf{j} + \mathbf{k}$; $\mathbf{p} = 2\mathbf{i} - \mathbf{j} - \mathbf{k}$

3 Find the vector equation of the plane in scalar product form which passes through a point with position vector, $\mathbf{a}$, and is perpendicular to the direction, $\mathbf{l}$.

(a) $\mathbf{a} = \mathbf{i} + 2\mathbf{j} + \mathbf{k}$; $\mathbf{l} = \mathbf{i} - 3\mathbf{j} + 4\mathbf{k}$ (b) $\mathbf{a} = 4\mathbf{j} + 5\mathbf{k}$; $\mathbf{l} = 2\mathbf{i} - 8\mathbf{j} + 3\mathbf{k}$

(c) $\mathbf{a} = 3\mathbf{i} + \mathbf{j} - 2\mathbf{k}$; $\mathbf{l} = 6\mathbf{j} - 2\mathbf{k}$

4 Find the Cartesian equation of the plane which contains the points A, B and C where

(a) $A(1, 2, -4)$, $B(3, 1, -1)$, $C(4, 0, 5)$ (b) $A(2, -3, 5)$, $B(5, 3, -2)$, $C(3, -2, -1)$

(c) $A(0, 3, 7)$, $B(5, 6, -1)$, $C(1, 4, 5)$

5 Find the distance of the following planes from the origin.

(a) $x + 3y - z = 4$ (b) $2x - 4y + 3z = 5$ (c) $3x - 2y + 5z = 2$ (d) $-x + y + 4z = 3$

6 Find the distance between the following pairs of planes.

(a) $x - y + 4z = 5$; $x - y + 4z = 3$ (b) $3x - 2y + z = 1$; $3x - 2y + z = -2$

7 Write down the parametric equation of the plane which contains the vectors $\mathbf{p}$ and $\mathbf{q}$ and passes through the point with position vector $\mathbf{s}$.

(a) $\mathbf{p} = \mathbf{a} - \mathbf{b}$; $\mathbf{q} = 2\mathbf{a} + \mathbf{b}$; $\mathbf{s} = 3\mathbf{a}$ (b) $\mathbf{p} = 3\mathbf{a} + 2\mathbf{b}$; $\mathbf{q} = 3\mathbf{a} - 2\mathbf{b}$; $\mathbf{s} = 4\mathbf{a} - \mathbf{b}$

(c) $\mathbf{p} = \mathbf{c} - 3\mathbf{a}$; $\mathbf{q} = \mathbf{b} - 2\mathbf{a}$; $\mathbf{s} = 5\mathbf{a}$

8 Find the Cartesian equation of the plane containing the direction vectors $\mathbf{l}$ and $\mathbf{m}$ and the point A.

(a) $\mathbf{l} = 3\mathbf{i} - 4\mathbf{j} - \mathbf{k}$; $\mathbf{m} = \mathbf{i} + 2\mathbf{j} - 3\mathbf{k}$; $A(1, 0, 1)$ (b) $\mathbf{l} = -2\mathbf{i} - \mathbf{j} + \mathbf{k}$; $\mathbf{m} = \mathbf{i} + 3\mathbf{j} - 2\mathbf{k}$; $A(-1, 1, 2)$

(c) $\mathbf{l} = \mathbf{i} + \mathbf{j} + \mathbf{k}$; $\mathbf{m} = 2\mathbf{i} + \mathbf{j} + 6\mathbf{k}$; $A(4, 3, 1)$

9 Find the Cartesian equation of the plane containing the point P, which is parallel to the given plane.

(a) $P(1, 0, -1)$; $2x + y - z = 4$ (b) $P(2, -1, 4)$; $x + 5y + z = -1$

(c) $P(3, -1, 2)$; $4x - y = 2$

10 Find the parametric equation of the plane which contains the following pairs of lines with equations:

(a) $\mathbf{r} = 2\mathbf{k} + \lambda(\mathbf{i} - \mathbf{j} + \mathbf{k}); \quad \mathbf{r} = \mathbf{i} - \mathbf{j} + 3\mathbf{k} + \mu(2\mathbf{i} - \mathbf{j} - \mathbf{k})$

(b) $\mathbf{r} = \mathbf{i} - 2\mathbf{j} + 3\mathbf{k} + \lambda(\mathbf{i} + 2\mathbf{j}); \quad \mathbf{r} = 3\mathbf{i} + \mu(2\mathbf{j} + 3\mathbf{k})$

11 Find, in parametric form, the vector equation of the plane which contains the lines:

$$\frac{x-2}{1} = \frac{2y-17}{2} = \frac{z}{1} \quad \text{and} \quad \frac{x+2}{1} = \frac{y-3}{2} = \frac{z+1}{-1}$$

12 Find the point of intersection of the line $\dfrac{x+1}{1} = \dfrac{y-1}{2} = \dfrac{z-3}{-1}$ and the plane $2x - 5y + 4z = 1$.

13 Find the vector equation of the plane through the origin and the point $(3, 0, -1)$, which is perpendicular to the plane $\mathbf{r} \cdot (\mathbf{i} - 2\mathbf{j} + 3\mathbf{k}) = 4$.

14 Find the vector equation of the plane through the points $A(1, 0, 2)$ and $B(1, -1, 3)$, which contains the line $\mathbf{r} = -2\lambda\mathbf{i} + (2\lambda - 1)\mathbf{j} + (3 - \lambda)\mathbf{k}$.

The angle between a line and a plane

Unless a line is parallel to a plane, clearly it intersects the plane in one point.

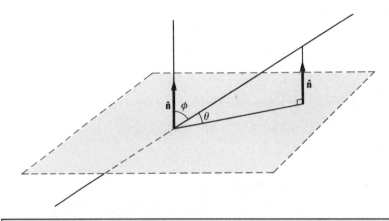

> The angle between a line and a plane is defined as the angle between the line and its projection on the plane.

Thus, in the figure, θ is the angle between the line and the plane.

Let the equation of the line be $\mathbf{r} = \mathbf{m} + \lambda\mathbf{a}$ and let $\hat{\mathbf{n}}$ be a unit vector normal to the plane.

Now, the angle between two vectors $\mathbf{a}$ and $\mathbf{b}$ is given by $\cos\theta = \dfrac{\mathbf{a} \cdot \mathbf{b}}{|\mathbf{a}||\mathbf{b}|}$.

Hence, $\cos\phi = \dfrac{\mathbf{a} \cdot \hat{\mathbf{n}}}{|\mathbf{a}|}$ since $\mathbf{a}$ is the direction vector of the line.

Now,
$$\theta + \phi = \tfrac{1}{2}\pi \quad \Rightarrow \quad \phi = \tfrac{1}{2}\pi - \theta$$
$$\Rightarrow \quad \cos\phi = \cos(\tfrac{1}{2}\pi - \theta) = \sin\theta$$
$$\sin\theta = \frac{\mathbf{a} \cdot \hat{\mathbf{n}}}{|\mathbf{a}|}$$

> The angle, θ, between a line with direction vector **a** and a plane with normal vector **n**, is given by
>
> $$\sin \theta = \frac{\mathbf{a} \cdot \mathbf{n}}{|\mathbf{a}||\mathbf{n}|}$$

WORKED EXAMPLE

Find the angle between the line $\mathbf{r} = 2\mathbf{i} + 3\mathbf{j} + \lambda(\mathbf{i} + \mathbf{j} - \mathbf{k})$ and the plane $\mathbf{r} \cdot (2\mathbf{i} + \mathbf{j} - 4\mathbf{k}) = 1$.

In this case, the direction vector of the line $\mathbf{a} = \mathbf{i} + \mathbf{j} - \mathbf{k}$ and the normal vector to the plane $\mathbf{n} = 2\mathbf{i} + \mathbf{j} - 4\mathbf{k}$.

If θ is the angle between the line and the plane, then

$$\sin \theta = \frac{\mathbf{a} \cdot \mathbf{n}}{|\mathbf{a}||\mathbf{n}|} = \frac{(\mathbf{i} + \mathbf{j} - \mathbf{k})}{\sqrt{3}} \cdot \frac{(2\mathbf{i} + \mathbf{j} - 4\mathbf{k})}{\sqrt{21}}$$

$$= \frac{7}{\sqrt{3}\sqrt{21}} = \frac{\sqrt{7}}{3}$$

The angle between the line and the plane, $\theta = \sin^{-1}\left(\dfrac{\sqrt{7}}{3}\right) = 61.9°$.

The angle between two planes

Unless two planes are parallel they will intersect in a straight line.

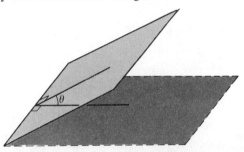

> The angle between two planes is defined as the angle between two lines, one in each plane, which are perpendicular to the line of intersection.

So, θ is the angle between the two planes in the figure. If these two lines are both rotated through 90° then they lie along the directions of the normals to the planes but still enclose an angle of θ.

> The angle between two planes is the angle between the normals to the planes.

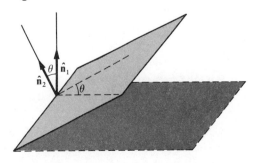

101

> If the unit vectors in the directions of the normals to the planes are $\hat{\mathbf{n}}_1$ and $\hat{\mathbf{n}}_2$ then the angle between the plane, θ, is given by
>
> $$\cos\theta = \hat{\mathbf{n}}_1 . \hat{\mathbf{n}}_2$$

WORKED EXAMPLE

Find the angle between the two planes whose equations are $\mathbf{r} . (\mathbf{i}+\mathbf{j}+3\mathbf{k}) = 2$ and $\mathbf{r} . (2\mathbf{i}+2\mathbf{j}-\mathbf{k}) = 1$.

If θ is the angle between the two planes, then

$$\cos\theta = \hat{\mathbf{n}}_1 . \hat{\mathbf{n}}_2 = \frac{(\mathbf{i}+\mathbf{j}+3\mathbf{k})}{\sqrt{11}} . \frac{(2\mathbf{i}+2\mathbf{j}-\mathbf{k})}{\sqrt{9}} = \frac{1}{3\sqrt{11}}$$

Hence, the angle between the planes is $\cos^{-1}\left(\frac{1}{3\sqrt{11}}\right) = 84.2°$.

The distance of a point from a plane

Consider a point P with position vector $\mathbf{a}$ and a given plane, π_1, whose equation is $\mathbf{r} . \hat{\mathbf{n}} = d$.

The plane π_2 through P parallel to the given plane π_1 is given by

$$\mathbf{r} . \hat{\mathbf{n}} = \mathbf{a} . \hat{\mathbf{n}}$$

since $OX = \mathbf{a} . \hat{\mathbf{n}} =$ the distance of the plane from the origin.

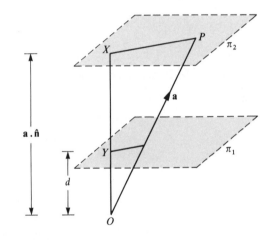

Since the distance of the plane π_1 from the origin is d, the distance of the point P from the plane π_1 is

$$XY = \mathbf{a} . \hat{\mathbf{n}} - d$$

if P and the origin are on opposite sides of the plane π_1.

Draw a diagram to show the case when P and the origin are on the same side of the plane π_1, when the result above will give a negative answer.

Investigation 15

Draw a simple diagram to show the plane $\mathbf{r} . \mathbf{k} = 4$. Use the result above to show that the distance of the point $(1, 1, -3)$ from the plane gives a value of -7.

Find the distance if the point $(1, 1, 5)$ is chosen. Note that in the first case the point and the origin are on the same side of the plane but in the second case they are on opposite sides.

Try out your observations with some other points and planes.

WORKED EXAMPLE

Find the equation of the line of intersection of two planes whose vector equations are $\mathbf{r} \cdot (\mathbf{i}+\mathbf{j}+2\mathbf{k}) = 6$ and $\mathbf{r} \cdot (2\mathbf{i}-\mathbf{j}+\mathbf{k}) = 1$.

The figure shows two intersecting planes. Since the line of intersection lies in both planes it is perpendicular to both $\mathbf{n}_1$ and $\mathbf{n}_2$, the normal vectors to each plane.

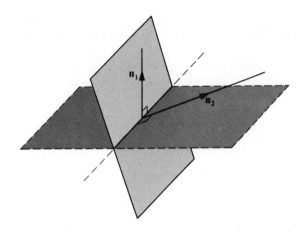

The Cartesian equations of the planes are

$$x+y+2z=6$$

$$\text{and} \quad 2x-y+z=1$$

If a general point of the line is $P(t, y, z)$ then the equations will be satisfied by these co-ordinates.

Hence,

$$t+y+2z=6 \qquad \text{............ (1)}$$

$$2t-y+z=1 \qquad \text{............ (2)}$$

Adding gives

$$3t+3z=7 \quad \Rightarrow \quad z = \frac{7-3t}{3}$$

Multiplying Equation **(2)** by 2 and subtracting from Equation **(1)** gives

$$-3t+3y=4 \quad \Rightarrow \quad y = \frac{4+3t}{3}$$

Thus, a general point of the line of intersection can be written as

$$\left(t, \frac{4+3t}{3}, \frac{7-3t}{3} \right)$$

Hence, the vector equation will be $\quad \mathbf{r} = (\tfrac{4}{3}\mathbf{j}+\tfrac{7}{3}\mathbf{k}) + t(\mathbf{i}+\mathbf{j}-\mathbf{k})$.

Thus the Cartesian equations will be $\quad \dfrac{x}{1} = \dfrac{3y-4}{3} = \dfrac{3z-7}{-3}$.

Exercise 4.6

1 Find the angle between each of the following pairs of lines and planes:

(a) line $\dfrac{x-2}{3} = \dfrac{y-1}{4} = \dfrac{z-4}{1}$; plane $5x - y + z = 10$

(b) line $\dfrac{x}{2} = \dfrac{y+3}{-1} = \dfrac{z-1}{3}$; plane $x + 3y - z = 4$

(c) line $\dfrac{x}{1} = \dfrac{y-2}{-2} = \dfrac{z+4}{3}$; plane $x - 2y + 4z = 1$

2 Find the angle between the line $\mathbf{r} = \mathbf{i} + 2\mathbf{j} - \mathbf{k} + \lambda(2\mathbf{i} - 3\mathbf{j} + 4\mathbf{k})$ and the plane $\mathbf{r} \cdot (\mathbf{i} + 3\mathbf{j} - \mathbf{k}) = 4$.

3 Find the sine of the angle between each line and plane, whose equations are,

(a) $\mathbf{r} = \mathbf{i} - 3\mathbf{j} + \mathbf{k} + \lambda(\mathbf{i} - 2\mathbf{k})$; $\mathbf{r} \cdot (\mathbf{i} - 2\mathbf{j} + 3\mathbf{k}) = 5$

(b) $\mathbf{r} = 3\mathbf{i} - \mathbf{j} + \mathbf{k} + \lambda(\mathbf{i} + \mathbf{j} + \mathbf{k})$; $\mathbf{r} = \mathbf{i} - 2\mathbf{j} + s(2\mathbf{i} + \mathbf{j}) + t(\mathbf{j} + 3\mathbf{k})$

4 Find the angle between the two given planes with equations:

(a) $\mathbf{r} \cdot (2\mathbf{i} - \mathbf{j} + \mathbf{k}) = 4$ and $\mathbf{r} \cdot (3\mathbf{i} - \mathbf{j} + 4\mathbf{k}) = 6$

(b) $\mathbf{r} \cdot (\mathbf{j} + 3\mathbf{k}) = 1$ and $\mathbf{r} \cdot (\mathbf{i} + 2\mathbf{k}) = 5$

(c) $x - 2y + 4z = 1$ and $2x - 3y + z = 1$

(d) $x - 3y + z = 2$ and $3x + y + 2z = 6$

5 Find the cosine of the angle between the two planes with equations:

(a) $\mathbf{r} \cdot (4\mathbf{i} - \mathbf{j} + \mathbf{k}) = 3$ and $\mathbf{r} \cdot (2\mathbf{i} - \mathbf{j} + 2\mathbf{k}) = 4$

(b) $\mathbf{r} = \mathbf{i} + \lambda(\mathbf{i} + \mathbf{j}) + \mu(2\mathbf{i} - \mathbf{j} + 2\mathbf{k})$ and $\mathbf{r} = (\mathbf{i} - \mathbf{j} + 2\mathbf{k}) + s(\mathbf{i} - \mathbf{j} + \mathbf{k}) + t(3\mathbf{i} + \mathbf{j} + \mathbf{k})$

6 Find the distance of the point $(1, 2, 3)$ from the following planes.

(a) $2x + 6y + 3z = 4$ (b) $\mathbf{r} \cdot (4\mathbf{i} + 7\mathbf{j} + 4\mathbf{k}) = 18$

7 Find the vector equation of the line of intersection of each of the following pairs of planes.

(a) $\mathbf{r} \cdot (\mathbf{i} - 2\mathbf{j} - \mathbf{k}) = 1$; $\mathbf{r} \cdot (3\mathbf{i} - \mathbf{j} - 2\mathbf{k}) = 2$

(b) $\mathbf{r} \cdot (2\mathbf{i} - 3\mathbf{j} + \mathbf{k}) = 5$; $\mathbf{r} \cdot (\mathbf{i} + \mathbf{j} + \mathbf{k}) = 1$

(c) $3x - 2y + 4z = 2$; $x + y + z = 6$

(d) $x + 4y - 3z = 4$; $2x - 3y + z = 5$

8 Find the position vector of the point of intersection of the line
$\mathbf{r} = (2\mathbf{i} + \mathbf{j} - \mathbf{k}) + \lambda(\mathbf{i} + 3\mathbf{j})$ and the plane $\mathbf{r} \cdot (\mathbf{i} + 2\mathbf{j} + 4\mathbf{k}) = 5$.

9 Find the equation of a line which passes through the point $A(2, 0, 1)$ and which is normal to the plane $2x - y + 3z = 4$. Calculate the co-ordinates of the point of intersection of this line and the plane. If $B(1, -1, 2)$ also lies in the plane, find the Cartesian equations of the line AB.

5 Rigid bodies

Introduction

Investigation 1

When using a ladder it is recommended that the angle of inclination to the horizontal should be $\tan^{-1} 4$ for safety.

Can you stand your pencil or ruler against a vertical surface at this sort of angle without it slipping down?

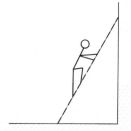

Investigation 2

Examine the stability of a ruler placed inside a hemispherical bowl?

Vary the length of ruler compared with the diameter of the bowl.

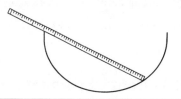

Investigation 3

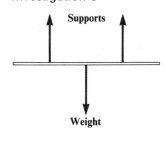

The striplight in my room is supported by two chains attached one-quarter and three-quarters of the way along the light. Why not at the ends?

If a weight equal to the weight of the light were attached to one end, where would the chains be attached for the best effect?

Consider equal weights at both ends.

Investigation 4

The inn-sign at my local hostelry hangs on the end of a horizontal pole, which is supported at its base on the wall and by a rope attached higher up the wall.

Where would you attach the rope to best effect?

Would you angle the pole?

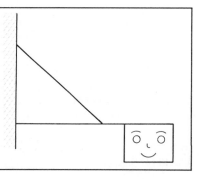

Investigation 5

Place two coins on one end of a ruler and one coin on the other. Where is the balance point?

Repeat the experiment for three coins on one end and one on the other. Try various combinations of coins on each end.

What is your result for *x* coins on one end and *y* coins on the other end?

How large is the force on the supporting pivot each time?

In these Investigations we are considering the resultants of forces and their turning effects (moments).

When opening a door or a gate, it is easier to push at a point furthest from the hinge (**axis of rotation**) because the force has a greater turning effect (**moment**) the further it is applied away from the hinge.

Think of everyday situations in which you apply a force to give a turning effect. Which turning effect is used in the following situations?

(a) using a screwdriver **(b)** unlocking and opening the front door

(c) using a socket set **(d)** changing a car wheel

(e) riding a bicycle **(f)** driving a car **(g)** stirring your coffee

(h) doing aerobics or workouts **(i)** standing up

(j) sitting down **(k)** writing a letter

Resultant of parallel forces

If my striplight (of weight 60 N) is supported by two chains placed symmetrically, then the tensions in the supports are both equal to 30 N.

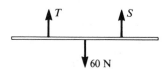

If T acts at the lefthand end and S three-quarters of the length from that end, $T \neq S$, but $T + S = 60$ and we find T and S by considering their turning effect about different points.

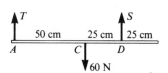

We know that a weight of 120 N will balance one of 80 N if their distances from the pivot are in the ratio 2:3.

The turning effect or moment of the forces must be equal, i.e.

$$120 \times 2 = 80 \times 3$$

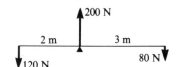

The moment of the force, 120 N, about the pivot is $120 \text{ N} \times 2 \text{ m} = 240$ Newton-metres.

Moreover, the pivot must support a force of 200 N, so we can deduce that the resultant of the 120 N and 80 N forces is a force of 200 N acting through the pivot.

In the middle diagram, above, T and S are in the inverse ratio of $AC:CD$ i.e. $T:S = 1:2 \Rightarrow T = 20 \text{ N}$ and $S = 40 \text{ N}$.

It may be easier to consider the moments of the forces about different points. Writing 'moments about C' as $\hat{C}$, and so on, then

$\hat{C} \Rightarrow T \times 50 = S \times 25 \Rightarrow 2T = S = 40$ (as $T + S = 60$)

$\hat{A} \Rightarrow 60 \times 50 = S \times 75 \Rightarrow 3S = 120 \Rightarrow S = 40 \text{ N}$

$\hat{D} \Rightarrow T \times 75 = 60 \times 25 \Rightarrow 3T = 60 \Rightarrow T = 20 \text{ N}$

The resultant of the parallel forces 20 N at A and 40 N at D, is a parallel force of 60 N at C, where $AC:CD = 2:1$.

> The resultant of two parallel forces, P acting through A and Q acting through B, is a parallel force of $P+Q$ acting through C, where $AC:CB = Q:P$.

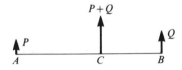

To prove this statement: we know that vertical forces of P at A and Q at B, will support a weight of $P+Q$ at C, where $AC:CB = Q:P$;   and that a weight of $P+Q$ at C is equal and opposite to a vertical force of $P+Q$ at C, which must therefore be the resultant of the forces at A and B.

Alternatively, the forces P at A and Q at B, must be equivalent to a total upwards force of $P+Q$, and the turning effect of P and Q about any point must equal the moment of $P+Q$ about that point.

$\Rightarrow \hat{A}$ give $(P+Q) \times AC = Q \times AB = Q \times (AC + CB) \Rightarrow P \times AC = Q \times CB$

$$\Rightarrow AC:CB = Q:P$$

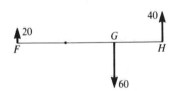

Looking at the balanced forces in another way, the fact that forces of $20\uparrow$ at F and $40\uparrow$ at H are balanced by $60\downarrow$ at G $\Rightarrow$ the resultant of the forces at F and H is 60 upwards at G, and leads us to find the resultant of two parallel forces in opposite directions.

$20\uparrow$ at F and $60\downarrow$ at G are balanced by $40\uparrow$ at H (the three forces are in equilibrium) $\Rightarrow$ the resultant of $20\uparrow$ at F and $60\downarrow$ at G is a force of $40\downarrow$ at H.

You can check the forces have the same effect by taking moments about any point.

Similarly the resultant of $40\uparrow$ at H and $60\downarrow$ at G will be $20\downarrow$ at F.

WORKED EXAMPLE

Find the moments of the forces shown, about A, B, C and D and say what this tells you about the position of the resultant. Find the resultant and its point of action.

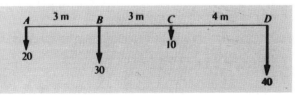

$\hat{A}$ $30 \times 3 + 10 \times 6 + 40 \times 10$ $= 550$ clockwise

$\hat{B}$ $10 \times 3 + 40 \times 7 - 20 \times 3$ $= 250$ clockwise

$\hat{C}$ $40 \times 4 - 30 \times 3 - 20 \times 6$ $= -50$ anticlockwise

$\hat{D}$ $-10 \times 4 - 30 \times 7 - 20 \times 10$ $= -450$ anticlockwise

The total moment of the forces about a point X decreases as X moves along AD.

Rigid bodies

At a certain point, G, between B and C, the total moment will equal zero, which indicates that the resultant, R, acts through G.

$\overset{\frown}{A}$ the moment of the four forces, will equal the moment of the resultant.

$R = 100$, so $100 \times AG = 550 \Rightarrow AG = 5.5$ m

Exercise 5.1

1 Find the balance point for the following weights (given in N) on a metre rule.

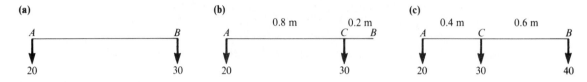

(a)   **(b)**   **(c)**

2 Specify the resultant forces in question **1**.

3 Find the balancing force and then the resultant of the forces given.

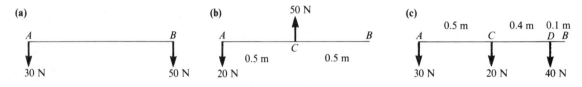

(a)   **(b)**   **(c)**

4 Find the moments, M, of the following forces about A, B and C, and the resultant in each case.

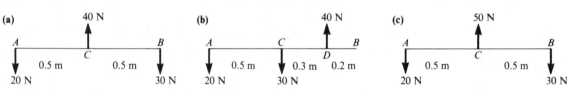

(a)   **(b)**   **(c)**

Moment of a force

The moment (turning effect) of a force $\mathbf{R} = \mathbf{AB}$ about the origin O, is given by the product of $|\mathbf{R}| = R$ and ON, where $ON \perp AB$.

If R makes an angle α with Ox then,

Moment $= R \times OC \sin \alpha$ (anticlockwise if $OC > 0$)

If $\mathbf{R} = X\mathbf{i} + Y\mathbf{j}$ then $R = \sqrt{X^2 + Y^2}$) and

$$\tan \alpha = \frac{Y}{X} \Rightarrow \sin \alpha = \frac{Y}{\sqrt{(X^2 + Y^2)}}$$

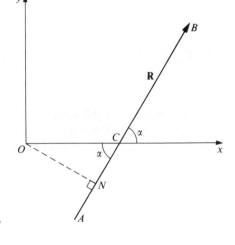

Moment $= R \times OC \times \sin \alpha = \sqrt{(X^2 + Y^2)} \times OC \times \dfrac{Y}{\sqrt{(X^2 + Y^2)}} = Y \times OC$

It may be more convenient when taking moments, to take the moments of the components.

108

Moment of a couple

The system of forces in Exercise 5.1, question **4(c)**, reduces to a couple where the forces in each direction are the same but do not act in the same line. Consequently, they will have a turning effect.

The resultant of 20 N and 30 N downwards is 50 N downwards, acting at D (0.4 m from B).

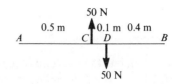

These two forces of 50 N form a **couple**, the two equal parallel forces acting in opposite directions whose lines of action are 10 cm apart. They have no resultant force except a turning effect equal to a moment of $50 \times 0.1 = 5$ N m clockwise.

It is interesting that the 'moment of a couple' is the same about any point, so that the moments about A, B, C and D are all 5 N m clockwise. For example,

$$\overset{\frown}{A} \quad \Rightarrow \quad \text{moment} = 50 \times 0.6 - 50 \times 0.5 = 30 - 25 = 5 \text{ N m clockwise.}$$

Also $\overset{\frown}{P}$, where $PC = x$, $\Rightarrow$ moment $= 50(x + 0.1) - 50 \times x = 5$ N m clockwise.

WORKED EXAMPLE

> In Exercise 5.1, question **1(a)**, insert a force at A to make the system into a couple, and find its moment.

The resultant of the 20 N and 30 N forces is a force of 50 N downwards acting at C, where $AC = 0.6$ m.

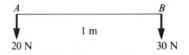

Insert an upwards force at A, equal to 50 N, which will give a couple of moment $= 50 \times 0.6 = 30$ N m clockwise.

Alternatively, the resultant of 20 N and 30 N is 50 N downwards, so insert a force of 50 N upwards at A and take moments about any point for the 20 N at A, the 30 N at B and the 50 N upwards at A.

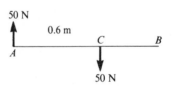

$$\overset{\frown}{A} \quad \Rightarrow \quad \text{moment} = 50 \times 0.6 = 30 \text{ N m clockwise}$$

Exercise 5.2

In each question of Exercise 5.1, insert a force at A to make the system into a couple, and find its moment, M.

For example: in **1(a)**, insert 50 N upwards at A; in **1(c)**, insert 90 N upwards at A;
in **3(b)** insert 30 N downwards at A.

Resultant of coplanar forces

Forces acting at a point

The resultant of the three forces here is clearly zero, as they balance (symmetry).

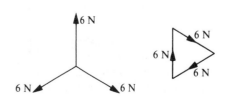

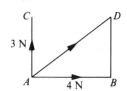

The resultant of **AB** + **AC** is a force of 5 N along *AD*. **AD** = 4**i** + 3**j** through *A*.

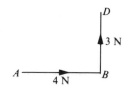

The resultant of **AB** + **BD** = 4**i** + 3**j** in direction *AD*, but acting through the point *B*.

Forces acting along the sides of a polygon

We shall find the resultant of the forces which act along *AB*, *BC*, *CD* and *AD* in the figure.

$\mathbf{R} = 4\mathbf{i} + 3\mathbf{j} - \mathbf{i} + 2\mathbf{j} = 3\mathbf{i} + 5\mathbf{j}$, which gives the direction and magnitude $(\sqrt{34})$ of the resultant.

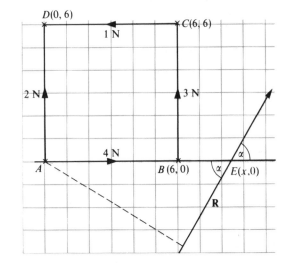

To find its line of action, take moments about $A(0, 0)$.

The moment of the resultant equals the sum of the moments of the constituent forces.

Assume $\mathbf{R} = 3\mathbf{i} + 5\mathbf{j}$ acts through $E(x, 0)$ where $AE = x$.

$\overset{\curvearrowright}{A}$ gives $Rx \sin\alpha = 3 \times 6 + 1 \times 6 \Rightarrow \sqrt{34}x \times \dfrac{5}{\sqrt{34}} = 24$

$$\Rightarrow \quad x = 4.8$$

So $\mathbf{R} = 3\mathbf{i} + 5\mathbf{j}$, acting through $E(4.8, 0)$.

This can be confirmed by taking moments about A, B and C to show that the resultant must pass between A and B and between B and C.

Alternatively

The resultant of the 2 N and 3 N forces is a force of $5\mathbf{j}$ along $x = 3.6$ and the resultant of $4\mathbf{i}$ along AB and $-\mathbf{i}$ along CD is $3\mathbf{i}$ along $y = -2$ (parallel forces).

So the total resultant is $3\mathbf{i} + 5\mathbf{j}$ through $(3.6, -2)$, acting along the line with gradient $5/3$ through $(3.6, -2)$, i.e.

$$y = \frac{5}{3}x + k \quad \text{where} \quad -2 = \frac{5 \times 3.6}{3} + k \quad \Rightarrow \quad k = -2 - 6 = -8$$

$$\Rightarrow \quad y = \frac{5}{3}x - 8 \quad \text{which cuts the } x\text{-axis at } y = 0 \quad \Rightarrow \quad \frac{5}{3}x = 8$$

$$\Rightarrow \quad x = 4.8$$

It is satisfying to be able to find the resultant without taking moments, as this proves that:

> The sum of the moments of the constituent forces is equal to the moment of the resultant.

You will usually find that it is easier to take moments to find the resultant, although it is satisfying to be able to use our results for parallel forces which we developed in the last section.

WORKED EXAMPLES

1 Find the resultant of the forces in the figure and show that the resultant has the same moment about A, B and C as its constituent forces.

Method 1

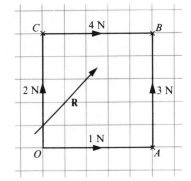

$2\mathbf{j}$ along OC, $+3\mathbf{j}$ along AB, $= 5\mathbf{j}$ along $x = 3$

$\mathbf{i}$ along OA, $+4\mathbf{i}$ along CB, $= 5\mathbf{i}$ along $y = 4$

$\Rightarrow \quad \mathbf{R} = 5\mathbf{i} + 5\mathbf{j}$ through $(3,4)$ i.e. along $y = x + 1$.

Method 2

$\mathbf{R} = X\mathbf{i} + Y\mathbf{j} = \mathbf{i} + 3\mathbf{j} + 4\mathbf{i} + 2\mathbf{j} = 5\mathbf{i} + 5\mathbf{j}$

$\widehat{O}$ gives $4 \times 5 - 3 \times 5 = 5$ N m clockwise, so $\mathbf{R}$ cuts the y axis between O and C, at D $(0,y)$, say.

$\widehat{O}$ gives $5 = Ry \cos 45° = 5\sqrt{2}y \times \dfrac{1}{\sqrt{2}} \quad \Rightarrow \quad y = 1$.

So, $\mathbf{R}$ acts through $(0, 1)$, i.e. along $y = x + 1$.

Method 3

Assume $\mathbf{R} = X\mathbf{i} + Y\mathbf{j}$ acts through (x, y).

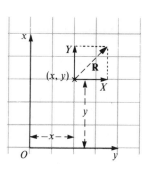

$\mathbf{R} = 5\mathbf{i} + 5\mathbf{j}$ as before, and $\widehat{O} \quad \Rightarrow \quad Xy - Yx = 5$

$X = 5$, $Y = 5 \quad \Rightarrow \quad 5y - 5x = 5 \quad \Rightarrow \quad y = x + 1$, the line of action of the resultant.

Check

(Clockwise +ve)	Moment of resultant $\mathbf{R} = X\mathbf{i} + Y\mathbf{j}$	Moment of constituents
$\widehat{A}$	$(X \times 4) + (Y \times 2) = \quad (5 \times 4) + (5 \times 2) = \quad 30$	$2 \times 5 + 4 \times 5 = \quad 30$
$\widehat{B}$	$(Y \times 2) - (X \times 1) = \quad (5 \times 2) - (5 \times 1) = \quad 5$	$2 \times 5 - 1 \times 5 = \quad 5$
$\widehat{C}$	$(-Y \times 3) - (X \times 1) = (-5 \times 1) - (5 \times 3) = -20$	$-1 \times 5 - 3 \times 5 = -20$

2 Forces 5 N, 2 N, 2 N and 4 N act along the sides AB, BC, CD and DE of a regular hexagon of side $2a$ as shown in the figure here. Find **(a)** the resultant and **(b)** the values of P and Q, if P along EF and Q along FA reduce the whole system to a couple.

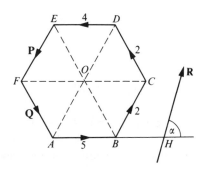

(a) If $\mathbf{R} = X\mathbf{i} + Y\mathbf{j}$, then resolving,

$\rightarrow \qquad X\mathbf{i} = 5\mathbf{i} - 4\mathbf{i} + 2\cos 60°\,\mathbf{i} - 2\cos 60°\,\mathbf{i} = \mathbf{i}$

$\uparrow \qquad Y\mathbf{j} = 2\cos 30°\,\mathbf{j} + 2\cos 30°\,\mathbf{j} = 4\dfrac{\sqrt{3}}{2}\,\mathbf{j} = 2\sqrt{3}\mathbf{j}$

$\Rightarrow \quad \mathbf{R} = \mathbf{i} + 2\sqrt{3}\mathbf{j} \;\Rightarrow\; R = \sqrt{13}$

If $\mathbf{R}$ acts through AB produced, moments about B

$\Rightarrow \qquad R \times BH\sin\alpha = 2a\sqrt{3} + 4 \times 2a\sqrt{3}$

$\Rightarrow \qquad \sqrt{13} \times BH \times \dfrac{2\sqrt{3}}{\sqrt{13}} = 10a\sqrt{3} \;\Rightarrow\; BH = 5a$

Or $\overset{\frown}{B}$ $\quad Y \times BH = 2a\sqrt{3} + 4 \times 2a\sqrt{3} \;\Rightarrow\; 2\sqrt{3}BH = 10\sqrt{3}a \;\Rightarrow\; BH = 5a$

(b) **Method 1** Forces must balance when resolving to give a couple.

$\nwarrow$ along $BF \;\Rightarrow\; 5\cos 30° + Q\cos 30° = 2\cos 30° + 4\cos 30° \;\Rightarrow\; 5 + Q = 6$

$\Rightarrow\; Q = 1$

$\nearrow$ along $AC \;\Rightarrow\; (5+2)\cos 30° = (4+P)\cos 30° \;\Rightarrow\; P = 3$

Method 2 For a couple, the moments about any two points are the same.

$\overset{\frown}{O} \quad (5+2+2+4+P+Q)\sqrt{3}a \qquad \overset{\frown}{A} \quad 2a\sqrt{3} + 2 \times 2a\sqrt{3} + 4 \times 2a\sqrt{3} + Pa\sqrt{3}$

$\Rightarrow\; 13 + P + Q = 2 + 4 + 8 + P \;\Rightarrow\; Q = 1$

$\overset{\frown}{E} \quad Qa\sqrt{3} + 5 \times 2a\sqrt{3} + 2 \times 2a\sqrt{3} + 2a\sqrt{3} \;\Rightarrow\; 13 + P + Q = Q + 10 + 4 + 2 \;\Rightarrow\; P = 3$

Exercise 5.3

1 In the figure, forces act as shown along the sides of the square $OABC$ of side 5 m

    (a) Find the resultant $\mathbf{R} = X\mathbf{i} + Y\mathbf{j}$.

    (b) Find the moment, M of the system about O, A, B and C.

    (c) Find the line of action of $\mathbf{R}$.

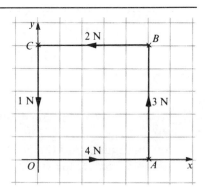

2 Repeat question **1**, adding a force of 3 N along BO.

3 Repeat question **1**, adding a force of $2\sqrt{2}$ N along BO.

4 Repeat question **1**, adding a force of $2\sqrt{2}$ N along CA.

5 Repeat question **1**, adding a force of $\sqrt{2}$ N along CA and 3 N along CO.

Questions **6—10** refer to the diagram of the hexagon in Worked Example **2** on page 111.

6 Find the resultant of forces 6, 5, 4, 3, 2, 1 along *AB*, *BC*, *CD*, *DE*, *EF*, *FA*.

7 Find the resultant of forces 6, 2, 4, 3, 5, 1 along *AB*, *BC*, *CD*, *DE*, *EF*, *FA*.

8 Find the resultant of forces 1 along each of *AB*, *BC*, *CD*, *DE*, *EF*, *FA*.

9 Find the resultant of forces 1 along *AB*, 2 along *BD*, 1 along *DE* and 2 along *EA*.

10 Find the resultant of forces 2 along each of *AB*, *BC*, *CD* and *AD*.

11 Forces: 1 along *OA*; 2 along *AB*; *P* along *BC*; and *Q* along *CO*; act along the sides of a square *OABC* where *O* is the origin of co-ordinates and *A* is the point (1, 0). Find *P* and *Q* if **(a)** the forces are in equilibrium **(b)** the forces form a couple **(c)** the resultant is 5**i** **(d)** the resultant is 5**j**.
   In **(c)** and **(d)** find the line of action of the resultant.

12 Three forces $\mathbf{F} = \mathbf{i} + 2\mathbf{j}$, $\mathbf{G} = 2\mathbf{i} - \mathbf{j}$, $\mathbf{H} = -3\mathbf{i} - \mathbf{j}$ act through points whose position vectors are $\mathbf{i} + \mathbf{j}$, $2\mathbf{i} + 3\mathbf{j}$, $4\mathbf{i} + 2\mathbf{j}$, respectively.

   (a) Show their resultant is a couple.

   (b) If **H** now acts through a point *A* in the same direction as before so that the three forces are in equilibrium, where is *A*?

   (c) What is the moment of **F** about the origin?

   (d) What is the moment of **G** about the origin?

   (e) What is the moment of **H** about the origin?

   (f) What is the moment of $\mathbf{F} + \mathbf{G} + \mathbf{H}$ about the origin?

   (g) If the three forces acted through the origin, what would be the resultant?

Equilibrium of rigid bodies

Until this chapter we have considered forces acting on bodies which can be regarded as particles, where the forces act through the centre of mass (gravity).

Certain larger bodies have forces acting at different points, such as in the situations in the Investigations at the beginning of this chapter.

We consider a **rigid** body as one which does not deform but keeps its shape under the action of the forces present. In reality, a ladder sags as soon as someone steps on to it, but we consider it to remain straight.

WORKED EXAMPLES

1 A uniform ladder of weight *W* leans against a smooth wall on rough ground. It is about to slip when the angle it makes with the horizontal is arctan 4. (See Investigation 1.)

   (a) Find the coefficient of friction μ between the ladder and the ground.
   (b) When $\mu = \frac{1}{4}$, find how far a man of weight 2*W* can climb up the ladder.

(a) Let the length of the ladder be 2*a*. The reaction of the ground on the ladder has two components, *R* vertically and friction *F* towards the wall. There is no friction at *B* as the wall is smooth.

$\uparrow \quad R=W; \qquad \rightarrow \quad F=N; \qquad \tan\alpha=4$

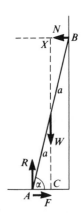

As the ladder is about to slip $\quad F=\mu R$

The forces on the ladder are in equilibrium $\Rightarrow$ the moments of the forces about any point are equal.

$\overset{\frown}{A} \quad Wa\cos\alpha=N\,2a\sin\alpha \quad \Rightarrow \quad N=\dfrac{W}{2}\cot\alpha=\dfrac{W}{8}$

$F=\mu R \quad \Rightarrow \quad \dfrac{W}{8}=\mu W \quad \Rightarrow \quad \mu=\dfrac{1}{8}$

We can regard F and R as combining to produce the total reaction, P. The three forces P, W and N balance and so must act through the same point, X, as three forces acting on a rigid body must be **concurrent** (act through the same point) or parallel. If they were not concurrent, taking moments about the intersection of any two and the third force would give a resultant turning effect, which contradicts equilibrium.

Thus, $\qquad \dfrac{R}{F}=\dfrac{XC}{AC}=\dfrac{2a\sin\alpha}{a\cos\alpha}=2\tan\alpha=8 \quad \Rightarrow \quad R=8F \quad \text{and} \quad F=\mu R \quad \Rightarrow \quad \mu=\dfrac{1}{8}$

(b) Let the man climb a distance $x=AM$ up the ladder to the point where the ladder is just about to slip, so it is in limiting equilibrium $\Rightarrow \quad F=\mu R$.

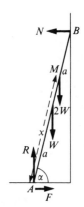

$\uparrow \quad R=3W; \qquad \rightarrow \quad F=N; \qquad \tan\alpha=4$

$\overset{\frown}{A} \quad Wa\cos\alpha+2Wx\cos\alpha=N\,2a\sin\alpha \qquad \text{............ (1)}$

$F=\mu R=\tfrac{1}{4}3W \quad \Rightarrow \quad F=N=\tfrac{3}{4}W$

Substituting for N and α in Equation **(1)** gives

$Wa+2Wx=\tfrac{3}{4}W\,2a\tan\alpha=6Wa$

$2Wx=5Wa \quad \Rightarrow \quad x=\dfrac{5a}{2} \qquad$ and the man can climb to the top of the ladder.

---

2 From Investigation 2, find the length of the rod resting inside the smooth-surfaced bowl.

The three forces R, N and W meet at X.

Surprisingly, the problem is best solved using geometry.

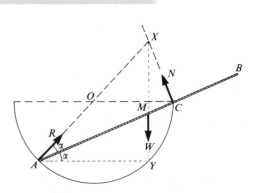

$A\widehat{C}X=90° \quad \Rightarrow \quad X$ lies on the circle with AX as diameter i.e. X lies on the circle including the bowl.

AY horizontal, XY vertical meet at Y.

$A\widehat{Y}X=90° \quad \Rightarrow \quad Y$ lies on the bowl.

If r is the radius of the bowl, $\quad AY=AX\cos 2\alpha=2r\cos 2\alpha$

$AM\cos\alpha=AY \quad \Rightarrow \quad AM=\dfrac{AY}{\cos\alpha}=\dfrac{2r\cos 2\alpha}{\cos\alpha} \quad \Rightarrow \quad AB=\dfrac{4r\cos 2\alpha}{\cos\alpha}$

Special cases

1 $\alpha = 0 \Rightarrow AB = 4r$ and the uniform rod balances at C (mid-point of the rod).

2 $\alpha = 45° \Rightarrow AB = \dfrac{4r \cos 90°}{\cos 45°} = 0$ (!) In fact, $\alpha = 45°$ is impossible, since R would act vertically and not intersect W and N.

3 $\alpha = 30° \Rightarrow \dfrac{4r \cos 60°}{\cos 30°} = 2.31r \Rightarrow$ stable equilibrium.

---

3 From Investigation 4: the inn-sign has length 0.5 m and the pole length 2 m; the pole's weight being negligible compared with the sign's weight of 20 N. The rope is at 45°, attached to the centre of the pole. Find the tension and the reaction at the foot of the pole.

Let the reaction at O be $\mathbf{R} = X\mathbf{i} + Y\mathbf{j}$

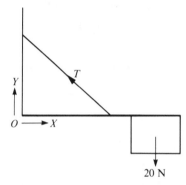

$\uparrow\ \ Y + T \cos 45° = 20 \quad \longrightarrow \quad X = T \cos 45°$

$\hat{O}\ \ T \times 1 \cos 45° = 20 \times 1.75 \quad \Rightarrow \quad T = 35\sqrt{2} = 49.5\ \text{N}$

$X = T \cos 45° = 35\ \text{N} \quad \text{and} \quad Y = 20 - T \cos 45° = -15$

So Y acts downwards and the pole pushes upwards against the wall.
The pole has to be fixed to prevent upward movement.

You could anticipate this by regarding X and Y as a single reaction $\mathbf{R}$ which, by considering the three forces $\mathbf{R}$, $\mathbf{T}$ and 20, have to intersect below the level of the pole.

Exercise 5.4

1 A uniform ladder of weight 200 N standing on rough horizontal ground leans against a smooth vertical wall.

   (a) Find the reactions on the ladder from the wall and ground and the angle of inclination of the ladder, if the coefficient of friction between the ladder and the ground is 0.5, and the ladder is about to slip.

   (b) Find the coefficient of friction if the ladder is about to slip when inclined at arctan 0.25 to the vertical.

2 A uniform ladder of weight 200 N leans against a rough vertical wall standing on rough horizontal ground, the coefficient of friction being 0.5 at both points of contact. Find the angle of inclination of the ladder, if the ladder is about to slip at both points of contact.

3 A uniform flagpole of weight 200 N and length 2 m is supported in a horizontal position against a vertical wall, with a rope of length 3 m attached to the tip of the pole and to a point vertically above where the flagpole base is supported on the wall. Find the tension in the rope and the reaction of the wall on the base of the flagpole. Can you lessen the tension in the rope by fixing it to a different point on the flagpole and/or attaching the other end to a point higher up or lower down the wall?

4 A uniform rod $ABCD$ of weight 50 N and length 1 m rests on supports at B and C. Find the supporting forces at B and C if (a) $AB = CD = 25$ cm (b) $AB = 40$ cm and $CD = 20$ cm (c) $AB = 20$ cm and $CD = 40$ cm (d) $AB = 10$ cm and $CD = 40$ cm.

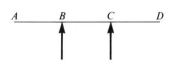

5 A uniform piano lid, AC, weighing 400 N is hinged at A and supported in the open position with a bar BC ($=1$ m). $AB = AC = 2$ m and the weight of the lid acts through D, where $AD = 0.8$ m. Find the thrust in BC and the reaction on the hinge at A.

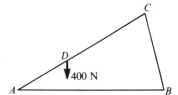

6 A uniform triangular lamina ABC has mass m and each side is of length $2a$. The lamina rests in a fixed vertical plane with A on a horizontal table and AC is vertical. Equilibrium is maintained by a force of magnitude T, which acts along the side BC. Show that $T = mg/3$. Find also the magnitude of the reaction at A.

Show that equilibrium is only possible when $\mu \geqslant \sqrt{3}/5$, where μ is the coefficient of friction between the lamina and the table.

7 A smooth horizontal rail is fixed at a height $3h$ above rough horizontal ground. A uniform rod AB of mass M and length $6h$ is placed in a vertical plane perpendicular to the rail, with A resting on the ground. $AC = 5h$, where C is the point of contact between the rail and the rod. Show that the force exerted by the rail on the rod is of magnitude $12\,Mg/25$.

Given that equilibrium is limiting, find the coefficient of friction between the rod and the ground and show that the force exerted by the ground on the rod is of magnitude $17\,Mg/25$.

Find in terms of M and g, the greatest magnitude of the horizontal force which can be applied to the rod at A without disturbing equilibrium.

6 Complex numbers

Introduction

A particular quadratic equation of the form $ax^2+bx+c=0$ can be solved by factorization, completing the square, or by using the standard formula,

$$x = \frac{-b \pm \sqrt{b^2 - 4ac}}{2a}$$

Hence, if $x^2+3x-4=0$ then $(x+4)(x-1)=0$

$$\Leftrightarrow \quad x=-4 \quad \text{or} \quad x=1$$

Thus, the equation has two real and distinct roots. The graph would confirm that the curve cuts the x-axis at $x=-4$ and $x=1$.

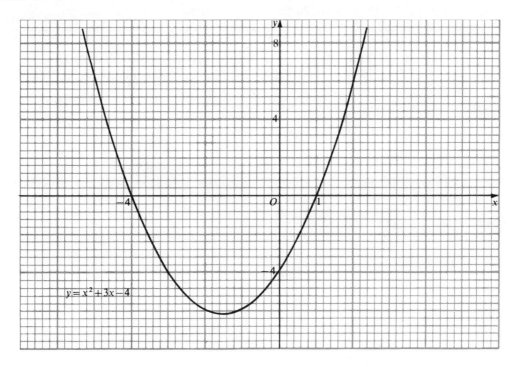

However, if $x^2-6x+9=0$ then $(x-3)^2=0$

$$\Leftrightarrow \quad x=3 \text{ twice}$$

Thus, in this case, the equation has a repeated real root, $x=3$. This means that the graph would show a curve which is tangential to the x-axis at $x=3$.

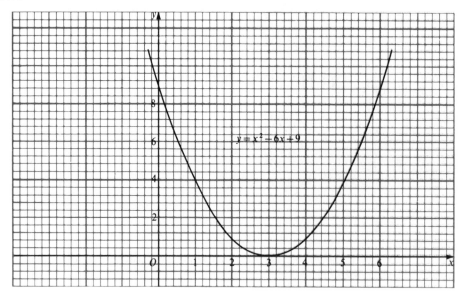

$y = x^2 - 6x + 9$

Investigation 1

Solve the equations   **(a)** $x^2 + 2x - 5 = 0$ **(b)** $x^2 + 4x = 0$

                                  (c) $4x^2 + 4x + 1 = 0$ **(d)** $2x^2 - x - 2 = 0$

In each case, draw a sketch of the curve to illustrate your answers.

You should have found that each of the equations in Investigation 1 produced real solutions. However, if we try to solve the equation $x^2 + 4 = 0$, we find that it is not possible to complete the solution using the real numbers.

So, $x^2 + 4 = 0 \iff x^2 = -4$

$$\iff x = \pm\sqrt{-4} \qquad \text{............ (1)}$$

Since the square of a non-zero real number is always positive there are no real solutions to this equation. To deal with this equation we must extend the number system to include the square roots of negative numbers.

This is done by defining $\boxed{i = \sqrt{-1} \ \text{ or } \ i^2 = -1}$

Hence, Equation **(1)** becomes,

$$x = \pm\sqrt{-4} = \pm\sqrt{-1 \times 4} = \pm\sqrt{-1} \times \sqrt{4} = \pm 2i$$

Investigation 2

Solve the quadratic equation $x^2 - 2x + 5 = 0$, using the fact that $i^2 = -1$.

Use the formula method to show that $x = 1 \pm \tfrac{1}{2}\sqrt{-16}$.

Let $i^2 = -1$ and show that $x = 1 \pm 2i$.

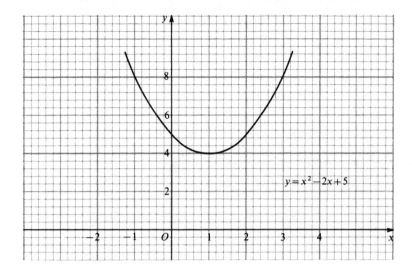

The equation in Investigation 2 has no real solutions and hence the curve $y = x^2 - 2x + 5$ will not cut the x-axis. The graph is shown in the figure.

WORKED EXAMPLE

Solve the equations **(a)** $x^2 + 2x + 2 = 0$ **(b)** $x^2 + x + 3 = 0$ using $i^2 = -1$.

The solution of the quadratic equation $ax^2 + bx + c = 0$ is given by

$$x = \frac{-b \pm \sqrt{b^2 - 4ac}}{2a}$$

(a) Hence, for $x^2 + 2x + 2 = 0$ we have,

$$x = \frac{-2 \pm \sqrt{4 - 4(1)(2)}}{2} = \frac{-2 \pm \sqrt{-4}}{2} = \frac{-2 \pm 2i}{2} = -1 \pm i$$

(b) For the equation $x^2 + x + 3 = 0$ we have,

$$x = \frac{-1 \pm \sqrt{1 - 4(1)(3)}}{2} = \frac{-1 \pm \sqrt{-11}}{2} = \frac{-1 \pm i\sqrt{11}}{2} = \tfrac{1}{2}(-1 \pm i\sqrt{11})$$

Algebraic form

The solutions to the equations in the Worked Example have produced numbers in the form $a + ib$ where a and b are real numbers (integers, rational or irrational).

The set of all such numbers of the form $a + ib$ is called the **set of complex numbers**, $\mathcal{C}$. We often use a single letter to denote a complex number, so we write $z = a + ib$.

119

The real numbers, a and b, are known as the **real** and **imaginary parts** of z and we write

$$\text{Real}(z) = \text{Re}(z) = a \quad \text{and} \quad \text{Imag}(z) = \text{Im}(z) = b$$

Complex numbers of this form are not just the result of the abstract ideas of mathematicians, but can be used as a tool to provide real solutions to real problems as, for example, in the theory of alternating current.

Investigation 3

Since we have defined a new type of number, we must also define the rules for combining these (c.f. vectors and vector algebra, matrices and matrix algebra etc.).

We assume the usual operations for real numbers apply to complex numbers, and apply the same rules of algebra.

For **addition**: $\quad (a+ib) + (c+id) = (a+c) + i(b+d)$

For **subtraction**: $\quad (a+ib) - (c+id) = (a-c) + i(b-d)$

For **multiplication**, $\quad$ we expand the brackets algebraically as usual. Show that if we use $i^2 = -1$,

$$(a+ib)(c+id) = (ac - bd) + i(ad + bc)$$

If $z_1 = 3 + 4i$ and $z_2 = -1 + 2i$, show that

(a) $z_1 + z_2 = 2 + 6i$ $\quad$ **(b)** $z_1 - z_2 = 4 + 2i$ $\quad$ **(c)** $z_1 z_2 = -11 + 2i$

Find $z_2 - z_1$ and then try some more examples of your own.

Investigation 4

Find the product of the complex numbers,

(i) $2 + 5i$ and $2 - 5i$ $\quad$ **(ii)** $-3 - 4i$ and $-3 + 4i$

These two products give real answers which suggests that the product of two complex numbers of the form $a + ib$ and $a - ib$ is real.

Let $z_1 = a + ib$ and $z_2 = a - ib$ be two complex numbers. Find the product $z_1 z_2$ and show that it is equal to $a^2 + b^2$.

So the product of two complex numbers $a + ib$ and $a - ib$ is real.

Such pairs of complex numbers are called **complex conjugates**.

Hence, if $z = a + ib$ is a complex number, then $a - ib$ is its complex conjugate denoted by $\bar{z}$ or z^* (we shall use the $\bar{z}$ form in this chapter).

Note that, if $z = a - ib$ then $z = a + ib$ is its complex conjugate.

Write down the complex conjugates of the following complex numbers and show that, in each case, $z\bar{z}$ is real.

(a) $z = 3 + i$ $\quad$ **(b)** $z = -6 + 2i$ $\quad$ **(c)** $z = 7 - 4i$ $\quad$ **(d)** $z = -5 - 3i$

Investigation 5

We can use the idea of the complex conjugate to consider the **division** of two complex numbers.

By multiplying the numerator and denominator of $\dfrac{3+4i}{1+2i}$ by the conjugate of $1+2i$ (i.e. $1-2i$) show that the result is $\dfrac{1}{5}(11-2i)$.

Repeat this technique to show that $\dfrac{5+i}{1-3i} = \dfrac{1}{5}(1+8i)$.

Show that, in general $\dfrac{(a+ib)}{(c+id)} = \left(\dfrac{ac+bd}{c^2+d^2}\right) + i\left(\dfrac{bc-ad}{c^2+d^2}\right)$

Now try some more examples of your own.

WORKED EXAMPLE

Express the following in the form $a+ib$.

(a) $(4-2i)+(3+5i)$ (b) $(1-3i)-(-3+2i)$

(c) $(2-5i)(-1+2i)$ (d) $\dfrac{3+2i}{2-i}$

(a) Using the result for addition, we take the sum of the real parts and the sum of the imaginary parts.

$$(4-2i)+(3+5i)=(4+3)+(-2i+5i)=7+3i$$

(b) Similarly, by subtracting the real parts and the imaginary parts,

$$(1-3i)-(-3+2i)=[1-(-3)]+(-3i-2i)=4-5i$$

(c) Expanding the brackets we have,

$$(2-5i)(-1+2i)=2(-1+2i)-5i(-1+2i)$$
$$=-2+4i+5i-10i^2$$
$$=-2+4i+5i+10=8+9i$$

(d) Multiply the numerator and denominator by the complex conjugate of $2-i$.

$$\frac{(3+2i)(2+i)}{(2-i)(2+i)} = \frac{3(2+i)+2i(2+i)}{4-i^2}$$
$$=\frac{6+3i+4i+2i^2}{5}=\frac{1}{5}(4+7i)$$

Equality of complex numbers

Two complex numbers $a+ib$ and $c+id$ are equal if

$$a+ib=c+id \quad \Rightarrow \quad (a-c)=i(d-b)$$

Since a, b, c and d are real, this can only be true if $a-c=0$ and $d-b=0$, i.e. if $a=c$ and $b=d$.

Hence, two complex numbers $a+ib$ and $c+id$ are equal if, and only if, $a=c$ and $b=d$.

> **WORKED EXAMPLE**

Find the square roots of $7-24i$ in the form $a+ib$.

Let the square root of $7-24i$ be $a+ib$ where a and b are real.

Thus, $$(a+ib)^2 = 7-24i \quad \Rightarrow \quad (a^2-b^2)+2abi = 7-24i$$

Equating the real and imaginary parts gives
$$a^2-b^2 = 7 \quad \text{and} \quad 2ab = -24$$

Solving by eliminating b gives, $b = -\dfrac{24}{2a} = -\dfrac{12}{a}$ and hence $a^2 - \left(-\dfrac{12}{a}\right)^2 = 7$

$$\therefore \quad a^2 - \frac{144}{a^2} = 7 \quad \Rightarrow \quad a^4 - 7a^2 - 144 = 0$$

$$\Rightarrow \quad (a^2-16)(a^2+9) = 0 \quad \Rightarrow \quad a^2 = 16 \quad \text{or} \quad a^2 = -9$$

Since a is real and $a^2 = -9$ does not produce real solutions, we use $a^2 = 16$ to provide the answers.

Hence, $\quad a^2 = 16 \quad \Rightarrow \quad a = \pm 4$

When $a=4$, $b=-3$ and when $a=-4$, $b=3$.

The square roots of $7-24i$ are $4-3i$ and $-4+3i$.

Exercise 6.1

1 Simplify **(a)** $\sqrt{-64}$ **(b)** i^2 **(c)** i^4 **(d)** i^5 **(e)** $\sqrt{-18}$

2 Solve the following equations.

    (a) $x^2-9=0$ **(b)** $x^2+9=0$ **(c)** $x^2+x+1=0$ **(d)** $x^2-2x+3=0$

    (e) $2x^2+5x+4=0$ **(f)** $x^2-4x+7=0$

3 Write down the complex conjugates of the following complex numbers.

    (a) $1-i$ **(b)** $2+i$ **(c)** $3+2i$ **(d)** $-4-3i$ **(e)** $5i$

4 Simplify **(a)** $(3+2i)+(-1+3i)$ **(b)** $(1-i)+(4-3i)$ **(c)** $(4-i)-(2-3i)$

    (d) $(-5+2i)-(1-4i)$

5 Express in the form $a+ib$,

    (a) $(2+i)(3-2i)$ **(b)** $(3-2i)(4+i)$ **(c)** $(-5+2i)(-2-i)$ **(d)** $(3-4i)(3+4i)$

    (e) $2i(5-3i)$ **(f)** $(3-2i)^2$

6 Simplify **(a)** i^3 **(b)** $\dfrac{1}{i}$ **(c)** $\dfrac{1}{i^2}$ **(d)** $\dfrac{1}{(1-i)(1+i)}$

7 Simplify **(a)** $(1-i)^2$ **(b)** $(2+i)^3$ **(c)** $(3+2i)^2$ **(d)** $(x+iy)^2$

8 Express in the form $a+ib$,

(a) $\dfrac{1}{3+i}$ (b) $\dfrac{2}{1-i}$ (c) $\dfrac{1}{\sqrt{3}-2i}$ (d) $\dfrac{3i}{2+5i}$

9 Simplify, giving your answers in the form $a+ib$,

(a) $\dfrac{1+i}{1-i}$ (b) $\dfrac{2-i}{3+i}$ (c) $\dfrac{1-i}{3+2i}$ (d) $\dfrac{4-5i}{1+2i}$

10 Show that the sum of any complex number and its conjugate is always real.

11 Find the square roots of the following complex numbers,

(a) $3+4i$ (b) $5-12i$ (c) $8-6i$ (d) $8i$

12 If $5(a+ib)+(3-2i)=6i$, find the values of a and b.

13 If $\dfrac{1}{3-i}+\dfrac{2}{1+2i}=\dfrac{1}{10}(x+iy)$, show that $x=7$ and $y=-7$.

14 Simplify $[z-(1+2i)][z+(1+2i)]=0$. Hence find the square roots of $-3+4i$.

> **Investigation 6**
>
> (a) Solve the equation $x^2-2x+3=0$ and show that its roots are given by $\alpha=1+i\sqrt{2}$ and $\beta=1-i\sqrt{2}$.
>
> (b) Find the sum of these roots, $\alpha+\beta$. Find also the product $\alpha\beta$ and compare your answers with the coefficients of the equation $x^2-2x+3=0$, showing that the relation between the sum and product of the roots of a quadratic equation and its coefficients is still true, even if the roots are complex.
>
> (c) Repeat the working above for the equations   (i) $x^2+x+4=0$ (ii) $2x^2-3x+2=0$.

It follows from the work in Investigation 6 that if a quadratic equation with integral coefficients has a complex root $a+ib$, then it must possess a second complex root $a-ib$, otherwise the sum of the roots cannot give a real integral coefficient of the x term.

> The complex roots of a quadratic equation must occur in complex conjugate pairs.

> **Investigation 7**
>
> (a) Use the remainder theorem to show that $(x-1)$ is a factor of $f(x)=x^3-2x^2-5x+6$. Hence solve completely the equation $f(x)=0$, showing that it has three real roots $x=1,\,-2$ and 3.
>
> (b) Use the remainder theorem to show that $(x-2)$ is a factor of $g(x)=x^3-x^2-x-2$. Hence show that the equation $g(x)=0$ has one real root and two complex roots. Find the two complex roots.
>
> *(cont.)*

Investigation 7—continued

(c) The graph of $f(x)=0$ is shown here and illustrates that a cubic equation may have three real roots.

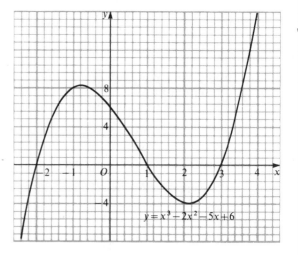

(d) The graph of $g(x)=0$ is shown in the figure here and illustrates a cubic equation which has one real root and two complex roots.

(e) Deduce that a cubic equation cannot possess two real roots and one complex root.

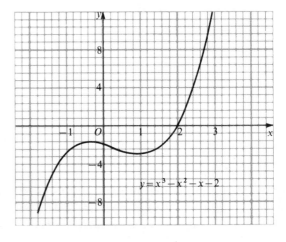

A cubic equation may have three real roots or only one real root. If it has any complex roots they must occur as complex conjugate pairs.

The cube roots of unity

The technique used in Investigation 7 can be applied to the equation $z^3-1=0$ to find the cube roots of $+1$.

Investigation 8

(a) Use the Remainder Theorem to show that $z-1$ is a factor of the equation z^3-1. Hence solve completely the equation $z^3-1=0$ showing that the roots are 1, $\frac{1}{2}(-1+i\sqrt{3})$ and $\frac{1}{2}(-1-i\sqrt{3})$.

(b) Let $\alpha=\frac{1}{2}(-1+i\sqrt{3})$ and $\beta=\frac{1}{2}(-1-i\sqrt{3})$ and show that $\alpha^2=\beta$ and $\beta^2=\alpha$.

(c) Deduce that the roots of the equation can be written as 1, w, w^2, where $w=\frac{1}{2}(-1+i\sqrt{3})$ or $\frac{1}{2}(-1-i\sqrt{3})$.
Show that $1+w+w^2=0$.

Investigation 9

Use the method of Investigation 8 to find the cube roots of -1 by solving the equation $z^3 + 1 = 0$.

Deduce, from your investigations, the cube roots of $+8$ and -8.

The Argand diagram

A complex number $z = x + iy$ is completely specified by the values of x and y. This suggests that there is a one-to-one correspondence between a complex number z and the ordered pair (x, y).

Hence, we may represent a complex number $z = x + iy$ by a point in a two dimensional plane with co-ordinates (x, y).

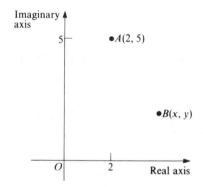

The diagram shows two complex numbers $z_1 = 2 + 5i$ represented by the point $A(2, 5)$ and $z_2 = x + iy$ represented by the point $B(x, y)$.

Diagrams which show complex numbers represented by points in a plane, are called **Argand diagrams**, after the Swiss mathematician, Jean Argand (1768–1822). The axes are usually called the real and imaginary axes, since points on the x-axis define real numbers and points on the y-axis define purely imaginary numbers.

Consider the following Argand diagram with points A, B and C representing the complex numbers $3 + 2i$, $-4 + i$ and $2 - 3i$.

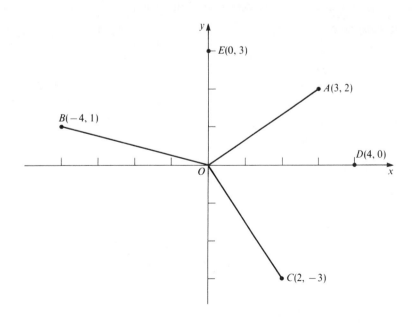

Note that a point $D(4, 0)$ would represent the complex number $4+0i$ (i.e. the real number 4) and the point $E(0, 3)$ would represent the complex number $0+3i$ (i.e. the imaginary number $3i$). So real numbers are represented by points on the x-axis and complex numbers, in which the real part is zero, are represented by points on the y-axis.

Also, each point has an associated vector often called the **radius vector** (i.e. **OA**, **OB**, **OC**, **OD** and **OE**).

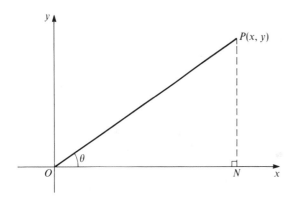

In general, a complex number $z = x + iy$ represented by a point $P(x, y)$ on the Argand diagram, will have a radius vector **OP** associated with it.

Modulus and argument

The magnitude of the vector (usually denoted by r) is called the **modulus of the complex number**, z. We write this as $|z|$ or $|x+iy|$.

Using the theorem of Pythagoras in triangle OPN gives

$$OP^2 = ON^2 + NP^2 \quad \Rightarrow \quad r^2 = x^2 + y^2$$

Hence, the modulus of z is the length of OP and

$$\boxed{|z| = |x+iy| = \sqrt{x^2+y^2}}$$

The angle, θ, which OP makes with the positive direction of the real axis, is called the **argument of the complex number**.

Values of θ measured anticlockwise from the real axis are positive, and values of θ measured clockwise are negative. We write

$$\boxed{\theta = \arg z \quad \text{or} \quad \arg(x+iy)}$$

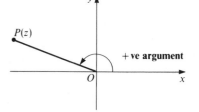

This is a function that can take many values (i.e. multiples of 2π) and so we take (by convention) the **principal value** of the argument to be in the range $-\pi$ to $+\pi$.

Thus, $\qquad\qquad\qquad -\pi < \arg z \leqslant \pi$

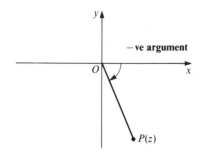

WORKED EXAMPLE

Find the modulus and argument of the following complex numbers.

(a) $z_1 = 1 + i$ **(b)** $z_2 = -1 + i\sqrt{3}$ **(c)** $z_3 = 3 - 2i$

(a) Since $z_1 = 1 + i$,

$$|z_1| = |\mathbf{OA}| = \sqrt{1^2 + 1^2} = \sqrt{2}$$

and $\arg z_1 = \alpha = \tan^{-1} 1$

$$\Rightarrow \quad \arg z_1 = \frac{\pi}{4}$$

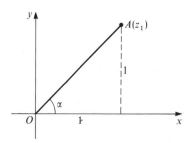

(b) Since $z_2 = -1 + i\sqrt{3}$,

$$|z_2| = |\mathbf{OB}| = \sqrt{(-1)^2 + (\sqrt{3})^2} = 2$$

and $\arg z_2 = \beta = \pi - \tan^{-1}\sqrt{3}$

$$= \pi - \frac{\pi}{3} = \frac{2\pi}{3}$$

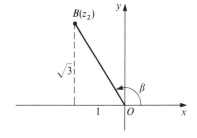

(c) Since $z_3 = 3 - 2i$,

$$|z_3| = |\mathbf{OC}| = \sqrt{3^2 + (-2)^2} = \sqrt{13}$$

and $\arg z_3 = -\gamma = -\tan^{-1}\left(\frac{2}{3}\right)$

$$= -33.5°$$

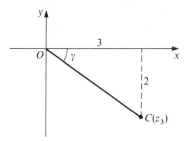

Representing addition

The **sum of two complex numbers** $z_1 = x_1 + iy_1$ and $z_2 = x_2 + iy_2$ can be shown on an Argand diagram by the following geometrical representation.

Consider the complex numbers z_1 and z_2 represented by the points P and Q respectively. Let R be the fourth vertex of the completed parallelogram, $OPRQ$.

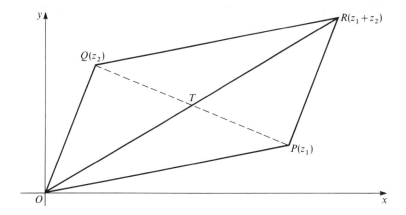

Since P is the point (x_1, y_1) and Q is the point (x_2, y_2), the point T has co-ordinates $(\frac{1}{2}(x_1 + x_2), \frac{1}{2}(y_1 + y_2))$, because the diagonals of a parallelogram bisect each other. Hence, the point R has co-ordinates $(x_1 + x_2, y_1 + y_2)$.

Thus, R represents the complex number $(x_1 + x_2) + i(y_1 + y_2) = z_1 + z_2$.

Note that $|z_1 + z_2| = OR$. Now, since OPR is a triangle, one side cannot be greater than the sum of the other two.

$$OR \leqslant OP + PR \quad \Rightarrow \quad OR \leqslant OP + OQ \qquad (\text{since} \quad PR = OQ)$$

$$\boxed{|z_1 + z_2| \leqslant |z_1| + |z_2|}$$

Under what circumstances is $|z_1 + z_2| = |z_1| + |z_2|$?

Representing subtraction

To **subtract two complex numbers** $z_1 = x_1 + iy_1$ and $z_2 = x_2 + iy_2$, we use the fact that $z_1 - z_2 = z_1 + (-z_2)$.

Consider the complex numbers z_1 and z_2 represented by points P and Q, respectively. Now $(-z_2)$ will be represented by the point R, as shown in the figure below. When the parallelogram $OPTR$ is formed, the point T will represent the complex number $z_1 - z_2$.

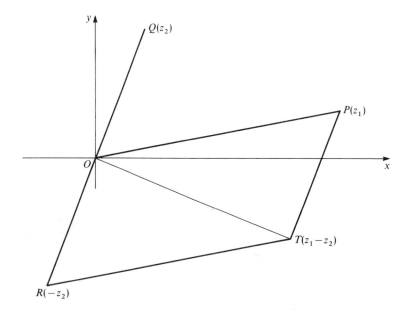

Note that $|z_1 - z_2| = OT = QP$ and gives the length of the line joining two complex numbers on the Argand diagram.

Using triangle OPT,

$$OP \leqslant OT + TP \quad \Rightarrow \quad OP \leqslant OT + OR$$

$$\therefore \quad |z_1| \leqslant |z_1 - z_2| + |z_2| \quad \text{and} \quad |z_1 - z_2| \geqslant |z_1| - |z_2|$$

Alternatively, using the same triangle,

$$PT \leqslant OT + OP \qquad OR \leqslant OT + OP$$

$$|z_2| \leqslant |z_1 - z_2| + |z_1| \quad \text{and} \quad |z_1 - z_2| \geqslant |z_2| - |z_1|$$

Combining these two results gives

$$\boxed{|z_1 - z_2| \geqslant \left| |z_1| - |z_2| \right|}$$

Exercise 6.2

1 Show the following complex numbers on an Argand diagram and find the modulus of each.

    (a) $2i$ *(b)* $1+i$ **(c)** $2-4i$ **(d)** $-1+3i$ **(e)** $-2-i$

2 Write down the complex numbers represented by the following points on an Argand diagram.

    (a) $(0,3)$ **(b)** $(1,0)$ **(c)** $(2,1)$ **(d)** $(-3,2)$ **(e)** $(2,-5)$

3 Find the modulus and principal value of the argument of the following complex numbers.

    (a) $1-i$ **(b)** $3+4i$ **(c)** $-3-4i$ **(d)** $2-5i$ **(e)** $-3+2i$

4 Find the modulus and principal value of the argument of each of the following complex numbers.

    (a) $-1-i$ **(b)** $\sqrt{3}+i$ **(c)** $1-4i$ **(d)** $-4-3i$ **(e)** $-1+\sqrt{3}i$

5 If $z=1+i$, show the following complex numbers and their associated radius vectors on the same Argand diagram.

    (a) z **(b)** z^2 **(c)** $\dfrac{1}{z}$ **(d)** $\dfrac{1}{z^2}$

6 If $z=3+4i$, show the following complex numbers and their associated radius vectors on the same Argand diagram.

    (a) z **(b)** $\bar{z}$ **(c)** iz **(d)** i^2z **(e)** $i\bar{z}$

7 If $z_1=2+i$ and $z_2=1+3i$, show on an Argand diagram the complex numbers,

    (a) z_1 **(b)** z_2 **(c)** z_1+z_2 **(d)** z_1-z_2 **(e)** z_1-2z_2

8 If $z_1=-2+i$ and $z_2=3-i$, find the complex numbers,

    (a) z_1+z_2 **(b)** z_1-z_2 **(c)** z_1z_2 **(d)** $\dfrac{z_1}{z_2}$ **(e)** $\dfrac{z_2}{z_1}$

9 If $z_1=1+2i$ and $z_2=-3+i$, find the modulus and argument of each of the following.

    (a) z_1+z_2 **(b)** z_1-z_2 **(c)** z_2-z_1 **(d)** z_1z_2 **(e)** $\dfrac{z_1}{z_2}$

10 If $z_1=3+2i$, find what form must z_2 take if,

    (a) $|z_1+z_2|=|z_1|+|z_2|$ **(b)** $|z_1+z_2|=|z_1|-|z_2|$

11 Find the distance between the points on the Argand diagram representing the following complex numbers.

    (a) $1+i$; $2+3i$ **(b)** $-3+2i$; $4-i$ **(c)** $2i$; 6 **(d)** $3+7i$; $1-2i$

12 Plot the points representing the following complex numbers on an Argand diagram and show the associated radius vectors.

    (a) 2 **(b)** $3i$ **(c)** $2+2i$ **(d)** $-1-i$

    Multiply each of these complex numbers by i and plot your results on the same Argand diagram showing the associated radius vectors. What geometrical effect is produced by multiplying a complex number by i?

13 Plot on an Argand diagram the points A and B representing the complex numbers $z_1=3+2i$ and $z_2=3+4i$, respectively. Draw the triangle OAB. If $z_3=1+2i$, find the complex numbers **(a)** z_1z_3 and **(b)** z_2z_3 and show your results as the points A' and B' on the Argand diagram. Draw the triangle $OA'B'$. Deduce that the triangles are similar and that the effect of multiplying by $1+2i$ is a rotation about O of $63.4°$, combined with an enlargement with scale factor $\sqrt{5}$.

Modulus-argument form for a complex number

Consider a complex number $z = x + iy$ represented by the point $P(x, y)$ in the Argand plane. If the modulus of z is denoted by r, and the argument by θ, then, using triangle OPN

$$x = r \cos \theta \quad \text{and} \quad y = r \sin \theta$$

Hence any complex number can be written in the form

$$z = r(\cos \theta + i \sin \theta)$$

where r is the modulus and θ is the argument.

This is known as the **modulus-argument form** of z.

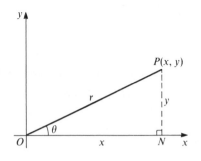

> **WORKED EXAMPLE**

> Express **(i)** $1 + i$ and **(ii)** $1 - \sqrt{3}i$ in modulus-argument form.

(i) If $z = 1 + i$, then $|z| = \sqrt{1^2 + 1^2} = \sqrt{2}$

and $\arg z = \theta = \tan^{-1} 1 = \dfrac{\pi}{4}$

Hence, $1 + i$ can be expressed as $\sqrt{2}\left(\cos \dfrac{\pi}{4} + i \sin \dfrac{\pi}{4}\right)$

(ii) If $z = 1 - \sqrt{3}i$ then $|z| = \sqrt{1^2 + (-\sqrt{3})^2} = 2$

and $\arg z = \theta = -\tan^{-1}\left(\dfrac{\sqrt{3}}{1}\right) = -\dfrac{\pi}{3}$

Hence, $1 - \sqrt{3}i = 2\left[\cos\left(-\dfrac{\pi}{3}\right) + i \sin\left(-\dfrac{\pi}{3}\right)\right]$

Note that, since $\cos\left(-\dfrac{\pi}{3}\right) = \cos\dfrac{\pi}{3}$ and $\sin\left(-\dfrac{\pi}{3}\right) = -\sin\dfrac{\pi}{3}$, this is often expressed as

$2\left(\cos\dfrac{\pi}{3} - i \sin\dfrac{\pi}{3}\right)$.

Investigation 10

Consider two complex numbers $z_1 = r_1(\cos \theta + i \sin \theta)$

and $z_2 = r_2(\cos \phi + i \sin \phi)$

Form the product $z_1 z_2 = r_1 r_2 (\cos \theta + i \sin \theta)(\cos \phi + i \sin \phi)$ and show that this is equivalent to

$$z_1 z_2 = r_1 r_2 [\cos(\theta + \phi) + i \sin(\theta + \phi)]$$

This shows that

> The product of two complex numbers can be obtained by finding the product of the moduli and the sum of the arguments.

WORKED EXAMPLES

---

1 If $z_1 = 2\left(\cos\dfrac{\pi}{3} + i\sin\dfrac{\pi}{3}\right)$ and $z_2 = 3\left(\cos\dfrac{\pi}{4} - i\sin\dfrac{\pi}{4}\right)$, find the product $z_1 z_2$.

Firstly, put z_2 into the correct modulus-argument form, $r(\cos\theta + i\sin\theta)$.

Hence,
$$z_1 = 2\left(\cos\frac{\pi}{3} + i\sin\frac{\pi}{3}\right) \quad \text{and} \quad z_2 = 3\left[\cos\left(-\frac{\pi}{4}\right) + i\sin\left(-\frac{\pi}{4}\right)\right]$$

$$z_1 z_2 = 6\left\{\cos\left[\frac{\pi}{3} + \left(-\frac{\pi}{4}\right)\right] + i\sin\left[\frac{\pi}{3} + \left(-\frac{\pi}{4}\right)\right]\right\}$$

$$= 6\left(\cos\frac{\pi}{12} + i\sin\frac{\pi}{12}\right)$$

---

2 Evaluate $z_1 z_2$ if $z_1 = 2\left(\cos\dfrac{\pi}{3} + i\sin\dfrac{\pi}{3}\right)$

and $z_2 = 5\left(\cos\dfrac{\pi}{6} + i\sin\dfrac{\pi}{6}\right)$.

Convert z_1 and z_2 to algebraic form and check the result.

Since z_1 and z_2 are both in the correct modulus-argument form,

$$z_1 z_2 = 10\left[\cos\left(\frac{\pi}{3} + \frac{\pi}{6}\right) + i\sin\left(\frac{\pi}{3} + \frac{\pi}{6}\right)\right]$$

$$= 10\left(\cos\frac{\pi}{2} + i\sin\frac{\pi}{2}\right) = 10i$$

If $z_1 = 2\left(\cos\dfrac{\pi}{3} + i\sin\dfrac{\pi}{3}\right)$ then $z_1 = 2\left[\dfrac{1}{2} + i\left(\dfrac{\sqrt{3}}{2}\right)\right] = 1 + i\sqrt{3}$

If $z_2 = 5\left(\cos\dfrac{\pi}{6} + i\sin\dfrac{\pi}{6}\right)$ then $z_2 = 5\left[\dfrac{\sqrt{3}}{2} + i\left(\dfrac{1}{2}\right)\right] = \dfrac{5}{2}(\sqrt{3} + i)$

Hence, $z_1 z_2 = \dfrac{5}{2}(1 + i\sqrt{3})(\sqrt{3} + i)$

$$= \frac{5}{2}\left(\sqrt{3} + i + 3i + i^2\sqrt{3}\right) = \frac{5}{2}(4i) = 10i$$

The product of two complex numbers on an Argand diagram

The product of two complex numbers $z_1 = r_1(\cos\theta + i\sin\theta)$ and $z_2 = r_2(\cos\phi + i\sin\phi)$ can be represented on an Argand diagram by finding a point whose radius vector makes an angle $(\theta + \phi)$ with the positive real axis and whose length is $r_1 r_2$.

In the figure, the points P and Q represent the complex numbers z_1 and z_2, respectively, with their associated radius vectors **OP** and **OQ**.

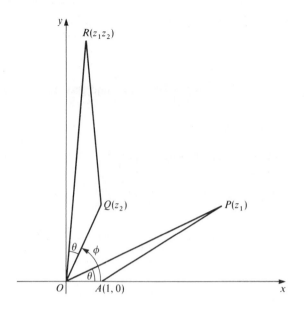

Take a point $A(1,0)$ on the real axis and complete the triangle OAP. Now construct triangle OQR similar to triangle OAP. The vertex R represents the product z_1z_2, since its argument is clearly $\theta + \phi$ ($\angle ROx$) and its modulus is given by $|OR|$, where

$$\frac{|OR|}{|OQ|} = \frac{|OP|}{|OA|} \quad \Rightarrow \quad |OR| = \frac{|OP| \times |OQ|}{|OA|} = \frac{|z_1| \times |z_2|}{1} = r_1 r_2$$

Hence, the point R represents the complex number

$$z_1z_2 = r_1r_2\left[\cos(\theta + \phi) + i\sin(\theta + \phi)\right]$$

Note that the diagram above, in fact, illustrates the product of

$$z_1 = 3(\cos 30° + i\sin 30°) \quad \text{and} \quad z_2 = 2(\cos 55° + i\sin 55°)$$

which gives

$$z_1z_2 = 6(\cos 85° + i\sin 85°)$$

Investigation 11

Consider the two complex numbers $z_1 = r_1(\cos\theta + i\sin\theta)$ and $z_2 = r_2(\cos\phi + i\sin\phi)$.

Form the quotient $\dfrac{z_1}{z_2}$ and by multiplying the numerator and denominator by $(\cos\phi - i\sin\phi)$, show that

the quotient can be expressed in the form $\dfrac{r_1}{r_2}\left[\cos(\theta - \phi) + i\sin(\theta - \phi)\right]$.

The working of Investigation 11 illustrates the following result.

> The quotient of two complex numbers can be obtained by dividing the moduli and subtracting the arguments.

WORKED EXAMPLE

Convert the complex numbers $z_1 = 1 - 2i$ and $z_2 = 2 + i$ to modulus-argument form, and hence find the quotient $\dfrac{z_1}{z_2}$.

If $z_1 = 1 - 2i$ then $|z_1| = \sqrt{5}$ and $\arg z_1 = -\tan^{-1} 2 = -63° \, 26'$.

If $z_2 = 2 + i$ then $|z_2| = \sqrt{5}$ and $\arg z_2 = \tan^{-1}\left(\tfrac{1}{2}\right) = 26° \, 34'$.

Hence, $$z_1 = \sqrt{5}[\cos(-63° \, 26') + i \sin(-63° \, 26')]$$

and $$z_2 = \sqrt{5}[\cos(26° \, 34') + i \sin(26° \, 34')]$$

$$\therefore \quad \frac{z_1}{z_2} = 1[\cos(-90°) + i\sin(-90°)] = -i$$

This can be checked by algebraic division,

$$\frac{z_1}{z_2} = \frac{1 - 2i}{2 + i} = \frac{(1 - 2i)(2 - i)}{(2 + i)(2 - i)} = \frac{2 - 5i + 2i^2}{4 - i^2} = -i$$

Exercise 6.3

---

1 Express the following in modulus-argument form.

   (a) 3   (b) $2i$ (c) $1 + i$ (d) $\sqrt{3} + i$ (e) $1 - 2i$

   (f) $-2 - 2i$ (g) $3 + 4i$ (h) $-2 + 3i$ (i) $-5 + 12i$

2 Convert the following complex numbers to the form $a + ib$.

   (a) $2(\cos 60° + i\sin 60°)$ (b) $4\left(\cos\dfrac{\pi}{6} + i\sin\dfrac{\pi}{6}\right)$ (c) $5\left(\cos\dfrac{2\pi}{3} + i\sin\dfrac{2\pi}{3}\right)$

   (d) $3\left[\cos\left(-\dfrac{\pi}{4}\right) + i\sin\left(-\dfrac{\pi}{4}\right)\right]$ (e) $2\left[\cos\left(-\dfrac{3\pi}{4}\right) + i\sin\left(-\dfrac{3\pi}{4}\right)\right]$

3 Write down the modulus and argument of the following.

   (a) $2\left(\cos\dfrac{\pi}{12} + i\sin\dfrac{\pi}{12}\right)$ (b) $\dfrac{3}{2}\left(\cos\dfrac{5\pi}{12} + i\sin\dfrac{5\pi}{12}\right)$ (c) $\cos 2\theta - i\sin 2\theta$

   (d) $\dfrac{1}{2}\left(\cos\dfrac{2\pi}{3} - i\sin\dfrac{2\pi}{3}\right)$

4 Simplify the following products.

   (a) $(\cos\tfrac{1}{2}\pi + i\sin\tfrac{1}{2}\pi)(\cos\tfrac{1}{2}\pi - i\sin\tfrac{1}{2}\pi)$

   (b) $(\cos\tfrac{1}{4}\pi + i\sin\tfrac{1}{4}\pi)(\cos\tfrac{3}{4}\pi + i\sin\tfrac{3}{4}\pi)$

   (c) $4(\cos 40° + i\sin 40°)(\cos 110° + i\sin 110°)$

   (d) $\left(\cos\dfrac{\pi}{12} - i\sin\dfrac{\pi}{12}\right)^2$

5 Find $\dfrac{z_1}{z_2}$ and $\dfrac{z_2}{z_1}$ when,

    (a) $z_1 = \cos\frac{1}{3}\pi + i\sin\frac{1}{3}\pi;\quad z_2 = \cos\frac{1}{4}\pi + i\sin\frac{1}{4}\pi$

    (b) $z_1 = \cos\dfrac{2\pi}{3} + i\sin\dfrac{2\pi}{3};\quad z_2 = 2\left(\cos\dfrac{5\pi}{12} + i\sin\dfrac{5\pi}{12}\right)$

    (c) $z_1 = 4\left(\cos\dfrac{3\pi}{5} + i\sin\dfrac{3\pi}{5}\right);\quad z_2 = \sqrt{2}\left(\cos\dfrac{2\pi}{5} - i\sin\dfrac{2\pi}{5}\right)$

6 If $z = \cos\theta + i\sin\theta$ where $0 < \theta < \pi$, find the modulus and argument of   **(a)** $\bar{z}$ **(b)** $\dfrac{1}{z}$ **(c)** z^2.

7 If $z = \cos\theta + i\sin\theta$ where $0 < \theta < \pi$, find the modulus and argument of   **(a)** $1 + z$ **(b)** $z - 1$ **(c)** $\dfrac{2z}{z-1}$.

8 If $z_1 = 1 + i$ and $z_2 = -1 + i$, plot the corresponding points on an Argand diagram. Construct the point which represents $z_1 z_2$.

    Check your result by converting z_1 and z_2 to modulus-argument form and evaluating the modulus and argument of the product.

9 Write $z = \sqrt{3} + i$ in modulus-argument form and hence find   **(a)** $(\sqrt{3} + i)^2$ **(b)** $(\sqrt{3} + i)^3$
    (c) $(\sqrt{3} + i)^5$ giving your answers in the form $a + ib$. Check your results for **(a)** and **(b)** by direct expansion.

Loci in the Argand plane

If $z = x + iy$ is a complex number which is represented on an Argand diagram by the point $P(x, y)$, then the position of P varies with the values of x and y. If z (i.e. the values of x and y) varies according to a given condition, then the set of all possible positions of P is called the locus of P.

Most conditions involve distances or angles, and the following are particularly important.

$|z|$ defines the distance of the point P from the origin.

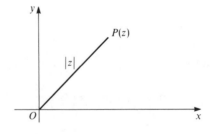

$|z - a|$ defines the distance of the point P from a point representing the complex number, a.

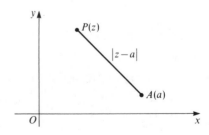

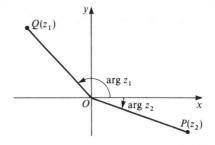

Arg z defines the angle that the radius vector makes with the positive increasing direction of the real axis.

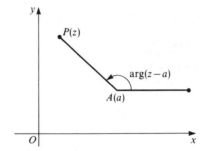

Arg $(z-a)$ defines the angle that the line joining the points representing the complex numbers z and a, makes with the positive increasing direction of the real axis.

Investigation 12

If z is a variable complex number, find points which satisfy the following conditions and show your results on an Argand diagram.

(a) $|z-3| = 2$ **(b)** $|z+i| = |z-3i|$ **(c)** $|z+1| = 2|z-1|$

By putting $z = x+iy$ and substituting into the equations, try to find the Cartesian equation of each of the loci.

WORKED EXAMPLES

1 Find the locus in the Argand plane if $|z| = 3$.

This equation states that the distance of any point representing z from the origin is constant and equal to 3. Thus, all the possible positions of z must lie on a circle whose centre is at the origin and whose radius is 3.

The Cartesian equation can be found as follows.

As $z = x+iy$, we have

$$|z| = 3 \implies |x+iy| = 3$$
$$\therefore \quad x^2 + y^2 = 3^2$$
$$\therefore \quad x^2 + y^2 = 9$$

i.e. a circle, centre at the origin and of radius 3 units.

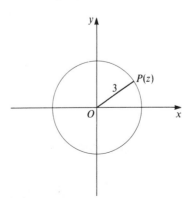

2 Find the locus in the Argand plane if $|z-2+i|=4$.

Since $|z-a|$ is the distance between the points representing the complex numbers z and a, $|z-(2-i)|$, is the distance between the points representing the variable complex number z and $2-i$.

Thus, the locus will be a circle of radius 4 units with its centre at the point $(2,-1)$.

The Cartesian equation will be found as follows.

$$|z-(2-i)|=4$$

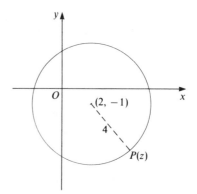

Let $z=x+iy$

$$\therefore \quad |(x+iy)-(2-i)|=4$$

$$\Rightarrow \quad |(x-2)+i(y+1)|=4$$

$$\Rightarrow \quad \sqrt{(x-2)^2+(y+1)^2}=4$$

$$\Rightarrow \quad (x-2)^2+(y+1)^2=16$$

i.e. a circle, centre at $(2,-1)$ and of radius 4 units.

3 Find the locus in the Argand plane if $|z-1|=|z+1+i|$.

In this case, the distance of the point representing z is equidistant from the point $(1,0)$ on the real axis and the point $(-1,-1)$. The locus will be the straight line bisecting the line joining $(1,0)$ to $(-1,-1)$ at right angles.

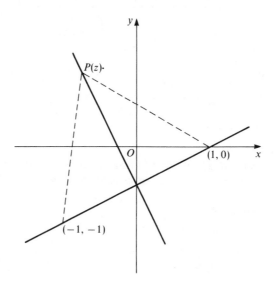

The Cartesian equation can be found by letting $z=x+iy$ and substituting in the given equation.

$$|z-1|=|z+1+i|$$

$$\therefore \quad |(x-1)+iy|=|(x+1)+i(y+1)|$$

$$\therefore \quad \sqrt{(x-1)^2+y^2}=\sqrt{(x+1)^2+(y+1)^2}$$

Squaring both sides gives

$$(x-1)^2 + y^2 = (x+1)^2 + (y+1)^2$$

$$\Rightarrow \quad x^2 - 2x + 1 + y^2 = x^2 + 2x + 1 + y^2 + 2y + 1$$

$$\Rightarrow \quad -2x + 1 = 2x + 2y + 2$$

$$\Rightarrow \quad 4x + 2y + 1 = 0$$

Check that this straight line is perpendicular to the line joining the points $(1, 0)$ and $(-1, -1)$ and that it passes through its mid-point.

4 Find the locus in the Argand plane if $\arg(z + 1 - i) = \frac{3}{4}\pi$.

Now, $\arg(z + 1 - i)$ is the angle which the line joining the points representing the complex numbers z and $(-1 + i)$ makes with the positive increasing direction of the real axis, and hence this angle has a constant value of $\frac{3}{4}\pi$.

Thus, the equation $\arg(z + 1 - i) = \frac{3}{4}\pi$ gives a half line (or ray) starting from $(-1, 1)$ and inclined at $\frac{3}{4}\pi$ to the positive real axis.

The Cartesian equation can be found as follows.

$$\arg(z + 1 - i) = \frac{3}{4}\pi$$

Let $z = x + iy$

$\therefore \quad \arg[(x+1) + i(y-1)] = \frac{3}{4}\pi$

$\therefore \quad \tan^{-1}\left(\dfrac{y-1}{x+1}\right) = \frac{3}{4}\pi$

$\therefore \quad \dfrac{y-1}{x+1} = \tan\frac{3}{4}\pi = -1$

$\therefore \quad y - 1 = -x - 1$

$\therefore \quad y + x = 0$

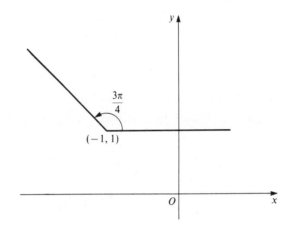

Note that the other half of the line would be given by $\arg(z + 1 - i) = -\frac{1}{4}\pi$.

5 Shade the region on an Argand diagram for which $|z - 1| < 2$ and $\mathrm{Re}(z) \geqslant 0$.

Since $|z - 1| = 2$ represents a circle of radius 2 units and centre $(1, 0)$, the inequality $|z - 1| < 2$ represents the area within this circle not including the boundary (shown dashed in the diagram).

However, $\mathrm{Re}(z) \geqslant 0$ means that the real part of z (i.e. x) is greater than or equal to zero.

Hence, the area produced by both of these conditions is shown by the shaded section within the circle.

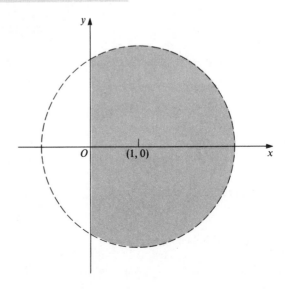

Exercise 6.4

1 Sketch on separate Argand diagrams the loci represented by the following equations and find the Cartesian equation in each case.

(a) $|z| = 1$ (b) $|z| = 4$ (c) $|z-1| = 1$ (d) $|z+2| = 3$ (e) $|z-i| = 1$

(f) $|z+i| = 2$ (g) $|z-1-i| = 2$ (h) $|z+1-i| = 3$ (i) $|z-4+3i| = 5$

2 Sketch, on separate Argand diagrams, the loci represented by the following equations.

(a) $\arg z = \frac{1}{4}\pi$ (b) $\arg z = \frac{3}{4}\pi$ (c) $\arg z = -\dfrac{\pi}{6}$

(d) $\arg(z-1) = \dfrac{\pi}{3}$ (e) $\arg(z+2) = -\dfrac{2\pi}{3}$

(f) $\arg(z-2+i) = \pi$ (g) $\arg(z+3-3i) = \pm\dfrac{\pi}{6}$

3 Sketch the following loci in the complex plane and find the Cartesian equation of each.

(a) $|z| = |z-1|$ (b) $|z-2| = |z+3|$ (c) $|z-1+i| = |z+2-i|$ (d) $|z+i| = |z-3-i|$

4 Shade the region defined by the following inequalities on an Argand diagram.

(a) $|z| \leqslant 3$ (b) $|z| > 4$ (c) $|z-1-i| < \sqrt{2}$ (d) $0 \leqslant \arg z \leqslant \frac{1}{2}\pi$

(e) $\frac{1}{3}\pi \leqslant \arg z \leqslant \frac{2}{3}\pi$ (f) $|z| < 5$ and $\mathrm{Re}(z) < 0$ (g) $|z+1| \leqslant 2$ and $\mathrm{Im}(z) \leqslant 0$

5 Find the Cartesian equation of each of the loci represented by the following equations.

(a) $|z+2i| = |z+1|$ (b) $\left|\dfrac{z}{z-4}\right| = 5$ (c) $|z-i| = 4$ (d) $|z-1| = 2|z-3|$

6 Prove that $z\bar{z} = |z|^2$. Show that the equation $2|z-1| = |z-4|$ can be expressed as $z\bar{z} = 4$.

7 Given that $|z-1+2i| = 3$, find the greatest and least values of $|z|$.

8 Given that the complex number z varies such that $|z-1| = 1$, find the greatest and least values of $|z+i|$.

9 Sketch the following loci on an Argand diagram.

(a) $\arg(z-1) - \arg(z+1) = \dfrac{\pi}{2}$ (b) $\arg(z-2) - \arg(z) = \dfrac{\pi}{4}$

De Moivre's theorem for an integral index

From Investigation 10 on page 130, we found that

$$(\cos\theta + i\sin\theta)(\cos\phi + i\sin\phi) = \cos(\theta+\phi) + i\sin(\theta+\phi)$$

If $\phi = \theta$, this becomes

$$(\cos\theta + i\sin\theta)^2 = \cos 2\theta + i\sin 2\theta$$

Investigation 13

Use this last result to show that

$$(\cos \theta + i \sin \theta)^3 = \cos 3\theta + i \sin 3\theta$$

and $\quad (\cos \theta + i \sin \theta)^4 = \cos 4\theta + i \sin 4\theta$

The results suggest that, in general, if n is a positive integer,

$$(\cos \theta + i \sin \theta)^n = \cos n\theta + i \sin n\theta$$

This can be proved by induction.

Assume that it is true for some value $n = k$. That is,

$$(\cos \theta + i \sin \theta)^k = \cos k\theta + i \sin k\theta$$

By considering

$$(\cos \theta + i \sin \theta)^{k+1} \quad \text{as} \quad (\cos \theta + i \sin \theta)^k (\cos \theta + i \sin \theta)$$

show that

$$(\cos \theta + i \sin \theta)^{k+1} = \cos (k+1)\theta + i \sin (k+1)\theta$$

Now, show that it is true for $n = 1$ and deduce that it is true for $n = 2$, $n = 3$, and for all positive integral values of n.

This result is known as **De Moivre's theorem.**

$$\boxed{(\cos \theta + i \sin \theta)^n = \cos n\theta + i \sin n\theta}$$

It can also be shown that the result is true if n is negative or fractional.

> WORKED EXAMPLES

1 Evaluate $(\cos \frac{1}{3}\pi + i \sin \frac{1}{3}\pi)^6$.

Using De Moivre's theorem,

$$(\cos \tfrac{1}{3}\pi + i \sin \tfrac{1}{3}\pi)^6 = \cos \left(6 \times \frac{\pi}{3} \right) + i \sin \left(6 \times \frac{\pi}{3} \right) = \cos 2\pi + i \sin 2\pi = 1$$

2 Evaluate $(1 - i)^{10}$.

Express $1 - i$ in modulus-argument form.

Since $\quad |1 - i| = \sqrt{2} \quad$ and $\quad \arg(1 - i) = -\frac{\pi}{4}, \quad$ it follows that

$$1 - i = \sqrt{2} \left[\cos \left(-\frac{\pi}{4} \right) + i \sin \left(-\frac{\pi}{4} \right) \right]$$

Hence,

$$(1-i)^{10} = (\sqrt{2})^{10}\left[\cos\left(-\frac{\pi}{4}\right) + i\sin\left(-\frac{\pi}{4}\right)\right]^{10}$$

$$= 2^5\left[\cos\left(-\frac{10\pi}{4}\right) + i\sin\left(-\frac{10\pi}{4}\right)\right]$$

$$= 32\left[\cos\left(-\frac{5\pi}{2}\right) + i\sin\left(-\frac{5\pi}{2}\right)\right]$$

$$= 32\left[\cos\left(-\frac{\pi}{2}\right) + i\sin\left(-\frac{\pi}{2}\right)\right] = -32i$$

3 Evaluate $\dfrac{1}{(-1+i\sqrt{3})^4}$.

Since $\ |-1+i\sqrt{3}| = 2\ $ and $\ \arg(-1+i\sqrt{3}) = \frac{2}{3}\pi,\ $ it follows that $\ -1+i\sqrt{3} = 2\left[\cos\left(\frac{2}{3}\pi\right) + i\sin\left(\frac{2}{3}\pi\right)\right]$

Thus,

$$\frac{1}{(-1+i\sqrt{3})^4} = (-1+i\sqrt{3})^{-4} = 2^{-4}\left[\cos\left(\frac{2}{3}\pi\right) + i\sin\left(\frac{2}{3}\pi\right)\right]^{-4}$$

$$= \frac{1}{16}\left[\cos\left(-\frac{8\pi}{3}\right) + i\sin\left(-\frac{8\pi}{3}\right)\right]$$

$$= \frac{1}{16}\left[\cos\left(\frac{8\pi}{3}\right) - i\sin\left(\frac{8\pi}{3}\right)\right]$$

$$= \frac{1}{16}\left[\cos\left(\frac{2\pi}{3}\right) - i\sin\left(\frac{2\pi}{3}\right)\right]$$

$$= \frac{1}{16}\left[-\frac{1}{2} - i\frac{\sqrt{3}}{2}\right] = \frac{1}{32}(-1-i\sqrt{3})$$

4 Prove that $\tan 3\theta = \dfrac{3\tan\theta - \tan^3\theta}{1 - 3\tan^2\theta}$.

We use De Moivre's theorem to express $\ \cos 3\theta + i\sin 3\theta\ $ in terms of powers of $\sin\theta$.

Hence, by De Moivre's theorem,

$$\cos 3\theta + i\sin 3\theta = (\cos\theta + i\sin\theta)^3$$

Expanding the right-hand side gives,

$$\cos 3\theta + i\sin 3\theta = (\cos\theta + i\sin\theta)^2\,(\cos\theta + i\sin\theta)$$

$$= (\cos^2\theta + 2i\sin\theta + \sin^2\theta)\,(\cos\theta + i\sin\theta)$$

$$= \cos^3\theta + 3i\cos^2\theta\sin\theta - 3\cos\theta\sin^2\theta - i\sin^3\theta$$

Equating the real and imaginary parts gives,

$$\cos 3\theta = \cos^3\theta - 3\cos\theta\sin^2\theta \quad\text{and}\quad \sin 3\theta = 3\cos^2\theta\sin\theta - \sin^3\theta$$

Hence, dividing

$$\tan 3\theta = \frac{\sin 3\theta}{\cos 3\theta} = \frac{3\cos^2\theta\sin\theta - \sin^3\theta}{\cos^3\theta - 3\cos\theta\sin^2\theta}$$

Dividing the numerator and denominator by $\cos^3 \theta$ gives,

$$\tan 3\theta = \frac{3 \tan \theta - \tan^3 \theta}{1 - 3 \tan^2 \theta}$$

Exercise 6.5

1 Evaluate **(a)** $\left(\cos \frac{3\pi}{4} + i \sin \frac{3\pi}{4} \right)^4$ **(b)** $\left(\cos \frac{\pi}{8} + i \sin \frac{\pi}{8} \right)^4$

   (c) $\left[\cos \left(-\frac{\pi}{3} \right) + i \sin \left(-\frac{\pi}{3} \right) \right]^6$ **(d)** $\left(\cos \frac{\pi}{6} - i \sin \frac{\pi}{6} \right)^3$

2 Evaluate **(a)** $(1+i)^5$ **(b)** $(1+i\sqrt{3})^8$ **(c)** $(-\sqrt{3}-i)^5$

3 Evaluate **(a)** $\dfrac{1}{\left[\cos \left(\frac{2\pi}{5} \right) + i \sin \left(\frac{2\pi}{5} \right) \right]^{10}}$ **(b)** $\left[\cos \left(\frac{2\pi}{3} \right) - i \sin \left(\frac{2\pi}{3} \right) \right]^{-4}$

4 Find expressions for **(a)** $\tan 2\theta$ **(b)** $\tan 5\theta$ in terms of $\tan \theta$.

5 Prove that $\cos 4\theta = 8 \cos^4 \theta - 8 \cos^2 \theta + 1$.

6 Show that $\sin 4\theta = \sin \theta \, (8 \cos^3 \theta - 4 \cos \theta)$.

7 Find an expression for $\cos 6\theta$ in terms of powers of $\cos \theta$ and show that
$\sin 6\theta = \sin \theta \, (32 \cos^5 \theta - 32 \cos^3 \theta + 6 \cos \theta)$.

Expressions for $\cos^n \theta$ and $\sin^n \theta$ in terms of multiple angles

In the previous section, Worked Example **4** illustrated how to use De Moivre's theorem to express a sine, cosine or tangent of a multiple angle in terms of powers of sine, cosine or tangent, i.e.

$$\tan 3\theta = \frac{3 \tan \theta - \tan^3 \theta}{1 - 3 \tan^2 \theta}$$

It is also possible to use De Moivre's theorem to express powers of sine, cosine or tangent in terms of the sine, cosine or tangent of multiple angles. The process, however, is rather different.

If $z = \cos \theta + i \sin \theta$, then by De Moivre's theorem

$$z^{-1} = (\cos \theta + i \sin \theta)^{-1} = \cos (-\theta) + i \sin (-\theta)$$
$$= \cos \theta - i \sin \theta$$

Also

$$z^n = (\cos \theta + i \sin \theta)^n = \cos n\theta + i \sin n\theta$$

and

$$z^{-n} = (\cos \theta + i \sin \theta)^{-n} = \cos (-n\theta) + i \sin (-n\theta)$$
$$= \cos n\theta - i \sin n\theta$$

Hence,

$$z + \frac{1}{z} = 2 \cos \theta \qquad z - \frac{1}{z} = 2i \sin \theta$$

$$z^n + \frac{1}{z^n} = 2 \cos n\theta \qquad z^n - \frac{1}{z^n} = 2i \sin n\theta$$

Using these results

$$2^n \cos^n \theta = \left(z + \frac{1}{z}\right)^n$$

which can be expanded to give cosines of multiple angles. Similarly,

$$(2i \sin \theta)^n = \left(z - \frac{1}{z}\right)^n$$

will give cosine terms if n is even and sine terms if n is odd.

WORKED EXAMPLE

Prove that $\cos^5 \theta = \dfrac{1}{16} (\cos 5\theta + 5 \cos 3\theta + 10 \cos \theta)$.

Using the relation $(2 \cos \theta)^n = \left(z + \dfrac{1}{z}\right)^n$ we have $(2 \cos \theta)^5 = \left(z + \dfrac{1}{z}\right)^5$

$$\Rightarrow \quad 32 \cos^5 \theta = z^5 + 5z^3 + 10z + \frac{10}{z} + \frac{5}{z^3} + \frac{1}{z^5}$$

$$= \left(z^5 + \frac{1}{z^5}\right) + 5\left(z^3 + \frac{1}{z^3}\right) + 10\left(z + \frac{1}{z}\right)$$

$$\therefore \quad 32 \cos^5 \theta = (2 \cos 5\theta) + 5(2 \cos 3\theta) + 10(2 \cos \theta)$$

$$= 2 \cos 5\theta + 10 \cos 3\theta + 20 \cos \theta$$

$$\therefore \quad \cos^5 \theta = \tfrac{1}{16}(\cos 5\theta + 5 \cos 3\theta + 10 \cos \theta)$$

The nth roots of unity

It is possible to find particular roots of unity by using De Moivre's theorem.

WORKED EXAMPLE

Find the roots of $z^5 - 1 = 0$ by using De Moivre's theorem.

Now, $z^5 - 1 = 0 \Rightarrow z^5 = 1$

Hence, $z = \sqrt[5]{1} = 1^{\frac{1}{5}}$ .... **(1)**

Now, it is possible to express the real number 1 as a complex number in many ways, for example,

$$\cos 0 + i \sin 0; \qquad \cos 2\pi + i \sin 2\pi; \qquad \cos 4\pi + i \sin 4\pi$$

In general, $\quad 1 = \cos 2n\pi + i \sin 2n\pi \quad$ where n is an integer

Thus, from Equation **(1)** $\qquad\qquad z = 1^{\frac{1}{5}} = (\cos 2n\pi + i \sin 2n\pi)^{\frac{1}{5}}$

Using De Moivre's theorem,

$$z = \cos\left(\frac{2n\pi}{5}\right) + i \sin\left(\frac{2n\pi}{5}\right)$$

Substituting integral values for n will produce the roots.

$$n=0 \quad \Rightarrow \quad z_1 = \cos 0 + i \sin 0 = 1$$

$$n=1 \quad \Rightarrow \quad z_2 = \cos \frac{2\pi}{5} + i \sin \frac{2\pi}{5}$$

$$n=2 \quad \Rightarrow \quad z_3 = \cos \frac{4\pi}{5} + i \sin \frac{4\pi}{5}$$

$$n=3 \quad \Rightarrow \quad z_4 = \cos \frac{6\pi}{5} + i \sin \frac{6\pi}{5}$$

$$n=4 \quad \Rightarrow \quad z_5 = \cos \frac{8\pi}{5} + i \sin \frac{8\pi}{5}$$

$$n=5 \quad \Rightarrow \quad z_6 = \cos 2\pi + i \sin 2\pi = 1$$

Notice that after $n=4$, the roots begin to repeat themselves. Hence there are just five principal roots:

$$\cos\left(\frac{2n\pi}{5}\right) + i \sin\left(\frac{2n\pi}{5}\right) \quad \text{for} \quad n=0, 1, 2, 3, 4$$

Exercise 6.6

1   Find expressions for   **(a)** $\cos^4 \theta$ **(b)** $\cos^7 \theta$ in terms of the cosines of multiples of θ.

2   Express   **(a)** $\sin^3 \theta$ in terms of the sines of multiples of θ;   **(b)** $\sin^4 \theta$ in terms of the cosines of multiples of θ.

3   Show that $\sin^5 \theta = \frac{1}{16}(\sin 5\theta - 5 \sin 3\theta + 10 \sin \theta)$.

4   Show that $\cos^3 \theta = \frac{1}{4}(\cos 3\theta + 3 \cos \theta)$.

5   Evaluate   **(a)** $\displaystyle\int_0^\pi \cos^4 \theta \, d\theta$ **(b)** $\displaystyle\int_0^{\pi/2} \sin^5 \theta \, d\theta$.

6   Show that $\cos^3 \theta - \sin^3 \theta = \cos 3\theta + \sin 3\theta - 3(\cos^2 \theta \sin \theta - \cos \theta \sin^2 \theta)$.

Hence evaluate $\displaystyle\int_0^{\pi/4} \cos^3 \theta - \sin^3 \theta \, d\theta$.

7   Use De Moivre's theorem to find the 4th roots of unity. Also show that $z^4 - 1 = (z^2 - 1)(z^2 + 1)$ and hence find the roots of $z^4 = 1$. Compare your results obtained by the two methods.

8   Use De Moivre's theorem to find the cube roots of **(a)** 1 and **(b)** -1. Simplify your answers into the form $a + ib$ and compare your results with the solutions on pages 124 and 125.

9   Solve the equation $z^5 = -1$ using De Moivre's theorem.

10   Find all the roots of the equation $z^4 + 1 = 0$ and plot your answers on an Argand diagram.

11   Write the complex number $4 + 4i$ in modulus-argument form, and hence show that the roots of the equation $z^5 = 4 + 4i$ are given by

$$z = \sqrt{2}\left[\cos\left(\frac{(8r+1)\pi}{20}\right) + i \sin\left(\frac{(8r+1)\pi}{20}\right)\right] \quad \text{for} \quad r=0, 1, 2, 3, 4.$$

Revision Exercises C

These Exercises should enable you to revise work covered in Book 1.

Forces

1 Identify the forces acting **(a)** on the 3 kg mass
   (b) on the 5 kg mass.

   (c) If both move with acceleration f, write down the equation of motion
   for each mass and find f and the tension in the string.

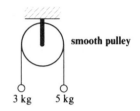

2 If the forces balance, find AB. Find the supporting force
   at B.

3 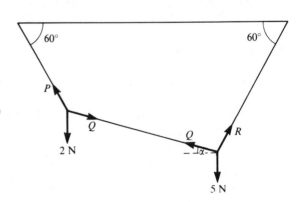 If $AC = 10$, find AB when the forces balance.

4 Alongside my table, I have a sloping work-surface inclined at an angle of 10 degrees. On this I have my calculator
   in its plastic cover and a sheet of paper. The calculator remains at rest on the work-surface but when I place the
   calculator on the sheet of paper, both slide down the surface. Why does this happen?

5 Find the coefficient of friction between *this* book and the surface on which it is resting.

6 Find the resultant of the forces $\mathbf{F}_1 = 3\mathbf{i} + 2\mathbf{j}$, $\mathbf{F}_2 = 2\mathbf{i} - 3\mathbf{j}$, $\mathbf{F}_3 = 4\mathbf{j}$, acting at a point.

7 Find the resultant of forces $\mathbf{F}_1 = 20\,\text{N}$ at $30°$ with Ox, and $\mathbf{F}_2 = 30\,\text{N}$ at $20°$ with Ox, both forces acting
   through the origin, O.

8 From the diagram find the tensions
   P, Q, R and the angle α.

144

Kinematics

1 A particle moves in a straight line so that its displacement in metres from its starting position, O $(t=0)$, is given by $s = t^4 - 2t^3 - t^2 + 2t$, ($t$ in seconds). Find when it is subsequently at O and when and where it is momentarily at rest. Find the distances it covers during the first, second and third seconds.

2 Repeat question 1 for $s = \sin t + \cos t$.

3 A ball is thrown vertically upwards from a window with velocity 6 m s^{-1}. Taking $g = 10 \text{ m s}^{-2}$, find **(a)** the time taken to reach the ground and **(b)** the velocity with which the ball hits the ground, if the window is **(i)** 3 m **(ii)** 5 m, above the ground.

4 Examine the motion represented by $\mathbf{r} = x\mathbf{i} + y\mathbf{j} = 20t\mathbf{i} - 5t^2\mathbf{j}$.

When is $x = 0$? What is the starting velocity? What situation could this equation represent?

5 Examine the motion for $\mathbf{r} = 4 \sin t\,\mathbf{i} + 4 \cos t\,\mathbf{j}$. Describe the path, velocity and acceleration.

6 Examine the motion for $\mathbf{r} = 4 \cos t\,\mathbf{i} + 3 \sin t\,\mathbf{j}$.

7 A particle travelling with constant acceleration covers distances of 2 m and 3 m in successive seconds. Find the velocities at the beginning and end of each second.

8 A particle is projected under gravity $(g = 10 \text{ ms}^{-2})$ with a starting velocity of $8\mathbf{i} + 6\mathbf{j}$. Find its velocity and displacement after **(a)** 1 s **(b)** 2 s **(c)** 3 s. **(d)** Find the x, y equation of its path.

Newton's laws

1 A book of mass 2 kg rests on a rough table inclined at $30°$ to the horizontal.

(a) Find the coefficient of friction between the book and the table if the book is about to move.

(b) Find the force acting up the slope parallel to the table which will just move the book up the slope.

(c) If this force acts down the slope, what will be the acceleration of the book?

2 The book in question 1 rests on the same table which is now horizontal.

(a) What horizontal force will give the book an acceleration of 3 m s^{-2}?

(b) Find the least force which will just move the book and the angle at which it acts.

3 A crate of mass 30 kg is lifted by a crane with an acceleration of 2 m s^{-2}. Find the tension in the cable.

4 A parcel of mass 20 kg is placed in a lift. Find the reaction between the parcel and the floor of the lift if the lift accelerates **(a)** upwards at 2 m s^{-2} **(b)** downwards at 3 m s^{-2}.

5 A force of 8 N acts on a particle of mass 2 kg for 5 seconds. Find the acceleration of the particle and the speed it attains, if its initial speed is 6 m s^{-1}.

6 A bullet of mass 0.01 kg is fired horizontally into a fixed vertical metal plate at a speed of 500 m s^{-1}. If the bullet penetrates the plate to a depth of 0.05 m, find the resistance of the plate, assuming this to be constant.

7 A car of mass 1200 kg tows a trailer of mass 800 kg. If the force exerted by the engine of the car is 4800 N and the resistances are 1.5 N per kg, find the acceleration of the car and the tension in the towbar.

8. A train is travelling at 54 km h^{-1} on level track when it begins to climb a hill of inclination $\sin^{-1}\left(\frac{1}{100}\right)$. If the tractive force developed by the engine is maintained at a constant value of $\frac{1}{200}$th of the weight of the train, and the resistances are $\frac{1}{40}$th of the weight of the train, find the distance moved up the slope before the train comes to rest.

9 A particle A of mass 2 kg lies on a rough horizontal table and is connected to a particle B of mass 4 kg which hangs freely by a light inextensible string passing over a smooth pulley at the edge of the table. If the system is released from rest with the string taut, find the acceleration of the masses if the coefficient of friction is $\frac{3}{4}$.

 If the 4 kg mass is initially 1 m above the floor, find the speed of the masses just before particle B strikes the floor. Find also the further time that elapses before A comes to rest, assuming that it does not reach the edge of the table.

10 A block of wood is in the form of a wedge with a triangular cross-section ABC and is fixed with AC horizontal. Angle BAC is 30° and angle BCA is 60°. A particle P of mass 2 kg lies on the smooth face containing AB and is connected by a light inextensible string passing over a smooth pulley at B to a particle Q of mass 3 kg on the smooth face containing BC. If the system is released from rest with the string taut, find the acceleration of the particles.

7 Centres of mass

Introduction

Investigation 1

Can you balance your ruler on your pencil? Where is the balancing point on the ruler?

Such a balancing point is called the **centre of gravity** and is the point at which the **weight** of a body is said to act. The point at which the **mass** is said to be concentrated is called the **centre of mass**. In these Books, where the force of gravity is constant over the whole body, the points coincide.

Find by experiment the centre of mass of **(a)** this book **(b)** a semi-circular protractor **(c)** yourself **(d)** an envelope **(e)** a triangular sheet of cardboard **(f)** two people on a seesaw.

Investigation 2

Which of the triangle centres of a triangular sheet of card is the centre of mass? **(a)** circumcentre **(b)** centroid **(c)** incentre **(d)** orthocentre.

Which is the same as three equal masses placed at the vertices of the triangle?

We must distinguish between **particles**, **laminae** and **rigid bodies**.

A **particle** is a body of small dimension whose mass can be regarded as concentrated at one point, i.e. its centre of mass. Theoretically a particle has no size, but sometimes we can regard a book sliding down a desk as a particle whose mass is concentrated at its centres of mass, although in reality its mass is spread throughout its shape.

A **lamina** (like a thin sheet of cardboard or metal) has no thickness (theoretically) but has a two-dimensional (rigid) shape.

Rigid bodies do not distort under the action of the forces acting on them. We regard a ladder as a rigid body, although in reality it sags a little when someone uses it.

We shall be considering the centres of mass of laminae (triangle, rectangle etc.), hollow shapes (cone, hemisphere) and solid shapes (pyramid, cone, hemisphere), whose mass distribution is uniform.

Investigation 3

Find the centres of mass of **(a)** a square **(b)** a rectangle **(c)** a parallelogram **(d)** a trapezium.

Investigation 4

Find the centre of mass of three particles, of mass as follows.

(a) 1 kg at (6, 0),   1 kg at (0, 6)   and   1 kg at (6, 6)

(b) 3 kg at (6, 0),   2 kg at (0, 6)   and   1 kg at (6, 6)

(c) 1 kg at (6, 0),   2 kg at (0, 6)   and   3 kg at (6, 6)

(d) 2 kg at (6, 0),   1 kg at (0, 6)   and   3 kg at (6, 6)

(e) 1 kg at (6, 0),   3 kg at (0, 6)   and   2 kg at (6, 6)

(f) m kg at (6, 0), m kg at (0, 6)   and m kg at (6, 6)

(g) m kg at (a, b), m kg at (c, d) and m kg at (e, f)

A practical way to find the centre of mass of a body is to suspend it from a point on its perimeter. When suspended, the centre of mass lies directly under the point of suspension. Two different points of suspension will give two lines on which the centre of mass lies, so their point of intersection is the centre of mass.

Sometimes consideration of symmetry will provide the centre of mass. For example, the centre of mass of a rectangle is at the centre of the rectangle (where the diagonals cross) since the mass is uniformly distributed (balanced) about that point.

It is obvious that the centre of mass (C of M) of a uniform rod (such as a ruler) is halfway along the rod. We can regard a rectangle as composed of many rods each having its C of M on XY, so the C of M of the rectangle is on XY.

Similarly, the C of M will be on AB, so it must be at the centre C.

Intuitively, we can accept that there is only one balance point i.e. **one** C of M, for if we move away from the balance point the mass ceases to be balanced about the new point.

What do we mean by **balance**?

Consider two children on a seesaw whose masses are 40 kg and 60 kg.

Their balance point, B, divides CD in the ratio x to y where

$$40x = 60y$$

as $40x$ gives the turning effect of C and $60y$ the turning effect of D, which must be equal.

$$40x = 60y \implies x:y = 3:2 \quad \text{so } B \text{ divides } CD \text{ in the ratio } 3:2$$

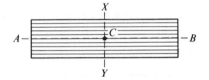

If CD were of length 5 m, B would be 3 m from C and 2 m from D.

B is also known as the centre of **gravity** because it supports the gravitational forces (weights) 400 N at C and 600 N at D. The total force downwards of 1000 N is balanced by a support force of 1000 N upwards acting at B. This shows that the resultant of two parallel forces, 400 N acting at C and 600 N acting at D, is a force of 1000 N acting through B parallel to the forces at C and D.

Considering the turning effects (moments) of the forces about the point B.

$$400x = 600y \implies x:y = 3:2 \quad \text{and} \quad x + y = 5 \implies x = 3 \quad \text{and} \quad y = 2$$

The turning effect of a force is called the **moment of a force** about a particular point (B in this case), and considering turning effects is called **taking moments about** (B in this case). As in previous chapters, 'taking moments about B' will be denoted by $\hat{B}$.

Knowing that the support force at B must be 1000 N, we can take moments about C and D.

$\hat{C}$ $\qquad 1000x = 600(x+y) \quad \Rightarrow \quad 5x = 3(x+y) \quad \Rightarrow 2x = 3y \quad \Rightarrow \quad x:y = 3:2$

$\hat{D}$ $\qquad 400(x+y) = 1000y \quad \Rightarrow \quad 2(x+y) = 5y \quad \Rightarrow \quad 2x = 3y \quad \Rightarrow \quad x:y = 3:2$

Taking moments gives a powerful method of finding forces and has an important application in the methods of finding centres of mass (gravity).

Centre of mass of point masses

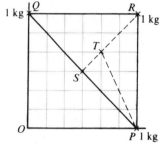

Referring to Investigation 4, part **(a)** the C of M of 1 kg at $P(6, 0)$ and 1 kg at $Q(0, 6)$ is at $S(3, 3)$, so we can regard the resultant of these as a mass of 2 kg at S.

Then the C of M of 2 kg at $S(3, 3)$ and 1 kg at $R(6, 6)$ is at T on RS where

$$ST:TR = 1:2 \quad \text{so} \quad T \text{ is } (4, 4)$$

For part **(b)** of Investigation 4, 2 kg at Q with 1 kg at R gives a resultant of 3 kg at $V(2, 6)$, which together with 3 kg at P gives a resultant 6 kg acting at W, where $VW = WP \quad \Rightarrow \quad W$ is $(4, 3)$

Alternatively, using co-ordinates, V divides QR in the ratio $1:2$,

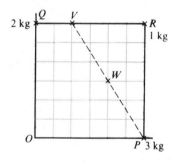

$\Rightarrow \quad \mathbf{V} = \frac{2}{3}\mathbf{Q} + \frac{1}{3}\mathbf{R}$

$\mathbf{V} = \frac{2}{3}(0, 6) + \frac{1}{3}(6, 6) = (0, 4) + (2, 2) = (2, 6)$

Using vectors, $\quad \mathbf{OV} = \frac{2}{3}(6\mathbf{j}) + \frac{1}{3}(6\mathbf{i} + 6\mathbf{j}) = 2\mathbf{i} + 6\mathbf{j}$

$\mathbf{OW} = \frac{1}{2}\mathbf{OV} + \frac{1}{2}\mathbf{OP} = \frac{1}{2}(2\mathbf{i} + 6\mathbf{j}) + \frac{1}{2}(6\mathbf{i}) = 4\mathbf{i} + 3\mathbf{j} \quad \Rightarrow \quad W \text{ is } (4, 3)$

Also $\quad \mathbf{OW} = \frac{1}{2}\mathbf{OV} + \frac{1}{2}\mathbf{OP} = \frac{1}{2}(\frac{2}{3}\mathbf{OQ} + \frac{1}{3}\mathbf{OR}) + \frac{1}{2}\mathbf{OP} = \frac{1}{6}(3\mathbf{OP} + 2\mathbf{OQ} + \mathbf{OR})$

This suggests that W, the C of M of m_1 at P, m_2 at Q, m_3 at R, is given by

$$\mathbf{OW} = \frac{m_1 \mathbf{OP} + m_2 \mathbf{OQ} + m_3 \mathbf{OR}}{m_1 + m_2 + m_3} \qquad \dots\dots\dots\dots \text{ (1)}$$

WORKED EXAMPLES

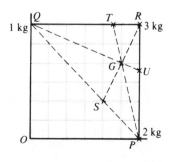

1 Referring to Investigation 4, part **(d)**, find the C of M of 2 kg at P, 1 kg at Q and 3 kg at R by **(a)** first pairing P and Q **(b)** then Q and R **(c)** then P and R **(d)** by Equation **(1)**.

(a) 2 kg at P + 1 kg at Q = 3 kg at $S(4, 2)$

3 kg at S + 3 kg at R = 6 kg at $(5, 4)$ the mid-point of SR

(b) 1 kg at Q + 3 kg at R = 4 kg at T, where $QT:TR = 3:1$ $\Rightarrow$ **T** is $(4\frac{1}{2}, 6)$

4 kg at T + 2 kg at P = 6 kg at G, where $TG:GP = 2:4$ $\Rightarrow$ $\mathbf{G} = \frac{1}{3}\mathbf{P} + \frac{2}{3}\mathbf{T}$

$$= (2, 0) + (3, 4) = (5, 4)$$

(c) 2 kg at P + 3 kg at R = 5 kg at U, where $PU:UR = 3:2$ $\Rightarrow$ $\mathbf{U} = \frac{2}{5}\mathbf{P} + \frac{3}{5}\mathbf{R}$

$$= \tfrac{1}{5}[(12, 0) + (18, 18)] = (6, 3.6)$$

5 kg at U + 1 kg at Q = 6 kg at H, where $UH:HQ = 1:5$ $\Rightarrow$ $\mathbf{H} = \frac{5}{6}\mathbf{U} + \frac{1}{6}\mathbf{Q}$

$$= \tfrac{1}{6}[(30, 18) + (0, 6)] = (5, 4)$$

(d) If centroid is at G, $\mathbf{OG} = \dfrac{2\mathbf{OP} + 1\mathbf{OQ} + 3\mathbf{OR}}{2+1+3} = \dfrac{2(6\mathbf{i}) + 6\mathbf{j} + 3(6\mathbf{i} + 6\mathbf{j})}{6}$

$$= \frac{30\mathbf{i} + 24\mathbf{j}}{6} = 5\mathbf{i} + 4\mathbf{j}$$

The formula in **(d)** gives the most direct way to achieve the result.

Note that SR, QU and PT are concurrent, meeting at G. This is **Ceva's theorem**, that if the product of the ratios in which S, T, U divide the sides is 1, the 'medians' are concurrent.

$$\frac{QS}{SP} \times \frac{PU}{UR} \times \frac{RT}{TQ} = \frac{2}{1} \times \frac{3}{2} \times \frac{1}{3} = 1 \quad \Rightarrow \quad SR, \quad QU, \quad PT \quad \text{concurrent}$$

---

2 Find the C of M of m at $A(3, 0)$, $2m$ at $B(6, 3)$, $3m$ at $C(3, 6)$ and $4m$ at $D(0, 3)$.

By extending the formula in Equation **(1)**, if G is the C of M,

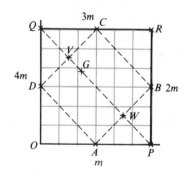

$$\mathbf{OG} = \frac{m(1\mathbf{OA} + 2\mathbf{OB} + 3\mathbf{OC} + 4\mathbf{OD})}{m(1+2+3+4)}$$

$$= \tfrac{1}{10}[3\mathbf{i} + 2(6\mathbf{i} + 3\mathbf{j}) + 3(3\mathbf{i} + 6\mathbf{j}) + 4(3\mathbf{j})]$$

$$= \tfrac{1}{10}(24\mathbf{i} + 36\mathbf{j}) = 2.4\mathbf{i} + 3.6\mathbf{j}$$

Notice that G lies on PQ $(x + y = 6)$ because we have $5m$ on the line AD and $5m$ on BC. So the C of M lies on the line $PWVQ$ halfway between AD and BC. Looking at DC ($7m$) and AB ($3m$) implies that G lies $\frac{3}{10}$ of the way between DC and AB, but this is too difficult to calculate at this stage.

Exercise 7.1

1 Find the balance point for the following masses on the seesaw $ABCDEF$, of length 5 m.

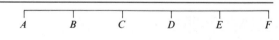

(a) 6 kg at A and 4 kg at F **(b)** 6 kg at B and 6 kg at F **(c)** 1 kg at A and 3 kg at E
(d) 2 kg at B and 1 kg at E **(e)** 1 kg at A, 2 kg at B, 3 kg at C and 4 kg at D
(f) 1 kg at A, 4 kg at B, 3 kg at E and 2 kg at F

2 Where would you place a mass of 2 kg on the 5 m seesaw in question **1** to make the centre of mass act through C in the following cases?

(a) 2 kg at A and 2 kg at F **(b)** 3 kg at A and 3 kg at E
(c) 4 kg at A and 1 kg at E **(d)** 3 kg at A, 2 kg at B and 3 kg at F

3 Find mass m to balance the systems about the point D, for the above seesaw.

    (a) 1 kg at A, 2 kg at B, m kg at F **(b)** 3 kg at B, 2 kg at C, m kg at E

    (c) 5 kg at A , 6 kg at C, m kg at F **(d)** 2 kg at A, m kg at C, m kg at F

4 If three masses, each m kg, balance about D, on the above seesaw, where could they be placed?

5 If four masses m, $2m$, $3m$ and $4m$ have a C of M at D, on the above seesaw, at which of the points A, B, C, E and F could they be attached **(a)** in ascending order **(b)** in descending order **(c)** in any order?

6 Find the C of M for the following distributions (all masses are in kg).

    (a) 3 at $A(3, 0)$,   2 at $B(6, 3)$,   1 at $C(3, 6)$

    (b) 3 at $A(3, 0)$,   2 at $B(6, 3)$,   1 at $C(3, 6)$ and 4 at $D(0, 3)$

    (c) 1 at $Q(0, 6)$,   2 at $R(6, 6)$,   3 at $P(6, 0)$

    (d) 1 at $Q(0, 6)$,   2 at $R(6, 6)$,   3 at $P(6, 0)$ and 4 at $O(0, 0)$

    (e) 2 at $Q(0, 6)$,   4 at $R(6, 6)$,   6 at $P(6, 0)$

    (f) 3 at $P(6, 0)$,   2 at $B(6, 3)$,   3 at $C(3, 6)$ and 2 at $Q(0, 6)$

    (g) 2 at $Q(0, 6)$,   1 at $E(6, 5)$,   3 at $F(5, 1)$ (Check by joining the C of M of E and F to Q, the C of M of F and Q to E, and the C of M of E and Q to F. The lines should meet at the C of M of all three.)

    (h) The distribution in **(g)** with 2 at $O(0, 0)$.

7 With the points in their positions of question **6**, find the C of M of the following.

    (a) 1 at A and 2 at B **(b)** 1 at A, 2 at B and 3 at C

    (c) 1 at A, 2 at B, 3 at C and 4 at D **(d)** As for **(c)**, with 5 at E

Alternative method by taking moments

Imagine four masses of 1 kg at $A(3, 0)$, 2 kg at $B(6, 3)$, 3 kg at $C(3, 6)$ and 4 kg at $D(0, 3)$ placed on a flat surface (the xy-plane) so that their weights act **into** the page of the book.

The centre of mass, $G(x, y)$, is the point through which the resultant force (a weight of 10 kg mass) acts. This resultant has the same effect as the constituents and so will have the same turning effect about the x and y axes.

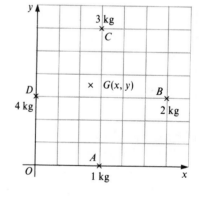

$\stackrel{\frown}{Oy}$ (taking moments about Oy)

$$\Rightarrow \quad 10x = (1 \times 3) + (2 \times 6) + (3 \times 3) + (4 \times 0) = 24 \quad \Rightarrow \quad x = 2.4$$

$\stackrel{\frown}{Ox}$ $10y = (1 \times 0) + (2 \times 3) + (3 \times 6) + (4 \times 3) = 36 \quad \Rightarrow \quad y = 3.6$

$$\Rightarrow \quad G \text{ is } (2.4, 3.6)$$

This method is virtually the same as in Worked Example **2** opposite, but may be easier to follow and it has important applications later in more difficult situations.

Centres of mass of laminae

In Investigation 2 you may have found the centre of mass of a triangle, i.e. a triangular lamina, a flat rigid sheet of negligible thickness.

By dividing $\triangle ABC$ into thin strips parallel to AB we can see that the centre of mass of each strip lies at its mid-point. So the C of M of all the strips together must lie on CM (locus of mid-points), the median from C to AB.

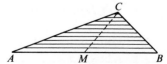

Centres of mass

By dividing the triangle into strips parallel to BC we can see that the C of M lies on AN (the median from A to BC).

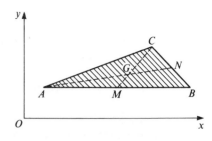

So the C of M lies at G, the intersection of the medians. G divides each median in the ratio $2:1$ and is called the **centroid** of $\triangle ABC$ (proof in Chapter 4, Vectors).

If the co-ordinates of A, B, and C are $A(x_1, y_1)$, $B(x_2, y_2)$, $C(x_3, y_3)$, then G is given by

$$x = \frac{x_1 + x_2 + x_3}{3}, \qquad y = \frac{y_1 + y_2 + y_3}{3}$$

If three equal masses are placed at A, B and C, then their C of M is at G, also.

The C of M of a square, parallelogram, rectangle and rhombus are clearly at their geometric centres (intersection of their diagonals).

Investigation 5

Locate the centres of mass of a kite and a trapezium. Are they at the intersection of the diagonals? Does it help to divide them into triangles?

Investigation 6

Find the centre of mass of $\triangle ACM$.

Find the centre of mass of $\triangle BCM$.

What are the relative masses of these two triangles?

Deduce the position of the centre of mass of the large triangle ABC.

Does this agree with the theory above (intersection of the medians)?

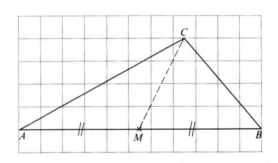

Investigation 7

To find the C of M of the trapezium $ABCE$.

(a) Find the C of M of the square $ABCD$.

(b) Find the C of M of the triangle CDE.

(c) What are the relative masses of the square $ABCD$ and the triangle CDE.

(d) Deduce the C of M of $ABCE$.

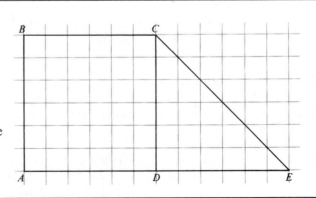

152

Composite bodies

The ideas of the last two Investigations enable us to find the centres of mass of shapes or bodies made up of simpler components whose centres of mass are known.

Consider the square $ODCB$ and triangle CDE, where D is $(6, 0)$, B $(0, 6)$ and E $(12, 0)$.

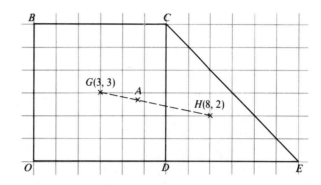

The C of M of ODCB is $G(3, 3)$.

The C of M of CDE is $H(8, 2)$. (This can be deduced as the centroid of the triangle lies on a horizontal line one-third of the distance from D to C, and on a vertical line one-third of the distance from D to E.)

As the square is double the mass of the triangle, the C of M of $OECB$ lies at A on GH such that $GA:AH = 1:2$ (inverse ratio of the masses).

A is $(3 + \frac{5}{3}, 3 - \frac{1}{3})$ i.e. $(4\frac{2}{3}, 2\frac{2}{3})$.

Alternatively, using moments

	Square	**Triangle**	**Trapezium** $OECB$
mass $\propto$ area	36	18	54
distance of G from OB	3	8	x
distance of H from OE	3	2	y

$\stackrel{\frown}{OB}$ $54x = 36 \times 3 + 18 \times 8$ $\Rightarrow$ $3x = 6 + 8$ $\Rightarrow$ $x = \frac{14}{3}$

$\stackrel{\frown}{OE}$ $54y = 36 \times 3 + 18 \times 2$ $\Rightarrow$ $3y = 6 + 2$ $\Rightarrow$ $y = \frac{8}{3}$

A is $(\frac{14}{3}, \frac{8}{3})$ i.e. $(4\frac{2}{3}, 2\frac{2}{3})$.

Exercise 7.2

For Questions **1** to **9**, find the centres of mass of the laminae.

1 $\triangle ABC$	**2** $\triangle PQR$	**3** $\triangle PRC$	**4** $\triangle PCQ$	**5** $\triangle QAR$
6 $\triangle ADR$	**7** $APCQ$	**8** $PBQD$	**9** $ODRB$	

For Questions **10** to **18** you may need to consider the laminae as composite bodies made up of simpler shapes (possibly using symmetry).

10 $APRQ$	**11** $OPBM$	**12** $OPBCQ$
13 $OABCQ$	**14** $OBRC$	**15** $OMPBCDO$
16 $OPRQMO$	**17** $ABRD$	**18** $ARDMA$

19 Find the centre of mass of a kite.

20 Find the centre of mass of a trapezium whose parallel sides are a and b $(b > a)$, the distance between them being h.

The following Worked Examples (pages 154 and 155), which you may have tried in Exercise 7.2, illustrate methods of approach which may be of use later.

Centres of mass

1 Find the centre of mass of a kite.

By symmetry the C of M lies on BD at J which is below O if $h > g$.

C of M of $\triangle ABC$ is at G, where $OG = \frac{1}{3}g$

C of M of $\triangle ADC$ is at H, where $OH = \frac{1}{3}h$

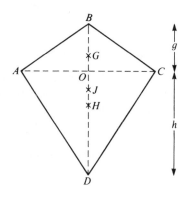

Mass of $\triangle ABC \propto \dfrac{g \times AC}{2}$;   Mass of $\triangle ADC \propto \dfrac{h \times AC}{2}$

$$\overset{\frown}{AC} \qquad \frac{hAC}{2} \times \frac{h}{3} - \frac{gAC}{2} \times \frac{g}{3} = \frac{(gAC + hAC)}{2} \times OJ$$

$$\Rightarrow \quad h^2 - g^2 = 3(g + h)OJ \quad \Rightarrow \quad OJ = \frac{h - g}{3}$$

2 Find the centre of mass of a trapezium.

Let the parallel sides $AB = b$, $CD = a$.

By dividing $ABCD$ into strips parallel to AB, the C of M lies at G, on the locus of the centres of the strips, LM.

If the height of $ABCD$ is h (the distance between AB and CD), then we need only to find how far (in terms of h) G is from AB, i.e. the perpendicular distance of G from AB.

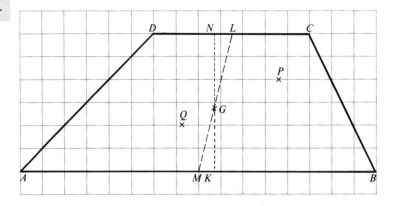

Here are three methods of doing this, each of which is instructive for following work.

Method A

The C of M of $\triangle BCD$ lies on BL at P, where $LP : PB = 1 : 2 \Rightarrow P$ is $\dfrac{h}{3}$ from CD.

The C of M of $\triangle ABD$ lies on DM at Q, where $MQ : QD = 1 : 2 \Rightarrow Q$ is $\dfrac{h}{3}$ from AB.

G lies on PQ, where $QG : GP = $ area $BCD :$ area $ABD = \frac{1}{2}ah : \frac{1}{2}bh = a : b$.

Measuring along a line through G perpendicular to AB, the distance of G from AB is

$$\frac{h}{3} + \frac{ah}{3(a + b)} = \frac{(a + b)h + ah}{3(a + b)} = \frac{(2a + b)h}{3(a + b)}$$

Method B

Take moments about AB for $\triangle ABD + \triangle BCD = $ trapezium $ABCD$.

$$\overset{\frown}{AB} \qquad GK \times \frac{(a + b)h}{2} = \left(\frac{h}{3} \times \frac{bh}{2} \right) + \left(\frac{2h}{3} \times \frac{ah}{2} \right) \quad \text{where } NGK \perp AB$$

Thus, $3(a + b)\,GK = bh + 2ah \Rightarrow GK = \dfrac{(2a + b)h}{3(a + b)} \Rightarrow GN = \dfrac{(a + 2b)h}{3(a + b)}$

The distance of the centre of mass from AB is $\dfrac{(2a+b)h}{3(a+b)}$.

Method C

Produce AD and BC to meet at T and take moments about AB for trapezium $ABCD + \triangle CDT \equiv \triangle ABT$.

Notice that TL produced meets AB at M, so that $DL = LC = \dfrac{a}{2}$ and $AM = MB = \dfrac{b}{2}$.

If the height of $\triangle CDT = k$, then by similar triangles,

$$\frac{k}{h+k} = \frac{a}{b} \quad \Rightarrow \quad kb = ah + ak \quad \Rightarrow \quad k = \frac{ah}{b-a}$$

$$A\tilde{B} \quad \Rightarrow \quad \left(\frac{(a+b)h}{2}\right)GK + \frac{ak}{2}\left(h+\frac{k}{3}\right) = \left(\frac{b(h+k)}{2}\right)\left(\frac{(h+k)}{3}\right)$$

Thus, $$[3h(a+b)]\,GK + ak(3h+k) = b(h+k)^2 \quad \text{and} \quad h+k = h + \frac{ah}{(b-a)} = \frac{bh}{(b-a)}$$

So, $$[3h(a+b)]\,GK = b(h+k)^2 - ak(3h+k) = \frac{b(bh)^2}{(b-a)^2} - \frac{a^2h}{(b-a)}\left(3h + \frac{ah}{(b-a)}\right)$$

$$[3h(a+b)(b-a)^2]\,GK = h^2[b^3 - a^2(3b-2a)] = h^2(b^3 - 3a^2b + 2a^3)$$

$$= h^2(b-a)(b^2 + ab - 2a^2) = h^2(b-a)(b-a)(b+2a)$$

$$\Rightarrow \quad GK = \frac{(2a+b)h}{3(a+b)}$$

Special Cases

$a = 0$ gives a triangle $\Rightarrow GK = \dfrac{h}{3}$, the C of M for a triangle.

$a = b$ gives a parallelogram $\Rightarrow GK = \dfrac{h}{2}$ i.e. half-way up.

The distance of G from CD is

$$h - \frac{(2a+b)h}{3(a+b)} = \frac{(3a+3b-b-2a)h}{3(a+b)} = \frac{(a+2b)h}{3(a+b)} \qquad \text{a symmetrical result}$$

Centres of mass by integration

Laminae

We can still use the method of dividing into strips for more complicated shapes. We shall start by confirming the C of M of a right-angled triangle by integration.

Consider $\triangle OAB$, where $OA = a$, $AB = b$.

We would expect to find G as $(\tfrac{2}{3}a, \tfrac{1}{3}b)$.

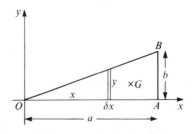

Divide $\triangle OAB$ into vertical strips, a typical strip having height y, width δx, and distant x from Oy.

If the mass per unit area is m, then the mass of the strip is $y\,\delta x\,m$, and its C of M is at $\left(x,\ \dfrac{y}{2}\right)$.

The sum of the moments of all the strips about Oy (or Ox) will equal the moment of the **whole** triangle about Oy (or Ox). If the C of M is at $G(X, Y)$,

$$\widehat{Oy}\quad \tfrac{1}{2}abm\,X = \lim_{\delta x \to 0} \sum_{x=0}^{x=a} my\,\delta x\,x = \int mxy\,dx \quad \text{where} \quad y = \frac{bx}{a} \quad \text{for this shape} \quad (\triangle OAB)$$

$$\Rightarrow \quad \tfrac{1}{2}abm\,X = m\int_{x=0}^{x=a} \frac{bx^2}{a}\,dx = \frac{mb}{a}\int_0^a x^2\,dx = \frac{mb}{a}\left[\frac{x^3}{3}\right]_0^a = \frac{mb}{a}\left(\frac{a^3}{3}\right) \quad \Rightarrow \quad X = \frac{2a}{3}$$

$$\widehat{Ox}\quad \tfrac{1}{2}abm\,Y = \lim_{\delta x \to 0} \sum_{x=0}^{x=a} my\,\delta x\,\frac{y}{2} = \int_0^a \frac{m}{2}y^2\,dx = \frac{m}{2}\int_0^a \left(\frac{bx}{a}\right)^2 dx = \frac{m}{2}\left(\frac{b}{a}\right)^2\left(\frac{a^3}{3}\right) \quad \Rightarrow \quad Y = \frac{b}{3}$$

> **WORKED EXAMPLES**

1 Find the centre of mass of a semicircular plate (lamina).

For convenience align the semicircle (radius r) with Ox as axis of symmetry, with C of M, $G(X, 0)$, on Ox.

First Method

Consider a typical strip parallel to Oy of height $2y$, width δx, distant x from Oy; mass per unit area m.

For each strip, $x^2 + y^2 = r^2$ (equation of circle)

$$\widehat{Oy}\quad m\frac{\pi}{2}r^2 X = \lim_{\delta x \to 0} \sum_{x=0}^{x=r} 2my\,\delta x\,x = \int_0^r 2mx\,y\,dx$$

$$= m\int_0^r 2x\sqrt{(r^2 - x^2)}\,dx$$

$$= m\left[\frac{-2(r^2 - x^2)^{3/2}}{3}\right]_0^r = 2m\left[-0 + \frac{r^3}{3}\right]$$

$$= \frac{2mr^3}{3} \quad \Rightarrow \quad X = \frac{4r}{3\pi}$$

So G is $\left(\dfrac{4r}{3\pi}, 0\right)$.

We could have considered the quadrant OAB as its C of M, $H(X, Y)$, will be the same same distance from Oy as G. Check by a similar method that you do obtain the same result.

$$\widehat{Ox}\quad m\frac{\pi}{4}r^2 Y = \lim_{\delta x \to 0} \sum_{x=0}^{x=r} my\,\delta x\,\frac{y}{2} = \int_0^r \frac{m}{2}y^2\,dx = \frac{m}{2}\int_0^r (r^2 - x^2)\,dx$$

$$= \frac{m}{2}\left[r^2 x - \frac{x^3}{3}\right]_0^r = \frac{m}{2}\left(\frac{2r^3}{3}\right) \quad \Rightarrow \quad Y = \frac{4r}{3\pi}$$

So H is $\left(\dfrac{4r}{3\pi}, \dfrac{4r}{3\pi}\right)$, which we would expect for the quadrant, by symmetry. Comparing the semicircle with the rectangle $DPQB$, C of M $\left(\dfrac{r}{2}, 0\right)$, we would expect X to be just less than $\dfrac{r}{2}$ and $\dfrac{4r}{3\pi} = 0.42r$.

Second Method, for the quadrant

Divide the quadrant into strips of sectors which are almost triangular, of angle $\delta\theta$, making an angle θ with Ox.

C of M of the sector is distant $\dfrac{2r}{3}\cos\theta$ from Oy and the mass of each sector is $\dfrac{r^2}{2}\delta\theta\,m$

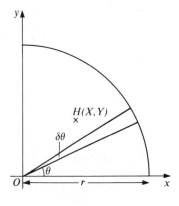

$$\hat{Oy} \quad \frac{\pi r^2}{4}\,m\,X = \lim_{\delta\theta\to 0}\ \sum_{\theta=0}^{\pi/2}\frac{r^2}{2}\delta\theta\,m\left(\frac{2r}{3}\cos\theta\right)$$

$$= m\int \frac{r^3}{3}\cos\theta\,d\theta = \frac{mr^3}{3}\left[\sin\theta\right]_{\theta=0}^{\pi/2} = \frac{mr^3}{3}$$

$$\therefore \quad X = \frac{4r}{3\pi} \quad \text{and by symmetry} \quad X = Y = \frac{4r}{3\pi}$$

---

2 Find the centre of mass of an arc of a circle of radius r making an angle of $2q$ at the centre of the arc.

Consider an element (of length δs) of the arc ABC subtending an angle $\delta\theta$ and making an angle θ with Ox. In this example the angular method (using θ) is easier than using x and y.

In our previous methods, each strip was taken as a rectangle neglecting the sloping top which was negligible compared with the whole strip.

For the second quadrant method, the narrow sector was regarded as a triangle neglecting the curved edge.

Here we are assuming that the whole of the elemental arc, δs, is straight and distance $x = r\cos\theta$ from Oy. If δs is small (as it is when we take the limit for integrating) this does not matter.

By symmetry, the C of M lies on Ox at $G(X, 0)$ and the top half of the arc BC will have the same X value for its C of M as AB and as the whole arc.

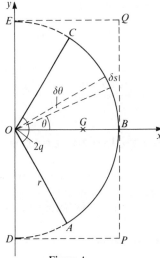

Figure A

Again, we can consider only the top half of the arc (but remember to halve the mass and take for limits $\theta = 0$ to $\theta = q$ with mass per unit length, m.

$$\hat{Oy} \quad rq\,m\,X = \lim_{\delta\theta\to 0}\ \sum_{\theta=0}^{q} m\,\delta s\,r\cos\theta = \lim_{\delta\theta\to 0}\ \sum_{\theta=0}^{q} mr\,\delta\theta\,(r\cos\theta)$$

$$= m\int_0^q r^2\cos\theta\,d\theta = mr^2\left[\sin\theta\right]_0^q = mr^2\sin q$$

$$X = \frac{r\sin q}{q} \quad \text{and for a semicircular arc,}\quad q = \frac{\pi}{2} \ \Rightarrow\ X = \frac{2r}{\pi}$$

Exercise 7.3

Find the centres of mass of the following:

1 A lamina in the shape of a sector of a circle radius r, angle $2q$. (This is $OABC$ in Figure A of Worked Example **2**, above.)

2 A lamina in the shape of a sector of a circle radius r, angle $2 \times 45°$. (Again this is a quadrant; with the same result?)

3 A lamina in the shape of three quadrants (total angle $270°$).

    (a) Arrange the lamina with Ox as line of symmetry (limits $-135° < \theta < +135°$)

    (b) Arrange three quadrants in 1st, 2nd and 4th trigonometrical quadrants so that $y = x$ is the line of symmetry and regard the total as made up of a quadrant and a semicircle.

    (c) Arrange as in **(b)** and take the 3rd quadrant from the whole circle.

4 The minor segment ABC, by taking $\triangle OAC$ from sector $OABC$ (Figure A).

5 The trapezium-like shape $DACE$, (Figure A) by taking the segment ABC from the semicircle $DABCE$.

6 A wire in the shape of a letter 'D' i.e. $ODABCEO$ (Figure A; semicircular arc + diameter).

7 A wire in the shape of $\triangle DBE$ (Figure A; three sides of a triangle).

8 A wire in the shape of the perimeter of the sector $OABCO$ (Figure A; arc + two radii).

9 A wire in the shape of $DPBQE$ (Figure A; three sides of a rectangle).

10 A lamina in the shape of 'moon' $BQECB$ (Figure A).

11 A lamina in the shape of semicircle DBE (Figure A) with isosceles $\triangle DEF$ to the left. (F is $(-r, 0)$.) What length must OF be to make the C of M at O?

Formulae for centres of mass of laminae

If the outline of a lamina can be described by the curve $y = f(x)$, then there are two formulae for X and Y, the co-ordinates of the centre of mass, which are worth remembering.

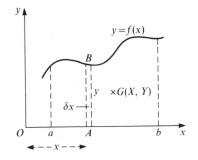

Let the lamina be described by $y = f(x)$, $x = a$, $x = b$ and Ox.

Consider a typical strip AB of width δx and height y distant x from Oy, so that the C of M of this strip is at $(x, y/2)$.

Let the C of M of the lamina be $G(X, Y)$, then the moment of the lamina about each axis will equal the sum of the moments of all the strips.

If the mass per unit area is m, then the mass of the lamina is $m \displaystyle\int y\, dx$.

$$O\hat{y} \qquad X m \int_a^b y\, dx = \lim_{\delta x \to 0} \sum_{x=a}^{x=b} m\, y\, \delta x\, x = m \int_a^b xy\, dx \quad \Rightarrow \quad X = \frac{\displaystyle\int_a^b xy\, dx}{\displaystyle\int_a^b y\, dx}$$

$$O\hat{x} \qquad Y m \int_a^b y\, dx = \lim_{\delta x \to 0} \sum_{x=a}^{x=b} m\, y\, \delta x\, \frac{y}{2} = m \int_a^b \frac{y^2}{2}\, dx \quad \Rightarrow \quad Y = \frac{\displaystyle\int_a^b \frac{y^2}{2}\, dx}{\displaystyle\int_a^b y\, dx}$$

WORKED EXAMPLES

1 The centre of mass of a triangle.

Let the triangle have sides $y = mx$, the x-axis and $x = h$.

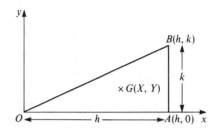

$$X = \frac{\int_0^h xy\,dx}{\int_0^h y\,dx} = \frac{\int_0^h mx^2\,dx}{\int_0^h mx\,dx} = \frac{\left[\frac{x^3}{3}\right]_0^h}{\left[\frac{x^2}{2}\right]_0^h} = \frac{2h}{3}$$

$$Y = \frac{\int_0^h y^2/2\,dx}{\int_0^h y\,dx} = \frac{\int_0^h m^2x^2/2\,dx}{\int_0^h mx\,dx} = \frac{\left[\frac{m^2x^3}{6}\right]_0^h}{\left[\frac{mx^2}{2}\right]_0^h} = \frac{mh}{3} = \frac{k}{3} \text{ since } k = mh$$

This is a special (right-angled) triangle, so we ought to generalize the result.

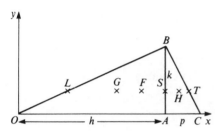

In the diagram, $OA = h$, $AC = p$, $AB = k$.

The centre of mass of $\triangle OAB$ is at $G\,(2h/3, k/3)$ where $LG = GS = h/3$.

The C of M of $\triangle ACB$ is at $H(h + p/3, k/3)$ where $SH = HT = p/3$.

So the C of M of $\triangle OCB$, the composite body, is at a distance $k/3$ from OC and lies on GH at F.

$$\frac{GF}{FH} = \frac{\text{area } ACB}{\text{area } OAB} = \frac{pk}{hk} = \frac{p}{h} \quad \text{and} \quad GH = \frac{h}{3} + \frac{p}{3} \;\Rightarrow\; GF = \frac{p}{3}; \quad FH = \frac{h}{3} \;\Rightarrow\; LF = LG + GF = \frac{h}{3} + \frac{p}{3} = FT$$

So F lies on BM, the median of $\triangle OBC$ from B to OC, and by a similar argument will lie on the medians from O to BC and from C to OB.

The C of M lies on the intersection of three lines which are concurrent in the ratio 2:1 along each line, and the medians are the only lines which satisfy this requirement, as in $\triangle PQR$.

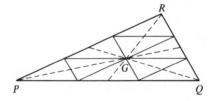

It is easy to see from geometric properties of enlargement that G, the intersection of the medians, divides each median in the ratio 2:1.

2 The centre of mass of a quadrant.

By symmetry, the C of M lies on OK at $G(X, Y)$ where $X = Y$.

The equation of the curve is $x^2 + y^2 = r^2$ and the formula for Y looks easier to integrate. The area of the quadrant is $\dfrac{\pi r^2}{4}$.

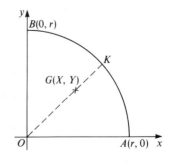

$$Y = \frac{\int_0^r y^2/2\,dx}{\pi r^2/4} = \frac{\int_0^r (r^2 - x^2)\,dx}{\pi r^2/2} = \frac{\left[r^2x - x^3/3\right]_0^r}{\pi r^2/2}$$

$$= \frac{2r^3/3}{\pi r^2/2} = \frac{4r}{3\pi} \quad \text{as before}$$

Centres of mass of hollow and solid shapes

Solid hemisphere

Let the hemisphere (radius r) have Ox as axis of symmetry, so the surface would be generated by rotating arc BC about Ox (out of the xy-plane) through 360°.

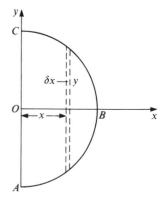

Consider a typical 'slice' (thin cylinder, such as a coin) of radius y, thickness δx (height of cylinder), distant x from base Oy.

For each slice $x^2 + y^2 = r^2$.

By symmetry, the C of M of each slice and of the hemisphere is on Ox at $(x, 0)$ and $G(X, 0)$ respectively.

Let the mass per unit volume be m.

$$\widehat{Oy} \quad \tfrac{2}{3}\pi r^3 m X = \lim_{\delta x \to 0} \sum_{x=0}^{r} \pi y^2\, \delta x\, m\, x = \pi m \int_{x=0}^{x=r} x y^2\, dx = m\pi \int_0^r x\,(r^2 - x^2)\, dx = m\pi \left[\frac{r^2 x^2}{2} - \frac{x^4}{4}\right]_0^r$$

$$\tfrac{2}{3}\pi r^3 m\, X = m\pi \left[\frac{r^4}{2} - \frac{r^4}{4}\right] = \frac{m\pi r^4}{4} \quad \Rightarrow \quad X = \frac{3r}{8}$$

> The centre of mass of a solid hemisphere, radius r, is $\dfrac{3r}{8}$ from O along the axis of symmetry.

Solid cone (right-circular)

With the notation in the figure, the cone is being generated by rotating the line OC given by the equation $y = rx/h$ about Ox. A typical slice is a disc radius y, thickness δx, distant x from Oy.

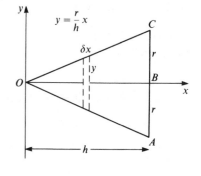

Mass per unit volume is m.

$$\widehat{Oy} \quad (\text{mass of cone})\, X = \lim_{\delta x \to 0} \sum_{x=0}^{h} m\pi y^2 \delta x\, x = \int_0^h m\pi x\, y^2\, dx$$

If you do not know the cone's volume: 'one third base area × height', you can find it by adding up the volumes of the slices.

Have you ever proved $V = \tfrac{1}{3}\pi r^2 h$?

$$\therefore \quad X \int_0^h m\pi y^2\, dx = m\pi \int_0^h x y^2\, dx \quad \Rightarrow \quad X = \frac{\displaystyle\int_0^h x y^2\, dx}{\displaystyle\int_0^h y^2\, dx} = \frac{\displaystyle\int_0^h x r^2 x^2/h^2\, dx}{\displaystyle\int_0^h r^2 x^2/h^2\, dx} = \frac{(r^2/h^2)\left[\dfrac{x^4}{4}\right]_0^h}{(r^2/h^2)\left[\dfrac{x^3}{3}\right]_0^h} = \frac{3h}{4}$$

> The centre of mass of a right-circular solid cone, height h, is $\dfrac{3h}{4}$ from the apex or $\dfrac{h}{4}$ from the base of the cone along the axis.

Hollow hemisphere

The **hollow** hemisphere is generated by rotating the arc CE about the axis of symmetry Ox so AOE represents the circular base.

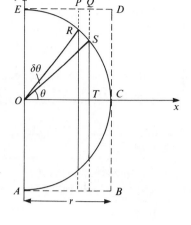

The arc RS generates a typical slice (like a dinner plate separator) and Archimedes discovered that the surface area of this slice is the same as the surrounding cylindrical section (generated by rotating PQ).

The C of M will be the same as for the surrounding hollow cylinder i.e. $\dfrac{r}{2}$ from the base.

However, using the θ method, the slice has area $RS.2\pi.ST$ and moment

$$RS.2\pi.ST.OT = r\,\delta\theta\,(2\pi r \sin\theta)\,(r\cos\theta)$$

By symmetry, the C of M is at $G(X, 0)$

$\widehat{Oy}$ $X\displaystyle\int_0^{\pi/2} 2\pi r^2 \sin\theta\,d\theta = \int_0^{\pi/2} 2\pi r^3 \sin\theta\cos\theta\,d\theta$

$$2\pi r^2 X\left[-\cos\theta\right]_0^{\pi/2} = 2\pi r^3\left[\frac{\sin^2\theta}{2}\right]_0^{\pi/2} \quad\Rightarrow\quad 2\pi r^2 X = \frac{2\pi r^3}{2} \quad\Rightarrow\quad X = \frac{r}{2}$$

> The centre of mass of a hollow hemisphere is $\dfrac{r}{2}$ from O along its axis of symmetry.

Hollow cone

This cone has no base; slant height k, radius r, height h, mass/unit area m.

The surface of the cone is generated by rotating the line $y = \dfrac{rx}{h}$ about Ox. By symmetry the C of M is at $G(X, 0)$.

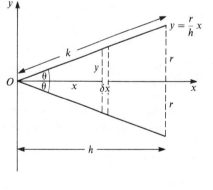

A typical slice has thickness δx, radius y, distant x from O. Its mass is $2\pi yL$, where L is its slant height and $L\cos\theta = \delta x$.

$\widehat{Oy}$ $X\pi rkm = \displaystyle\lim_{\delta x\to 0}\sum_{x=0}^{h} m\,2\pi yxL = \sum m\,2\pi\left(\frac{rx}{h}\right)x\,\delta x\,\frac{1}{\cos\theta}$

$\cos\theta = \dfrac{h}{k} \quad\Rightarrow\quad \dfrac{1}{\cos\theta} = \dfrac{k}{h}$

$\Rightarrow\quad X\pi rkm = 2\pi m\displaystyle\int_0^h \frac{rx^2}{h}\left(\frac{k}{h}\right)dx = \frac{2\pi mrk}{h^2}\int_0^h x^2\,dx = \frac{2\pi mrk}{h^2}\left[\frac{x^3}{3}\right]_0^h = \frac{2\pi mrkh}{3}$

$\Rightarrow\quad X = \dfrac{2h}{3}$ or $\dfrac{h}{3}$ from the base.   This can easily be demonstrated by dividing

the cone into small triangles each of which has centre of mass $\dfrac{h}{3}$ from the base.

> The centre of mass of a hollow cone, height h, is $\dfrac{h}{3}$ from the base along its axis of symmetry.

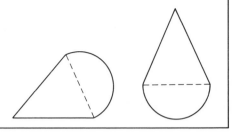

Investigation 8

My magic ball-point pen consists of a cone (ball-point at the apex) with a hemispherical base which always 'stands up' even when I lay it down. Why is this?

Find the combined centre of mass if cone and base are
(a) solid **(b)** hollow.

Exercise 7.4

1 For the pen in Investigation 8, find the combined centre of mass if both cone and hemisphere are **(a)** hollow and $h = 2r$ **(b)** solid and $h = 2r$ **(c)** hollow and $h = r$ **(d)** solid and $h = r$ where h is the cone height.

   In each case comment on the stability of the pen.

   How is the pen made to 'stand up' from any position?

2 Find the centre of mass of a solid hemisphere of radius r, joined to a solid cylinder radius r and height h.

   Find the value of h (in terms of r) for which, if h is less than this value, the body always 'rights' itself from any position with the curved surface in contact with a horizontal plane.

3 If **(a)** a hollow cone with $h = r$ **(b)** a solid cone with $h = r$ **(c)** a solid hemisphere **(d)** a hollow hemisphere, is suspended from a point on the rim of its circular base, what angle does the base make with the vertical?

4 I once saw a two-year-old drinking from a cup in the shape of a hollow cylinder with a solid hemispherical base which 'righted' itself if at all displaced from the upright position. Obviously the base was heavier than the cup (cylinder) but was this a good design?

5 Find the centre of mass of a lamina whose boundary is the parabola $y^2 = 4ax$ and the line $x = b$.

6 Find the centre of mass of a solid paraboloid with the dimensions of question **5**.

7 Find the centre of mass of a solid hemisphere of radius a with a similar hemisphere of radius b $(b < a)$ removed from its centre. (Both have the same axis of symmetry.) Deduce the centre of mass of a hollow hemisphere by letting $b \rightarrow a$.

8 Repeat question **7** for a solid cone of height H with a similar smaller cone of height h removed from its centre, both cones having the same base plane. Deduce the centre of mass of a hollow cone by letting $h \rightarrow H$.

9 Find the centre of mass of

   (a) letter W, made of wire

   (b) letter E, a lamina

   (c) letter F, a lamina.

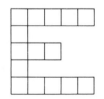

10 Find the resultant centre of mass when an isosceles triangle AEF is cut away from the square $ABCD$, of side $2a$.

   If $x = a$ when $\triangle AEF$ is cut away, the rest will not topple over but if $x = 2a$, clearly the $\triangle BCD$ left will topple over.

   Find the critical value of x (in terms of a) for which the shape remaining is just about to topple about F.

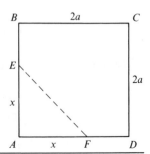

8 Work and power

Introduction

In Chapter 11 of Book 1 we considered the motion of a particle and the relationship between its acceleration, velocity, displacement and time in one dimension. In particular, we derived and used the equations of motion for constant acceleration:

$$v = u + at$$

$$v^2 = u^2 + 2as$$

$$s = ut + \tfrac{1}{2}at^2$$

$$s = \frac{u + v}{2}t$$

These can be applied when **u**, **v**, **a** and **s** are vectors.

We also investigated the effect of forces acting on bodies subject to Newton's laws of motion. The second law, $\mathbf{F} = m\mathbf{a}$, was applied to a variety of different situations.

In this chapter we shall take another look at moving bodies and consider the effect on these bodies of forces which act on them whilst they move through a specified displacement.

Work done by a force

If you push a car along a road we say that you do work. In general, **work is done when a force moves its point of application**.

If the force is increased or the distance moved is greater, then more work will be done.

We define the work done by a force, mathematically, as the product of the force and the distance it moves in the direction of the force.

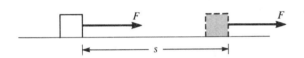

> The work done by a force is the product of the magnitude of the force and the distance it moves in the direction of the force.
>
> Work done $= F \times s$

In a more general situation, when the displacement is not in the direction of the force we have,

Work done $= F \times (AB \cos \theta)$

This can be written,

$\qquad = (F \cos \theta) \times AB$

Hence, the definition can be expressed in a different way, namely

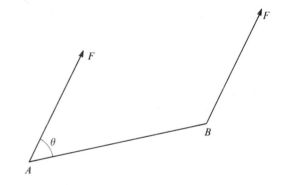

> The work done when a body moves under the action of a constant force, is the product of the component of the force in the direction of motion and the distance moved by the point of application of the force.

When the force is measured in Newtons and the distance in metres, the unit of work is the Newton-metre. This is called a **Joule (J)**.

Unit of work

One Joule is the amount of work done when a force of 1 Newton moves through a distance of 1 metre in the direction of the force.

WORKED EXAMPLES

---

1 A small object of mass 2 kg falls from a shelf which is 1.5 m above the floor. Find the work done by the weight of the object as it falls to the floor.

The weight of the object is $2g$ N and this force moves 1.5 m in the direction of the force.

$$\text{Work done} = \text{force} \times \text{distance moved in the direction of the force}$$
$$= 2g \times 1.5 \text{ N m} = 3g \text{ N m}$$
$$= 29.4 \text{ J}$$

---

2 A boy pulls a sledge through 20 m on horizontal ground by a rope inclined at 30° to the ground. Find the work done by the horizontal component of the tension in the rope if the magnitude of the tension is 8 N.

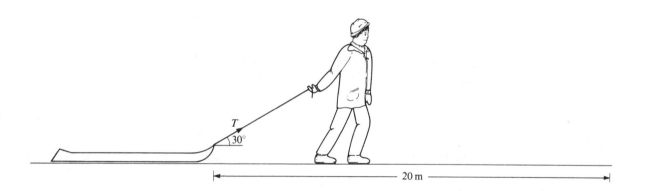

$$\text{Work done} = \text{component of force} \times \text{distance moved}$$
$$= (8 \cos 30°) \times 20 \text{ J}$$
$$= 80\sqrt{3} \text{ J} = 138.6 \text{ J}$$

Since the work done by a force is $(F \times s) \cos \theta$, where s is the distance moved and θ is the angle between the directions of F and s, it is possible to express the work done by a force as a scalar product.

In general, when a force $\mathbf{F}$ N moves its point of application through a displacement $\mathbf{r}$ m, the work done by $\mathbf{F}$ is given by

$$|\mathbf{F}| \cos \theta \times |\mathbf{r}|$$
$$= |\mathbf{F}||\mathbf{r}| \cos \theta = \mathbf{F} . \mathbf{r}$$

Thus,

$$\boxed{\text{Work done} = \mathbf{F} . \mathbf{r}}$$

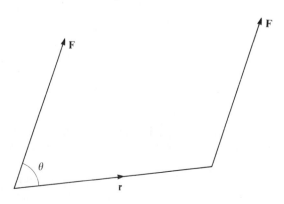

WORKED EXAMPLE

> A particle moves from a point A with position vector $\mathbf{i}+\mathbf{j}-2\mathbf{k}$ to a point B with position vector $2\mathbf{i}+3\mathbf{j}-\mathbf{k}$ under the action of a force $\mathbf{F}=5\mathbf{i}-2\mathbf{j}+\mathbf{k}$. Find the work done by the force.

The displacement $\quad \mathbf{AB}=\mathbf{r}=(2\mathbf{i}+3\mathbf{j}-\mathbf{k})-(\mathbf{i}+\mathbf{j}-2\mathbf{k})$

$$=\mathbf{i}+2\mathbf{j}+\mathbf{k}$$

$\therefore$ Work done $=\mathbf{F}.\mathbf{r}$

$$=(5\mathbf{i}-2\mathbf{j}+\mathbf{k}).(\mathbf{i}+2\mathbf{j}+\mathbf{k})$$

$$=5-4+1=2 \text{ J}$$ 　　　　　(assuming that the force is measured in Newtons and the displacement in metres.)

Zero or negative work

It is possible that the work done by a force may be negative or even zero.

A particle moves along a horizontal table through a distance s m. If the normal reaction is R N, then the work done by the reaction is

$$R\times s\cos 90^\circ = 0$$

A particle is pushed along a rough horizontal surface against a frictional force F N through s m.

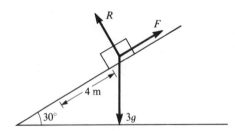

Work done by $F = F\times s\cos 180^\circ$

$$= -Fs$$

Investigation 1

A mass of 3 kg slides 4 m down a rough plane inclined at 30° to the horizontal. If the coefficient of friction is $\frac{1}{4}$, find the work done by　**(a)** the normal reaction　**(b)** the frictional force and　**(c)** the weight.

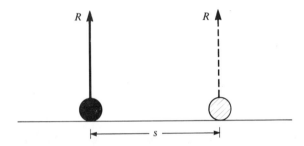

Now resolve the weight into its components parallel and perpendicular to the plane and show that the work done by these component forces is　$12g\sin 30^\circ$　(58.8 J)　and　0 J,　respectively.

Compare the sum of these results with your answer to part **(c)**.

Your working in Investigation 1 should have produced 0 J for part **(a)** and $-\dfrac{3\sqrt{3}}{2}g$ J $(-25.46$ J$)$ for part **(b)**.

When the work done is negative, we often say that the **work is done against the force**. So, in the Investigation, we say, 'the work done against friction is 25.46 J', instead of, 'the work done by the friction is -25.46 J'.

You should note from the Investigation that the work done by the weight in part **(c)** could be found from the sum of work done by any two components of the weight.

So, work done $= (3g\sin 30° \times 4) + (3g\cos 30° \times 0) = 3g \times 4\sin 30° = 58.8$ J

Total work

It is generally true that, **if a body is subject to a number of forces, the total work done by all the forces is the sum of the work done by each force.**

If a number of forces $F_1, F_2, \ldots F_n$ act on a body which moves so that the forces have displacements $r_1, r_2, \ldots r_n$ respectively, then

$$\text{the total work done} = F_1 . r_1 + F_2 . r_2 + \ldots F_n . r_n = \sum_{i=1}^{n} F_i . r_i$$

Special cases exist when the force is constant or when all the forces have the same displacement.

If a **constant force** moves through different displacements $r_1, r_2, \ldots r_n$ then,

$$\text{The total work done} = F . r_1 + F . r_2 + \cdots + F . r_n$$

$$= F . (r_1 + r_2 + \cdots + r_n) = F . \left(\sum_{i=1}^{n} r_i\right)$$

i.e. the scalar product of the force and the resultant displacement.

If a number of different forces $F_1, F_2, \ldots F_n$ all move through the **same displacement r** then,

$$\text{The total work done} = F_1 . r + F_2 . r + \cdots + F_n . r$$

$$= (F_1 + F_2 + \cdots + F_n) . r = \left(\sum_{i=1}^{n} F_i\right) . r$$

i.e. the scalar product of the resultant force and the displacement.

WORKED EXAMPLE

Three forces given by $F_1 = 2i + 3j + 4k$, $F_2 = -i + 2j + k$ and $F_3 = -2i - j - k$ act on a particle causing it to move from A to B whose position vectors are $i + j - k$ and $2i + 3j + k$, respectively. Find the total work done by the forces if the displacements are measured in metres and the forces in Newtons.

The displacement $AB = r_b - r_a$

$\therefore$ $AB = r = (2i + 3j + k) - (i + j - k)$

$= i + 2j + 2k$

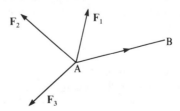

Work and power

The total work done is $(\mathbf{F_1} + \mathbf{F_2} + \mathbf{F_3}) . \mathbf{r}$

$$= [(2\mathbf{i} + 3\mathbf{j} + 4\mathbf{k}) + (-\mathbf{i} + 2\mathbf{j} + \mathbf{k}) + (-2\mathbf{i} - \mathbf{j} - \mathbf{k})] . (\mathbf{i} + 2\mathbf{j} + 2\mathbf{k})$$

$$= (-\mathbf{i} + 4\mathbf{j} + 4\mathbf{k}) . (\mathbf{i} + 2\mathbf{j} + 2\mathbf{k}) = -1 + 8 + 8 = 15 \text{ J}$$

Check that the same result is obtained by calculating the sum of the work done by each force.

i.e. work done $= \mathbf{F_1} . \mathbf{r} + \mathbf{F_2} . \mathbf{r} + \mathbf{F_3} . \mathbf{r}$

Exercise 8.1

1 A crate of mass 300 kg is raised by a crane to a height of 12 m. Find the work done by the crane against gravity.

2 A block of wood of mass 3 kg is pushed 2 m along a smooth table by a horizontal force of 4 N. Find the work done by

  (a) the pushing force (b) the normal reaction and (c) the weight of the block

3 A parcel of mass 2 kg lies on a work bench of length 2 m and is pushed from one end to the other. If the coefficient of friction is $\frac{1}{4}$, find the work done against friction.

4 Two forces $\mathbf{F_1} = 4\mathbf{i} + 2\mathbf{j} - \mathbf{k}$ and $\mathbf{F_2} = 3\mathbf{i} - \mathbf{j} - \mathbf{k}$ act on a particle which moves from a point A with position vector $\mathbf{i} + \mathbf{j} - 3\mathbf{k}$ to a point B with position vector $\mathbf{i} - \mathbf{j} + \mathbf{k}$. Find the work done by each force and their resultant, if the displacements are measured in metres and the forces in Newtons.

5 A man cycles 15 km against an average resistance of 20 N. How much work does he do against the resistance?

6 A particle of mass 2 kg lies on a plane inclined at $30°$ to the horizontal. The particle is pulled 6 m up the plane by a force of 15 N applied through a string which lies along the line of greatest slope of the plane. If the coefficient of friction is $\frac{1}{5}$, find the total work done by all the forces.

7 A particle is pulled along a rough horizontal surface at a constant speed of 3 m s^{-1} by a horizontal string in which there is a tension of 5 N. Find the work done by the tension in 4 seconds.

8 A climber of mass 75 kg reaches the top of a mountain 2300 m high. If he starts from sea level, find how much work he does against gravity.

9 Three forces, $\mathbf{F_1} = \mathbf{i} + \mathbf{j} + \mathbf{k}$, $\mathbf{F_2} = 2\mathbf{i} + 3\mathbf{j} - 4\mathbf{k}$ and $\mathbf{F_3} = \mathbf{i} - 4\mathbf{j} + 2\mathbf{k}$, act on a particle which moves from a point A with position vector $\mathbf{i} + \mathbf{j}$ to a point B with position vector $-\mathbf{i} - \mathbf{j} + \mathbf{k}$ and then to a point C with position vector $-\mathbf{i} + 2\mathbf{j} - 3\mathbf{k}$. Find the work done by (a) each force in moving from A to B (b) each force in moving from B to C and (c) the resultant of the forces in moving from A to B and from B to C.
  Assume that the forces are measured in Newtons and the displacements in metres.

10 A block is pulled up a smooth slope inclined at $\sin^{-1}\left(\frac{1}{28}\right)$ to the horizontal at a constant speed of 4 m s^{-1}. Find the mass of the block, if the work done against gravity in 2 seconds is $4g$ J.

Power

If a family saloon car and a sports car of the same weight climb a hill, then they do the same amount of work against gravity.

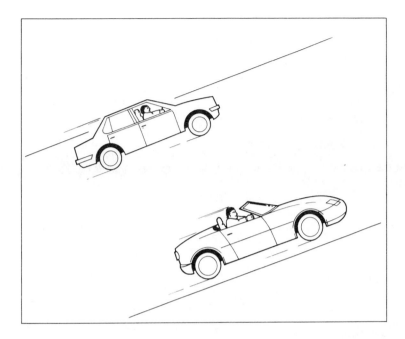

However, if the sports car gets to the top in a shorter time, then we say that this car has a **greater power** than the saloon car.

> Power is the rate at which a force does work.

Thus power is the rate of working, or, $\text{power} = \dfrac{\text{work done}}{\text{time taken}}$

Unit of power

Since work done is measured in Joules and time in seconds, the **unit of power** is a Joule per second (J s^{-1}), which we call a **Watt** (W).

WORKED EXAMPLE

A crane lifts a container of mass 250 kg through a height of 15 m. If the operation takes 20 seconds, find the power of the crane.

The weight of the container $= 250g$ N.

Hence, the work done by the crane against gravity $= 250g \times 15$ J $= 36\,750$ J

Thus, the power of the crane $= \dfrac{36\,750}{20} = 1837.5$ W

Larger units of power

When powerful engines are being used, large amounts of work are done in short intervals of time. It is convenient to use larger units of power.

$$1 \text{ kW} = 1 \text{ kilowatt} = 1000 \text{ Watts}$$

$$1 \text{ MW} = 1 \text{ megawatt} = 1000 \text{ kW} = 1\,000\,000 \text{ W}$$

The former unit of power was the horsepower, which was used to rate engines. One horsepower is equivalent to 746 W.

WORKED EXAMPLE

A car engine is rated at 7460 W. If the car, of mass 1200 kg, climbs a hill inclined at $\sin^{-1}(\frac{1}{40})$ to the horizontal at a steady speed, find the shortest time taken to travel 600 m up the slope, if the resistance to motion is 800 N. (Take g as 9.8 m s^{-1}.)

Let the pulling force of the engine be P N, the resistance to motion be F N and the normal reaction R N.

Since the car climbs the hill at a steady speed, there will be no resultant force on the car in the direction of motion.

$$P = F + 1200g \sin \alpha$$

$$= 800 + \frac{1200g}{40} = 1094 \text{ N}$$

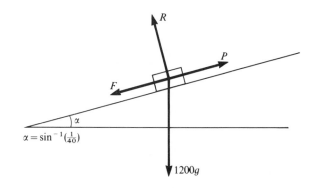

Since the normal reaction R and the component of weight $1200g \cos \alpha$ do no work,

the total work done by the engine $= 1094 \times 600 \text{ J} = 656\,400 \text{ J}$

If the car takes t s to travel 600 m then,

$$\text{Power} = \text{rate at which work is done} = \frac{656\,400}{t} \text{ J s}^{-1}$$

Hence, $$7460 = \frac{656\,400}{t} \quad \Rightarrow \quad t = \frac{656\,400}{7460} = 88 \text{ s}$$

Thus the time taken to travel 600 m up the slope is 88 s.

Investigation 2

Calculate the approximate power developed by a boy of mass 50 kg when he runs up a flight of 20 steps, each 24 cm high, in 3.7 s.

Show that the work done by the boy in gaining the vertical height is 2352 J. Hence, show that the power developed is 0.64 kW.

Why is this an approximate value? What factors have been overlooked?
Remember that this is a short-term estimate of the boy's power and such a value could not be maintained over a long interval of time.

Find a friend to time you while you run up some stairs, to find the power that you can develop.

Investigation 3

The concept of power can be applied to any machine which does work in converting energy from one form to another.

For example, a 100 W light bulb does 100 J of work per second in converting electrical energy to heat and light.

Find some typical values for the power ratings of the following,

(a) a kettle

(b) a sports car

(c) a modern electric locomotive

(d) the hose pump on a fire engine

(e) an aircraft engine

(f) a television set

(g) a space rocket

(h) a juggernaut engine

Think of some more everyday articles and find their power ratings. For example, an iron, an electric drill, a pump for a garden pond, etc.

The power of a moving vehicle

The Worked Example on page 170 considered a car climbing a hill and doing work in the process. We can form a general result for the power of a moving engine as follows.

Consider an engine working at P Watts and exerting a pulling force $\mathbf{F}$ N on the car. Assume that this force remains constant over a small interval of time δt and that the car moves through a displacement $\delta \mathbf{r}$ in this time.

The work done in time $\delta t \simeq \mathbf{F} . \delta \mathbf{r}$

Hence, the average rate of working $\simeq \dfrac{\mathbf{F} . \delta \mathbf{r}}{\delta t}$

Taking the limit as $\delta t \rightarrow 0$ will give an expression for the power at time t.

$$\therefore \qquad \text{Power} = \lim_{\delta t \to 0} \left[\frac{\mathbf{F} . \delta \mathbf{r}}{\delta t} \right] = \mathbf{F} . \left[\lim_{\delta t \to 0} \frac{\delta \mathbf{r}}{\delta t} \right] = \mathbf{F} . \mathbf{v}$$

where $\mathbf{v}$ is the velocity of the car at time t.

Hence, if the point of application of a force, $\mathbf{F}$, is moving with velocity $\mathbf{v}$, then

$$\boxed{\begin{array}{c} \text{The power developed} = \mathbf{F} . \mathbf{v} \\ \text{or,} \qquad P = \mathbf{F} . \mathbf{v} \end{array}}$$

Notice that if the force and the velocity are in the same direction, the result becomes

$$\boxed{\text{Power} = Fv}$$

It is clear that if the velocity is not constant, then this result gives the power at a particular instant of time and this will vary with the velocity.

> WORKED EXAMPLES ⟩

1 A force $\mathbf{F} = 3\mathbf{i} + \mathbf{j} + 2\mathbf{k}$ is applied to a mass of 2 kg which is free to move on a smooth horizontal surface. If the mass starts with a velocity $\mathbf{i} + \mathbf{j}$, find the velocity of the mass at time t s and hence deduce the power of the force $\mathbf{F}$ at time t s.

Using Newton's second law, $\mathbf{F} = m\mathbf{a}$, gives

$$3\mathbf{i} + \mathbf{j} + 2\mathbf{k} = 2\mathbf{a} \quad \Rightarrow \quad \mathbf{a} = \tfrac{1}{2}(3\mathbf{i} + \mathbf{j} + 2\mathbf{k})$$

Since the acceleration is constant, we can use $\mathbf{v} = \mathbf{u} + \mathbf{a}t$ to determine the velocity at time t.

$$\therefore \qquad \mathbf{v} = \mathbf{i} + \mathbf{j} + \tfrac{1}{2}(3\mathbf{i} + \mathbf{j} + 2\mathbf{k})t = (1 + \tfrac{3}{2}t)\mathbf{i} + (1 + \tfrac{1}{2}t)\mathbf{j} + t\mathbf{k}$$

Now the power at time t s is $\mathbf{F} . \mathbf{v}$

$$\therefore \qquad \text{Power} = (3\mathbf{i} + \mathbf{j} + 2\mathbf{k}) . [(1 + \tfrac{3}{2}t)\mathbf{i} + (1 + \tfrac{1}{2}t)\mathbf{j} + t\mathbf{k}] = (3 + \tfrac{9}{2}t) + (1 + \tfrac{1}{2}t) + 2t$$

$$\therefore \quad \text{Power} = 4 + 7t$$

2 A car has a maximum speed of 180 km h^{-1} on a level road when the resistances amount to 820 N. Find the maximum power developed by the engine.

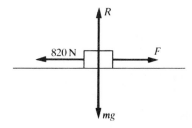

At maximum speed there is no acceleration and hence the resultant force in the direction of motion is zero. Thus, the pulling force of the engine, F, is 820 N.

Since F is constant, the maximum power of the engine will occur when the speed is greatest.

Now,   180 km h^{-1} = $\dfrac{180 \times 1000}{3600}$ m s^{-1} = 50 m s^{-1}

Using   Power = $F \times v$,   we have

Power = 820×50 W = 41 kW

The maximum power of the engine is 41 kW.

3 A car of mass 1200 kg can maintain a steady speed of 30 m s^{-1} when travelling up a hill inclined at $\sin^{-1}(\frac{1}{20})$ to the horizontal, against resistances which are proportional to the speed and when the engine is working at a constant rate of 60 kW.

Find the acceleration of the car when it descends this hill at 25 m s^{-1} if the engine is then working at half its original power.

Let the pulling force of the engine be F N and the resistances at speed v m s^{-1} be R N.

Since the resistances are proportional to the speed, it follows that

$$R = kv \qquad \text{where } k \text{ is a constant}$$

$$\Rightarrow \quad R = 30k \qquad\qquad \text{............ (1)}$$

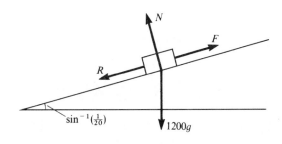

Work and power

Now, since the speed is constant, there is no resultant force in the direction of motion.

Thus,   Pulling force = Resistances + Weight component down the slope

$$\therefore \quad F = R + 1200g \times \left(\tfrac{1}{20}\right)$$

$$\Rightarrow \quad F = R + 60g \qquad\qquad\qquad\qquad\qquad \text{............ (2)}$$

Using Equation **(1)** and substituting for R, we have,

$$F = 30k + 60g \qquad\qquad\qquad\qquad\qquad \text{............ (3)}$$

Since   Power $= F \times v$

$$60\,000 = F \times 30 \quad \Rightarrow \quad F = 2000 \text{ N}$$

Substituting into Equation **(3)** gives,

$$2000 = 30k + 60g$$

$$\Rightarrow \quad 30k = 2000 - 60g \quad \Rightarrow \quad k = \tfrac{1412}{30}$$

Now consider the case when the car descends the hill. The resistances act up the slope. Let the pulling force of the engine be F_1 and the resistances be R_1 N.

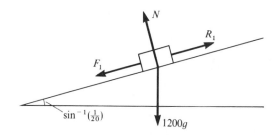

Since   Power $= F_1 \times v$

$$30\,000 = F_1 \times 25 \quad \Rightarrow \quad F_1 = 1200 \ N$$

In this case, the forces along the slope do not balance since the car has an acceleration.

The effective force down the slope is given by

$$F_1 + \left[(1200g \times \left(\tfrac{1}{20}\right)\right] - R_1 = 1200 + 60g - R_1 \qquad \text{............ (4)}$$

Since the resistances are proportional to the speed,

$$R_1 = k_1 v \quad \Rightarrow \quad R_1 = \frac{1412}{30} \times 25 = 1176.7 \text{ N}$$

Substituting into Equation **(4)** gives,

$$\text{Effective force down the slope} = 1200 + 60g - 1176.7 = 611.3 \text{ N}$$

Using Newton's second law, $F = ma$,

$$611.3 = 1200a \quad \Rightarrow \quad a = 0.51 \text{ m s}^{-1}$$

The acceleration of the car down the slope when the speed is 25 m s^{-1} is 0.51 m s^{-1}.

Investigation 4

A car of mass 1000 kg starts from rest on a level road and accelerates at 1.5 m s^{-2} for 12 seconds. If the resistances to motion are 400 N, find the power developed by the engine of the car at 2 s intervals and display your results on the table below.

Time (s)	2	4	6	8	10	12
Speed (m s^{-1})	3					
Power (kW)	5.7					

If, instead, the same car moves from rest with the engine working at a constant rate of 20 kW, show that the pulling force of the engine, D N, is given by $\dfrac{20\,000}{v}$ where v is the speed of the car in m s^{-1}.

Draw a graph showing how D varies with v.

Show that the equation of motion is given by $\dfrac{dv}{dt} = \dfrac{20}{v} - \dfrac{2}{5}$.

Find the maximum speed attained by the car.

Exercise 8.2

1 A train has a maximum speed of 180 km h^{-1} when travelling on level track against resistances of 30 kN. Find the power developed by the engine.

2 A car has a maximum power output of 42 kW. Find the greatest speed of the car on a level road if the resistances to motion are 1400 N.

3 Find the power of an engine which would pull a train of mass 180 Mg up a slope of inclination $\sin^{-1}\left(\frac{1}{120}\right)$ at a uniform speed of 60 km h^{-1}, if the resistance due to friction etc. is $3000g$ N.

4 A particle of mass 2 kg moves under the action of a force $\mathbf{F} = \mathbf{i} + 2\mathbf{j}$. Find an expression for the velocity, $\mathbf{v}$, after time t seconds, given that $\mathbf{v} = \mathbf{j} + \mathbf{k}$ when $t = 0$. Hence deduce the power of the force $\mathbf{F}$ at time t seconds.

5 A particle of unit mass is acted upon by a force $\mathbf{F} = 4\mathbf{i} + 2t\,\mathbf{k}$ at time t seconds. If the velocity of the particle is $2\mathbf{i} + \mathbf{j}$ when $t = 0$, find expressions for the velocity and the power of $\mathbf{F}$ at time t s.

6 A cyclist working at 250 W descends a hill inclined at $\sin^{-1}\left(\frac{1}{25}\right)$ to the horizontal. If he can maintain a steady speed of 30 km h^{-1}, find the resistance to motion due to friction if the mass of the cyclist and his bicycle is 100 kg.

7 A car of mass 1000 kg accelerates at 1.2 m s^{-2} up a slope inclined at $\sin^{-1}\left(\frac{1}{50}\right)$ to the horizontal. If the resistance due to friction is 400 N, find the power developed by the car when its speed is 36 km h^{-1}.

8 An engine working at 800 kW pulls a train of mass 200 tonnes up a slope of inclination $\sin^{-1}\left(\frac{1}{150}\right)$, against frictional resistances of 45 N per tonne. Find the greatest uniform speed that it can maintain up the slope.
   If, when it reaches the top of the slope, the track becomes level, find the acceleration of the train at that instant if it maintains the same power output.

9 A car of mass 1200 kg has a maximum speed of 15 m s^{-1} up a slope inclined at sin^{-1} ($\frac{1}{10}$) to the horizontal and a maximum speed of 40 m s^{-1} when travelling down the same slope with the same power developed by the engine. Given that the resistance varies as the speed, find the maximum power developed by the car.

10 If a van of mass 5000 kg experiences frictional resistances of 800 N when climbing a hill inclined at sin^{-1} ($\frac{1}{20}$) to the horizontal at a maximum speed of 24 km h^{-1}, find the power of the engine of the car.
    If the resistance varies as the square of the speed, find the maximum speed of the car on a level road.

11 A train of mass 120 Mg coasts down a slope of inclination sin^{-1} ($\frac{1}{40}$) with the steam shut off. If it maintains a constant speed of 144 km h^{-1} down the slope, and the frictional resistances are proportional to the speed, find the maximum speed of the train up the slope if the engine can develop 90g kW.

12 A train of mass 240 tonnes travels up an incline whose gradient is sin^{-1} ($\frac{1}{120}$). If the engine develops 400 kW and the resistance is 80 N per tonne, find the acceleration of the train at the instant when the speed is 18 km h^{-1}.
    What is the maximum speed of the train up the slope if the engine continues to develop the same power?
    What is the maximum speed down a slope of inclination sin^{-1} ($\frac{1}{240}$), if the resistances remain the same and the engine works at one-half of the original power?

Work done by a variable force

When a force **F** varies in magnitude or direction, it is still possible to calculate the work done by the force. In such cases we consider the path over which the point of application of the force moves as a series of small displacements, $\delta\mathbf{r}_1, \delta\mathbf{r}_2, \ldots, \delta\mathbf{r}_n$.

Let the curve between A and B be divided into n parts by points $P, Q, R, \ldots$.

Let $\delta\mathbf{r}_1, \delta\mathbf{r}_2, \ldots$ represent the displacements **AP, PQ**, ...

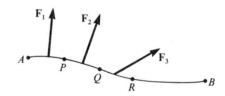

If we assume that the force **F** takes constant values $\mathbf{F}_1, \mathbf{F}_2, \ldots$ over these intervals then,

$$\text{Work done by } \mathbf{F} \simeq \mathbf{F}_1 . \delta\mathbf{r}_1 + \mathbf{F}_2 . \delta\mathbf{r}_2 + \cdots + \mathbf{F}_n . \delta\mathbf{r}_n$$

$$= \sum_{i=1}^{n} \mathbf{F}_i . \delta\mathbf{r}_i$$

In the limit, as $n \to \infty$, $\delta\mathbf{r}_i \to 0$ and thus,

$$\text{Work done by } \mathbf{F} = \lim_{n \to \infty} \sum_{i=1}^{n} (\mathbf{F}_i . \delta\mathbf{r}_i) = \int_{A}^{B} \mathbf{F} . d\mathbf{r}$$

The integral is known as a line integral along the curve and the evaluation of the integral is beyond the scope of the Advanced level syllabus unless the path AB is a straight line.

So, in the case when the direction of the force **F** is in the same direction as the displacement, i.e. along $d\mathbf{r}$,

> **The work done by F through a displacement from x_1 to x_2 is**
> $$\int_{x_1}^{x_2} |\mathbf{F}| \, dx$$

WORKED EXAMPLE

> Find the work done when a particle is moved through a displacement of 5 m along a smooth horizontal surface by a force **F** of magnitude $(8 - \frac{1}{2}x)$ N, where x m is the displacement of the particle from its initial position.

Since the force acts along the direction of motion we have,

$$\text{Work done} = \int_0^5 |\mathbf{F}|\, dx$$

$$= \int_0^5 (8 - \tfrac{1}{2}x)\, dx = \left[8x - \tfrac{1}{4}x^2 \right]_0^5 = 40 - 6.25 = 33.75 \text{ J}$$

Elastic strings and springs

A common example is the movement of an elastic string since, when a force is applied, it causes the string to change its length and creates a variable tension in the string.

A spring is similar to an elastic string except that it can also be compressed.

So, if a mass of 2 kg is attached to an elastic string and hung vertically, the string will increase its length and work will be done. Similarly, a boy with chest expanders will have to do a considerable amount of work if he is to stretch them a significant distance. On the other hand, the spring inside a ball-point pen can be compressed.

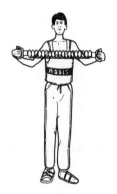

If the string or spring is not stretched or compressed, we say that it is at its **natural length.** Remember that the tension in a stretched elastic string acts towards the centre from each end, but in a compressed spring it acts towards the ends, and is usually called a **thrust**.

The figures show how the tension acts in a stretched elastic string and a compressed spring.

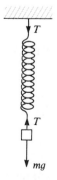
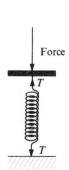

Investigation 5

In an experiment, various masses are hung on the end of an elastic spring which is hanging vertically with its other end attached to a beam. Use the data given to plot a graph to show how the extension (in cm) varies with the weight of the masses (in Newtons) called the **load**.

Load (N)	0	4	8	12	16	20	24	28
Extension (cm)	0	1	2	3	4	5.8	8.2	14.1

What do you notice about the graph for a load of less than 16 N? What happens as the load is increased above 16 N?

Discuss what might happen to the spring if the maximum load applied was

(a) 16 N   **(b)** 24 N   **(c)** 35 N.

Hooke's law

The observations from Investigation 5 should have led you to conclude that,

> When a load is applied to an elastic string or spring, the extension is proportional to the load, provided the elastic limit is not exceeded.

This fact was discovered by Robert Hooke in the seventeenth century when much research was being undertaken into the strength of materials used for building. This result is known as **Hooke's law** and applies equally to the loading of horizontal beams fixed at one end.

The law is the result of observation and cannot be proved mathematically.

In general, graphs produced from experiments similar to the one in Investigation 5 will follow a standard pattern and will show several critical points.

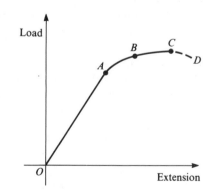

A is called the **elastic limit** and, if the string is not loaded beyond this point, the string will return to its original length when the load is removed.

If the string is loaded beyond this point, then it may stretch permanently and not return to its original length when it is unloaded.

B is called the **yield point**, where an increase in load produces a much larger extension. Point *C* defines the maximum load that the string can support, and *D* will be the breaking point.

For OA, the relationship can be written

$$\text{Load} \propto \text{extension}$$

or $\quad \text{Load} = k \times (\text{extension}) \quad$ where k is a constant

In equilibrium, the load will equal the tension in the string,

$$\text{Tension} = k \times (\text{extension}) = kx$$

We use the fraction $\dfrac{\lambda}{l}$ as the constant k where l is the natural length of the string and λ is the **modulus of elasticity** of the string.

Hence, **Hooke's law** can be written as $\quad \boxed{T = \dfrac{\lambda x}{l}}$

The modulus of elasticity of the string is a measure of the stiffness of the string. If the extension is equal to the natural length we have,

$$T = \frac{\lambda x}{l} \quad \Rightarrow \quad T = \frac{\lambda l}{l} \quad \Rightarrow \quad T = \lambda$$

> λ is a constant for a given elastic string and is equal to the tension that would double the length of the string. It is independent of the length of the string.
>
> The units for the modulus are those of force, i.e. Newtons.

WORKED EXAMPLES

1   A light elastic string of natural length 0.7 m is fixed to the ceiling and hangs vertically with a mass of 4 kg attached to the other end. If the modulus of elasticity of the string is 48 N, find the extension of the string when the mass is in equilibrium.

Let the extension of the string be x m in the equilibrium position.

Since there is equilibrium, the tension in the string will equal the weight of the mass.

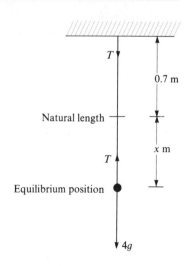

Hence, $\quad T = 4g$

Now, $\quad T = \dfrac{\lambda x}{l} \quad$ from Hooke's law

$\Rightarrow \quad 4g = \dfrac{48x}{0.7}$

$\Rightarrow \quad x = \dfrac{4g \times 0.7}{48} = 0.57$

Thus, the extension of the string is 0.57 m.

179

2 An elastic string of natural length $2l$ and modulus of elasticity $3g$ N is stretched between two points A and B on the same horizontal level, where $AB = 3l$. A particle of mass m kg is attached to the mid-point C of the string and allowed to rest in equilibrium with the portions of the string making an angle of $25°$ with the horizontal. Find the mass of the particle.

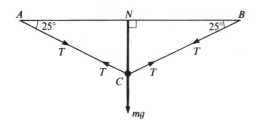

Since the mass is attached to the string, we can regard the mass as being supported by two identical strings of natural length l and modulus of elasticity $3g$.

Also, by considering the horizontal components of the tensions at C, it is clear that the tension in the portions AC and BC are equal.

Resolving vertically at C,

$$2T \sin 25° = mg \quad \Rightarrow \quad T = \tfrac{1}{2}mg \operatorname{cosec} 25° = 1.18\, mg$$

Considering the string BC, we have, using triangle BCN,

$$BC = 1.5l \sec 25°$$

Since the natural length is l, the extension is

$$1.5l \sec 25° - l = 0.655l$$

Using Hooke's law, $\quad T = \dfrac{\lambda x}{l}$

$$\therefore \quad 1.18mg = \frac{3g(0.655l)}{l} \quad \Rightarrow \quad m = \frac{3 \times 0.655}{1.18} = 1.66$$

Thus, the mass of the particle is 1.66 kg.

3 Two identical elastic strings, AB and BC, of natural length a and modulus of elasticity $4mg$, are fastened together at B and their other ends fixed to two points $6a$ apart in a vertical line with A above C. If a particle of mass $3m$ is attached at B, find the distance AB when the particle is in equilibrium.

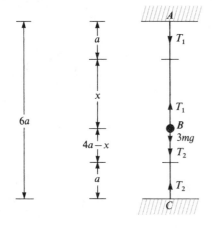

Let the tensions in the strings AB and BC be T_1 and T_2 respectively, as shown.

If the extension in AB is x, then the extension in BC is

$$(6a - 2a - x) = (4a - x)$$

Since the particle is in equilibrium,

$$T_1 = T_2 + 3mg \qquad \dots\dots (1)$$

Using Hooke's law for AB gives,

$$T_1 = \frac{\lambda x}{a} = \frac{4mgx}{a} \qquad \dots\dots (2)$$

Using Hooke's law for BC gives,

$$T_2 = \frac{\lambda(4a-x)}{a} = \frac{4mg(4a-x)}{a}$$ .... (3)

Substituting Equations (2) and (3) into Equation (1) we have,

$$\frac{4mgx}{a} = \frac{4mg(4a-x)}{a} + 3mg \quad \Rightarrow \quad \frac{4x}{a} = \frac{4(4a-x)}{a} + 3$$

$$\Rightarrow \quad 4x = 4(4a-x) + 3a \quad \Rightarrow \quad 8x = 19a \quad \Rightarrow \quad x = \frac{19a}{8}$$

Hence, the extension of the string AB is $2\frac{3}{8}a$ and the particle lies $3\frac{3}{8}a$ below A.

Exercise 8.3

1 In an experiment, different masses were attached to an elastic spring and the total length of the spring measured. Draw a graph to show the relation between the load and the total length. Comment upon your results.

Load (N)	0	5	10	15	20	25	30	35	40	41
Length (mm)	50	50.1	50.2	50.3	50.4	50.5	50.6	50.8	51.2	51.4

2 An elastic string of natural length 2.5 m is stretched to 3.1 m. If the modulus of elasticity is 18 N, find the tension in the string.

3 An elastic string of natural length 3 m is fixed at one end and a mass of 0.5 kg is attached to the other end. It hangs vertically in equilibrium. What is the total length of the string if it has a modulus of elasticity of 15 N?

4 A spring of natural length 20 cm is compressed to a length of 16 cm. Find the thrust in the spring if its modulus of elasticity is 10 N.

5 A mass of weight 10 N is hung from one end of an elastic string whose other end is fixed. If the string has a modulus of elasticity of 20 N and the length of the string when stretched is 1.8 m, find its natural length.

6 A mass is pulled along a smooth horizontal surface by a horizontal force $\mathbf{F}$ of magnitude $\left(1 - \dfrac{x}{4}\right)$ N, where x m is the displacement of the mass from its starting point. If the mass moves through 6 m, find the total work done.

7 A boy pulls a sledge 100 m along level ground by a force $\mathbf{F}$ of magnitude $\left(8 - \dfrac{x}{10}\right)$ N, where x is the distance of the sledge from its starting point. If the direction of $\mathbf{F}$ is inclined at $60°$ to the horizontal, find the work done by $\mathbf{F}$.

8 A mass of 2 kg rests on a smooth plane inclined at $45°$ to the horizontal. It is kept in equilibrium by a light elastic string of natural length 1.2 m attached to the mass and to a point on the plane. If the modulus of elasticity of the string is 50 N and it lies parallel to the plane along the line of greatest slope, find its extension.

9 An elastic string of natural length 2 m has one end fixed to the ceiling and hangs vertically in equilibrium. It supports a mass of 3 kg at its other end, 3.2 m below the ceiling. If a horizontal force of 15 N is then applied to the mass, find the extension of the string when the mass is again in equilibrium.

10 A particle of mass m is attached to an elastic string of natural length l at its mid-point. The ends of the string are fixed to two points a distance $2l$ apart on the same horizontal level. If the mass rests in equilibrium with the two portions of the string each making an angle θ with the vertical, find the modulus of elasticity.

11 A particle of mass 2 kg is attached to the mid-point of an elastic string of modulus of elasticity 15 N and natural length 1.5 m. If the other ends of the string are fixed to the ceiling and the floor of a room of height 3 m, find the distance of the particle from the ceiling when it is in equilibrium.

The work done in stretching an elastic string

If a force is applied to one end of an elastic string whose other end is fixed, then the string will extend and create a varying tension in the string which, in equilibrium, will be equal to the pulling force.

Consider an elastic string fixed at one end, O, resting in equilibrium with an extension x beyond its natural length l.

If the string is extended a further distance δx, from P to Q, we can assume that the tension remains constant provided that δx is small.

Hence, the work done by the pulling force over this interval as $\delta x \to 0$ is $T\,dx$.

Thus, the work done in stretching the string from x_1 to x_2 is

$$\int_{x_1}^{x_2} T\,dx = \int_{x_1}^{x_2} \frac{\lambda x}{l}\,dx = \left[\frac{\lambda x^2}{2l}\right]_{x_1}^{x_2} = \frac{\lambda}{2l}(x_2{}^2 - x_1{}^2)$$

$$= \frac{\lambda}{2l}(x_2 + x_1)(x_2 - x_1)$$

If T_1 and T_2 represent the tensions at extensions x_1 and x_2 respectively, then

$$T_1 = \frac{\lambda x_1}{l} \quad \text{and} \quad T_2 = \frac{\lambda x_2}{l}$$

> The work done in stretching an elastic string from x_1 to x_2 is
>
> $$\tfrac{1}{2}(T_1 + T_2)(x_2 - x_1) \quad \text{or} \quad \frac{\lambda}{2l}(x_2{}^2 - x_1{}^2)$$

If $x_1 = 0$, we can deduce that the work done in stretching an elastic string from its natural length to an extension x, is given by the following result.

> The work done in stretching an elastic string by an extension x, is $\dfrac{\lambda x^2}{2l}$.

Investigation 6

Show that the tension in an elastic string which is extended a distance x can be represented by a straight line graph of gradient $\dfrac{\lambda}{l}$.

If T_1 and T_2 represent the tensions at extensions x_1 and x_2 respectively, deduce that the work done when the extension is increased from x_1 to x_2 is given by the area under the graph.

Hence show that the work done
$$= \tfrac{1}{2}(T_1 + T_2)(x_2 - x_1)$$
$$= \frac{\lambda}{2l}(x_2{}^2 - x_1{}^2) \quad \text{as before.}$$

Consider an elastic string of natural length 0.4 m which extends 0.01 m under the action of a force of 12 N. If one end of this string is fixed draw a graph to show that the work done in stretching the string to a length of 0.5 m is 60 J.

WORKED EXAMPLE

An elastic spring of natural length 2 m is fixed at one end and stretched from its natural length by a force of 15 N applied at the other end until equilibrium is maintained. Find the modulus of elasticity, if the work done is 60 J.

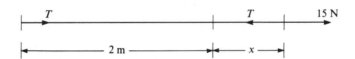

Let the extension in the equilibrium position be x m. The tension in the string will be 15 N.

Using Hooke's law, $\quad T = \dfrac{\lambda x}{l}$

$$\therefore \quad 15 = \frac{\lambda x}{2} \quad \Rightarrow \quad \lambda x = 30 \qquad \text{............ (1)}$$

Now, the work done in stretching the spring from its natural length to an extension x is $\dfrac{\lambda}{2l}x^2$.

Hence, $$60 = \frac{\lambda}{4}x^2 \quad \Rightarrow \quad \lambda x^2 = 240 \qquad \text{............ (2)}$$

Dividing Equation (2) by Equation (1) gives,

$$\frac{\lambda x^2}{\lambda x} = \frac{240}{30} \quad \Rightarrow \quad x = 8$$

Using Equation (1) and substituting $x = 8$ gives $\quad 8\lambda = 30 \quad \Rightarrow \quad \lambda = 3.75$.

Thus the modulus of elasticity is 3.75 N.

Exercise 8.4

1 An elastic string of natural length 4 m and modulus of elasticity 10 N is stretched by a force of 6 N until equilibrium is reached. Find the extension of the string and the work done.

2 An elastic string of natural length 0.4 m is stretched 5 cm by a mass of 2 kg when it is hanging vertically from a fixed point. Find the work done in stretching the string from a length of 0.5 m to 0.55 m.

3 A spring of natural length 1.8 m will stretch 2 cm under the action of a force of 8 N. Find the work done in stretching the string from **(a)** 1.8 m to 1.9 m and **(b)** 2 m to 4.1 m.

4 An elastic spring of natural length 2 m is fixed at one end and is slowly stretched from its natural length by a force of 20 N applied to the other end. The work done in reaching the equilibrium position is 25 J. Find the modulus of elasticity of the spring.

5 An elastic string of modulus of elasticity λ and natural length l is fixed at one end and has a mass m attached at the other end. Find the work done when the string is hanging vertically and it is stretched from its natural length to a position of equilibrium, a distance $(3 + l)$ below the fixed end.

6 Prove that the work done in stretching an elastic string of modulus of elasticity λ and natural length l a distance x, is $\lambda x^2/2l$.

   A spring of natural length $\frac{3}{2}l$ and modulus of elasticity mg is compressed to a length of l. Find the work done. If a second spring of natural length a and modulus of elasticity $2mg$ is compressed to a length l, find the natural length of this spring in terms of l, if the work done remains the same.

9 Energy

Introduction

Anything which is capable of doing work, is said to possess **energy**.

> Energy is the capacity for doing work.

Thus energy has the same units as work, i.e. **Joules**. Energy exists in many different forms.

Chemical energy is stored in the food we eat and provides the energy we need in our daily lives.

A burning fire provides heat, light and sound energy.

A person playing a violin produces sound energy.

What other forms of energy are illustrated on the next page?

185

You should have mentioned nuclear and electrical energy, elastic potential energy, kinetic energy and gravitational potential energy.

One of the laws of physics states that, 'The amount of energy in the universe is constant'. It cannot be created or destroyed but may be converted from one form to another. So when we strike a match we convert chemical energy into heat, light and sound energy. A television set also converts electrical energy into heat, light and sound energy, and a falling stone converts potential energy into kinetic energy.

In mathematics, we concentrate on one of the forms of energy, known as **mechanical energy**, which includes just **potential** and **kinetic** energy. This is the energy due to motion, the energy due to the position of a body, and the energy stored in an elastic string or spring.

Kinetic energy

> The kinetic energy (K.E.) of a body is its capacity to do work by virtue of its motion.

We measure this by calculating the work done by a force in bringing the body from rest to a velocity v.

Consider a particle of mass m subject to a constant force F. The acceleration, a, is given by $F = ma$.

If it starts from rest and reaches a velocity v in a distance s, then using $v^2 = u^2 + 2as$, we have,

$$v^2 = 0 + 2as \quad \Rightarrow \quad v^2 = 2as$$

So, as the work done is $Fs = mas = \frac{1}{2}mv^2$, we conclude that,

> The kinetic energy of a body of mass m, moving with velocity v, is $\frac{1}{2}mv^2$.

Note that m and v are always positive and hence the kinetic energy is always positive and does not depend upon the direction of motion of the body.

Gravitational potential energy

> The gravitational potential energy (P.E.) of a body is the energy it possesses due to its position.

If the body is in such a position that, if it were released from rest it would move, then it possesses **gravitational potential energy**. A hydro-electric power station makes good use of the potential energy of the water.

We measure the **potential energy** by calculating the work done in moving from an actual position to a standard position.

If a mass m is at a height h above a standard level (often ground level), then the work done in moving from its present position to the standard level is mgh.

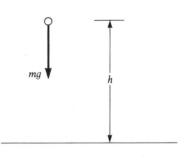

> The gravitational potential energy of a body of mass m, at a height h, is mgh.

Note that if the body is below the standard level, the potential energy will be negative since $h < 0$. In answering questions, it must be made clear where the standard level (i.e. the level of zero potential energy) is, since it can be chosen arbitrarily.

Elastic potential energy

An elastic string can be stretched and will store energy. A catapult makes use of this energy which is known as **elastic potential energy**.

> The elastic potential energy of an elastic string is stored by the string when it is stretched.

Remember that a spring can be compressed as well as stretched. This energy is measured by the work done in stretching an elastic string to an extension, x, from its natural length, l. This was deduced on page 182 of Chapter 8 and produced the expression $\dfrac{\lambda x^2}{2l}$.

> The elastic potential energy stored in an elastic string of natural length l, when it has an extension x, is $\dfrac{\lambda x^2}{2l}$.

WORKED EXAMPLES

1 A mass of 15 g slides from rest down a smooth slide of length 4 m inclined at 30° to the horizontal. Find the loss in potential energy and the gain in kinetic energy experienced by the mass.

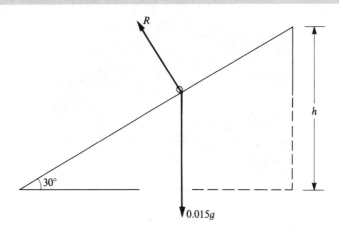

Taking the bottom of the slide as the zero level of potential energy of the mass, the potential energy of the mass at the top of the slide is *mgh*. Note the use of the vertical height for *h*.

P.E. at the top of the slide $= 0.015g \times 4 \sin 30° = 0.294$ J

The loss in P.E. of the mass $= 0.294$ J

To calculate the kinetic energy, we must find the speed of the mass after sliding 4 m down the slope under the action of its weight component, $0.015g \sin 30°$.

Using Newton's second law of motion, $F = ma$, we have

$$0.015g \sin 30° = 0.015a \quad \Rightarrow \quad a = \tfrac{1}{2}g$$

The acceleration down the slope is $\frac{1}{2}g = 4.9$ m s^{-2}.

Using $v^2 = u^2 + 2as$ gives,

$$v^2 = 0 + 2(4.9) \times 4$$
$$v^2 = 39.2 \quad \Rightarrow \quad v = 6.26 \text{ m s}^{-1}$$

The speed at the bottom of the slope is 6.26 m s^{-1}.

The kinetic energy at the foot of the slope is $\frac{1}{2}mv^2$.

Hence, the K.E. $= \frac{1}{2}(0.015) \times (6.26)^2 = 0.294$ J

Thus, the gain in K.E. of the mass is 0.294 J.

---

2 An elastic string of natural length 2 m and of modulus $3g$ N is suspended from the ceiling and a mass of 2 kg is attached at the other end. Find the elastic potential energy stored in the string when it rests in equilibrium.

Let the string rest in equilibrium with an extension x beyond the natural length of the string.

In equilibrium, the tension in the string will be $2g$ N.

Using Hooke's law, $\quad T = \dfrac{\lambda x}{l}$

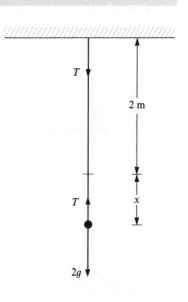

Hence, $\qquad T = \dfrac{3gx}{2} \quad \Rightarrow \quad x = \dfrac{4}{3}$ m

The elastic potential energy in the string is given by $\dfrac{\lambda x^2}{2l}$.

Elastic P.E. $= \dfrac{1}{2}\left(\dfrac{3g}{2}\right)\left(\dfrac{4}{3}\right)^2 = 13.07$ J.

Exercise 9.1

1 What forms of energy are stored in the following examples?

   (a) a biscuit   **(b)** a shell in flight   **(c)** a stretched bow   **(d)** a torch battery   **(e)** a rotating flywheel.

2 What changes of energy take place when,

   (a) using your mouth to blow up a balloon   **(b)** a steam engine moves along the track   **(c)** a stone is dropped from a tower   **(d)** a boy slides down a slide.

3 A book of mass 1 kg rests on a shelf 1.5 m above the floor. Calculate its potential energy relative to the floor level.

4 A body of mass 8 kg moves with a speed of 5 m s^{-1} and is brought to rest by a constant force in 10 m. Find the loss in kinetic energy of the body.

5   A pendulum of length 3 m is pulled aside until the string makes an angle of 25° with the vertical. If the bob has a mass of 1.5 kg, find the potential energy of the bob relative to the lowest position of the bob.

6   An elastic string is suspended from the ceiling and supports a mass of 5 kg in equilibrium. If the modulus of the string is $2g$ N and its natural length is 0.7 m, find the elastic potential energy stored in the string in the equilibrium position.

7   Calculate the gravitational potential energy of,

(a) a mass of 5.2 kg raised to a height of 5 m
(b) a mass of 100 kg at a height of 30 cm.

8   A body of mass 2 kg slides 3 m down a smooth slope from rest. If the slope is inclined at 20° to the horizontal, find the kinetic energy of the body when it reaches the foot of the slope.

9   A particle of mass 2 kg is attached to an inextensible string of length 1.5 m. If the string rotates in a vertical circle, find the potential energy of the mass at the highest and lowest positions of the mass   (a) relative to the centre of the circle and   (b) relative to the lowest point of the circle.

10   A pebble of mass 2 kg is thrown so that it just clears the top of a breakwater 1.5 m high with a speed of 3.2 m s^{-1}. Calculate the total mechanical energy of the pebble as it passes over the breakwater.

Continuous change of energy

When water is raised from a reservoir or is pumped from a hosepipe the change in energy that it experiences is continuous, but we can find the change per second.

Investigation 1

A pump is used to raise water from a reservoir 10 m below ground level and deliver it through a hosepipe of radius 0.04 m at a speed of 6 m s^{-1}.

Imagine that the water issues from the pipe as does toothpaste from a tube. Calculate the volume of water leaving the pipe in one second (i.e. the volume of a cylinder 6 m long).

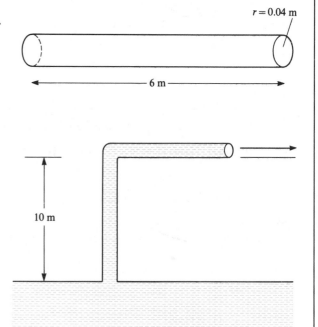

Show that, since 1 m^3 of water has a mass of 1000 kg, the mass of water delivered per second is 30 kg.

Show that the P.E. gained by this mass of water in one second (mgh) is 2940 J.

Show that the K.E. gained by this mass of water in one second ($\frac{1}{2}mv^2$) is 540 J.

The total mechanical energy gained per second is a measure of the work done by the pump per second. Hence show that the power of the pump is 3.48 kW.

Exercise 9.2

1   2 m³ of water are pumped from a depth of 15 m every minute and are delivered through a pipe of cross-sectional area 0.1 m². Find the total gain in mechanical energy each second.

2   0.01 m³ of water are delivered through a pipe every second. If the pipe has a radius of 0.02 m, calculate the velocity of the water as it leaves the pipe. Find the gain in kinetic energy of the water and hence find the power of the pump required.

3   A pump is required to deliver 4 m³ of water per minute through a circular pipe of radius 6 cm. If it is also to be raised from a depth of 15 m, find the power of the pump.

4   Water is to be pumped at a rate of 1.5 m³ per minute from ground level to a height of 6 m and is delivered through a pipe of cross-sectional area 0.002 m². If the pump is only 80% efficient, calculate the power of the pump required.

5   A pump is used to fill a swimming pool from a reservoir 10 m below the level of the pool. The pool measures 35 m by 10 m, slopes uniformly from a depth of 1 m to a depth of 2.5 m, and the process of filling must not take more than 6 hours. Calculate the power of the pump if it is assumed to be 60% efficient, and if the water is delivered through a pipe of diameter 8 cm.

Conservation of mechanical energy

It was stated on page 186 that the total amount of energy in the universe is constant although it can be changed from one form to another. Thus, in mechanics, the total mechanical energy (i.e. the potential and kinetic energy) of a body or group of bodies will change if some of the energy is converted into other forms (e.g. heat, light, sound, etc.).

This will happen, for example, when there is an impact creating sound, when two rough surfaces rub together producing heat (and light and sound?) or when an external force is applied to the system. Remember that the weight of a body is included through the potential energy and is not regarded as an external force.

Investigation 2

When a space capsule is returning to earth through the atmosphere, it has a speed of 60 m s⁻¹ when it is 3 km above the ocean and is being retarded by a parachute. If the capsule has a mass of 12 tonnes and its speed just before splashdown is 6 m s⁻¹, find the loss in mechanical energy of the space capsule.

Show that the sum of the P.E. and K.E. at a height of 3 km is 374 400 kJ.

Show that the total mechanical energy at splashdown is 216 kJ. Hence deduce the loss in mechanical energy. How is it dissipated?

Investigation 3

A child of mass 60 kg slides from rest down a straight water chute into a swimming pool. If the vertical height of the chute is 6 m and the inclination to the horizontal is 60°, show that the speed of the child just before entering the water is $\sqrt{12g} = 10.84 \text{ m s}^{-1}$, assuming that any friction or air resistance can be neglected.

Hence, find the sum of the P.E. and K.E. at the bottom of the chute and deduce that this sum has the same value at the top of the chute.

Repeat this calculation if the chute is inclined at 45° to the horizontal but the vertical height is maintained at 6 m.

Although the slide is longer in this case, you should find that the speed at the bottom of the chute is the same as in the first example.

Similarly, the sum of the P.E. and K.E. at the top and the bottom of the chute are equal to the values in the previous case. This suggests that the sum of the P.E. and K.E. in this motion is always constant.

Now consider the child has reached a point C some way down the chute where C is at a distance h above the water level.

Show that the speed of the child, v, at point C, is given by

$$v^2 = 2g(6-h)$$

Show also that the P.E. of the child at the point C is $60gh$ and that the K.E. at C is $60g(6-h)$.

Deduce that the sum of the P.E. and K.E. at point C is $360g$, which is the value obtained for this at both the top and the bottom of the chute.

We say that the mechanical energy is conserved in this example. Would this be true in the chute shown in the picture if friction and air resistance are neglected?

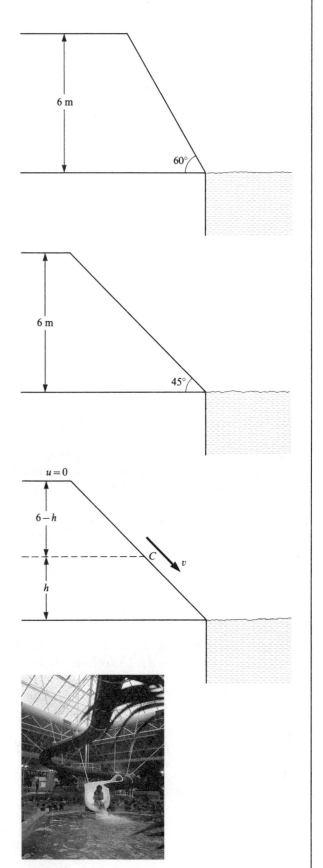

These two Investigations show that the mechanical energy of a body may be conserved. When does this happen? Mechanical energy will be conserved if no work is done by external forces and none of the mechanical energy is converted into another form.

This is called the **Principle of Conservation of Mechanical Energy**:

> The total mechanical energy of a body or system of bodies is constant provided no work is done by any external forces acting on the body or system of bodies.

In the water chute Investigation there are two forces acting on the child: the weight of $60g$ N and the normal reaction R N.

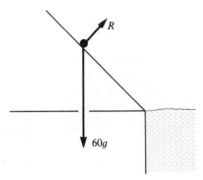

The normal reaction is always perpendicular to the direction of travel and hence the work done by R is zero. The work done by the weight is potential energy and is already included in the total mechanical energy.

So the external forces do not do any work and hence the total mechanical energy is conserved.

Clearly in the examples of the water chute, the work done in moving down the chute depends only on the initial and final positions and *not* on the path taken. When this is the case we say that the **forces are conservative**.

> Conservative forces are such that the work done by them in moving a body depends only on the initial and final positions and not on the path taken.

Conservative forces include gravitational forces but not frictional and impulsive forces. These are said to be **non-conservative**. It is possible to restate the **Principle of Conservation of Mechanical Energy**:

> If the forces acting on a body or system of bodies are conservative then the total mechanical energy remains constant.

WORKED EXAMPLES

1   A small bead of mass 2 kg is threaded on a smooth wire in the form of a circle of radius 0.2 m. If the circular wire is fixed in a vertical plane and the bead is projected with a speed of 1.5 m s^{-1} from the highest point of the wire, find the angle that the radius to the bead makes with the vertical when its speed has doubled.

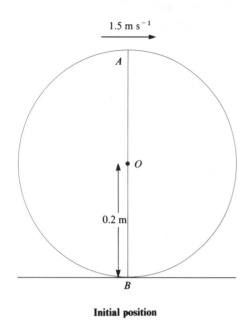

Initial position

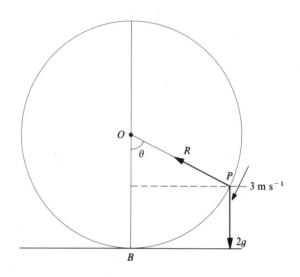

Position when speed is 3 m s^{-1}

The diagram, above left, shows the initial position, A, of the bead when it is given a speed of 1.5 m s^{-1}. The diagram, above right, shows a general position of the bead when the speed has doubled (i.e. 3 m s^{-1}). We have assumed that this is below the level of the centre of the circle. Let B be the level of zero potential energy.

Let the angle that the radius makes with the vertical be θ and the normal reaction between the bead and the wire be R. The only forces acting on the bead are the weight $2g$ N and the normal reaction R N. Since this normal reaction is always along the radius, it will be perpendicular to the direction of motion and hence does no work.

Thus, mechanical energy will be conserved.

$$\text{Total P.E. and K.E. at } A = \text{Total P.E. and K.E. at } P \quad \ldots\ldots\ldots\ldots(1)$$

At the highest point, $\text{P.E.} = mgh = 2g(0.4) = 0.8g \text{ J}$

$\text{K.E.} = \tfrac{1}{2}mv^2 = \tfrac{1}{2}(2)(1.5)^2 = 2.25 \text{ J}$

At point P, $\text{P.E.} = mgh = 2g(0.2)(1 - \cos\theta)$

$= 0.4g(1 - \cos\theta) \text{ J}$

$\text{K.E.} = \tfrac{1}{2}mv^2 = \tfrac{1}{2}(2)(3)^2 = 9 \text{ J}$

Using the Energy Equation **(1)**, we have

$$0.8g + 2.25 = 0.4g(1 - \cos\theta) + 9$$

$$\Rightarrow \quad 7.84 + 2.25 - 9 = 0.4(9.8)(1 - \cos\theta)$$

$$\Rightarrow \quad 1.09 = 3.92(1 - \cos\theta)$$

$$\therefore \quad 1 - \cos\theta = \frac{1.09}{3.92} \quad \Rightarrow \quad \cos\theta = 1 - \frac{1.09}{3.92}$$

$$\Rightarrow \quad \cos\theta = 0.7219 \quad \Rightarrow \quad \theta = 43.8°$$

The radius to the bead makes an angle of $43.8°$ to the vertical when the speed has doubled.

---

2 A light elastic string of natural length 3 m and modulus of elasticity $24g$ N has one end fixed to a point A of the ceiling. To its other end is attached a mass of 2 kg which is held at A and released from rest.
 Find **(a)** the speed of the mass when the length of the string is 4 m and **(b)** the depth below A at which the mass comes instantaneously to rest.

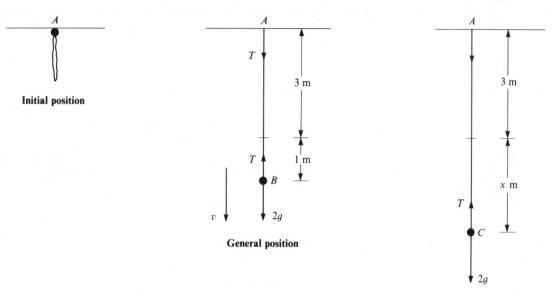

In this example, the forces acting on the 2 kg mass are the tension in the string and its own weight. Since the tension in the string is an internal force, no work is done by external forces and hence energy is conserved.

Remember that the total energy will include the elastic potential energy stored in the string.

(a) Using conservation of energy,

$$\text{total energy at } A = \text{total energy at } B$$

At A, the mass has gravitational potential energy but no kinetic or elastic potential energy. At B, however, the mass possesses kinetic energy and elastic potential energy stored in the string, but no gravitational potential energy (if this is taken as the zero level).

$$\text{P.E. at } A = \text{K.E. at } B + \text{elastic P.E. at } B$$

$$\therefore \quad mgh = \tfrac{1}{2}mv^2 + \frac{\lambda x^2}{2l}$$

$$\therefore \quad 2g(4) = \tfrac{1}{2}(2)v^2 + \frac{24g(1)^2}{6}$$

$$\therefore \quad v^2 = 4g \quad \Rightarrow \quad v = \sqrt{4g} = 6.26 \text{ m s}^{-1}$$

The speed of the mass at point B is 6.26 m s^{-1}.

Energy

(b) Let the mass come to rest at a point C, a distance $(x+3)$ m below A, i.e. the extension of the string is x m.

In this case the gravitational potential energy at A is converted to the elastic potential energy of the string at C. There is no kinetic energy at A or C. Let C be the level of zero potential energy.

$$\text{P.E. at } A = \text{elastic P.E. at } C$$

$$\therefore \quad mgh = \frac{\lambda x^2}{2l}$$

$$\therefore \quad 2g(x+3) = \frac{24gx^2}{6} \quad \Rightarrow \quad x+3 = 2x^2 \quad \Rightarrow \quad 2x^2 - x - 3 = 0$$

$$\Rightarrow \quad (2x-3)(x+1) = 0 \quad \Rightarrow \quad x = -1 \quad \text{or} \quad x = 1\tfrac{1}{2}$$

Since $x > 0$, the only solution possible is $x = 1\tfrac{1}{2}$. Hence the mass will come to rest instantaneously $4\tfrac{1}{2}$ m below A.

Investigation 4

A boy puts a conker of mass m on the end of an inelastic string and then spins the string so that the conker completes vertical circles of radius 0.5 m. If the speed of the conker at the lowest point of the circle is u m s^{-1}, deduce that mechanical energy is conserved and that the speed at the top of the circle is given by $\sqrt{u^2 - 2g}$.

By considering the total downward force on the conker at the highest point, show that the tension in the string at this point is given by $m(2u^2 - 5g)$. Deduce that the conker will only make complete vertical circles if the speed at the lowest point, u, is such that $2u^2 \geqslant 5g$.

Investigation 5

A small bead of mass m is placed at the highest point of a smooth fixed sphere of radius 2 m and is slightly displaced from rest so that the bead slides down the surface of the sphere.

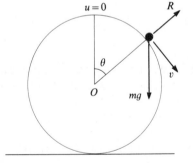

By considering conservation of energy and the equations of circular motion, show that the normal reaction between the bead and the sphere is given by

$$R = mg(3\cos\theta - 2)$$

Deduce that the bead will leave the surface when $\cos\theta = \tfrac{2}{3}$.

What difference does it make to this problem if the bead is threaded on to a fixed smooth vertical circular wire? What is the speed of the bead at the lowest point? What can you discover about the normal reaction? In which direction does it act?

Exercise 9.3

1 A ball falls freely from rest at the top of a tower 50 m high. Find the speed of the ball at the base of the tower, assuming that air resistance can be neglected.

2 A child slides down a water chute 15 m high at a theme park. If the child starts with a speed of 0.5 m s^{-1}, find the child's speed at the bottom of the chute.

3 A girl on a garden swing moves in an arc of a circle of radius 2 m. If her speed at the lowest point is 4 m s^{-1}, find the height above the ground when she comes instantaneously to rest.

4 A particle of mass 2 kg moves down a smooth slope of length 5 m inclined at an angle of $\tan^{-1}\frac{3}{4}$ to the horizontal. If the particle starts from rest, find its speed at the foot of the slope.

5 A boy whirls a conker of mass 0.2 kg on the end of a string in a vertical circle of radius 1 m. If the speed of the conker at the highest point is 2 m s^{-1}, find its speed at the lowest point of the circle.

6 A man of mass 65 kg jumps from a balcony 10 m above the ground by swinging on the end of a rope of length 15 m attached to a point 7 m above the level of the balcony.

If his initial speed is 1 m s^{-1}, find **(a)** his speed as he passes through the lowest point of the path and **(b)** the height above the ground when he comes to rest instantaneously.

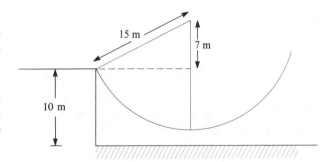

7 A particle of mass 0.5 kg lies inside a smooth fixed hollow sphere of radius 2 m and is projected with a speed of $\sqrt{12g}$ m s^{-1} from the lowest point of the sphere so as to move in a vertical circle. Show that the normal reaction between the particle and the sphere at the top of its path is $\frac{1}{2}g$ N.

8 An elastic string of natural length 1.5 m and modulus of elasticity $20g$ N, has one end fixed to a point A. A particle of mass 2 kg is attached to the other end and held at A. If it is released from rest, find **(a)** the speed of the particle when the string has an extension of 0.5 m and **(b)** the depth below A when the mass comes instantaneously to rest.

9 A mass m is attached to one end of an inextensible string of length l. The other end is fastened to a fixed point O and the mass hangs at rest with the string vertical. If the mass is then projected with a speed u, show that the speed v at any point of the circular path is given by

$$v^2 = u^2 - 2gl(1 - \cos \alpha)$$

where α is the angle that the string makes with the downward vertical.

Show that the tension in the string at this point is given by

$$T = \frac{mu^2}{l} + (3 \cos \alpha - 2)mg$$

10 One end of a light elastic string of natural length 2 m and modulus of elasticity 8 N is attached to a fixed point A on a smooth horizontal table. A mass of 3 kg is attached at the other end of the string and the system is released from rest with the mass at a point 5 m from A.

Find the speed of the mass when it reaches point A.

11 A particle of mass m is attached to a light inextensible string of length a, the other end of which is attached to a fixed point O. If the particle is then projected with speed u from its lowest point to perform vertical circles, show that if it comes to rest whilst the string is still taut then $u^2 < 2ga$ and the particle will perform oscillations below the level of O.

Show also that, if the string becomes slack above the level of O, then $2ga < u^2 < 5ga$.

10 Impulse, momentum and impact

Introduction

When moving bodies collide, equal and opposite forces act on the two bodies at the moment of impact. The effect of such interactions can be quite different and dramatic.

A car in collision with a lamp post comes to rest very quickly. Why do the passengers suffer injuries?

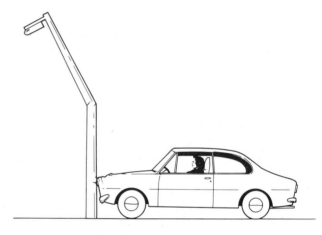

A snooker player strikes the white ball with his cue with various forces, causing it to move with different speeds. The white ball then collides with another ball so that this travels in the right direction and with the correct speed towards the pocket.

A batsman in a cricket match makes the ball change direction and speed by striking it with the bat at different angles, different forces and for different lengths of time.

When a boy kicks a plastic ball it will travel further than a heavier leather football, even though the same force is applied to both. Why is this?

In this chapter we shall consider how forces which act for a finite interval of time, when bodies come into contact, cause different effects.

Sometimes these forces are external to the bodies but in other situations they may be internal. For example, consider two masses *A* and *B* which are connected by a light inextensible string and placed on a smooth table. If an external force, *P*, acting at 45° to the string is applied to *A*, it will cause an internal momentary tension in the string. What force causes *B* to move? Can you decide the approximate direction in which *A* and *B* move initially?

Investigation 1

A particle of mass 2 kg has a force of 10 N applied to it. Use Newton's second law to find the acceleration of the particle. If its speed increases from 12 m s^{-1} to 20 m s^{-1}, find the time for which the force acts.

Definitions of impulse and momentum

In Investigation 1, the force applied to the 2 kg mass caused it to increase its speed from 12 m s^{-1} to 20 m s^{-1} in 1.6 seconds. It is possible to obtain this answer without actually calculating the value of the acceleration.

Consider a particle of mass *m*, to which a constant force **F** is applied causing an acceleration **a**. If the velocity increases from **u** to **v** in time *t*, then,

$$\mathbf{v} = \mathbf{u} + \mathbf{a}t \qquad \text{............ (1)}$$

Using Newton's second law of motion,

$$\mathbf{F} = m\mathbf{a} \qquad \text{............ (2)}$$

Eliminating **a** from Equations **(1)** and **(2)** gives

$$\mathbf{F} = m\left(\frac{\mathbf{v} - \mathbf{u}}{t}\right)$$

This can be expressed in the form,

$$\boxed{\mathbf{F}t = m\mathbf{v} - m\mathbf{u}}$$

The result is known as the **impulse-momentum equation**. The product **F**t is called the **impulse of the force F**, and the product *m***v** is known as the **momentum** of the mass *m* at velocity **v**. Thus,

$$\boxed{\text{Impulse} = \text{change of momentum}}$$

Definitions

1 When a constant force **F** acts for a time *t*, it is said to exert an impulse **F***t*.

2 The momentum of a particle of mass *m* moving with velocity **v**, is defined as the product *m***v**.

The **S.I. unit of impulse** is the **Newton-second** (N s). Momentum can be measured in kg m s^{-1} but, as it is equivalent to impulse, we usually use the **Newton-second** (N s) as the **unit of momentum**.

So, in Investigation 1, we could obtain the solution by using the impulse-momentum equation.

$$\mathbf{F}t = m\mathbf{v} - m\mathbf{u}$$

$$\Rightarrow \quad 10t = 2(20) - 2(12) \quad \Rightarrow \quad t = \frac{40 - 24}{10} = 1.6 \text{ s}$$

WORKED EXAMPLES

1   A constant force $\mathbf{F} = \mathbf{i} + 2\mathbf{j} - 4\mathbf{k}$ acts on a particle for 1.3 seconds. Find the impulse of the force $\mathbf{F}$.

Since,   Impulse $= \mathbf{F}t$,   it follows that,

   Impulse $= (\mathbf{i} + 2\mathbf{j} - 4\mathbf{k}) \times 1.3 = 1.3\mathbf{i} + 2.6\mathbf{j} - 5.2\mathbf{k}$

2   A ball of mass 0.2 kg travelling at 4 m s^{-1} hits a wall and rebounds at a speed of 3 m s^{-1}. Find the impulse exerted by the wall on the ball.

Remember that impulse and momentum are **vector** quantities.

Let the impulse exerted by the wall on the ball be $\mathbf{I}$. Hence,

   Impulse $=$ change in momentum $= m(\mathbf{v} - \mathbf{u})$

Now, all the motion takes place in a straight line and thus,

$$|\mathbf{I}| = 0.2[3 - (-4)] = 1.4 \text{ N s}$$

Hence, the impulse exerted on the ball is 1.4 N s in the direction shown.

3   In a tennis match a player attempts to return a service, but the ball hits the edge of the racquet and flies off at a speed of 50 m s^{-1} at an angle of 60° to its original direction and at 30° above the horizontal.
   If the ball has a mass of 0.02 kg and was originally travelling horizontally at 40 m s^{-1}, find the magnitude of the impulse given to the ball by the racquet.

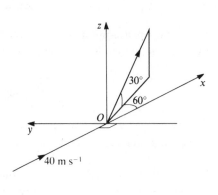

Clearly, the motion takes place in three dimensions. If the point of contact of the ball on the racquet is taken as the origin O, we can choose horizontal axes Ox and Oy along and perpendicular to the original direction of the ball and a vertical axis Oz, to form a mutually perpendicular set of axes.

Let $\mathbf{i}$, $\mathbf{j}$ and $\mathbf{k}$ be the unit vectors along Ox, Oy and Oz, respectively.

Writing the given velocities in terms of $\mathbf{i}$, $\mathbf{j}$ and $\mathbf{k}$, gives,

the initial velocity $\quad \mathbf{u} = 40\mathbf{i}$

the final velocity $\quad \mathbf{v} = 50 \cos 30° \cos 60° \, \mathbf{i} + 50 \cos 30° \sin 60° \, \mathbf{j} + 50 \sin 30° \, \mathbf{k}$

$$= \frac{25\sqrt{3}}{2}\mathbf{i} + \frac{75}{2}\mathbf{j} + 25\,\mathbf{k}$$

If $\mathbf{I}$ is the impulse given to the ball then, using the impulse-momentum equation,

$$\mathbf{I} = m\mathbf{v} - m\mathbf{u}$$

$$= 0.02\left(\frac{25\sqrt{3}}{2}\mathbf{i} + \frac{75}{2}\mathbf{j} + 25\mathbf{k} - 40\mathbf{i}\right)$$

$$= 0.02(21.65\mathbf{i} + 37.5\mathbf{j} + 25\mathbf{k} - 40\mathbf{i})$$

$$= 0.02(-18.35\mathbf{i} + 37.5\mathbf{j} + 25\mathbf{k})$$

Hence $\qquad |\mathbf{I}| = 0.02\sqrt{(-18.35)^2 + (37.5)^2 + (25)^2} = 0.02 \times \sqrt{2367.97} = 0.97 \text{ N s}$

The magnitude of the impulse given to the ball is 0.97 N s.

Investigation 2

(a) A boy stubs his foot against a rock of mass 0.25 kg and causes it to move from rest with a speed of 10 m s^{-1}. Since the rock is solid, it does not deform under the impact.

If his foot is in contact with the rock for only 0.01 s, calculate the force, $\mathbf{F}$, which is applied to the rock and deduce the force on his foot.

If, instead, the boy kicks a plastic football of the same mass, the ball will deform under the impact and, in addition, he is likely to follow through (keep his leg swinging after making contact with the ball) which means that the force will be applied for longer, say 0.1 seconds.

Calculate the force on his foot in this case, if the ball again moves with a speed of 10 m s^{-1}.

Which kick hurts more?

From these two examples deduce, using the impulse-momentum equation, the inverse relation which exists between the force and the time for which it acts when the change of momentum is constant.

(cont.)

Investigation 2—continued

(b) Explain:

    (i) why a tennis player sometimes 'follows through' with the racquet when making a shot;

    (ii) why a parachutist bends his knees and rolls over on landing;

    (iii) why some lorries have a sprung collision bar across the back of the vehicle;

    (iv) how a fielder in cricket reduces the 'sting' in catching a ball.

Clearly, from the impulse-momentum equation we can see that, for a given change in momentum, the product Ft can be achieved in many ways.

If $Ft = 200$ N s, then, $F = 100$ N and $t = 2$ s, $F = 50$ N and $t = 4$ s or $F = 400$ N and $t = \frac{1}{2}$ s, and so on.

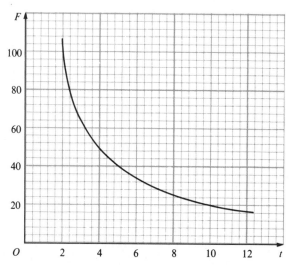

The longer the time for which the force acts, the smaller the force.

Since, in this case, the relation is

$$Ft = 200$$

We can write this as,

$$F = \frac{200}{t}.$$

The graph is shown here.

Impulse of a variable force

In the working so far we have considered the force to be constant. In many situations this is unlikely to be the case. For example, a car experiences a great variation in the applied force during a crash and when a ball is kicked the force applied varies during the time of contact.

In these cases we consider the total impulse as the sum of a large number of impulses over short intervals of time. So, if forces $\mathbf{F}_1, \mathbf{F}_2, \mathbf{F}_3, \ldots \mathbf{F}_n$ act for intervals of time $\delta t_1, \delta t_2, \delta t_3, \ldots \delta t_n$ respectively, then the total impulse $\mathbf{I}$ is given by,

$$\mathbf{I} \simeq \mathbf{F}_1\, \delta t_2 + \mathbf{F}_2\, \delta t_2 + \cdots + \mathbf{F}_n\, \delta t_n = \sum_{i=1}^{n} \mathbf{F}_i\, \delta t_i$$

In the limit, as $\delta t \to 0$,

$$\mathbf{I} = \lim_{\delta t \to 0} \sum_{i=1}^{n} \mathbf{F}_i\, \delta t_i = \int_{t_1}^{t_2} \mathbf{F}\, dt$$

This can be seen graphically in the case when the force $\mathbf{F}$ is **one-dimensional** by plotting the **magnitude of the force** against time. Since the total impulse is the sum of the separate impulses over the small intervals of time, it will be represented by the area under the curve.

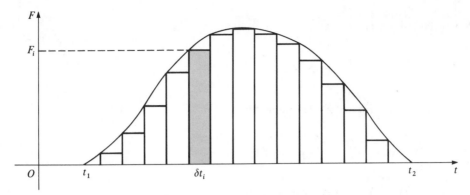

So, in this case, since $F = m\dfrac{dv}{dt}$

Then,
$$\int_{t_1}^{t_2} F \, dt = \int_{t_1}^{t_2} m \frac{dv}{dt} \, dt = \left[mv \right]_{t=t_1}^{t=t_2}$$

Now, if u is the initial speed at time t_1 and v is the final speed at time t_2, then,

$$\text{Impulse} = mv - mu$$

Thus the impulse is again equal to the change in momentum.

Since the working above for the one-dimensional case can apply equally to each component in two or three dimensions it follows that, in general, the previous result is still true if the force is variable, i.e.

$$\text{Impulse} = \int_{t_1}^{t_2} \mathbf{F} \, dt = m\mathbf{v} - m\mathbf{u}$$

WORKED EXAMPLE

A hockey player strikes a ball of mass 0.16 kg which is at rest with a force, $\mathbf{F}$, of magnitude $30\,000t \, e^{-100t}$. If the stick is in contact with the ball for 0.05 seconds, find the impulse given to the ball and the speed with which it begins to move.

In this case the force is variable and one-dimensional and hence the impulse will be represented by the area under the curve of the magnitude of $\mathbf{F}$ against t, which is drawn below.

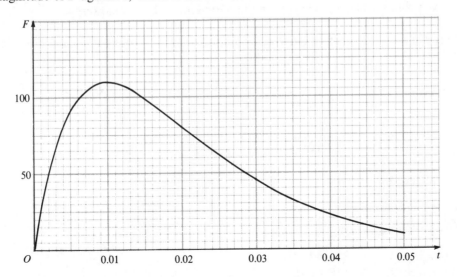

The impulse of a variable force is $\displaystyle\int_{t_1}^{t_2} |\mathbf{F}|\, dt$

$$\text{Impulse} = \int_{0}^{0.05} 30\,000t\, e^{-100t}\, dt$$

This function must be integrated by parts (see page 44).

Using $\displaystyle\int_{t_1}^{t_2} u\frac{dv}{dt}\, dt = \Big[uv \Big]_{t_1}^{t_2} - \int_{t_1}^{t_2} v\frac{du}{dt}\, dt$ and letting

$$u = t \;\Rightarrow\; \frac{du}{dt} = 1 \quad\text{and}\quad \frac{dv}{dt} = e^{-100t} \;\Rightarrow\; v = -\frac{1}{100}e^{-100t}$$

gives,

$$\text{Impulse} = \left[-\frac{30\,000}{100}t\, e^{-100t} \right]_{0}^{0.05} + \int_{0}^{0.05} \frac{30\,000}{100}e^{-100t}\, dt$$

$$= \left[-300t\, e^{-100t} \right]_{0}^{0.05} - \left[3e^{-100t} \right]_{0}^{0.05}$$

$$= (-0.1011 + 0) - (0.0202 - 3) = 2.88 \text{ N s}$$

Using the impulse-momentum equation,

$$\text{Impulse} = mv - mu$$

$$\therefore \qquad 2.88 = 0.16(v-0) \;\Rightarrow\; v = 18 \text{ m s}^{-1}$$

The impulse given to the ball is 2.88 N s, which produces an initial speed of 18 m s^{-1}.

Exercise 10.1

1   A particle of mass 2 kg moving with a speed of 15 m s^{-1} is brought to rest when it strikes a barrier. Find the magnitude of the impulse exerted by the barrier on the particle.

2   Find the magnitude of the impulse when the force F acts for t seconds,

   (a) $\mathbf{F} = \mathbf{i} + 2\mathbf{j}$ and $t = 4$ (b) $\mathbf{F} = -3\mathbf{i} + 4\mathbf{j} + \mathbf{k}$ and $t = 2$

3   A pebble of mass 0.25 kg falls vertically and has a speed of 12 m s^{-1} when it strikes the ground. If it is brought to rest by the impact, find the magnitude of the impulse exerted by the ground on the pebble.

4   A railway truck of mass 1500 kg moving at 4 m s^{-1} is brought to rest by the buffers in 1.5 seconds. What is the average force exerted by the buffers on the truck?

5   A particle of mass 3 kg is moving in a straight line with a speed of 2 m s^{-1} when a force $\mathbf{F}$ acts on it, in the direction of travel, for 3 seconds. If the final speed is 10 m s^{-1}, find the magnitude of the force $\mathbf{F}$.

6   A ball of mass 0.15 kg is moving horizontally at 30 m s^{-1} when it is struck by a bat. Find the impulse given to the ball if its final speed is 20 m s^{-1} in the opposite direction.

7   Find the impulse of the variable force $\mathbf{F}$ if,

   (a) $\mathbf{F} = 2t\mathbf{i} + 4\mathbf{j}$ and $0 \leqslant t \leqslant 1.5$

   (b) $\mathbf{F} = e^{-2t}\mathbf{i} + t\mathbf{j} + \mathbf{k}$ and $0 \leqslant t \leqslant 1$

   (c) $\mathbf{F} = \cos t\,\mathbf{i} + 2\sin t\,\mathbf{j} + \tan t\,\mathbf{k}$ and $0 \leqslant t \leqslant \frac{1}{4}\pi$

8 A squash ball of mass 0.02 kg is moving horizontally at 10 m s^{-1} when it strikes a vertical wall at an angle of 60° to the normal to the wall. If it leaves the wall at an angle of $\tan^{-1} 2\sqrt{3}$ to the normal find the magnitude of the impulse on the ball, assuming that the component of the velocity parallel to the wall remains constant.

9 A particle of mass 4 kg moving with velocity $2\mathbf{i} + 3\mathbf{j} + \mathbf{k}$ receives an impulse $-\mathbf{i} + 2\mathbf{j} + 3\mathbf{k}$. Find the final velocity of the particle.

10 A cricket ball of mass 0.2 kg is moving at 25 m s^{-1} horizontally towards the batsman. Find the magnitude of the impulse given to the ball if, after the batsman plays his shot, the ball moves **(a)** horizontally at 30 m s^{-1} at right angles to the original direction and **(b)** at 40 m s^{-1} at an angle of 45° to the original direction and at 45° above the horizontal.

11 A golf player strikes a stationary ball of mass 0.05 kg with a force $F = ke^{-500t}$, where $k = 1.6 \times 10^3$. If the contact lasts for 0.005 seconds, find the impulse applied to the ball and the speed with which it begins to move.

12 A ball of mass 0.8 kg is moving horizontally at a speed of $6\sqrt{3}$ m s^{-1}. It is given an impulse **I** which causes it to move at 120° to its original direction and at 30° above the horizontal. If its final speed is 4 m s^{-1}, find the magnitude of the impulse given to the ball.

Continuous change of momentum

We can also use the impulse-momentum equation when a continuous flow of water changes its velocity.

Consider a garden tap which is left running. A mass of water strikes the ground every second and, provided it does not splash back, this mass loses its momentum.

It is this momentum that is destroyed every second which gives a measure of the impulse on the surface. Since the change in momentum occurs in one second, it is possible to determine the average force on the surface.

Similar reasoning can be applied when rain falls on the ground or when a water jet from a hose is directed against the side of a car.

When water flows round a bend in a pipe or issues from a nozzle, it is given an impulse (i.e. a change in momentum) and this will cause an equal and opposite impulse on the pipe or hose. If the hose is being held by hand it may take a considerable effort to hold it still.

Can you explain why a garden sprinkler rotates although it does not have a motor?

Investigation 3

Consider the case of water running from a tap at a rate of 25 litres per second on to level ground. If the speed of the water just before it strikes the ground is 8 m s^{-1} and it does not rebound from the surface, find the momentum destroyed in 1 second. Remember that the mass of 1 litre of water can be taken as 1 kg.

Deduce that the impulse exerted by the ground on the water in 1 second is 200 N s. Hence, show that the water exerts an average force of 200 N on the ground.

WORKED EXAMPLE ⟩

Water issues horizontally from a pipe of diameter 6 cm at a speed of 20 m s^{-1} and strikes a vertical wall from which it does not rebound. What force is exerted by the water on the wall?

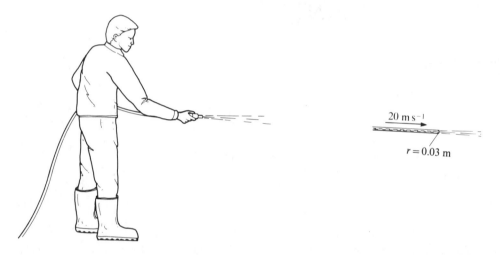

In one second a cylinder of water 20 m long will leave the hose. Think of it as toothpaste from a tube!

Hence, the volume of water issuing in 1 second $= \pi(0.03)^2 \times 20$ m^3

Since 1 m^3 of water has a mass of 10^3 kg,

$$\text{the mass of water issuing in 1 second} = \pi(0.03)^2 \times 20 \times 10^3 \text{ kg}$$

$$= 56.55 \text{ kg}$$

As the speed is 20 m s^{-1}, the momentum lost per second $= 56.55 \times 20$ N s

$$= 1130.97 \text{ N s}$$

Hence, the force exerted on the wall is 1.131 kN.

Exercise 10.2

1 A jet of water from a hosepipe, 2 cm in diameter, strikes a fixed vertical wall at right angles. If the water leaves the pipe at 12 m s^{-1} and does not rebound from the wall, find the force exerted on the wall.

2 Water leaves a pipe at a rate of 250 litres per second and strikes a vertical plane from which it does not rebound. If the speed of the water on leaving the pipe is 15 m s^{-1}, find the force exerted by the water on the plane.

3   A fireman holds a hosepipe and directs the jet of water horizontally so that it delivers 4 m³ of water per minute. If the hose has a diameter of 6 cm, find the force exerted on a vertical wall if the water does not splash back.

4   A jet of water issues horizontally from a hose at a speed of 10 m s⁻¹, strikes a vertical wall at right angles and rebounds at 2 m s⁻¹. Find the force on the wall if the hose has a diameter of 1.5 cm.

5   Coal dust falls from a hopper on to level ground with a speed of 8 m s⁻¹ and does not rebound. If 50 kg of coal dust falls per second, find the force exerted on the ground.

6   Water flows over a waterfall at a rate of 100 kg per second and exerts a force of 1200 N on the ground below. Assuming that the water falls vertically and does not rebound, find the speed at which the water hits the ground.

7   On a production line in a factory, small nails fall vertically on to a conveyor belt without bouncing at a rate of 40 kg per minute. If the belt stops whilst the nails are still falling, find the vertical force exerted by the belt on the nails if the speed of the nails just before striking the belt is 1.5 m s⁻¹.

Impact of two bodies

We now consider what happens when two movable objects collide, as in a game of bowls or a game of marbles. By considering the impulses and the change in momentum of each of the bodies, we can establish an important result concerning the impact of two bodies.

Whenever two bodies collide they exert equal and opposite forces on each other (by Newton's third law). Since the time for which they act is the same, it follows that they exert equal and opposite impulses on each other.

Now the impulse-momentum equation states that the impulse is equal to the change in momentum. Hence, if the two bodies experience equal and opposite impulses, they must also experience equal and opposite changes in momentum.

Thus, the **total change in momentum of the two bodies is zero**. This result is known as **the principle of conservation of momentum.**

> The total momentum in a given direction of a system of particles which interact remains constant provided no external forces act on the system in that direction.

This can be deduced mathematically as follows.

Consider two particles, A and B, of masses m_1 and m_2 respectively, which are moving in a straight line with speeds u_1 and u_2 where $u_1 > u_2$. When the particles collide there will be an equal and opposite impulse, **I**, between them, acting along the line of centres.

Let the velocities of the masses after impact be v_1 and v_2 respectively.

On diagrams, we usually show speeds before impact above the spheres and speeds after impact below the spheres.

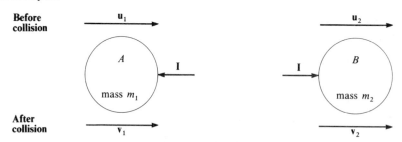

Using the impulse-momentum equation along the direction of motion,

On particle A, $\quad -\mathbf{I} = m_1(\mathbf{v}_1 - \mathbf{u}_1)$

On particle B, $\quad \mathbf{I} = m_2(\mathbf{v}_2 - \mathbf{u}_2)$

Remember that impulse is a vector quantity and hence particle A experiences a negative impulse from B which will cause it to slow down.

Adding these equations gives,

$$0 = m_1(\mathbf{v}_1 - \mathbf{u}_1) + m_2(\mathbf{v}_2 - \mathbf{u}_2) \quad \Rightarrow \quad m_1\mathbf{u}_1 + m_2\mathbf{u}_2 = m_1\mathbf{v}_1 + m_2\mathbf{v}_2$$

This result gives the **principle of conservation of momentum**. Notice that it is in a slightly different format from that stated on the previous page, i.e.,

> The total momentum before impact = the total momentum after impact.

This principle applies to any interaction between two or more bodies and hence could be used where an explosion occurs. An explosion happens when two or more bodies separate after an internal impulse.

So, just before a gun is fired the total momentum is zero and hence it must also be zero after the explosion.

The momentum given to the bullet is equal and opposite to the momentum given to the gun. This is felt as a recoil against one's shoulder.

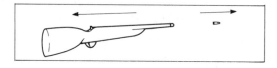

Investigation 4

A particle of mass m moving at 6 m s^{-1} horizontally meets a particle of mass $2m$ moving horizontally along the same straight line and in the same direction at a speed of 3 m s^{-1}. On impact the particles join together to form a single particle of mass $3m$.

Use the principle of conservation of momentum to show that the final speed of the combined mass is 4 m s^{-1}.

Calculate the kinetic energy of each of the particles before the impact and show that the kinetic energy lost during the impact is $3m$ J.

This illustrates an important point: **the total mechanical energy of a system is reduced by a collison**.

Why does the impact cause a loss of mechanical energy?

> In problems involving impacts, the mechanical energy is not conserved.

WORKED EXAMPLES

1 Two particles A and B of masses 5 kg and 4 kg, respectively, are moving along the same straight line such that the 5 kg mass catches and collides with the 4 kg mass. If the speeds of A and B before the impact are 2 m s^{-1} and 1.5 m s^{-1}, respectively, and the speed of A after the collision is 1 m s^{-1} in the same direction, find the final speed of B.

Use a diagram to show the speeds of the particles before and after the impact.

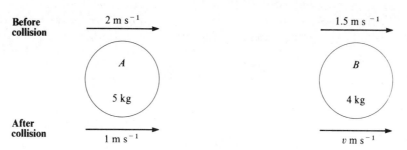

Since all the velocities shown are in the same direction we shall take this direction as positive. As the motion takes place in a straight line we can consider the momentum equation in a linear form.

Using the principle of conservation of momentum,

$$m_1 u_1 + m_2 u_2 = m_1 v_1 + m_2 v_2$$
$$\therefore \quad (5 \times 2) + (4 \times 1.5) = (5 \times 1) + (4 \times v)$$
$$\therefore \quad\quad 11 = 4v \quad \Rightarrow \quad v = 2\tfrac{3}{4}$$

The final speed of B is $2\tfrac{3}{4} \text{ m s}^{-1}$.

2 A sphere of mass 4 kg moving at 10 m s^{-1} impinges directly on another similar sphere of mass 3 kg moving in the opposite direction at 8 m s^{-1}. If, as a result of the impact, the 3 kg sphere has the direction of its motion reversed but its magnitude is unchanged, find the final velocity of the 4 kg sphere.

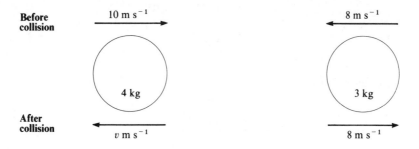

Let $v \text{ m s}^{-1}$ be the speed of the 4 kg sphere after the impact. We have assumed that the velocity of this sphere is reversed.

In this case the directions of motion vary, so we must be careful to assign the correct sign to each velocity.

Assume that velocity in the original direction of motion of the 4 kg sphere is positive.

By the principle of conservation of momentum,

$$m_1u_1 + m_2u_2 = m_1v_1 + m_2v_2$$

$$\therefore \quad (4 \times 10) + [3 \times (-8)] = [4 \times (-v)] + (3 \times 8)$$

$$\therefore \quad 40 - 24 = -4v + 24$$

$$\therefore \quad 4v = 8 \quad \Rightarrow \quad v = 2$$

The final velocity of the 4 kg sphere is 2 m s^{-1} in the opposite sense to its original direction.

3 A gun of mass 4 kg fires a bullet of mass 25 g at a speed of 600 m s^{-1}. Find the speed with which the gun recoils and the constant force required to bring it to rest in 0.4 seconds.

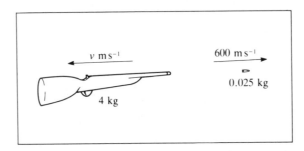

Let the velocity of the gun after firing be v m s^{-1} as shown in the diagram. The momentum before firing is zero and hence the total momentum after firing is also zero.

Using the principle of conservation of momentum,

$$m_1u_1 + m_2u_2 = m_1v_1 + m_2v_2$$

$$\therefore \quad 0 = (600 \times 0.025) + [4 \times (-v)]$$

$$\therefore \quad 4v = 15 \quad \Rightarrow \quad v = 3.75 \text{ m s}^{-1}$$

The speed at which the gun recoils is 3.75 m s^{-1}.

If F N is the force required to bring the gun to rest in 0.4 s then, using the impulse-momentum equation we have,

$$\text{Impulse} = \text{change in momentum}$$

$$F \times 0.4 = (4 \times 3.75) - (4 \times 0)$$

$$F = \frac{15}{0.4} = 37.5 \text{ N}$$

The force required to bring the gun to rest is 37.5 N.

Exercise 10.3

1 A particle A, of mass 3 kg moving at 4 m s^{-1}, overtakes a particle B, of mass 2 kg, moving in the same direction at 2 m s^{-1}. The particles collide and coalesce (join together). Find the speed of the combined mass after the impact.

2 A particle moving at 4 m s^{-1} collides directly with another particle of half its mass which is at rest. If the particles combine together on collision, find their common speed after the impact.

3 A truck *A* of mass 20 Mg travelling at 0.75 m s^{-1} collides with another truck *B* of mass 15 Mg moving in the opposite direction with a speed of 0.5 m s^{-1}. If the trucks couple together on impact, find their common velocity after the impact.

4 A bullet of mass 20 g travelling with a velocity of 900 m s^{-1} enters a block of wood of mass 8 kg which rests on a smooth horizontal plane. If the bullet becomes embedded in the wood, find the speed with which the block begins to move.

5 A gun of mass 10 kg fires a shot of mass 0.05 kg with a velocity of 1500 m s^{-1}. Find the speed with which the gun recoils. If the gun comes to rest in 0.5 seconds, find the constant force required to stop the gun's movement.

6 In a game of bowls, one wood of mass 1 kg hits a stationary jack and, as a result of the collision, the speed of the wood is reduced from 3 m s^{-1} to 2 m s^{-1}. If the jack moves with a speed of 4 m s^{-1} after the impact, find the mass of the jack.

7 The white cue ball in a game of snooker moves at a speed of 2 m s^{-1} and strikes a similar red ball which is moving at 1 m s^{-1} in the opposite direction.
  If the final speed of the white ball is 0.6 m s^{-1} in an unchanged direction, show that the red ball will move with a speed of 0.4 m s^{-1}.

8 A pile driver of mass 4 Mg falls onto a pile of mass 1 Mg with a speed of 8 m s^{-1} just before the impact. Immediately after the impact the pile and the pile driver move as one body. Find their common speed and the average resistance of the ground if they come to rest in 10 cm.

9 A particle of mass 2 kg moving with a speed of 5 m s^{-1} meets another particle of mass 3 kg moving with a speed of 4 m s^{-1} in the opposite direction. If the particles coalesce into one body upon impact, find the common speed after impact and the loss in kinetic energy.

10 A cannon of mass 8 tonnes fires a shot of mass 50 kg with a speed of 250 m s^{-1} relative to the ground when the barrel is **(a)** horizontal **(b)** inclined at 30° to the horizontal. Calculate the speed with which the cannon recoils in each case. (**Hint**: momentum will only be conserved horizontally.)

11 Two particles *A* and *B* of masses 3 kg and 2 kg respectively are moving in the same straight line but in opposite directions when they collide. Just before the impact the speeds of *A* and *B* are 6 m s^{-1} and 4 m s^{-1}, respectively, and after the collision *A* has a speed of 3 m s^{-1}. Find the velocity of *B* after the collision, if *A* and *B* move **(a)** in the same direction and **(b)** in the opposite direction, after the impact.

12 A sphere *A* of mass 10 kg moving at 8 m s^{-1} impinges directly on a similar sphere *B* of mass 40 kg moving in the opposite direction at 5 m s^{-1}. If the speed of *B* after the collision is $\frac{2}{3}$ m s^{-1} in an unchanged direction, find the speed of *A* after the impact.

13 A smooth sphere *A*, moving with speed *u*, impinges on a stationary sphere *B*, of half the mass. If, after the impact, *A* and *B* move in the same direction and the speed of *B* is twice the speed of *A*, find the loss in kinetic energy during the impact.

14 A particle *A*, of mass 3*m*, is suspended by a light inextensible string from a beam and hangs in equilibrium with the string vertical. A particle *B*, of mass *m* moving horizontally with speed *u*, collides with *A* and the combined mass moves in a circular arc. Find the height above the original position of *A* when the combined mass comes instantaneously to rest.

Impulsive tensions in strings

When a string suddenly tightens, an instantaneous impulse acts along the string giving equal and opposite forces acting towards the centre of the string.

When a car is being towed, such an impulsive force is often experienced as the tow rope tightens after being slack.

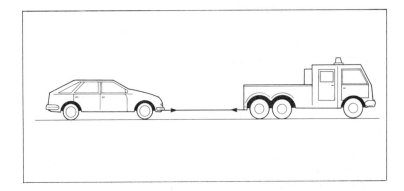

If a particle of mass m lies at rest on a smooth table and has an inelastic string attached to it, then any pull on the string will cause an impulsive tension. This, in turn, will cause a change in momentum equal to the impulse and will take place in the direction of the string.

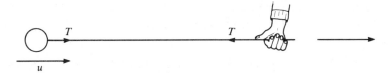

Notice that the initial velocity of the particle will be in the direction of the string.

Using the impulse-momentum equation gives,

$$T = mu$$

where T is the impulsive tension generated in the string and u is the initial velocity of the particle.

Investigation 5

A particle A, of mass 2 kg, is connected by a light inextensible string passing over a smooth pulley to a particle B, of mass 3 kg.

Calculate the acceleration of the system if it is released from rest.

After 5 seconds, particle B strikes the floor and does not rebound. Calculate the velocity of the system after 5 seconds and deduce, from the motion of A, its speed just before the string becomes taut again.

If I is the impulsive tension in the string as B is jerked off the floor, show that, by applying the impulse-momentum equation to particles A and B separately, the common velocity V, after the particle B is in motion again, is $2g/5$.

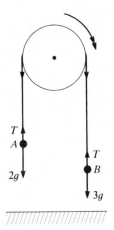

WORKED EXAMPLE

Two particles, A and B, of masses 2 kg and 3 kg respectively, are connected by a taut light inextensible string and lie at rest on a smooth horizontal table. If B is given an external impulse of magnitude I in a direction making an angle of 60° to the string AB away from A, find the velocities of A and B immediately after the impulse has been applied.

In this problem, the external impulse I applied to B causes an impulsive tension T in the string AB. Hence, A will move as a result of the impulsive tension in the direction of the string and B will move at an angle to AB as a result of the two impulses applied to it.

Let the initial components of the velocity of B be u in the direction AB and v in the direction perpendicular to AB. As the string is taut initially particle A will also move with velocity u along AB.

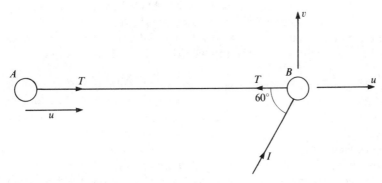

In solving these problems we use the impulse-momentum equation in suitable directions. It is possible to apply the equation to A in the direction of AB and to B in directions along and perpendicular to AB. Using the impulse-momentum equation gives, in general,

$$\text{Impulse} = \text{change in momentum}$$

On particle A, $T = 2u$ .... **(1)**

On particle B, along AB, $I \cos 60° - T = 3u$ .... **(2)**

On particle B, perpendicular to AB, $I \sin 60° = 3v$ .... **(3)**

From Equations **(1)** and **(2)**,

$$I \cos 60° - 2u = 3u \implies u = \frac{I}{10}$$ .... **(4)**

From Equation **(3)**, $v = \dfrac{I\sqrt{3}}{6}$

The velocity of A is $\dfrac{I}{10}$ along AB.

The velocity of B is $\sqrt{\left(\dfrac{I}{10}\right)^2 + \left(\dfrac{I\sqrt{3}}{6}\right)^2} = 0.31I$ at an angle $\tan^{-1}\dfrac{v}{u} = \tan^{-1}\left(\dfrac{I\sqrt{3}}{6} \times \dfrac{10}{I}\right) = 70.9°$ to AB.

Notice that if we consider the system as a whole, the internal impulsive tensions T have no resultant effect. The total momentum change is thus due to the external impulse I.

Applying the impulse-momentum equation to the whole system gives,

Along AB, $I\cos 60° = 3u + 2u$ i.e. producing Equation **(4)** in one step.

Perpendicular to AB, $I\sin 60° = 3v + 0 = 3v$ i.e. giving Equation **(3)**.

This provides an alternative approach to the solution.

Exercise 10.4

1   A particle of mass 0.5 kg lies at rest on a smooth horizontal table. A light taut inextensible string is attached to the particle and lies along the table. A boy tugs the string along its length with an impulse of 10 N s. Find the speed with which the particle begins to move.

2   Two particles of masses 6 kg and 4 kg are connected by a light inextensible string which passes over a smooth pulley. The system is released from rest and after moving 0.5 m the 6 kg mass hits the floor without rebounding. Find the further time that elapses until the system is again at rest instantaneously with the string taut.

3   Two particles of masses 1 kg and 2 kg are connected by a light inextensible string of length 2.5 m, which passes over a smooth peg 2 m above a horizontal table. The 2 kg mass lies on the table and the 1 kg mass is held next to the peg and released from rest. Find the common velocity of the two masses immediately after the string has become taut. Find the height above the table at which the 1 kg mass comes to rest.

4   Two particles, A and B, of masses 5 kg and 3 kg, respectively, lie at rest on a smooth horizontal table and are connected by a taut light inextensible string. If A is given an impulse of 40 N s in a direction inclined at 45° to AB and away from B, find the initial velocities of A and B and the impulsive tension in the string.

5   Two particles, A and B, each of mass 2 kg, lie at rest on a smooth horizontal table 3 m apart. They are connected by a light inextensible string of length 5 m. A is given an impulse of 10 N s in a direction perpendicular to AB. Find the speed of A immediately after the impulse of 10 N s has been applied and determine the initial speeds of A and B after the string becomes taut.

6   A particle A, of mass 1 kg, lies at the edge of a table and is connected by a light inextensible string of length 1.5 m to a particle B, of mass 3 kg, which is 0.5 m from the edge of the table. If AB is perpendicular to the edge of the table, find the velocity with which B begins to move after A is gently pushed over the edge of the table.

7   Three particles A, B and C, of mass m, $2m$ and m, respectively, lie on a smooth horizontal table with B connected to A and C by two light taut inextensible strings, such that the angle ABC is 60°. Particle B is given an impulse I in the direction AB away from A. Find the impulsive tensions in the strings AB and BC and the speeds with which the particles A, B and C begin to move.

Direct impact of elastic bodies

When two bodies collide they either coalesce (as discussed earlier in the chapter), or they bounce and separate again.

When they coalesce we say, in mathematics, that the impact is **inelastic**; and when they bounce and separate we say that there is an **elastic impact**.

It should be noted that this could be slightly confusing because, as those who study physics will know, the property of a body to recover its shape and rebound after a collision is called **elasticity**. If a body does not recover its shape it is said to be inelastic.

In this section, we shall consider the elastic impact of smooth spheres which separate after collision and where the mutual impulse between them will be along the line joining their centres. Such impacts are said to be **direct**. In these cases, the velocities just before the collision will be along the line of centres.

On diagrams, we show the initial and final velocities above and below the spheres as before.

In the last section, on impact, we found that momentum is conserved. This is still true. However, Newton discovered a further relationship between the velocities when he was investigating the rebound properties of different elastic bodies.

Investigation 6

Take a tennis ball and drop it from rest on to a hard floor. Measure the height H from which it was dropped and the height h to which is rebounds.

Repeat this for a golf ball, a squash ball, a super bouncing ball, a ball of putty etc. Make sure that you drop them all from the same height, H.

For each example form the ratio $\sqrt{\dfrac{h}{H}}$.

What conclusions can you draw about the value of this ratio and the material of the ball?

Show that the speed with which they reach the floor is $\sqrt{2gH}$.

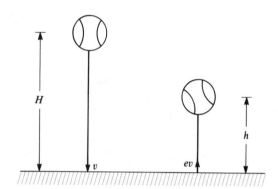

On the return upward flight, they must start with a lower speed since, moving under gravity, they do not reach the original height.

Let the return velocity be ev upwards, where $e < 1$. Show that $e = \sqrt{\dfrac{h}{H}}$.

The fraction e, defined in Investigation 6, is called the **coefficient of restitution** and is a constant for the particular materials in collision. For putty, which does not rebound, $e = 0$ and this is an inelastic collision. For a rubber bouncing ball, e might be as high as 0.9.

If $e = 1$, the collision is said to be **perfectly elastic**.

> In general, $0 \leqslant e \leqslant 1$.

The same principle applies if both bodies are moving, but in this case we use the relative velocities before and after impact.

Newton's law of restitution (or Newton's experimental law) states that,

> When two bodies impinge directly, the relative velocity after impact is in a constant ratio to the relative velocity before impact and is in the opposite direction.

Impulse, momentum and impact

This gives the formula, $\dfrac{v_2 - v_1}{u_2 - u_1} = -e$

$$\boxed{v_2 - v_1 = -e(u_2 - u_1)}$$

Remember that this is an experimental law and cannot be proved by mathematical logic. Since we are only considering velocities in one dimension, we use this result in a non-vector form, remembering to assign the correct sign to the velocities.

Two examples are shown to illustrate the possible directions for the final velocities and the correct use of signs.

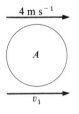

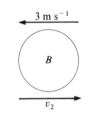

Using, $\qquad\qquad v_2 - v_1 = -e(u_2 - u_1)$

Case 1 $\qquad\qquad v_2 - v_1 = -e(-3 - 4)$

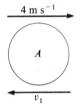

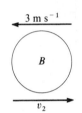

Case 2 $\qquad\qquad v_2 - (-v_1) = -e(-3 - 4)$

WORKED EXAMPLES

1 A sphere A, of mass $4m$ moving with a speed of 6 m s^{-1}, collides directly with a similar sphere, B, of mass $2m$ which is at rest. If the coefficient of restitution between the spheres is $\frac{1}{2}$, find the speeds of the spheres after the impact. Find also the loss of kinetic energy.

Let the speeds of spheres A and B after impact be v_1 and v_2 respectively, as shown in the diagram. Consider the initial direction of motion of A as the positive sense.

Using the principle of conservation of momentum gives,

$$m_1 u_1 + m_2 u_2 = m_1 v_1 + m_2 v_2$$

$$\therefore \qquad (4m \times 6) + (2m \times 0) = 4m v_1 + 2m v_2$$

$$\Rightarrow \quad 4v_1 + 2v_2 = 24$$

$$\Rightarrow \quad 2v_1 + v_2 = 12 \qquad\qquad \text{............ (1)}$$

Using Newton's law of restitution gives,

$$v_2 - v_1 = -e(u_2 - u_1)$$

$$\therefore \quad v_2 - v_1 = -\tfrac{1}{2}(0 - 6)$$

$$\Rightarrow \quad v_2 - v_1 = 3 \qquad \text{............ (2)}$$

Subtracting Equation **(2)** from Equation **(1)** gives,

$$3v_1 = 9 \quad \Rightarrow \quad v_1 = 3$$

Substituting in Equation **(2)** gives, $\quad v_2 = 6$

Hence, the speed of A is 3 m s^{-1} and the speed of B is 6 m s^{-1} (in the original direction of A).

Now, the K.E. before impact $= \tfrac{1}{2}(4m)6^2 + \tfrac{1}{2}(2m)0 = 72m \text{ J}$

and the K.E. after impact $= \tfrac{1}{2}(4m)3^2 + \tfrac{1}{2}(2m)6^2 = 54m \text{ J}$

Thus, the loss of kinetic energy due to the impact $= 18m \text{ J}$.

2 A sphere A, of mass $3m$ moving with speed $2u$, impinges directly on another similar sphere B, of mass m moving in the opposite direction with speed u. If, after the collision, the speed of A is half the speed of B and they move in the same direction, find the coefficient of restitution.

Let the speeds of A and B after the impact be v_1 and v_2 respectively, in the directions shown in the diagram. Note that you can choose the directions for v_1 and v_2 differently and the same solutions will be obtained. However, you must use the correct sign for v_1 and v_2 according to your choice.

Using the principle of conservation of momentum,

$$m_1 u_1 + m_2 u_2 = m_1 v_1 + m_2 v_2$$

$$\therefore \quad 3m(2u) + m(-u) = 3mv_1 + mv_2$$

$$\Rightarrow \quad 3v_1 + v_2 = 5u \qquad \text{............ (1)}$$

Using Newton's law of restitution,

$$v_2 - v_1 = -e(u_2 - u_1)$$

$$\therefore \quad v_2 - v_1 = -e(-u - 2u)$$

$$\Rightarrow \quad v_2 - v_1 = 3eu \qquad \text{............ (2)}$$

Now, since the speed of A is half the speed of B, $\quad v_2 = 2v_1$.

Impulse, momentum and impact

So Equations **(1)** and **(2)** become,

$$5v_1 = 5u \quad \Rightarrow \quad v_1 = u \qquad \text{.......... (3)}$$

$$\text{and} \quad 2v_1 - v_1 = 3eu \quad \Rightarrow \quad v_1 = 3eu \qquad \text{.......... (4)}$$

Clearly, from Equations **(3)** and **(4)**, $e = \frac{1}{3}$.

Thus, the coefficient of restitution between the spheres is $\frac{1}{3}$.

Exercise 10.5

1 A ball is dropped from a height of 2 m above horizontal ground. If the coefficient of restitution between the ball and the ground is $\frac{3}{4}$, find the height to which it rebounds.

2 A ball moves on a smooth horizontal plane and passes a point P with a speed of 10 m s^{-1}. It strikes a wall which is 5 m from P and at right angles to the path of the ball. If the coefficient of restitution between the ball and the wall is $\frac{3}{5}$, calculate the time that elapses before the ball returns to P.

3 A sphere of mass 4 kg moving at 3 m s^{-1} overtakes another sphere of mass 2 kg moving in the same direction at 2 m s^{-1}. Find the velocities of the spheres after the collision if the coefficient of restitution is $\frac{1}{2}$.

4 A sphere of mass 5 kg moving at 4 m s^{-1} impinges directly on another sphere of mass 4 kg moving in the opposite direction at 3 m s^{-1}. If the coefficient of restitution between the spheres is $\frac{1}{5}$, find the velocities of the spheres after the collision.

5 Two spheres each of mass m impinge directly when moving in opposite directions. If their speeds before impact are 6 m s^{-1} and 3 m s^{-1} and half the kinetic energy is lost in the collision, show that $e = \frac{2}{3}$.

6 A ball of mass 1.5 kg moving at 10 m s^{-1} collides directly with another similar ball of mass 3 kg moving at 5 m s^{-1} in the same direction. Find the loss in kinetic energy during the impact if $e = \frac{1}{2}$.

7 A sphere, A, of mass 0.5 kg is moving with a speed of 4 m s^{-1} when it collides directly with a similar stationary sphere, B. If A is brought to rest by the impact and the coefficient of restitution is $\frac{3}{4}$, find the speed of B after the collision and its mass.

8 A sphere of mass m_1 moving with speed $5u$ collides directly with a similar sphere of mass m_2 moving in the same direction with speed u. After the collision the spheres move in the same direction with speeds $2u$ and $3u$, respectively. Find the coefficient of restitution and the ratio $m_1 : m_2$.

9 Two spheres of equal mass lie on a smooth table. One sphere is given a speed u so as to collide directly with the other sphere. Show that the velocities after the collision are $\frac{1}{2}u(1-e)$ and $\frac{1}{2}u(1+e)$, where e is the coefficient of restitution.

10 A sphere A of mass $2m$ moving at 5 m s^{-1} collides directly with a sphere B of mass m which is at rest. Find the velocity of B after the impact, if the coefficient of restitution between the spheres is $\frac{1}{5}$. B then strikes a wall perpendicular to its path and, after the subsequent collision, A has a final speed of 1 m s^{-1} in an unchanged direction. Find the coefficient of restitution between the wall and the sphere.

11 Three perfectly elastic spheres A, B and C have masses of $6m$, $4m$ and $2m$, respectively. They lie in a straight groove in a smooth table and A is given a speed of 4 m s^{-1} towards B so that they collide directly. B subsequently collides with C. Find the speeds of each sphere after two collisions have taken place and deduce that there are no more impacts.

218

12 Three spheres A, B and C of equal mass lie in a straight line on a smooth table. If A is projected with speed u so that it collides directly with B, which then collides directly with C, show that the speeds of the spheres after two collisions are $\frac{1}{2}(1-e)u$, $\frac{1}{4}(1-e^2)u$ and $\frac{1}{4}(1+e)^2u$, where e is the coefficient of restitution.

13 Two beads, A and B, of masses $2m$ and m respectively, are threaded on to a smooth circular wire of radius a, which is fixed in a vertical plane. The beads are released from rest at the top of the wire and collide at the lowest point. If the coefficient of restitution is $\frac{1}{3}$, find the speeds of the beads immediately after the collision.

14 Three identical spheres A, B and C lie at rest in a straight smooth groove. Sphere A is projected towards B with speed u. Show that there will be exactly three impacts and find the final speed of each sphere if the coefficient of restitution is $\frac{1}{2}$.

Revision Exercises D

These Exercises should enable you to revise work covered in Book 1.

Permutations and combinations

1 Without using a calculator, find the values of the following.

     (a) $4!$ **(b)** $\dfrac{6!}{3!}$ **(c)** $(3!)^2$ **(d)** $\dfrac{18!}{16!3!}$ **(e)** $5! - 2(4!)$

2 Express in factorial notation,

     (a) $7 \times 6 \times 5 \times 4 \times 3 \times 2$ **(b)** $17 . 16 . 15 . 14 . 13$ **(c)** $\dfrac{9 . 8 . 7 . 6}{4 . 3 . 2}$

3 Simplify the following, leaving your answers in factorials.

     (a) $7! + 8!$ **(b)** $14! - 2(12!)$ **(c)** $\dfrac{6!}{2!} + \dfrac{6!}{3!}$

     (d) $4(9!) - 3(8!)$ **(e)** $n! + n(n-2)!$ **(f)** $\dfrac{8!}{5!3!} + \dfrac{8!}{6!2!}$

4 Find the number of permutations of the letters of the word TRIANGLE.

5 Six different shrubs are to be planted in a straight line in a garden. In how many ways can the shrubs be arranged?

6 In how many ways can eight different bedding plants be planted in a circle?

7 A girl buys seven different postcards. In how many ways can she select four of them and send one to each of her four friends?

8 How many 4-digit numbers can be made from the digits 2, 4, 5, 7, 8, 9, if no digit may be repeated?

9 Find the number of arrangements of the letters of the following words.

     (a) EXIST   **(b)** EXAMPLE   **(c)** EXPONENT   **(d)** STATISTICS

10 Three boys and two girls sit on a bench. In how many ways can they sit if the two girls **(a)** sit next to each other and **(b)** do not sit next to each other?

11 How many 4-digit odd numbers can be formed from the digits 2, 3, 5, 7, 8, 9?

12 How many even 5-digit numbers greater than 60 000 can be made from the digits 3, 4, 6, 7, 8?

13 A girl writes 8 different invitations to a party and addresses 8 envelopes. In how many ways can she place one invitation in each envelope if she fails to get every invitation in the correct envelope?

14 Find the number of arrangements of 3 letters chosen from the word HOTTER.

15 How many different 4-digit numbers greater than 5000 can be formed from the digits $1, 2, 4, 5, 6, 7$ if **(a)** no digit may be repeated **(b)** any of the digits may be repeated?

16 Show that $^{10}C_4 + {}^{10}C_3 = {}^{11}C_4$.

17 In how many ways can

(a) a group of 5 people be chosen from a group of 8 people?

(b) a committee of 4 people be chosen from 10 candidates?

(c) 7 pencils be chosen from a box of 12 pencils of different colours?

18 How many different cricket teams of 11 players can be formed from a squad of 16 players?

19 If 9 people are to travel in two cars each able to carry 5 passengers, find the number of ways in which the group can be divided.

20 If 10 people are waiting to visit an exhibition which can admit, at most, 4 people at a time, find the number of ways in which the 10 can be divided if the first two groups each contain 4 people.

21 A committee consists of 4 men and 3 women. If 8 men and 5 women are eligible to serve, find the number of different committees that can be formed.

22 How many different groups of 3 letters can be made from the letters of the word PARALLELOGRAM? How many of these contain the letter G?

23 Show that the total number of combinations of 6 items taken one or more at a time is 63.

24 By using the binomial expansion of $(1+x)^n$, show that
$$^nC_1 + {}^nC_2 + {}^nC_3 + \cdots + {}^nC_n = 2^n$$
Deduce that the total number of combinations of n items taken one or more at a time is $2^n - 1$.

25 Two boats can each hold 6 people. In how many ways can a group of 6 boys and 6 girls travel so that there are equal numbers of boys and girls in each boat?

Series

1 Find the first three terms and the 15th term of the following sequences, defined as follows.

(a) $u_n = 2^{n-1}$ for $n = 1, 2, 3, \ldots$ (b) $u_n = \dfrac{n}{(n+1)^2}$ for $n = 1, 2, 3, \ldots$

(c) $u_n = \dfrac{n^2}{(n-1)(n+2)}$ for $n = 2, 3, 4, \ldots$

2 For the first four terms of the following sequences.

(a) $u_{r+1} = 4(u_r - 1);$ $u_1 = 5$ (b) $u_{r+1} = \frac{1}{2}u_r;$ $u_1 = 100$ (c) $u_{r+1} = (r+2)u_r;$ $u_1 = 1$

3 Determine the behaviour of the following sequences as n becomes large.

(a) $u_n = \dfrac{2}{n+1}$ (b) $u_n = \dfrac{2n}{n+3}$ (c) $u_n = 5 - \dfrac{1}{n^2}$ (d) $u_n = \dfrac{4n-1}{n+1}$ (e) $u_n = 3 + (-1)^n$

4 Write down the terms of the following series.

    (a) $\displaystyle\sum_{r=3}^{7}(r^2-2)$ (b) $\displaystyle\sum_{r=1}^{4}\frac{r}{r+1}$ (c) $\displaystyle\sum_{r=2}^{8}(-1)^r(2^r-1)$

5 Write the following series using the Σ notation.

    (a) $1+3+5+7+\ldots+51$

    (b) $2(1)^2+3(2)^2+4(3)^2+\ldots$ to n terms

    (c) $(2\times5)+(3\times6)+(4\times7)+\ldots+(15\times18)$

6 Find the 50th term and the sum of the first 20 terms of the arithmetical progression $5+8+11+14+\ldots$

7 How many terms are there in the series $40+34+28+\ldots-134$?

8 Find three consecutive terms of an arithmetical progression such that their sum is 51 and their product is 4301.

9 The 10th term of an arithmetical progression is 22 and the sum of the first 10 terms is -95. Find the first term and the common difference.

10 Find the 18th term and the sum of the first 8 terms of the geometrical progression $20-10+5-2\frac{1}{2}+\ldots$

11 Find the sum of the geometrical progression $\dfrac{3}{4}-\dfrac{9}{8}+\dfrac{27}{16}-\ldots+\dfrac{2187}{256}$

12 The second term of a geometrical progression is 6 and the fifth term is 162. Find the common ratio and the sum of the first 10 terms.

13 In a geometrical progression, the sum of the second and third terms is 12 and the sum of the third and fourth terms is -48. Find the common ratio and the sum of the first 5 terms.

14 The 2nd, 5th and 12th terms of an arithmetical progression form part of a geometrical progression. If the 8th term is 66, find the first term and the common difference of the arithmetical progression.

15 Find the sum to infinity of the geometrical progression $3+\dfrac{9}{4}+\dfrac{27}{16}+\dfrac{81}{64}+\ldots$

16 Express the following recurring decimals as fractions in the lowest terms.

    (a) $0.\dot{4}$ (b) $0.\dot{3}\dot{2}$ (c) $0.\dot{5}1\dot{3}$

17 Find the range of values of x for which the following series converges.

$$1+\frac{(x-2)}{3}+\frac{(x-2)^2}{9}+\frac{(x-2)^3}{27}+\ldots$$

18 Expand the following functions as series in ascending powers of x.

    (a) $(1+x)^6$ (b) $(2+x)^5$ (c) $(1-x^2)^8$

19 Expand $(1-x)^{11}$ as far as the term in x^4 and hence find the value of $(0.975)^{11}$, correct to 3 decimal places.

20 Find the expansion of the following in ascending powers of x as far as the term in x^3.

    (a) $(1-2x)^{-2}$ (b) $\sqrt{1+4x}$ (c) $\dfrac{1}{(2+x)}$

21 Expand the following in ascending powers of x, giving the first 4 terms.

    (a) $\dfrac{3+x}{1+x}$ **(b)** $(1+x)^2(2+x)^{-3}$ **(c)** $\dfrac{2-x}{\sqrt{1+5x}}$

22 Expand $\dfrac{1-2x}{1+3x}$ in ascending powers of x as far as the term in x^3, stating the range for which it is valid.

23 Find the sums of the following series.

    (a) $1^2+2^2+3^2+4^2+\ldots+(n+1)^2$

    (b) $1^3+3^3+5^3+7^3+\ldots+(2n-1)^3$

    (c) $(2\times3)+(3\times4)+(4\times5)+\ldots$ to $(2n)$ terms

    (d) $(1\times3\times5)+(2\times4\times6)+(3\times5\times7)+\ldots$ to n terms.

24 Use the method of differences to find the sum of the following series to n terms. Deduce the sum to infinity.

    (a) $\dfrac{1}{1\times3}+\dfrac{1}{2\times4}+\dfrac{1}{3\times5}+\ldots$ **(b)** $\dfrac{1}{1\times2\times5}+\dfrac{1}{2\times3\times6}+\dfrac{1}{3\times4\times7}+\ldots$

25 If $f(r)=\dfrac{1}{r!}$, write down an expression for $f(r)-f(r+1)$. Use this to find $\displaystyle\sum_{r=1}^{n}\dfrac{r}{(r+1)!}$.

26 Show that, if x^3 and higher powers of x can be neglected, $\sqrt{\dfrac{1-x}{1+x}}=1-x+\tfrac{1}{2}x^2$. Use this result to show that the value of $\sqrt{14}$ is approximately 3.742.

11 Probability

Introduction

Investigation 1

What is the probability that it will rain tomorrow?

On which factors will it depend?

Is this a reasonable question?

Investigation 2

What is the probability that I will win the main prize in the draw of tickets in my ONE HUNDRED club?

Each of the 100 members pays in £1 per month, and there are prizes of £15, £10 and £5 per month and in addition a prize of £100 in June and £240 in December. £1200 is paid in by members of which £700 is paid out in prizes. Does this mean that I have a better than even chance of winning money?

Investigation 3

When we play bridge, four players cut to decide who deals. Does this mean I have a 1 in 4 chance of dealing?

When cards are dealt, I naturally want a good hand. What is the chance of being dealt all 4 aces?

To open the bidding I need at least 12 points (counting 4 for an ace, 3 for a king, 2 for a queen, 1 for a jack). Is my chance of bidding better than 1 in 4?

Investigation 4

When we play **Ludo** we throw a pair of dice and we can start if we throw a **six**. How many throws will I need to start?

Investigation 5

At the start of many games, players toss a coin to see who starts. Why do we use a coin? Do we use other methods? What happens if we have no coin?

Investigation 6

What is the chance of winning a prize with a **Premium Bond**?

If the probability of winning is so small, why do we buy them?

Investigation 7

Why do we enter the **football pools**, or bet on **horses**?

Why do we gamble at all?

Investigation 8

What is the probability of **(a)** catching a cold   **(b)** being involved in a road accident   **(c)** having a heart attack?

Investigation 9

What sayings involve a qualitative statement of probability, e.g. 'nine times out of ten' or 'a stitch in time saves nine'?

Investigation 10

With what value of probability do you associate   **(a)** ninety-nine times out of a hundred   **(b)** never **(c)** always   **(d)** invariably   **(e)** a good chance   **(f)** odds-on favourite   **(g)** evens   **(h)** 10 to 1?

What is probability?

What is the **difference** between **climate** and **weather**?

A fourteen-year-old gave this answer:

"Climate is what **should** happen and weather is what **does** happen."

In other words, the fourteen-year-old is saying that climate is a theoretical forecast based on years of observation of the weather. We can consider the relationship between probability and statistics in the same way. **Probability** is the theoretical prediction of results which resemble closely the statistical results which are observed.

Some of the answers to the Investigations will depend on previous events, while others can be analysed mathematically. Some will depend on factors beyond the control of the particular situation (such as the weather), while others can be worked out theoretically from the situation described (such as cards).

Situations like **football pools** and **Premium Bonds** involve so many possibilities that a generalization may be required.

Any analysis of the **weather** will require the **statistics** collected over many years, but even then results can be unpredictable.

When throwing a **die** (plural **dice**) the results are **random** (if the die is fair) and we would expect the number 6 to appear one time in six, but we could roll the die twenty times and not produce a 6.

The **theoretical probability** of throwing a 6 is one in six, or $\frac{1}{6}$, since each of the six results $(1, 2, 3, 4, 5$ or $6)$ is **equally likely**.

When we cut a pack of cards which has been thoroughly shuffled, each card has a equal chance of being cut as the cards have been **randomly** distributed throughout the pack. The drawing of one particular card is a **random event**.

When tickets are drawn in a raffle, the organizers try to ensure that each ticket has an equal chance of being selected. Nowadays, large competitions are decided by computers. Do you know how the program is devised to ensure **impartiality**?

When we roll a die there are **six possible outcomes** or **possibilities**.

The **event** of rolling the number 2 has a **probability** of one in six, or $\frac{1}{6}$.

The **event** of rolling an **even** number has a **probability** of three in six, or $\frac{3}{6} = \frac{1}{2}$, as each of 2, 4 or 6 is an even number.

The **event** of NOT rolling the number 2, $p \, (\text{not } 2) = \frac{5}{6} = 1 - p \, (2)$.

Exercise 11.1

1 When rolling a fair die, what is the probability of obtaining **(a)** a 3 **(b)** an odd number **(c)** a prime number **(d)** a number greater than 4 **(e)** a 7?

2 When cutting a pack of 52 playing cards, what is the probability of cutting **(a)** a red card **(b)** a diamond **(c)** a court card (J, Q or K) **(d)** an ace **(e)** a red ace **(f)** a king or a club **(g)** the king of clubs **(h)** a joker **(i)** a card with 12 hearts on it?

3 In a draw of 100 tickets, what is the probability of winning the **(a)** first prize **(b)** second prize **(c)** third prize?

4 What is the probability of passing your driving test first time?

5 What is the probability that a letter chosen at random from this Exercise will be **(a)** an 'a' **(b)** a 'b' **(c)** a 'c'?

6 Of the figures (digits 0 to 9) which are printed on this page, which occurs **(a)** the most **(b)** the least? **(c)** Why?

Trials and events

When we throw a die we regard this throw as a **trial** in which there are six possible results or **outcomes**.

We can describe one or more **events** as possible results arising from one **trial**, e.g. the event E of throwing an even number, which with one throw of a die has **probability** $\frac{3}{6} = \frac{1}{2}$, since the 2, 4 or 6 are equally likely outcomes of the six possible results.

If S is the event of throwing a 6, then $p(S) = \frac{1}{6}$.

If F is the event of throwing a 5, then $p(F) = \frac{1}{6}$.

Adding probabilities

The probability of throwing a 5 OR a 6 is

$$p(F \text{ OR } S) = p(F) + p(S) = \frac{1}{6} + \frac{1}{6} = \frac{1}{3}$$

In this case, the events S and F are **mutually exclusive**; they cannot both happen simultaneously, one event excludes the other, i.e.

$$\boxed{p(S \text{ AND } F) = 0 \quad \Rightarrow \quad p(S \text{ OR } F) = p(S) + p(F)}$$

$$p(E) = p(\text{even}) = \frac{1}{2} \quad \text{and} \quad p(S) = \frac{1}{6} \quad \text{but} \quad p(E \text{ OR } S) = \frac{3}{6} \quad \text{not} \quad \frac{1}{2} + \frac{1}{6} \quad \text{since} \quad p(E \text{ and } S) = \frac{1}{6}$$

Here we are concerned with **two events within one trial** with the possibility that both events can happen simultaneously. We shall deal with this more fully later.

The same trial repeated

Tossing two coins

When two coins are tossed there are four possible results, two heads, two tails, a head and a tail or a tail and a head. These four possible outcomes are best illustrated by using a **tree diagram**.

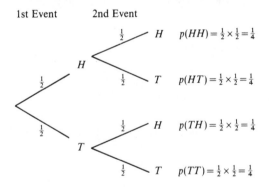

As the four outcomes are equally likely, the probability of each outcome: HH, HT, TH and TT, is equal to $\frac{1}{4}$. For each event, the probability of an H or T equals $\frac{1}{2}$ and if these are assigned to each branch, then the combined probability of HH, i.e. a head followed by a head, is obtained by **multiplying** along the branches, so

$$p(HH) = \frac{1}{2} \times \frac{1}{2} = \frac{1}{4}$$

The probability of a head and a tail in any order is obtained by **adding** the probabilities for HT and TH, so

$$p(\text{head and tail}) = p(HT) + p(TH) = \frac{1}{4} + \frac{1}{4} = \frac{1}{2}$$

Notice, in the tree diagram, that the sum of the probabilities in the final column is 1, showing that one of the four outcomes must occur.

WORKED EXAMPLE

> If the probability of having a girl is the same as having a boy at any birth, what is the probability of a family of three children being two girls and a boy?

The three births will produce three events in the tree diagram. Look along the branches for the outcomes *BGG, GBG, GGB* and add the probabilities.

$$p(BGG) + p(GBG) + p(GGB) = \frac{1}{8} + \frac{1}{8} + \frac{1}{8} = \frac{3}{8}$$

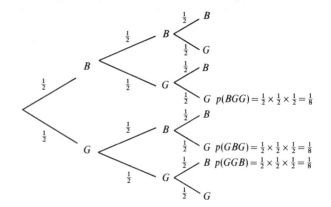

A quicker way to achieve the result, especially with more events, is to look at the Binomial Expansion.

$$(b+g)^3 = b^3 + 3b^2g + \underline{3bg^2} + g^3$$

where the first term represents 3 boys, the second 2 boys and a girl, the third a boy and 2 girls and the fourth 3 girls.

For this example we require the third term and, if we substitute $b = \frac{1}{2}$ and $g = \frac{1}{2}$,

$$3bg^2 = 3 \times \frac{1}{2} \times \frac{1}{2} \times \frac{1}{2} = \frac{3}{8}$$

Bear in mind that this includes *BGG, GBG, GGB*; in other words make sure this order does not matter.

Extension   For four children the Binomial method is so much quicker. To find the probability of three boys and a girl in a family of four children, we expand $(b+g)^4$.

$$(b+g)^4 = b^4 + \underline{4b^3g} + 6b^2g^2 + 4bg^3 + g^4$$

The term in b^3g gives the probability of three boys and a girl, so

$$p(BBBG, \text{ in any order}) = 4\left(\frac{1}{2}\right)^3 \left(\frac{1}{2}\right) = \frac{4}{16} = \frac{1}{4}$$

Throwing two dice

Noting that each die can land in 6 ways gives $6 \times 6 = 36$ possible results for 2 dice, as illustrated by the **possibility space** diagram.

Throwing 2 sixes can occur in only 1 way, so

$$p(\text{double six}) = \frac{1}{36}$$

$$p(\text{one six only}) = \frac{10}{36}$$

$$p(\text{no sixes}) = \frac{25}{36}$$

$$p(\text{total score of seven}) = \frac{6}{36} = \frac{1}{6}$$

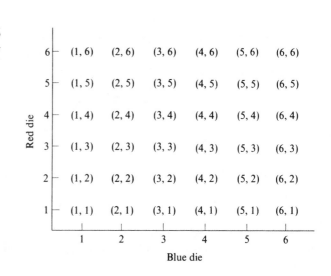

The same trial repeated

If we are concerned with the event of throwing sixes on each die, we can draw a **tree diagram** to describe the probabilities.

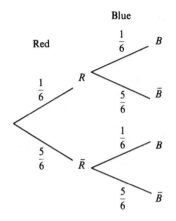

Along the branches, 4 routes lead to 4 results

$R \cap B$ a red six and a blue six, $p(R \cap B) = \dfrac{1}{6} \times \dfrac{1}{6} = \dfrac{1}{36}$

$R \cap \bar{B}$ a red six and not a blue six, $p(R \cap \bar{B}) = \dfrac{1}{6} \times \dfrac{5}{6} = \dfrac{5}{36}$

$\bar{R} \cap B$ not a red six and a blue six, $p(\bar{R} \cap B) = \dfrac{5}{6} \times \dfrac{1}{6} = \dfrac{5}{36}$

$\bar{R} \cap \bar{B}$ not a red six and not a blue six, $p(\bar{R} \cap \bar{B}) = \dfrac{5}{6} \times \dfrac{5}{6} = \dfrac{25}{36}$

In this case we are considering throwing the red die first and then the blue die. The two trials are **independent**, the result on the blue die not depending on the result on the red die. When two dice are rolled simultaneously we can analyse the results in this way, as though one die were being rolled after the other.

We know from the possibility space diagram for throwing two dice that

$$p(R \cap B) = \frac{1}{36}, \quad p(R \cap \bar{B}) = \frac{5}{36}, \quad p(\bar{R} \cap B) = \frac{5}{36}, \quad p(\bar{R} \cap \bar{B}) = \frac{25}{36}$$

and these results justify our method of multiplying along the branches of the tree diagram.

Remember that 'one six only' can occur in two ways, so

$$p(\text{one six only}) = p(R \cap \bar{B}) + p(\bar{R} \cap B) = \frac{5}{36} + \frac{5}{36} = \frac{10}{36} = \frac{5}{18}$$

> **WORKED EXAMPLE**

Find the probability of throwing 3 sixes with   **(a)** 3 dice   **(b)** 4 dice   **(c)** 5 dice.

(a)   If S is the event of throwing one six, we can use the Binomial Expansion for 3 dice (looking for sixes).

$(S + \bar{S})^3 = S^3 + \cdots$ to give $p(SSS) = \dfrac{1}{6} \times \dfrac{1}{6} \times \dfrac{1}{6} = \left(\dfrac{1}{6}\right)^3 = \dfrac{1}{216}$

(b)   For 3 sixes from 4 dice, use $(S + \bar{S})^4 = S^4 + 4S^3\bar{S} + \ldots$ to give

$p(SSS\bar{S}, \text{ in any order}) = 4 \times \left(\dfrac{1}{6}\right)^3 \times \dfrac{5}{6} = \dfrac{20}{1296} = \dfrac{5}{324}$

(c)   For 3 sixes from 5 dice, we expand $(S + \bar{S})^5$,   looking for the $S^3\bar{S}^2$ term.

$p(S^3\bar{S}^2, \text{ in any order}) = {}^5C_3 \times \left(\dfrac{1}{6}\right)^3 \times \left(\dfrac{5}{6}\right)^2$

$\qquad\qquad = \dfrac{5 \times 4}{2} \times \dfrac{1}{216} \times \dfrac{25}{36} = \dfrac{125}{3888} \simeq 0.032$

Binomial expansion

$$(a+b)^5 = a^5 + 5a^4b + 10a^3b^2 + 10a^2b^3 + 5ab^4 + b^5$$

If in throwing 5 dice we are looking for sixes, let $\ b = p(\text{six}) = \dfrac{1}{6} \ \Rightarrow \ a = \dfrac{5}{6}$

$$\left(\frac{5}{6}+\frac{1}{6}\right)^5 = \underset{p(\text{no sixes})}{\left(\frac{5}{6}\right)^5} + \underset{p(\text{1 six})}{5 \times \left(\frac{5}{6}\right)^4 \times \left(\frac{1}{6}\right)} + \underset{p(\text{2 sixes})}{10 \times \left(\frac{5}{6}\right)^3 \times \left(\frac{1}{6}\right)^2} + \underset{p(\text{3 sixes})}{10 \times \left(\frac{5}{6}\right)^2 \times \left(\frac{1}{6}\right)^3} + \underset{p(\text{4 sixes})}{5 \times \frac{5}{6} \times \left(\frac{1}{6}\right)^4} + \underset{p(\text{5 sixes})}{\left(\frac{1}{6}\right)^5}$$

Exercise 11.2

1 Find the probabilities of the following events in tossing 3 coins,

(a) 3 heads (b) 2 heads and 1 tail (in that order)
(c) 2 heads and 1 tail (in any order) (d) at least one head.

2 "There are 3 results in tossing 2 coins: (a) 2 heads (b) 2 tails (c) one of each; so that the probability of each of (a), (b), and (c) is $\dfrac{1}{3}$." Criticize!

3 I won the toss for my cricket team last Saturday for the first time in the last five matches. What is the probability of winning (a) only once in 5 games (b) twice in 5 games (c) more often than not?

4 What is the probability of throwing 2 heads with (a) 4 coins (b) 5 coins?

5 What is the probability of throwing an equal number of heads and tails with:

(a) 2 coins (b) 4 coins (c) 6 coins (d) 8 coins (e) 9 coins?

6 If the probability of passing my driving test each time I take it is $\dfrac{1}{3}$, what is the probability of passing it

(a) first time (b) second time (c) third time (d) nth time?

7 The probability that a car fails its MOT test on brakes is $\dfrac{1}{4}$. What is the probability that both my wife's car and mine will pass?

8 On a multi-choice examination paper, each question has 5 possible answers only one of which is correct. What is the probability of gaining exactly 40% by guessing your answers to (a) the first ten questions (b) the first twenty questions (c) the whole paper of 30 questions?

9 On each throw in a game of Ludo a player throws 2 dice. What is the probability of throwing a total of 4 sixes in 4 'throws' (i.e. 4 throws of 2 dice)?
Is this the same as throwing 4 sixes with 8 single throws of one die (a) in theory (b) in practice?

10 Last year, in 18 cricket games, our captain won the toss 12 times. What is the probability of this happening this year?

11 A die in the shape of an octahedron is numbered 1 to 8. What is the probability of throwing a total greater than 6 with (a) one throw (b) two throws (c) three throws?

12 What is the probability that, in your class (of 20 students), two students were born on (a) the same day of the week, e.g. a Monday (b) in the same month (c) in the same week?

Combined events (within one trial)

Rolling a die

There are six outcomes from one throw of a die and we have seen that the probability of a four is $\frac{1}{6}$ or the probability of an even number is $\frac{3}{6} = \frac{1}{2}$.

Also, $p(\text{five OR six}) = p(\text{five}) + p(\text{six}) = \frac{1}{6} + \frac{1}{6} = \frac{1}{3}$

as the events 'rolling a five' and 'rolling a six' cannot happen simultaneously.

However, $p(\text{even OR six}) \neq p(\text{even}) + p(\text{six})$ because the two events 'even' and 'six' are not **mutually exclusive**, i.e. they do not exclude each other.

In general, $\qquad p(E \text{ OR } S) = p(E) + p(S) - p(E \text{ AND } S)$

$$= \frac{1}{2} + \frac{1}{6} - \frac{1}{6} \qquad = \frac{1}{2}$$

The symbols used for the combined events,

$$E \text{ or } S \text{ (or both)}, \quad E \cup S \quad \text{and} \quad E \text{ and } S, \quad E \cap S$$

come from the connection with **sets** and the comparable result,

$$n(A \cup B) = n(A) + n(B) - n(A \cap B)$$

(Note that some books and examination boards use $E \vee S$ and $E \wedge S$.)

For instance, $p(\text{prime number}) = \frac{3}{6} = \frac{1}{2}$ since 2, 3, 5 are prime.

Venn diagram for the events E and P (even and prime).

For sets, $\quad n(E \cup P) = n(E) + n(P) - n(E \cap P)$

$$5 \quad = 3 \ + \ 3 \ - \quad 1$$

Probability $\quad \dfrac{5}{6} \ = \dfrac{1}{2} + \dfrac{1}{2} - \dfrac{1}{6}$

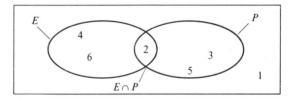

Combined events

The events, 'E AND P' denoted by $E \cap P$ and 'E OR P' denoted by $E \cup P$, are examples of **combined events**, which can be illustrated in different ways.

The symbols suggest a Venn diagram. There are four regions representing:

$E \cap P \quad$ E AND P both occur

$E \cap \bar{P} \quad$ E AND not P

$\bar{E} \cap P \quad$ not E AND P

$\bar{E} \cap \bar{P} \quad$ not E AND not P

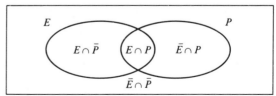

and these four regions correspond to the four branches of the **tree diagram** for the two events E and P.

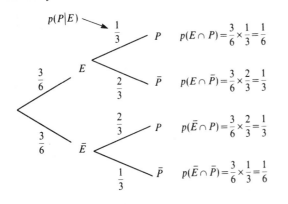 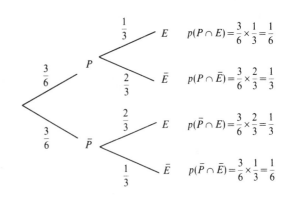

Writing *E* as the first event   Writing *P* as the first event

We can see that with two events described within one trial, the order of consideration makes no difference to the combined probability $E \cap P$ i.e.

$$p(E \cap P) = p(P \cap E)$$

and in general

$$p(A \cap B) = p(B \cap A)$$

With *E* written first, the probability that *P* happens is called 'the probability of *P* given *E*' and is written $p(P|E) = \frac{1}{3}$; and the probability of $\bar{P}$ given *E*, $p(\bar{P}|E) = \frac{2}{3}$.

From the second diagram, $p(E|P) = \frac{1}{3}$ and $P(\bar{E}|P) = \frac{2}{3}$.

In general, $p(A|B)$ is the **conditional probability** of event *A* given *B* has happened.

This is all reasonably obvious when one event precedes the next (as you will see later in the example on drawing beads), but not so easy to understand when the two events occur within one trial.

WORKED EXAMPLE

> With one throw of a die what is the probability of 'an odd number given that it is prime' and 'a prime number given it is odd'?

There are three primes of which two, 3 and 5, are odd so $p(O|P) = \frac{2}{3}$.

There are three odd numbers, 1, 3, 5, two of which are prime, so $p(P|O) = \frac{2}{3}$.

It will probably help at this stage to draw out the two tree diagrams for the events Prime and Odd, starting with *P* first then *O*, and also with *O* first then *P*, just as we have done for *P* and *E* earlier in this section.

Note   Remember that on every pair of branches emanating from an event, the probabilities add up to 1 (because one of them must happen).

The sum of the final column must also be 1, as that represents the total sum of the four possible combined results with two events.

Conditional probability

Formal definition for p(A/B)

Writing event B first in the tree diagram and reading along the top branches

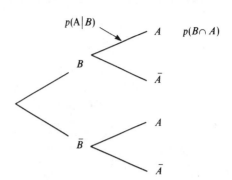

$$\Rightarrow \quad p(B) \times p(A|B) = p(B \cap A)$$

$$\Rightarrow \quad p(A|B) = \frac{p(B \cap A)}{p(B)}$$

Similarly, $\quad p(B|A) = \frac{p(A \cap B)}{p(A)}$

although to illustrate this, draw event A first.

The following example will provide clarification of **conditional probability**.

Drawing two beads from a bag

Let the bag contain 2 red beads and 3 blue beads.

With replacement	**Without replacement**
In this case, each bead chosen is returned to the bag.	In this case, each bead chosen is NOT returned.

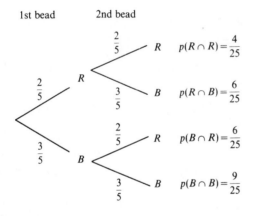 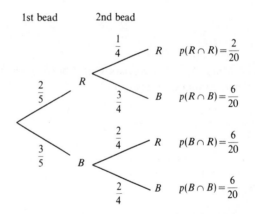

Here the same trial is repeated so the probabilities of drawing red or blue for the second bead are the same as for the first.

$$p(2 \text{ reds}) = p(R \cap R) = 0.16$$

$$p(\text{one of each}) = p(R \cap B) + p(B \cap R) = 0.48$$

$$p(2 \text{ blues}) = 0.36$$

Here the probabilities for the second bead depend on the result for the first bead. If a red bead is drawn first, one red and 3 blue are left; but if a blue is drawn first, 2 of each colour remain.

$$p(2 \text{ reds}) = p(R \cap R) = 0.10$$

$$p(\text{one of each}) = \frac{12}{20} = 0.60$$

$$p(2 \text{ blues}) = 0.30$$

Conditional probability

$p(B|R) = \dfrac{3}{5} = p(B|B)$ so drawing a blue second does not depend on the result of the first bead.

The second event is **independent** of the first.

$p(B|R) = \dfrac{3}{4}$ and $p(B|B) = \dfrac{2}{4}$ so the second event DOES depend on the result of the first bead.

The two events are **not independent**.

Exercise 11.3

1 One card is drawn from a pack of 52 playing cards. The following events are defined:

D = drawing a diamond; R = a red card; K = a king; C = a court card (J, Q or K).

Find the probabilities of the following events,

(a) D (b) R (c) K (d) C (e) $D \cap R$ (f) $D \cap \bar{R}$ (g) $D \cup R$ (h) $D \cap K$ (i) $D \cup K$

(j) $D \cup C$ (k) $D \cap C$ (l) $R \cap K$ (m) $R \cup K$ (n) $R \cap C$ (o) $R \cup C$ (p) $K \cap C$ (q) $K \cup C$ (r) $K | C$

(s) $C | K$ (t) $D | R$ (u) $R | D$ (v) $C | R$ (w) $R | C$ (x) $D | K$ (y) $K | D$ (z) $R | K$

2 Now two cards are drawn from the pack **without replacement**. Find the probabilities of the following events, $A \cap B$ and $A \cup B$ meaning A first B second, but $A | B$ meaning A second B first.

(a) $D \cap R$ (b) $R \cap K$ (c) $\bar{R} \cap D$ (d) $K \cap C$ (e) $\bar{K} \cap \bar{C}$ (f) $K \cup C$ (g) $\bar{D} \cap R$ (h) $D \cap \bar{R}$

(i) $\bar{D} \cup \bar{R}$ (j) $D \cup R$ (k) $\bar{D} \cap \bar{R}$ (l) $D | R$ (m) $R | \bar{D}$ (n) $K | C$ (o) $K | \bar{C}$ (p) $R | K$

3 Repeat question **2** for two cards drawn **with replacement**.

4 With the same events as defined in question **1**, find the probabilities of the following combinations of three successive events of drawing three cards from a pack successively **without replacement**, the order of drawing being the order of the letters.

(a) $R \cap D \cap C$ (b) $D \cap R \cap C$ (c) $C \cap R \cap K$ (d) $R \cap K \cap C$ (e) $K \cap R \cap C$ (f) $K \cap C \cap R$

5 With the same four events, D, R, K and C, and drawing **only one card**, can you find probabilities for the following compound events,

(a) $R \cap D \cap C$ (b) $D \cap R \cap K$ (c) $R \cap K \cap C$ (d) $D \cap K \cap C$ (e) $R \cup D \cup C$ (f) $D \cup R \cup K$

(g) $D \cup K \cup C$ (h) $R \cup K \cup C$ (i) $(D \cap R) \cup K$ (j) $D \cap (R \cup K)$ (k) $R \cup (K \cap C)$ (l) $(R \cup K) \cap C$

6 Find the probabilities of drawing 3 beads from a bag containing 2 red beads and 3 blue beads if the 3 beads are drawn **without replacement**.

(a) 3 blues   (b) 3 reds   (c) RRB in that order   (d) BRR (e) BRR in any order

(f) BRB (g) BRB in any order   (h) not 3 of the same colour.

7 In my 100 Club draw, each of the 100 members has one ticket and no one is allowed to win more than one prize, i.e. tickets are not replaced. What is the probability of my winning the   (a) first prize   (b) second prize   (c) third prize?

8 The following events are defined for throwing dice:
S = a six; F = a five; T = total of seven; E = even number on first throw.

Find the following probabilities for one die.

(a) $p(S)$ (b) $p(\bar{S})$ (c) $p(S \cup F)$ (d) $p(S \cap F)$ (e) $p(\bar{T})$ (f) $p(T)$ (g) $p(S \cap E)$

(h) $p(F \cup E)$ (i) $p(S | E)$ (j) $p(E | S)$ (k) $p(F | E)$ (l) $p(E | F)$ (m) $p(F | \bar{E})$ (n) $p(\bar{F} | \bar{E})$

Find the following probabilities for **two dice**.

(o) $p(S \cap F)$ (p) $p(S \cap S)$ (q) $p(S | F)$ (r) $p(S | \bar{F})$ (s) $p(T)$ (t) $p(T \cap S)$ (u) $p(F)$

(v) $p(S | E)$ (w) $p(F | E)$ (x) $p(T | E)$ (y) $p(T | F)$ (z) $p(S | T)$

9 If events A and B are mutually exclusive and events A and C are independent given $p(A) = \dfrac{1}{4}$, $p(B) = \dfrac{1}{5}$,

$p(A \cup C) = \dfrac{1}{2}$ and $p(B \cap C) = \dfrac{1}{15}$,   find

(a) $p(A \cup B)$ (b) $p(A \cap \bar{B})$ (c) $p(A \cap C)$ (d) $p(B \cap \bar{C})$ (e) $p(C)$

Theory of combined events

A pair of events can

  (a) happen simultaneously, e.g. rolling 2 dice, tossing 2 coins;

or **(b)** clearly be seen to happen one after the other, e.g. rolling a die twice, drawing 2 beads one after the other;

or **(c)** be defined within the same trial, e.g. Prime and Even with one die.

It may help (as in the case of 2 dice or 2 beads) to consider the two events successively rather than simultaneously.

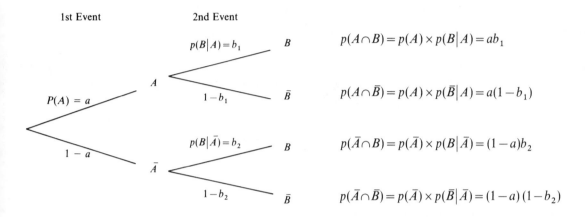

If $b_1 = b_2 = b$ then $p(B|A) = p(B|\bar{A})$ and B is **independent** of A.

In this case, $p(B) = p(A \cap B) + p(\bar{A} \cap B)$

becomes $p(B) = ab + (1-a)b = b$ and $p(B|A) = p(B|\bar{A}) = p(B)$

Also, $p(A \cap B) = ab = p(A) \times p(B)$ .... **(1)**

$$\Rightarrow \quad p(A|B) = \frac{p(A \cap B)}{p(B)} = \frac{ab}{b} = a = p(A) \quad \text{and} \quad p(A|\bar{B}) = \frac{p(A \cap \bar{B})}{p(\bar{B})} = \frac{a(1-b)}{(1-b)} = a = p(A)$$

and A is **independent** of B, in which case the events A and B are **statistically independent**.

$$\boxed{A \text{ being independent of } B \quad \Leftrightarrow \quad B \text{ is independent of } A.}$$

Be careful that events which appear, intuitively, to be independent, are in fact **statistically** independent.

From the nature of the events it may be possible to see that A and B are independent, but in most cases you will have to show that either

$$p(A \cap B) = p(A) \times p(B) \quad \text{or} \quad p(B|A) = p(B|\bar{A})$$

In general, if A appears first in the tree diagram, then

$$p(B) = p(A \cap B) + p(\bar{A} \cap B)$$

It is possible to prove that, if B is independent of A, then A is independent of B, by reversing the tree diagram and putting B first.

1 Given $p(A) = \frac{1}{3}$, $p(B|A) = \frac{1}{4}$, and B is independent of A, show that A is independent of B.

Tree diagram with A first

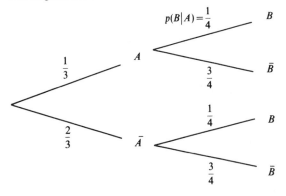

$$p(A \cap B) = p(A) \times p(B|A) = \frac{1}{3} \times \frac{1}{4} = \frac{1}{12}$$

B independent of A $\Rightarrow$ $p(B) = p(B|A) = \frac{1}{4}$

We can now start the reversed tree diagram with B first, as we know $p(B) = \frac{1}{4}$.

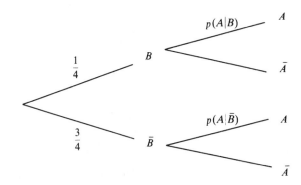

We need to show that $p(A|B) = p(A|\bar{B})$.

$$p(A|B) = \frac{p(A \cap B)}{p(B)} = \frac{\frac{1}{12}}{\frac{1}{4}} = \frac{1}{3} \qquad p(A|\bar{B}) = \frac{p(A \cap \bar{B})}{p(\bar{B})} = \frac{\frac{1}{3} \times \frac{3}{4}}{\frac{3}{4}} = \frac{1}{3}$$

So, A is independent of B.

2 Given, $p(A) = \frac{1}{3}$, $p(B|A) = \frac{1}{4}$, $p(B|\bar{A}) = \frac{1}{5}$, find $p(A|B)$ and $p(A|\bar{B})$.

Draw the tree diagram putting A first, and find $p(B)$.

$$p(A \cap B) = \frac{1}{3} \times \frac{1}{4} = \frac{1}{12}$$

$$p(\bar{A} \cap B) = \frac{2}{3} \times \frac{1}{5} = \frac{2}{15}$$

$$p(B) = P(A \cap B) + p(\bar{A} \cap B)$$

$$= \frac{1}{12} + \frac{2}{15} = \frac{13}{60}$$

Now reverse the tree diagram starting with B, $\qquad\qquad p(B \cap A) = p(A \cap B) = \dfrac{1}{12}$

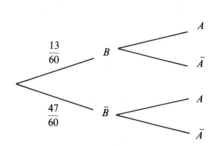

$p(B \cap A) = \dfrac{1}{12} \quad\Rightarrow\quad p(A \mid B) = \dfrac{p(B \cap A)}{P(B)} = \dfrac{\frac{1}{12}}{\frac{13}{60}} = \dfrac{5}{13}$

$p(A \mid \bar{B}) = \dfrac{p(A \cap \bar{B})}{p(\bar{B})} = \dfrac{\frac{1}{3} \times \frac{3}{4}}{\frac{47}{60}} = \dfrac{15}{47}$

Bayes theorem gives a formula for reversing the tree diagram, but it is very complicated to remember, and it is thought that you will find it more straightforward to use the method above.

Exercise 11.4

1 For the throw of one die E = the event of throwing an even number, P = a prime number (2, 3 or 5), and D an odd number,

(a) find $p(P \mid E)$ and $p(P \mid \bar{E})$ and say whether P is independent of E;

(b) find $p(E \mid P)$ and $p(E \mid \bar{P})$ and say whether E is independent of P;

(c) find $p(P \mid D)$ and $p(P \mid \bar{D})$ and say whether P is independent of D;

(d) find $p(D \mid P)$ and $p(D \mid \bar{P})$ and say whether D is independent of P.

(e) Are E and D independent?

2 These events are defined for cutting a pack of 52 cards once,

H = a heart; $\quad K$ = a king; $\quad R$ = a red card; $\quad C$ = a court card.

Decide whether the following pairs of events are independent,

(a) H and K (b) H and R (c) H and C (d) K and R (e) K and C (f) R and C.

3 I have 4 five pence pieces and 3 Deutschmarks in my pocket which are indistinguishable to the touch. Find the probability of drawing (a) 2 DMs (b) 2 five pence pieces (c) one of each when two coins are drawn from my pocket.

Does it matter which of the coins you consider to be drawn first?

4 I am dealt 2 cards from a well-shuffled pack of 52 cards. What is the probability of them being (a) both red (b) both black (c) one of each (d) 2 diamonds (e) 2 aces (f) an ace and a king (g) Pontoon (an ace with a king, queen, jack or ten) (h) Malta (a king and queen of the same suit)?

5 In my Maths class of sixteen: 10 students study Physics, P; 5 study Chemistry, C; and 2 neither. Use the symbols to describe the following events and find their probabilities.

(a) Physics and Chemistry (b) Physics or Chemistry (c) Physics only (d) Chemistry only (e) neither (f) Physics given Chemistry (g) Chemistry given Physics (h) Chemistry given not Physics (i) Physics given not Chemistry (j) Maths only (k) Physics (l) Chemistry. (m) Are Physics and Chemistry independent?

Expected values

In my Cricket 100 Club draw, what are my expected winnings during one year? I pay a total of £12, with prizes of £15, £10 and £5 per month and in addition a prize of £100 in June and £240 in December.

p(win £15) in 1 month $= \dfrac{1}{100}$, so in a year my expected winnings $= \dfrac{12 \times 15}{100} = £1.80$

p(win £10) in 1 month $= \dfrac{99}{100} \times \dfrac{1}{99}$, so in a year my expected winnings $= \dfrac{12 \times 10}{100} = £1.20$

p(win £5) in 1 month $= \dfrac{99}{100} \times \dfrac{98}{99} \times \dfrac{1}{98}$, so in a year my expected winnings $= \dfrac{12 \times 5}{100} = £0.60$

p(win £100 in June) $= \dfrac{1}{100}$, so my expected winnings $= \dfrac{1}{100} \times 100 = £1.00$

p(win £240 in December) $= \dfrac{1}{100}$, so my expected winnings $= \dfrac{1}{100} \times 240 = £2.40$

TOTAL EXPECTED WINNINGS = £7.00

The expected winnings of 100 members $= 100 \times £7 = £700$, the total prizes.

WORKED EXAMPLE

Each time I take my driving test the probability of passing is $\dfrac{1}{3}$.

(a) How many times do I expect to take the test, to be certain of passing?
(b) Which test gives me the highest probability?
(c) If the test costs £10, what is my expected outlay before I pass?

p(passing 1st time) $= \dfrac{1}{3}$; p(passing 2nd time) $= \dfrac{2}{3} \times \dfrac{1}{3}$; p(passing 3rd time) $= \dfrac{2}{3} \times \dfrac{2}{3} \times \dfrac{1}{3}$

(a) Total probability of passing $= \dfrac{1}{3} + \dfrac{2}{3} \times \dfrac{1}{3} + \dfrac{2}{3} \times \dfrac{2}{3} \times \dfrac{1}{3} + \ldots$

$$= \dfrac{1}{3}\left[1 + \dfrac{2}{3} + \left(\dfrac{2}{3}\right)^2 + \ldots \text{ an infinite series} \right]$$

Sum $= \dfrac{a}{1-r}$ $\Rightarrow$ Total p(passing) $= \dfrac{\frac{1}{3}}{1 - \frac{2}{3}} = 1$

So I am certain to pass after an infinite number of tests (!)

(b) p(passing 1st time) $= \dfrac{1}{3}$; p(2nd) $= \dfrac{2}{9}$; p(3rd) $= \dfrac{4}{27}$;   so the probability goes down each time, but each time the probability of passing is $\dfrac{1}{3}$ (!!)

At the outset, the probability of passing 3rd time is $\dfrac{4}{27}$. When you take the test 3rd time, the probability has increased to $\dfrac{1}{3}$. It depends on your point of view, or time of viewing.

(c) If I pass 1st time my cost is £10 with $p = \frac{1}{3}$ ⇒ expected outlay = £$\frac{10}{3}$

If I pass 2nd time my cost is £10 with $p = \frac{2}{9}$ ⇒ expected outlay = £$\frac{20}{9}$

If I pass 3rd time my cost is £10 with $p = \frac{4}{27}$ ⇒ expected outlay = £$\frac{40}{27}$

Total expected outlay $= 10 \times \frac{1}{3} + 10 \times \frac{2}{9} + 10 \times \frac{4}{27} + \ldots$

$$= \frac{10}{3}\left[1 + \frac{2}{3} + \frac{4}{9} + \ldots \right] \quad \text{an infinite series 1st term} = 1; \quad \text{common ratio} = \frac{2}{3}$$

$$= \frac{10}{3} \times \frac{1}{(1 - \frac{2}{3})} = £10 \quad \text{the cost of one test (!!)}$$

This seems a strange result. Discuss the assumptions, false or otherwise, that we have made.

Exercise 11.5

1 A player pays £1 to roll a die. If he rolls a six he wins £6, otherwise he loses his £1 stake. How much will he win in the long run?

2 A person takes a test until he passes. His probability of passing each time is $\frac{4}{5}$. If he has to pay £5 each time to take the test, what is his expected outlay?

3 Investigate the expression 'double or quits'. Is it ever worth it? How do you decide when to stop?

4 Analyse the game 'Yahtzee'. With 5 dice, each turn consists of three throws with the option after the first and second to keep as many of your dice still as you want to. On each turn you must score in one of 10 categories: ones, twos, threes, fours, fives, sixes, low straight, high straight, full house, total dice. For example, if in a turn you get 3 sixes you score 18. It will be helpful if one of the class has the game and you work on one category.
  For fives, what is the probability of rolling 1, 2, 3, 4, 5 fives in one throw? Find probabilities for totals of 0, 1, 2, 3, 4, 5 fives after 3 throws. Does it help your strategy in choosing which category to score? Is 3 sixes a good score or should you score the single 'two' you rolled with them? Or is it all based on luck?

5 What is the probability of **(a)** being dealt a pair in a poker hand of 5 cards **(b)** getting 4 aces in a bridge hand of 13 cards **(c)** getting a five in a cribbage hand of 6 cards **(d)** solving clock patience?

6 What is your chance of winning the football pools?
  If there are 10 scoring draws out of the 50 matches on Saturday, how many lines (selections of 8 matches) must you pay for to guarantee a winning line, i.e. a line with 8 scoring draws?

7 The probability that my car will fail its MOT: on lights is $\frac{1}{4}$; on brakes $\frac{1}{5}$; and on steering $\frac{1}{6}$. What is the probability that it passes its MOT?

8 A horse called *Hotrod* is quoted at 2–1 on. Is it worth backing? What do you win on a £1 stake? Is it worth backing 'each way'? Or is it all a mug's game?

9 Why did the Government introduce Premium Bonds? Does it make money if we can cash them in at any time? Why can't we use a lottery to finance the Health Service? Isn't it the same as Stocks and Shares? And the same morality?

12 Further series

Introduction

In Book 1, we considered the arithmetical and geometrical progressions as well as the binomial series.

So, if we expand the function $(1+x)^{-1}$ by the Binomial theorem, we obtain

$$(1+x)^{-1} = 1 - x + x^2 - x^3 + x^4 - \ldots \qquad \text{where} \quad -1 < x < 1$$

The graph of the function $y = (1+x)^{-1} = \dfrac{1}{(1+x)}$ is shown below.

Note that the line $x = -1$ is an asymptote to the curve and that $y \to 0$ as $x \to \pm\infty$.

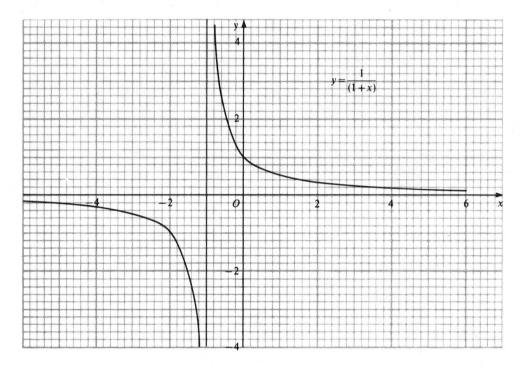

For a given value of x in the range $-1 < x < 1$, say $x = \frac{1}{8}$, the value of $(1+x)^{-1}$ is $\frac{8}{9} = 0.88\dot{8}$.

By substituting $x = \frac{1}{8}$ into the terms of the right-hand side of the expansion, an approximation to $\frac{8}{9}$ will be obtained. The more terms that are used, the better the approximation.

So, when $x = \frac{1}{8}$, we obtain, using 4 terms,

$$1 - x + x^2 - x^3 = 1 - \frac{1}{8} + \frac{1}{64} - \frac{1}{512} = 0.888\,672$$

If we use 6 terms,

$$1 - x + x^2 - x^3 + x^4 - x^5 = 1 - \frac{1}{8} + \frac{1}{64} - \frac{1}{512} + \frac{1}{4096} - \frac{1}{32768} = 0.888\,885\,5$$

Thus, the expansion of $(1+x)^{-1}$ gives an approximation to this function and we can make the fit as close as we like by choosing enough terms. It is possible to expand many other functions as series and in this chapter we shall consider some of these together with the techniques for dealing with them.

In Chapter 1 we considered the function e^x and the graph is shown below with the graphs of $y = 1 + x$ and $y = 1 + x + \frac{1}{2}x^2$ superimposed, using the same axes.

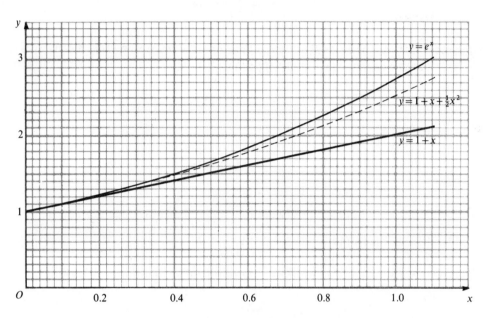

It is clear that, within the range $0 \leqslant x \leqslant 1$, the graphs are very similar. So the function e^x could be represented by a linear approximation $1 + x$ or by a quadratic function $1 + x + \frac{1}{2}x^2$, which gives a closer fit.

So, if x is small, $\quad e^x \simeq 1 + x + \frac{1}{2}x^2$

Investigation 1

Use graph paper to plot the graph of $y = \ln(1+x)$ for $-\frac{1}{2} \leqslant x \leqslant 2$.

On the same axes plot the graphs of $y = x$, $y = x - \frac{1}{2}x^2$ and $y = x - \frac{1}{2}x^2 + \frac{1}{3}x^3$, showing that they give a progressively better approximation to the function $\ln(1+x)$.

Deduce that, if x is small, $\quad \ln(1+x) \simeq x - \frac{1}{2}x^2 + \frac{1}{3}x^3$

By choosing values of x such that $-\frac{1}{2} \leqslant x \leqslant 2$ and substituting into the four functions above, show that the cubic expression gives the best approximation to $\ln(1+x)$.

Power series

A power series is an infinite series in the form

$$a_0 + a_1 x + a_2 x^2 + a_3 x^3 + \ldots a_r x^r + \ldots$$

where the coefficients $a_0, a_1, a_2, \ldots$ are rational constants.

So, from the work in Investigation 1 and the approximations for e^x, we have established the first few terms of the power series representing the functions $\ln(1+x)$ and e^x. It would suggest that, by adding further terms, better approximations will be obtained.

The power series $1 - x + x^2 - x^3 + x^4 - \ldots$ is the expansion of the function $\dfrac{1}{(1+x)}$, but the series only converges to the limit $(1+x)^{-1}$ when $-1 < x < 1$.

In general, power series only converge to a finite limit for a certain range of values of x. We say that the expansion is valid only for this range of x.

Maclaurin's theorem

Assume that a function $f(x)$ can be expanded as an infinite power series in x and that it converges to a finite limit. We must also assume that the function and its derivatives are all finite and continuous in the range of convergence.

Thus,
$$f(x) = a_0 + a_1 x + a_2 x^2 + a_3 x^3 + a_4 x^4 + a_5 x^5 + \ldots$$

Assuming that a power series which converges in a given interval may be differentiated term by term and that the resultant series also converges in the interval. Then we have,

$$f'(x) = a_1 + 2a_2 x + 3a_3 x^2 + 4a_4 x^3 + 5a_5 x^4 + \ldots$$
$$f''(x) = 2a_2 + (3 \times 2)a_3 x + (4 \times 3)a_4 x^2 + (5 \times 4)a_5 x^3 + \ldots$$
$$f'''(x) = (3 \times 2)a_3 + (4 \times 3 \times 2)a_4 x + (5 \times 4 \times 3)a_5 x^2 + \ldots$$
$$f^{(4)}(x) = (4 \times 3 \times 2)a_4 + (5 \times 4 \times 3 \times 2)a_5 x + \ldots$$

By substituting $x = 0$ in these equations we obtain values for the coefficients $a_0, a_1, a_2, \ldots$.

Hence, $a_0 = f(0)$; $\quad a_1 = f'(0)$; $\quad a_2 = \dfrac{f''(0)}{2!}$; $\quad a_3 = \dfrac{f'''(0)}{3!}$; $\quad$ and, in general, $\quad a_r = \dfrac{f^{(r)}(0)}{r!}$

Thus,
$$f(x) = f(0) + xf'(0) + \frac{x^2}{2!}f''(0) + \frac{x^3}{3!}f'''(0) + \ldots + \frac{x^r}{r!}f^{(r)}(0) + \ldots$$

This result is known as **Maclaurin's series** and it is used to give the expansion of a function as a power series in the neighbourhood of $x = 0$.

Note that the Maclaurin series for a function $f(x)$ exists only if $f(x)$ and all its derivatives are finite when $x = 0$.

WORKED EXAMPLE ▷

Use Maclaurin's series to find the expansion of $\sin x$ in ascending powers of x.

Let
$$
\begin{aligned}
f(x) &= \sin x &\Rightarrow\quad f(0) &= \sin 0 = 0 \\
f'(x) &= \cos x &\Rightarrow\quad f'(0) &= \cos 0 = 1 \\
f''(x) &= -\sin x &\Rightarrow\quad f''(0) &= -\sin 0 = 0 \\
f^{(3)}(x) &= -\cos x &\Rightarrow\quad f^{(3)}(0) &= -\cos 0 = -1 \\
f^{(4)}(x) &= \sin x &\Rightarrow\quad f^{(4)}(0) &= \sin 0 = 0
\end{aligned}
$$

Clearly, a pattern is established and the values are repeated in cycles.

Using Maclaurin's series,

$$f(x) = f(0) + xf'(0) + \frac{x^2}{2!}f''(0) + \frac{x^3}{3!}f'''(0) + \ldots$$

$$\therefore \quad \sin x = x - \frac{x^3}{3!} + \frac{x^5}{5!} - \frac{x^7}{7!} + \ldots$$

Exercise 12.1

1 Use graph paper to plot the graph of $y = \cos x$ for $0 \leqslant x \leqslant \frac{1}{2}\pi$. On the same axes draw the graphs of $y = 1$, $y = 1 - \frac{1}{2}x^2$ and $y = 1 - \frac{1}{2}x^2 + \frac{1}{24}x^4$.
Deduce that, when x is small, $\cos x \simeq 1 - \frac{1}{2}x^2 + \frac{1}{24}x^4$.

2 Use graph paper to plot the graphs of the following functions for the interval $0 \leqslant x \leqslant \frac{3}{8}\pi$.

(a) $y = \tan x$ (b) $y = x$ (c) $y = x + \frac{1}{3}x^3$

Deduce a power series approximation for $\tan x$.

3 State Maclaurin's series for the expansion of $f(x)$ in ascending powers of x. Use this to find the expansion of $\sec x$ as far as the term in x^4.

4 Use Maclaurin's series to find the expansions of the following functions in ascending powers of x as far as the term in x^4.

(a) $\sin 4x$ (b) $\sin^2 x$ (c) e^{2x} (d) $\sin^{-1} x$ (e) $\tan^{-1} x$

(f) $(1+x)^n$ (g) $\ln(\cos x)$ (h) $\ln(1-x)$ (i) $\ln(1+\sin x)$

5 Use Maclaurin's series to find the expansions of the following functions in ascending powers of x as far as the term indicated.

(a) $e^x \sin x$ (x^5) (b) $e^{\sin x}$ (x^4)

6 Use Maclaurin's series to show that,

(a) $\ln(1+x^2) = x^2 - \frac{1}{2}x^4 + \frac{1}{3}x^6 - \ldots$

(b) $\ln(1+\tan x) = x - \frac{1}{2}x^2 + \frac{2}{3}x^3 - \ldots$

The exponential series

The expansion of the function e^x is of particular interest and can be shown to be valid for all values of x.

Let, $f(x) = e^x \quad \Rightarrow \quad f(0) = e^0 = 1$

$f'(x) = e^x \quad \Rightarrow \quad f'(0) = e^0 = 1$

$f''(x) = e^x \quad \Rightarrow \quad f''(0) = e^0 = 1$

In fact, for this function, all the successive differentials are e^x, and hence,

$$f^{(n)}(x) = e^x \quad \Rightarrow \quad f^{(n)}(0) = 1$$

Further series

Maclaurin's series gives

$$f(x) = f(0) + xf'(0) + \frac{x^2}{2!}f''(0) + \frac{x^3}{3!}f'''(0) + \ldots$$

$$e^x = 1 + x + \frac{x^2}{2!} + \frac{x^3}{3!} + \frac{x^4}{4!} + \ldots + \frac{x^r}{r!} + \ldots$$

This series is **valid for all values of** x and is sometimes written as $\exp x$.

Investigation 2

In the expansion of e^x, let $x = 1$ and show that

$$e = 1 + 1 + \frac{1}{2!} + \frac{1}{3!} + \frac{1}{4!} + \frac{1}{5!} + \ldots$$

By evaluating a sufficient number of terms, show that the value of e is 2.718 281 8 correct to 7 decimal places.

---

e **is irrational** and thus cannot be expressed as a ratio of two integers. It contains an infinite number of decimal places and never repeats itself. It does **not** satisfy any polynomial equation with rational coefficients of the form,

$$a_0 + a_1 x + a_2 x^2 + \ldots + a_n x^n = 0$$

Such a number is called **transcendental**.

WORKED EXAMPLE

Expand **(a)** e^{2x} **(b)** $e^{(x-x^2)}$ as a series in ascending powers of x as far as the term in x^4.

In cases like these we use the basic expansion for e^x and replace x by the appropriate index, i.e. $2x$ and $(x - x^2)$.

(a) Since $e^x = 1 + x + \frac{x^2}{2!} + \frac{x^3}{3!} + \frac{x^4}{4!} + \ldots$ for all values of x,

$$e^{2x} = 1 + (2x) + \frac{(2x)^2}{2!} + \frac{(2x)^3}{3!} + \frac{(2x)^4}{4!} + \ldots$$

$$\Rightarrow \quad e^{2x} = 1 + 2x + 2x^2 + \frac{4}{3}x^3 + \frac{2}{3}x^4 + \ldots$$

(b) In this case the index is $(x - x^2)$.

Now, $e^x = 1 + x + \frac{x^2}{2!} + \frac{x^3}{3!} + \frac{x^4}{4!} + \ldots$

$$\therefore \quad e^{(x-x^2)} = 1 + (x - x^2) + \frac{(x-x^2)^2}{2!} + \frac{(x-x^2)^3}{3!} + \frac{(x-x^2)^4}{4!} + \ldots$$

All further terms will contain powers of x higher than x^4 and so there is no need to consider more terms.

Expand the powers of $(x-x^2)$ but include only those powers of $x \leqslant 4$.

$$\therefore \quad e^{(x-x^2)} = 1 + x - x^2 + \frac{1}{2}(x^2 - 2x^3 + x^4) + \frac{1}{6}(x^3 - 3x^4 + \ldots) + \frac{1}{24}(x^4 + \ldots)$$

$$\Rightarrow \quad e^{(x-x^2)} = 1 + x - \frac{1}{2}x^2 - \frac{5}{6}x^3 + \frac{1}{24}x^4 - \ldots$$

Investigation 3

Consider part **(b)** of the Worked Example above, and write $e^{(x-x^2)}$ as the product $e^x \times e^{-x^2}$.

Show that $\quad e^{-x^2} = 1 - x^2 + \frac{1}{2}x^4 - \ldots \quad$ up to the term in x^4.

Multiply this expansion by the expansion of e^x, including only those terms up to and including x^4.

Verify that $\quad e^{(x-x^2)} = 1 + x - \frac{1}{2}x^2 - \frac{5}{6}x^3 + \frac{1}{24}x^4 - \ldots \quad$ as before.

WORKED EXAMPLES

1 Find the first three non-zero terms in the expansion of $\dfrac{e^x}{1+x}$ in ascending powers of x.

Since $\dfrac{e^x}{1+x} = e^x(1+x)^{-1}$, we can expand e^x and $(1+x)^{-1}$ separately and multiply the results. Since only the first three non-zero terms are required, we shall ignore powers of x higher than 3.

Now, $$e^x = 1 + x + \frac{x^2}{2!} + \frac{x^3}{3!} + \ldots$$

and, using the Binomial theorem,

$$(1+x)^{-1} = 1 - x + \frac{(-1)(-2)}{2!}x^2 + \frac{(-1)(-2)(-3)}{3!}x^3 + \ldots$$

$$= 1 - x + x^2 - x^3 + \ldots$$

Hence, $$\frac{e^x}{1+x} = (1 + x + \tfrac{1}{2}x^2 + \tfrac{1}{6}x^3 + \ldots)(1 - x + x^2 - x^3 + \ldots)$$

$$= 1 + x + \tfrac{1}{2}x^2 + \tfrac{1}{6}x^3 - x - x^2 - \tfrac{1}{2}x^3 + x^2 + x^3 - x^3 + \ldots$$

$$= 1 + \tfrac{1}{2}x^2 - \tfrac{1}{3}x^3 + \ldots$$

2 Find the term in x^3 and the general term in the expansion of $(1+x)e^{-x}$ in ascending powers of x.

Now, $$e^{-x} = 1 + (-x) + \frac{(-x)^2}{2!} + \frac{(-x)^3}{3!} + \ldots + \frac{(-x)^{r-1}}{(r-1)!} + \frac{(-x)^r}{r!} + \ldots$$

$$= 1 - x + \frac{x^2}{2!} - \frac{x^3}{3!} + \ldots + \frac{(-1)^{r-1}x^{r-1}}{(r-1)!} + \frac{(-1)^r x^r}{r!} + \ldots$$

So, $(1+x)e^{-x} = (1+x)\left(1 - x + \tfrac{1}{2}x^2 - \dfrac{1}{6}x^3 + \ldots \dfrac{(-1)^{r-1}x^{r-1}}{(r-1)!} + \dfrac{(-1)^r x^r}{r!} + \ldots\right)$

The product of the brackets produces two terms containing x^3, i.e. $1 \times (-\tfrac{1}{6}x^3)$ and $x(\tfrac{1}{2}x^2)$.

Thus, the term in x^3 is $\quad -\tfrac{1}{6}x^3 + \tfrac{1}{2}x^3 = \tfrac{1}{3}x^3$.

Similarly, the product of the brackets will produce two terms in x^r.

Thus, the term in x^r is $\quad \dfrac{(-1)^{r-1}x^r}{(r-1)!} + \dfrac{(-1)^r x^r}{r!} = \dfrac{(-1)^{r-1}(r-1)}{r!}x^r$.

Exercise 12.2

1 Find the first four non-zero terms and the general term in the expansions of the following functions in ascending powers of x.

(a) e^{-x} (b) e^{3x} (c) e^{x^2} (d) $\sqrt{e^x}$ (e) $e^{(1+x)}$ (f) $(1-x)e^x$

2 Find the first three non-zero terms in the expansions of the following functions in ascending powers of x.

(a) $e^{2x} + e^{-2x}$ (b) $\dfrac{e^{\frac{1}{2}x}}{(1-x)}$ (c) $(1-x)^2 e^x$ (d) $e^{3x} \cdot e^{5x}$ (e) e^{1-x^2} (f) $(1+2x-5x^2)e^{-2x}$

3 Find the sum to infinity of the following series.

(a) $1 + 4x + 8x^2 + \dfrac{32}{3}x^3 + \dfrac{32}{3}x^4 + \ldots$ (b) $1 - \dfrac{x}{2} + \dfrac{x^2}{2^2 \cdot 2!} - \dfrac{x^3}{2^3 \cdot 3!} + \ldots$

4 Find the first four terms of the expansion of $(1+2x)e^x$ and hence deduce the sum to infinity of the series,

$$1 + \dfrac{3x}{1!} + \dfrac{5x^2}{2!} + \dfrac{7x^3}{3!} + \ldots$$

5 By writing down the series for e^x and e^{-x}, find the sum to infinity of the series

$$x + \dfrac{x^3}{3!} + \dfrac{x^5}{5!} + \ldots$$

6 Find the limits of the following functions as x tends to zero.

(a) $\dfrac{e^x - 1}{e^{2x} - 1}$ (b) $\dfrac{e^x + e^{-x} - 2}{x^2}$

7 Find the sum to infinity of each of the following series.

(a) $1 - x + \dfrac{x^2}{2!} - \dfrac{x^3}{3!} + \ldots$ (b) $1 + \dfrac{x^2}{2!} + \dfrac{x^4}{4!} + \ldots$ (c) $1 + \dfrac{2x}{1!} + \dfrac{3x^2}{2!} + \dfrac{4x^3}{3!} + \ldots$

(d) $2x + 2x^2 + x^3 + \dfrac{1}{3}x^4 + \ldots$

8 Show that the series $\dfrac{3}{1!} + \dfrac{4}{2!} + \dfrac{5}{3!} + \ldots$ can be written as $\displaystyle\sum_{r=1}^{\infty} \dfrac{r+2}{r!}$.

Hence show that the sum of the series is $3e - 2$.

9 Find the general term when each of the following functions is expanded in ascending powers of x.

(a) $e^x + e^{-x}$ (b) $(1+e^x)^2$ (c) $(1-x+x^2)e^x$

The logarithmic series

The function $\ln x$ is not finite and is not defined for $x = 0$, as can be seen by the graph.

Hence, it is not possible to use Maclaurin's theorem to expand this function.

Instead, we consider the function $\ln(1 + x)$.

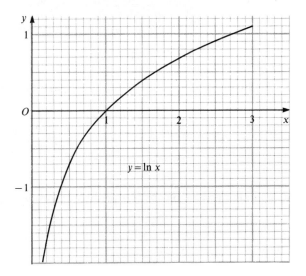

Investigation 4

Let $f(x) = \ln(1 + x)$ and form the successive differentials $f'(x), f''(x), f'''(x), \ldots$ Hence, find the values of $f'(0), f''(0), \ldots$

Use Maclaurin's theorem to show that,

$$\ln(1 + x) = x - \frac{x^2}{2} + \frac{x^3}{3} - \frac{x^4}{4} + \ldots + \frac{(-1)^{r-1}}{r}x^r + \ldots$$

Since the terms alternate between positive and negative and the denominators do not contain factorials (as in the exponential series), the logarithmic series converges very slowly unless x is small.

The series is only valid for $-1 < x \leqslant 1$ and the graphs, below, of $y = \ln(1 + x)$ and $y = x - \frac{x^2}{2} + \frac{x^3}{3} - \frac{x^4}{4} + \frac{x^5}{5}$ show how the approximation is useful only for $-1 < x \leqslant 1$.

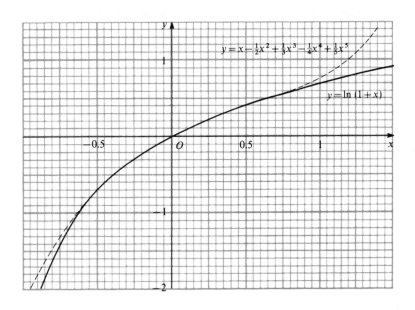

Hence,

$$\ln(1+x) = x - \frac{x^2}{2} + \frac{x^3}{3} - \frac{x^4}{4} + \ldots + \frac{(-1)^{r-1}}{r}x^r + \ldots \qquad \text{for } -1 < x \leqslant 1.$$

Note that $x \neq -1$ since $\ln 0$ is not defined.

If we replace x by $-x$ in the expansion, then we obtain the expansion of $\ln(1-x)$ which is valid for $-1 \leqslant x < 1$.

$$\ln(1-x) = -x - \frac{x^2}{2} - \frac{x^3}{3} - \frac{x^4}{4} - \ldots - \frac{x^r}{r} - \ldots \qquad \text{for } -1 \leqslant x < 1.$$

By combining these results it is possible to produce a series for $\ln\left(\dfrac{1+x}{1-x}\right)$.

$$\ln(1+x) = x - \frac{x^2}{2} + \frac{x^3}{3} - \frac{x^4}{4} + \ldots + \frac{x^{2r-1}}{2r-1} - \frac{x^{2r}}{2r} + \frac{x^{2r+1}}{2r+1} - \ldots$$

$$\ln(1-x) = -x - \frac{x^2}{2} - \frac{x^3}{3} - \frac{x^4}{4} - \ldots - \frac{x^{2r-1}}{2r-1} - \frac{x^{2r}}{2r} - \frac{x^{2r+1}}{2r+1} - \ldots$$

Subtracting gives

$$\ln(1+x) - \ln(1-x) = 2x + \frac{2x^3}{3} + \frac{2x^5}{5} + \ldots + \frac{2x^{2r-1}}{2r-1} + \frac{2x^{2r+1}}{2r+1} + \ldots$$

$$\ln\left(\frac{1+x}{1-x}\right) = 2\left(x + \frac{x^3}{3} + \frac{x^5}{5} + \ldots + \frac{x^{2r-1}}{2r-1} + \ldots\right)$$

So,

$$\frac{1}{2}\ln\left(\frac{1+x}{1-x}\right) = \ln\sqrt{\frac{1+x}{1-x}} = x + \frac{x^3}{3} + \frac{x^5}{5} + \ldots + \frac{x^{2r-1}}{2r-1} + \ldots$$

This series will only be valid when x satisfies $-1 \leqslant x < 1$ and $-1 < x \leqslant 1$. Thus the expansion is valid for $-1 < x < 1$.

This form is particularly useful for calculating logarithms of numbers outside the range $-1 < x < 1$.

WORKED EXAMPLES

1 Show that if $\dfrac{1+x}{1-x} = 4$ then $x = 0.6$. Hence deduce that $\ln 4 = 1.386$ correct to 4 significant figures.

Now,

$$\frac{1+x}{1-x} = 4 \quad \Rightarrow \quad 1+x = 4(1-x) = 4-4x \quad \Rightarrow \quad 5x = 3 \quad \Rightarrow \quad x = \frac{3}{5} = 0.6$$

Since

$$\ln\left(\frac{1+x}{1-x}\right) = 2\left(x + \frac{x^3}{3} + \frac{x^5}{5} + \frac{x^7}{7} + \frac{x^9}{9} + \ldots\right) \qquad \text{for } -1 < x < 1$$

then, if $x = 0.6$

$$\ln 4 = 2\left(0.6 + \frac{(0.6)^3}{3} + \frac{(0.6)^5}{5} + \frac{(0.6)^7}{7} + \frac{(0.6)^9}{9} + \frac{(0.6)^{11}}{11}\right)$$

$$= 1.2 + 0.144 + 0.031\ 104 + 0.007\ 998\ 2 + 0.002\ 394\ 8 + 0.000\ 659\ 6$$

$$= 1.386\ 157$$

The last term of the series evaluated here suggests that the value of $\ln 4$ is 1.386 to 4 significant figures. Notice that, even with a relatively high value of x, the series converges fairly rapidly.

2 Find the first three terms and the general term of the expansion of $\ln \sqrt{x^2 + 3x + 2}$.

Since $x^2 + 3x + 2 = (x+1)(x+2)$ it follows that

$$\ln \sqrt{x^2 + 3x + 2} = \ln (x^2 + 3x + 2)^{\frac{1}{2}} = \tfrac{1}{2} \ln (x^2 + 3x + 2)$$

$$= \tfrac{1}{2} \ln (x+1)(x+2) = \tfrac{1}{2} \ln (x+1) + \tfrac{1}{2} \ln (x+2)$$

$$= \tfrac{1}{2} \ln (1+x) + \tfrac{1}{2} \ln \left[2\left(1 + \frac{x}{2}\right)\right]$$

$$= \tfrac{1}{2} \ln (1+x) + \tfrac{1}{2} \ln \left(1 + \frac{x}{2}\right) + \tfrac{1}{2} \ln 2$$

$$= \tfrac{1}{2} \ln 2 + \frac{1}{2}\left[x - \frac{x^2}{2} + \frac{x^3}{3} - \frac{x^4}{4} + \ldots + \frac{(-1)^{r-1} x^r}{r} + \ldots\right]$$

$$+ \frac{1}{2}\left[\frac{x}{2} - \frac{1}{2}\left(\frac{x}{2}\right)^2 + \frac{1}{3}\left(\frac{x}{2}\right)^3 - \frac{1}{4}\left(\frac{x}{2}\right)^4 + \ldots + \frac{(-1)^{r-1}}{r}\left(\frac{x}{2}\right)^r + \ldots\right]$$

$$= \tfrac{1}{2} \ln 2 + \frac{1}{2}\left[\frac{3x}{2} - \frac{5x^2}{8} + \frac{9x^3}{24} + \ldots + (-1)^{r-1}\left\{1 + \left(\frac{1}{2}\right)^r\right\}\frac{x^r}{r} + \ldots\right]$$

$$\ln \sqrt{x^2 + 3x + 2} = \tfrac{1}{2} \ln 2 + \frac{3x}{4} - \frac{5x^2}{16} + \frac{9x^3}{48} + \ldots + (-1)^{r-1}\left\{1 + \left(\frac{1}{2}\right)^r\right\}\frac{x^r}{2r} + \ldots$$

3 Expand the function $\dfrac{\ln (1+2x)}{1-x}$ as far as the term in x^3 and state the range of values of x for which the expansion is valid.

Since $\dfrac{\ln (1+2x)}{1-x} = (1-x)^{-1} \ln (1+2x)$, the expansion will be formed from the product of the separate expansions of $(1-x)^{-1}$ and $\ln (1+2x)$.

Using the Binomial expansion,

$$(1-x)^{-1} = 1 + x + \frac{(-1)(-2)}{2!}(-x)^2 + \frac{(-1)(-2)(-3)}{3!}(-x)^3 + \ldots$$

$$= 1 + x + x^2 + x^3 + \ldots$$

$$\ln (1+2x) = 2x - \frac{(2x)^2}{2} + \frac{(2x)^3}{3} - \ldots$$

$$= 2x - 2x^2 + \frac{8x^3}{3} - \ldots$$

Further series

$$\therefore \quad \frac{\ln(1+2x)}{1-x} = (1+x+x^2+x^3+\ldots)\left(2x-2x^2+\frac{8x^3}{3}-\ldots\right)$$

Multiplying the brackets, neglecting x^4 and higher powers, gives

$$\frac{\ln(1+2x)}{1-x} = 2x-2x^2+\frac{8x^3}{3}+2x^2-2x^3+2x^3+\ldots$$

$$= 2x+\frac{8x^3}{3}+\ldots$$

The expansion of $\ln(1+2x)$ is valid for $-1<2x\leqslant 1$ i.e. $-\frac{1}{2}<x\leqslant\frac{1}{2}$, and the expansion of $(1-x)^{-1}$ is valid for $-1<x<1$.

Hence, the resulting product is valid only if $-\frac{1}{2}<x\leqslant\frac{1}{2}$.

Exercise 12.3

1 Find the first three non-zero terms in the expansions of the following functions in ascending powers of x.

(a) $\ln(1-2x)$ (b) $\ln(1+\frac{1}{3}x)$ (c) $\ln(3-x)$

2 Expand the following functions in ascending powers of x, giving the first four terms and the general term.

(a) $\ln(1-x^2)$ (b) $\ln(4-\frac{1}{2}x)$ (c) $\ln\left(\frac{4+3x}{1-x}\right)$

3 Expand the following functions in ascending powers of x, giving the first three non-zero terms and state the range of validity.

(a) $\ln(x^2-5x+4)$ (b) $(1+x)\ln(1+x)$ (c) $\dfrac{1+\ln(1+x^2)}{\sqrt{1-x}}$

(d) $e^x\ln(1-\frac{1}{2}x)$ (e) $(1+x+x^2)\ln\sqrt{1+3x}$ (f) $\dfrac{\ln(1-x)^2}{e^x}$

4 By writing the expansion of $\ln\left(\dfrac{1+x}{1-x}\right)$ in ascending powers of x and by letting $x=\frac{1}{5}$, show that $\ln 3\simeq 1.0986$.

5 By writing the expansions of $\ln(1+3x)$ and $\ln(1-x)$ in ascending powers of x show that, if $-\frac{1}{3}<x\leqslant\frac{1}{3}$,
$$\ln(1+2x-3x^2) = 2x-5x^2+\tfrac{26}{3}x^3+20x^4+\ldots$$

6 By substituting a suitable value of x in the expansion of $\ln\left(\dfrac{1+x}{1-x}\right)$ in ascending powers of x, find the value of $\ln\frac{5}{3}$ correct to four significant figures.

7 Show that $1-x^3\equiv(1-x)(1+x+x^2)$ and hence find the expansion of $\ln(1+x+x^2)$ in ascending powers of x as far as the term in x^6.

8 By writing $(1+x+x^2)$ as $1+(x+x^2)$, find the expansion of $\ln(1+x+x^2)$ in ascending powers of x as far as the term in x^6.
   Compare the method with that used in question **7**.

9 If x is small such that x^6 and higher powers can be neglected, show that $\ln(\cos x) = -\dfrac{x^2}{2} - \dfrac{x^4}{8}$.

10 Write down the expansion of $\ln\left(\dfrac{1+x}{1-x}\right)$ in ascending powers of x. Use this to show that,

(a) $\ln m = 2\left[\left(\dfrac{m-1}{m+1}\right) + \dfrac{1}{3}\left(\dfrac{m-1}{m+1}\right)^3 + \dfrac{1}{5}\left(\dfrac{m-1}{m+1}\right)^5 + \cdots\right]$

(b) $\ln\dfrac{m}{n} = 2\left[\left(\dfrac{m-n}{m+n}\right) + \dfrac{1}{3}\left(\dfrac{m-n}{m+n}\right)^3 + \dfrac{1}{5}\left(\dfrac{m-n}{m+n}\right)^5 + \cdots\right]$

Use these results to evaluate $\ln 2$ and $\ln\frac{5}{4}$.

11 Show that $\ln\left(1+\dfrac{1}{n}\right)^n = 1 - \dfrac{1}{2n} + \dfrac{1}{3n^2} - \dfrac{1}{4n^3} + \cdots$ if $n>1$.

Hence, show that $\left(1+\dfrac{1}{n}\right)^n = e\left(e^{-1/2n}\right)\left(e^{1/3n^2}\right)\left(e^{-1/4n^3}\right)$

Use the expansion of e^x to show that $\left(1+\dfrac{1}{n}\right)^n = e\left(1 - \dfrac{1}{2n} + \dfrac{11}{24n^2} - \dfrac{7}{16n^3} + \cdots\right)$

Deduce that $\lim\limits_{n\to\infty}\left(1+\dfrac{1}{n}\right)^n = e$.

The method of proof by induction

Sometimes we know that a statement is true for given values of the variable. Thus the statement,

$$3^{n+1} - 2n - 3 \quad \text{is divisible by 4 for any integral value of } n$$

can be verified for $n=1$ and $n=2$ by substitution.

So, if $f(n) = 3^{n+1} - 2n - 3$
then, $f(1) = 3^2 - 2 - 3 = 4$ and $f(2) = 3^3 - 4 - 3 = 20$

However, knowing that the statement is true for $n=1$ and $n=2$ does not prove that it is true for all values of n. To do this we use a technique similar to the method employed for defining sequences inductively.

To define a sequence inductively we give,

(a) a rule for obtaining each term from the previous one; and
(b) an initial value.

To prove a statement, we show that

(a) if it is valid for some general value of n, say $n=k$, then it must also be true for the next integral value of n, i.e. $n=k+1$; and
(b) it is true for some initial value of n.

This is known as **proof by induction**.

So assume that $f(n) = 3^{n+1} - 2n - 3$ is divisible by 4 when $n=k$.

Now, when $n = k+1$ (i.e. the next integral value of n),

$$f(k+1) = 3^{k+2} - 2(k+1) - 3$$

and $\quad f(k+1) - 3f(k) = 3^{k+2} - 2(k+1) - 3 - 3(3^{k+1} - 2k - 3)$

$$= -2k - 2 - 3 + 6k + 9$$

$$= 4k + 4 = 4(k+1)$$

Thus, $\qquad f(k+1) = 3f(k) + 4(k+1)$

Clearly, if $f(k)$ is divisible by 4, then $f(k+1)$ must be also, since $4(k+1)$ is obviously divisible by 4.

So if $f(k)$ is divisible by 4, then so is $f(k+1)$. Now we saw at the beginning of this section that $f(1) = 4$ (i.e. divisible by 4) and hence, the statement is true for $n = 2, n = 3, n = 4, \ldots$

Thus, $3^{n+1} - 2n - 3$ is divisible by 4 for all positive integral values of n.

Investigation 5

Prove by the method of induction that

$$1 + 3 + 5 + 7 + 9 + \ldots + (2n-1) = n^2 \qquad \ldots\ldots\ldots \text{ (1)}$$

Firstly, assume that this result is true for a value $n = k$, so

$$1 + 3 + 5 + 7 + 9 + \ldots + (2k-1) = k^2$$

Now add the next term in the series to both sides and show that

$$1 + 3 + 5 + 7 + 9 + \ldots + (2k-1) + (2k+1) = k^2 + (2k+1)$$

Now rearrange the right-hand side and show that

$$1 + 3 + 5 + 7 + 9 + \ldots + (2k-1) + (2k+1) = (k+1)^2$$

Since this is the correct form for the summation of the series **(1)** when $n = k+1$, we have shown that if the result is true for $n = k$, then it is also true for $n = k+1$.

Now show that the result **(1)** is true for $n = 1$ and deduce that it will therefore be true for all positive integral values of n.

WORKED EXAMPLE ⟩

Prove by induction that $\displaystyle\sum_{r=1}^{n} \frac{1}{r(r+1)} = \frac{n}{n+1}$.

Assume that the result is true for $n = k$,

$$\sum_{r=1}^{k} \frac{1}{r(r+1)} = \frac{k}{k+1}$$

The next term of the series is $\dfrac{1}{(k+1)(k+2)}$ (i.e. when $n = k+1$)

Adding this to both sides gives,

$$\sum_{r=1}^{k+1} \frac{1}{r(r+1)} = \frac{k}{k+1} + \frac{1}{(k+1)(k+2)}$$

Simplify the right-hand side by finding common factors,

$$\sum_{r=1}^{k+1} \frac{1}{r(r+1)} = \frac{1}{k+1}\left(k + \frac{1}{k+2}\right)$$

$$= \frac{1}{k+1}\left(\frac{k(k+2)+1}{(k+2)}\right)$$

$$= \frac{1}{k+1}\left(\frac{k^2+2k+1}{k+2}\right) = \frac{(k+1)^2}{(k+1)(k+2)} = \frac{k+1}{k+2}$$

This is the correct form for the right-hand side when $n = k+1$. Hence, if the result is true for $n = k$, then it is also true for $n = k+1$.

Now, when $n = 1$, $\qquad \sum_{r=1}^{n} \frac{1}{r(r+1)} = \frac{1}{1(2)} = \frac{1}{2}$ and $\dfrac{n}{n+1} = \dfrac{1}{2}$

Thus, the result holds for $n = 1$. Hence, by induction, it is true for $n = 2$, $n = 3$, $n = 4, \ldots$

Thus it is true for all positive integral values of *n*.

Investigation 6

In Book 1 we saw that the Binomial expansion for an integral index is given by

$$(a+b)^n = a^n + {}^nC_1 a^{n-1}b + {}^nC_2 a^{n-2}b^2 + \ldots + {}^nC_r a^{n-r}b^r + \ldots + b^n$$

By assuming that this is true for $n = k$ and multiplying by $(a+b)$, show that

$$(a+b)^{k+1} = a^{k+1} + ({}^kC_1+1)a^kb + ({}^kC_2+{}^kC_1)a^{k-1}b^2 + ({}^kC_3+{}^kC_2)a^{k-2}b^3 + \ldots$$
$$+ ({}^kC_r+{}^kC_{r-1})a^{k-r+1}b^r + \ldots + b^{k+1}$$

Using the results $\quad {}^kC_1 + 1 = k+1 = {}^{k+1}C_r$

and $\quad {}^kC_r + {}^kC_{r-1} = {}^{k+1}C_r \qquad$ (see Book 1, page 226)

show that

$$(a+b)^{k+1} = a^{k+1} + {}^{k+1}C_1 a^kb + {}^{k+1}C_2 a^{k-1}b^2 + \ldots + {}^{k+1}C_r a^{k-r+1}b^r + \ldots + b^{k+1}$$

Deduce that the result is true for all positive integral values of *n*.

Exercise 12.4

Prove the following results (questions **1** to **6**) by induction.

1 $\quad 1 + 2 + 3 + \ldots + n = \frac{1}{2}n(n+1)$

2 $\quad 1^2 + 2^2 + 3^2 + \ldots + n^2 = \frac{1}{6}n(n+1)(2n+1)$

3 $\quad 1^3 + 2^3 + 3^3 + \ldots + n^3 = \frac{1}{4}n^2(n+1)^2$

4 $1^2 + 3^2 + 5^2 + \ldots + (2n-1)^2 = \frac{1}{3}n(4n^2 - 1)$

5 $(1 \times 4) + (2 \times 5) + \ldots + n(n+3) = \frac{1}{3}n(n+1)(n+5)$

6 $\dfrac{1}{(1 \times 3)} + \dfrac{1}{(3 \times 5)} + \dfrac{1}{(5 \times 7)} + \ldots + \dfrac{1}{(2n-1)(2n+1)} = \dfrac{n}{2n+1}$

7 Show that $3^n + 1$ is even for all positive integral values of n.

8 Show that $5^n + 4n + 3$ is divisible by 4 for all positive integral values of n.

9 Show that $4^n + 3^n + 2$ is a multiple of 3 for all positive integral values of n.

Prove the following (questions **10** to **13**) by induction.

10 $\displaystyle\sum_{r=1}^{n} r(r+1) = \frac{1}{3}n(n+1)(n+2)$

11 $\displaystyle\sum_{r=1}^{n} 3r(r-1) = n(n^2 - 1)$

12 $\displaystyle\sum_{r=1}^{n} r(r+1)(r+2) = \frac{1}{4}n(n+1)(n+2)(n+3)$

13 $\displaystyle\sum_{r=1}^{n} \frac{1}{r(r+1)(r+2)} = \frac{1}{4} - \frac{1}{2(n+1)(n+2)}$

14 By assuming that $\dfrac{d}{dx}(x^k) = kx^{k-1}$, where k is a positive integer, and writing $x^{k+1} = x(x^k)$, show that
$\dfrac{d}{dx}(x^{k+1}) = (k+1)x^k$.

Deduce that $\dfrac{d}{dx}(x^n) = nx^{n-1}$ for all positive integral values of n.

15 Show, by induction, that $n^5 - n^3$ is divisible by 24 for all $n \in N$.

16 Prove, by induction, that $1(1!) + 2(2!) + 3(3!) + \ldots \quad n(n!) = (n+1)! - 1$.

17 Show by induction that the sum of the cubes of three consecutive integers is divisible by 9.

13 Differential equations

Introduction

Certain radioactive elements disintegrate at a rate proportional to the number of atoms present. This can be expressed approximately in mathematical terms by the equation

$$\frac{dN}{dt} = -kN$$

where N is the number of undecayed atoms, k is a constant and t is the time, measured in seconds.

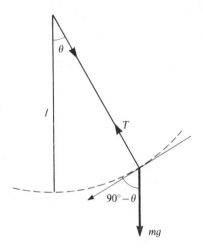

A particle of mass m attached to a light inextensible string whose other end is fixed, can swing in a circular arc in the vertical plane.

Applying Newton's second law along the direction of the tangent gives the equation of motion as,

$$ml\frac{d^2\theta}{dt^2} = -mg\sin\theta$$

$$\Rightarrow \quad \frac{d^2\theta}{dt^2} = -\frac{g}{l}\sin\theta$$

These two equations both contain the differential of a function and are known as **differential equations**. Such equations also occur in many other situations, for example, electrical circuits and damped oscillatory motion.

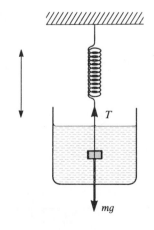

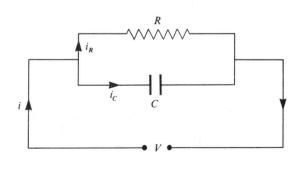

Order and degree

In general, any equation in which at least one term contains a differential coefficient is called a **differential equation**. The following equations give some examples.

(a) $\dfrac{dy}{dx} = 3x^2 - 1$ (b) $e^x \dfrac{dy}{dx} = \sin x$ (c) $(1+x)\dfrac{dy}{dx} = y(x^2 + 2x - 3)$

(d) $\dfrac{dy}{dx} + 4y = e^{3x}$ (e) $\dfrac{d^2y}{dx^2} = 4y$ (f) $\dfrac{d^2x}{dt^2} + \dfrac{dx}{dt} - 2x = 0$

Differential equations are grouped according to their order and degree.

> The **order** of a differential equation is the order of the highest derivative involved.

So Equations (a) to (d) above are all first order equations, whereas Equations (e) and (f) are second order.

> The **degree** of a differential equation is the power of the highest derivative involved.

So, (i) $\left(\dfrac{dx}{dt}\right)^2 = 4 - x^2$ is of second degree.

(ii) $\dfrac{d^2y}{dx^2} + \left(\dfrac{dy}{dx}\right)^2 + 4y = 0$ is of first degree.

General solution

Investigation 1

(a) Consider a simple first order differential equation $\dfrac{dy}{dx} = 4x$.

Integrate both sides with respect to x to obtain y in terms of x (do not forget the arbitrary constant, C).

The process of finding this relation between x and y from the original equation is called **solving the differential equation**.

Since the result $y = 2x^2 + C$ contains an arbitrary constant, it is known as the **general solution** of the equation and represents a **family of curves**, all of which have a gradient of $4x$.

If further data is given (say, a point on the solution curve), then the constant C can be evaluated which produces a **particular solution** of the equation, i.e. one particular member of the family of curves.

(b) Find the particular solution, given that $y = 5$ when $x = 1$.

(c) Sketch the family of curves for different values of C.

Investigation 2

Find the general solution of the differential equation $\dfrac{dx}{dt} = 2x$, if $x > 0$.

In this case it is not possible to integrate the equation immediately since the right-hand side is not a function of t. However, by changing the form of the equation, we can effect a solution.

Use the result $\dfrac{dt}{dx} = \dfrac{1}{\dfrac{dx}{dt}}$ to show that $\dfrac{dt}{dx} = \dfrac{1}{2x}$.

Integrate with respect to x and show that

$$t = \tfrac{1}{2} \ln x + C$$

By writing C as $\tfrac{1}{2} \ln k$ (i.e. choosing a convenient form for C), show that $2t = \ln kx$.

Show that this can be written as $x = \dfrac{1}{k} e^{2t}$ or $x = A e^{2t}$, where $A = \dfrac{1}{k}$.

Check, by differentiating this result, that it satisfies the original equation. Since the solution contains one arbitrary constant, A, it is the general solution of the differential equation for which $x > 0$.

Hence, equations of the form,

(a) $\dfrac{dy}{dx} = f(x)$ can be solved by direct integration to give $\quad y = \displaystyle\int f(x)\, dx + C$

and of the form,

(b) $\dfrac{dy}{dx} = f(y)$ can be written $\dfrac{dx}{dy} = \dfrac{1}{f(y)}$ and then integrated to give $\quad x = \displaystyle\int \dfrac{1}{f(y)}\, dy + C$

WORKED EXAMPLES

1 Solve the differential equation $\dfrac{dy}{dx} = 6x^2 - 2$ and illustrate the solution curves on a graph.

By direct integration with respect to x, the general solution is

$$y = 2x^3 - 2x + C$$

The solution curves can be drawn for specific values of C and the figure overleaf shows the appropriate curves when $C = -4, 0, 3$ and 5.

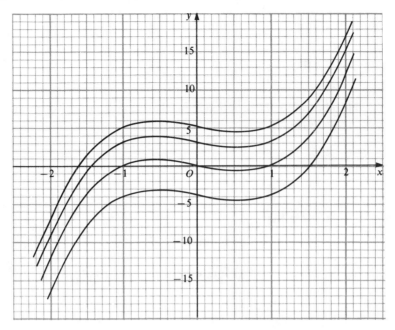

2 Solve the equation $\sec x \dfrac{dy}{dx} = \sin^3 x$ given that $y = 1$ when $x = \frac{1}{4}\pi$.

Rearranging, gives $\dfrac{1}{\cos x} \dfrac{dy}{dx} = \sin^3 x \Rightarrow \dfrac{dy}{dx} = \sin^3 x \cos x$

Integrating with respect to x gives, $y = \displaystyle\int \sin^3 x \cos x \, dx$

Thus, the general solution is $y = \frac{1}{4}\sin^4 x + C$ \qquad (by inspection)

Now, if $y = 1$ when $x = \frac{1}{4}\pi$, then $1 = \frac{1}{4}(\sin\frac{1}{4}\pi)^4 + C \Rightarrow C = 1 - \dfrac{1}{4}\left(\dfrac{1}{\sqrt{2}}\right)^4 = 1 - \frac{1}{16} = \frac{15}{16}$

The particular solution is $y = \frac{1}{4}\sin^4 x + \frac{15}{16}$.

3 Solve the equation $e^{-x}\dfrac{dy}{dx} + 2x = 2$.

Rearranging gives, $\dfrac{dy}{dx} + 2x\,e^x = 2e^x \Rightarrow \dfrac{dy}{dx} = 2e^x - 2x\,e^x = 2e^x(1-x)$

Integrating with respect to x gives, $y = \displaystyle\int 2e^x(1-x)\,dx$

This function is a product and hence we integrate by parts (see page 44).

Let $u = (1-x) \Rightarrow \dfrac{du}{dx} = -1$

and $\dfrac{dv}{dx} = 2e^x \Rightarrow v = 2e^x$

Using,
$$\int u \frac{dv}{dx}\, dx = uv - \int v \frac{du}{dx}\, dx$$

$$\therefore \quad \int 2e^x(1-x)\, dx = 2(1-x)\, e^x + \int 2e^x\, dx$$

$$= 2(1-x)\, e^x + 2e^x + C$$

Hence, the general solution is $\quad y = 2(1-x)\, e^x + 2e^x + C$

$$y = 2(2-x)\, e^x + C$$

4 Solve the equation $\dfrac{d^2x}{dt^2} = \cos t$ given that $\dfrac{dx}{dt} = 3, \quad x=0 \quad$ when $\quad t = \tfrac{1}{2}\pi.$

This equation can be solved by integrating twice with respect to t.

$$\frac{dx}{dt} = \sin t + A \qquad \text{where } A \text{ is the arbitrary constant}$$

Now, when $\dfrac{dx}{dt} = 3, \quad t = \tfrac{1}{2}\pi.$

Substituting gives, $\quad 3 = \sin\tfrac{1}{2}\pi + A \quad \Rightarrow \quad A = 2$

Hence,
$$\frac{dx}{dt} = \sin t + 2$$

Integrating again with respect to t gives $\quad x = -\cos t + 2t + B$

Now, when $\quad x=0, \quad t = \tfrac{1}{2}\pi.$

Substituting gives, $\qquad 0 = -\cos\tfrac{1}{2}\pi + \pi + B \quad \Rightarrow \quad B = -\pi$

Thus, the particular solution is $\quad x = 2t - \pi - \cos t.$

5 Solve the differential equation $\dfrac{dy}{dx} = 4 + y^2.$

Since $\quad \dfrac{dx}{dy} = \dfrac{1}{\dfrac{dy}{dx}}, \quad$ the equation can be written $\quad \dfrac{dx}{dy} = \dfrac{1}{4+y^2}.$

Integrating with respect to y gives $\quad x = \displaystyle\int \frac{1}{4+y^2}\, dy$

$$\Rightarrow \quad x = \tfrac{1}{2}\tan^{-1}(\tfrac{1}{2}y) + C$$

This can be written as $\qquad 2x - \alpha = \tan^{-1}(\tfrac{1}{2}y) \qquad$ where $\quad \alpha = 2C$

$$\Rightarrow \quad y = 2\tan(2x-\alpha)$$

Exercise 13.1

1 Show by differentiation that

(a) $y = 4 - \dfrac{1}{2x}$ is a particular solution of the equation $\dfrac{dx}{dy} = 2x^2$.

(b) $y = e^{3x}$ is a particular solution of $\dfrac{dy}{dx} = 3y$.

2 Show by differentiation that $y = 2\cos x + 3\sin x$ is a particular solution of the differential equation $\dfrac{d^2y}{dx^2} = -y$.

3 Show that $y = Ax\,e^{4x}$ is the general solution of $\dfrac{dy}{dx} - 4y = 3e^{4x}$.

4 Verify that the equation $\dfrac{d^2y}{dx^2} = 3 - y$ has particular solutions

(a) $y = 3$ (b) $y = 3 + 2\sin x$ (c) $y = 3 - 4\cos x$

5 Find the general solutions of the following differential equations and sketch the family of solution curves.

(a) $\dfrac{dy}{dx} = 2x - 1$ (b) $\dfrac{dy}{dx} = \cos x$ (c) $\dfrac{dy}{dx} = e^{2x}$

6 Solve the following differential equations,

(a) $\dfrac{dy}{dx} = 3x^2 - 6x + 1$ (b) $\dfrac{dy}{dx} = \cos\tfrac{1}{2}x + \sin\tfrac{1}{2}x$ (c) $\dfrac{dy}{dx} = x\sqrt{x^2 + 1}$

(d) $\dfrac{dy}{dx} = \sec^2 x$

7 Find the general solutions of the following differential equations

(a) $x\dfrac{dy}{dx} = x^2 - 1$ (b) $e^x\dfrac{dy}{dx} = 2 - e^{3x}$ (c) $\dfrac{d^2y}{dx^2} = 2\sin 3x$

(d) $\dfrac{d^2y}{dx^2} = 2e^{2x} - e^x$

8 Differentiate the following functions to form $\dfrac{dy}{dx}$. Use the result to form a differential equation by eliminating the constant A.

(a) $y = Ax$ (b) $y = \dfrac{A}{x}$ (c) $y = A\,e^x$ (d) $y = A\,e^{3x}$ (e) $y = x\ln x + Ax^2$

9 Solve the differential equations to give the general solutions

(a) $\dfrac{dy}{dx} = y^2$ (b) $\dfrac{dy}{dx} = 1 + y^2$ (c) $\dfrac{dy}{dx} = \cos^2 y$ (d) $\dfrac{dy}{dx} - e^{-y} = 0$

(e) $\dfrac{dy}{dx} = y + 1$ (f) $\dfrac{dy}{dx} = 1 - y^2$

10 Find the particular solutions of the following equations under the given conditions.

(a) $\dfrac{dy}{dx} = (2x-1)^2$; $y=2$ when $x=1$

(b) $\dfrac{dy}{dx} - \dfrac{1}{\sqrt{1-x^2}} = 0$; $y=0$ when $x=0$

(c) $\dfrac{dy}{dx} - 4y = 0$; $y=2$ when $x=0$

(d) $\dfrac{dy}{dx} = e^{2y}$; $y=0$ when $x=1$

11 Use the substitution $\dfrac{dy}{dx} = p$ to show that the differential equation $\dfrac{d^2y}{dx^2} - \dfrac{dy}{dx} = 1$ can be written as $\dfrac{dp}{dx} = 1 + p$. Solve this equation to find p as a function of x. Hence find a solution giving y in terms of x.

Equations with separable variables

Sometimes a first order differential equation contains a function of x and a function of y. If these are such that they can be separated, then the equation can be solved.

Investigation 3

Consider the first order equation $\dfrac{dy}{dx} = 2xy^2 + y^2$ and show that, by factorizing and rearranging, it can be written as $\dfrac{1}{y^2}\dfrac{dy}{dx} = 2x + 1$.

Now integrate with respect to x and show that $y(x^2 + x + k) = -1$ where k is an arbitrary constant.

In general, if a first order differential equation can be written in the form

$$f(y)\frac{dy}{dx} = g(x)$$

then we say that the **variables x and y are separable** and the solution will be obtained by integrating both sides with respect to x.

$$\therefore \quad \int f(y)\frac{dy}{dx}\,dx = \int g(x)\,dx$$

$$\Rightarrow \quad \int f(y)\,dy = \int g(x)\,dx$$

Differential equations

1 Solve the differential equation $\dfrac{dy}{dx} = \dfrac{y^2 - y}{x^2}$.

Separating the variables,

$$\frac{dy}{dx} = \frac{y(y-1)}{x^2} \quad \Rightarrow \quad \frac{1}{y(y-1)} \frac{dy}{dx} = \frac{1}{x^2}$$

Integrating both sides with respect to x gives,

$$\int \frac{1}{y(y-1)} \frac{dy}{dx} dx = \int \frac{1}{x^2} dx$$

$$\Rightarrow \quad \int \frac{1}{y(y-1)} dy = \int \frac{1}{x^2} dx$$

Using partial fractions on the left-hand side,

$$\int \left(\frac{1}{(y-1)} - \frac{1}{y} \right) dy = \int \frac{1}{x^2} dx$$

$$\Rightarrow \quad \ln|y-1| - \ln|y| = -\frac{1}{x} - \ln k$$

Note that the arbitrary constant has been written as $-\ln k$ to enable us to combine it with the other logarithmic functions. The constant may be written in different forms which will assist in simplifying the solution.

$$\therefore \quad \ln k \left| \frac{y-1}{y} \right| = -\frac{1}{x} \quad \text{or} \quad k \left(\frac{y-1}{y} \right) = e^{-1/x}$$

Writing $A = \dfrac{1}{k}$ we have $y - 1 = A y e^{-1/x}$

The general solution is $y(1 - A e^{-1/x}) = 1$.

2 Find the solution of $3\dfrac{dy}{dx} = e^{x-3y}$ given that $y = 1$ when $x = 3$.

Rearranging and separating the variables gives,

$$3\frac{dy}{dx} = e^x \times e^{-3y} \quad \Rightarrow \quad 3 e^{3y} \frac{dy}{dx} = e^x$$

Integrating with respect to x gives,

$$3\int e^{3y} \frac{dy}{dx} dx = \int e^x dx \quad \Rightarrow \quad 3\int e^{3y} dy = \int e^x dx \quad \Rightarrow \quad e^{3y} = e^x + C$$

Since $y = 1$ when $x = 3$, $e^3 = e^3 + C \quad \Rightarrow \quad C = 0$

Hence, $e^{3y} = e^x$ which clearly gives $3y = x$.

Thus the particular solution is $y = \frac{1}{3}x$.

3 Find the general solution of $x \sin y \dfrac{dy}{dx} = 3 - \cos y$.

Rearranging and separating the variables gives,

$$\frac{x \sin y}{3 - \cos y} \frac{dy}{dx} = 1 \quad \Rightarrow \quad \frac{\sin y}{3 - \cos y} \frac{dy}{dx} = \frac{1}{x}$$

Integrating with respect to x gives,

$$\int \frac{\sin y}{3 - \cos y} \frac{dy}{dx} \, dx = \int \frac{1}{x} \, dx \quad \Rightarrow \quad \int \frac{\sin y}{3 - \cos y} \, dy = \int \frac{1}{x} \, dx$$

$$\ln (3 - \cos y) = \ln |x| + \ln k$$

$$\ln \left| \frac{3 - \cos y}{x} \right| = \ln k$$

Hence,

$$3 - \cos y = kx$$

The general solution is $\cos y = 3 - kx$.

Exercise 13.2

Find the general solutions of the following first order differential equations.

1 $3x \dfrac{dy}{dx} = y$ **2** $\dfrac{dy}{dx} = xy^2$ **3** $x^2 \dfrac{dy}{dx} = y$ **4** $\dfrac{dy}{dx} = e^x \sec y$

5 $\dfrac{dy}{dx} = 4 \cos^2 y$ **6** $(1-x) \dfrac{dy}{dx} = 3y$ **7** $x \cos y \dfrac{dy}{dx} = 2 - \sin y$ **8** $\dfrac{dy}{dx} = \dfrac{2y}{x^2 - 1}$

9 $\dfrac{dy}{dx} - xe^{-y} = 0$ **10** $(1+x) \dfrac{dy}{dx} - xy = 0$ **11** $(1-x^2) \dfrac{dy}{dx} = 2xy$ **12** $e^x y \dfrac{dy}{dx} = x$

13 $xy \dfrac{dy}{dx} = \ln x$ **14** $\cot x \dfrac{dy}{dx} + 2y^2 \operatorname{cosec}^2 x = 0$

Find the particular solutions of the following differential equations under the given conditions.

15 $y^2 \dfrac{dy}{dx} = x + 1;$ $y = 3$ when $x = 2$ **16** $\dfrac{dy}{dx} = y \sin x;$ $y = 1$ when $x = \frac{1}{2}\pi$

17 $\dfrac{dy}{dx} = 2x(1 + y^2);$ $y = 1$ when $x = 0$ **18** $\dfrac{dy}{dx} = y - y \cos x;$ $y = 2$ when $x = 0$

19 $(1 + x^2) \dfrac{dy}{dx} - y^2 = 1;$ $y = 0$ when $x = 1$

20 If the gradient of a curve is given by $\dfrac{dy}{dx} = 2xy$ and the curve passes through the point $(2, 1)$, find the equation of the curve.

21 The acceleration of a body is given by the equation $\dfrac{dv}{dt} = -te^{-v}$.

     Find an expression for its velocity in terms of t, given that $v = 0$ when $t = 10$.

Formulation of differential equations

So far we have seen how to solve various types of first order differential equations. Such equations occur when we use a mathematical model to represent a physical situation.

For example, consider a parachutist falling under gravity against a resistance which is proportional to the velocity at any instant.

If the resistance is denoted by mkv then, by Newton's second law, the differential equation to model this situation is

$$m\frac{dv}{dt} = mg - mkv \quad \Rightarrow \quad \frac{dv}{dt} = g - kv$$

where k is a constant of proportionality. The motion described by this equation is discussed fully on page 283.

When a liquid is heated and then left to cool it is found that the rate of change of temperature of the liquid at time t is proportional to the difference in the temperatures of the surroundings and the liquid at that instant.

If θ is the temperature of the liquid at time t and θ_1 is the temperature of the surroundings, then the differential equation to model this situation is

$$\frac{d\theta}{dt} = k(\theta - \theta_1)$$

where k is a constant of proportionality.

Investigation 4

A cross country runner attempts to run a 10 km race and decides to run at a speed which is proportional to $(10 - 2x)$ where x km is his distance from the start after t hours.

Write down an expression for the remaining distance and show that the differential equation which models this situation is given by $\dfrac{dx}{dt} = k(10 - 2x)$, where k is a constant of proportionality.

Using the boundary condition that his speed is 15 km h^{-1} at the start, show that $k = \frac{3}{2}$. Hence deduce that the equation is $\dfrac{dx}{dt} = 15 - 3x$ and show that its solution is given by $t = -\frac{1}{3}\ln|15 - 3x| + c$.

Using the initial condition $x = 0$ when $t = 0$, show that $t = \dfrac{1}{3}\ln\left|\dfrac{5}{5-x}\right|$.

Find the time to complete one quarter of the race.

Although this model is suitable for the first part of the race it is not acceptable for the whole distance. Decide why it becomes unsuitable and why the runner does not finish the race. How far does he run?

WORKED EXAMPLE

In a chemical reaction a substance B is formed from a substance A such that the sum of the masses of A and B is a, where a is a constant. If x is the mass of B at time t, and the rate at which this mass is increasing at time t is proportional to the product of the two masses at that instant, show that $\dfrac{dx}{dt} = kx(a-x)$ where k is a constant of proportionality.

If $x = \frac{1}{4}a$ when $t = 0$ and $x = \frac{1}{2}a$ when $t = \ln 9$, show that $k = \dfrac{1}{2a}$.

As the total mass of A and B is a and the mass of B at time t is x, it follows that the mass of A at time t is $(a-x)$.

The rate of change of the mass of B is $\dfrac{dx}{dt}$, which is proportional to the product of the masses of A and B at time t.

Hence,

$$\frac{dx}{dt} \propto x(a-x) \quad \Rightarrow \quad \frac{dx}{dt} = kx(a-x)$$

where k is the constant of proportionality.

To solve this equation we rearrange it to give,

$$\frac{1}{x(a-x)} \frac{dx}{dt} = k$$

Integrating with respect to t gives,

$$\int \frac{1}{x(a-x)} \frac{dx}{dt} \, dt = \int k \, dt$$

$$\Rightarrow \quad \int \frac{1}{x(a-x)} \, dx = \int k \, dt$$

Using partial fractions it is possible to integrate the left-hand side.

$$\frac{1}{a} \int \left(\frac{1}{x} + \frac{1}{(a-x)} \right) dx = \int k \, dt \quad \Rightarrow \quad \frac{1}{a} \left[\ln x - \ln (a-x) \right] = kt + C$$

Note that modulus signs are unnecessary for $\ln (a-x)$ and $\ln x$, since $(a-x) > 0$ and $x > 0$.

Now, when $x = \frac{1}{4} a$, $t = 0 \quad \Rightarrow \quad C = \frac{1}{a} \ln \left[\dfrac{\frac{1}{4}a}{(a - \frac{1}{4}a)} \right] = \frac{1}{a} \ln \frac{1}{3}$

$$\therefore \quad \frac{1}{a} \left[\ln x - \ln (a-x) \right] = kt + \frac{1}{a} \ln \tfrac{1}{3}$$

$$\Rightarrow \quad \frac{1}{a} \left[\ln x - \ln (a-x) - \ln \tfrac{1}{3} \right] = kt \quad \Rightarrow \quad \frac{1}{a} \ln \left(\frac{3x}{a-x} \right) = kt$$

Now use the other condition, i.e. $x = \frac{1}{2}a$ when $t = \ln 9$

$$\therefore \quad \frac{1}{a} \ln \left[\frac{3(\frac{1}{2}a)}{a - \frac{1}{2}a} \right] = k \ln 9 \quad \Rightarrow \quad \ln 3 = ak \ln 9 \quad \Rightarrow \quad \ln 3 = 2ak \ln 3$$

$$\Rightarrow \quad 2ak = 1 \quad \text{or} \quad k = \frac{1}{2a}$$

Investigation 5

In a first order chemical reaction in which a substance A decomposes, the rate of reaction $\dfrac{dx}{dt}$, where x moles per litre decompose in time t, is proportional to the concentration of A at time t.

If the original concentration of A is a moles per litre, show that the rate of reaction is given by the equation $\dfrac{dx}{dt} = k(a-x)$ where k is a constant.

Solve this differential equation and show that $\quad t = -\dfrac{1}{k} \ln (a-x) + C$.

Use the condition that $x = 0$ when $t = 0$ (i.e. no part of A has decomposed at $t = 0$) to show that,

$$t = \frac{1}{k} \ln \left(\frac{a}{a-x} \right)$$

Use this result to show that the half life (the time for half the reaction to be completed, i.e. for a to become $\frac{1}{2}a$) is independent of the initial concentration.

Rewrite the solution above and show that

$$a - x = a\, e^{-kt}$$

Choose values for a and k and verify that by plotting $(a-x)$ against t (i.e. the concentration of A against time), the curve shown here is obtained.

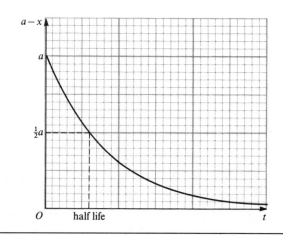

Investigation 6

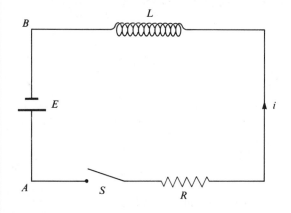

A simple electrical circuit shows a resistance R and an inductance L connected to a battery which maintains a constant potential difference E between A and B.

When switch S is closed, the growth of the current i in the circuit is given by $\dfrac{di}{dt}$ and is opposed by the inductance which has a potential difference $-L\dfrac{di}{dt}$ across its terminals.

The potential difference across the resistance is $-iR$.

(cont.)

Investigation 6—continued

Using the fact that the total potential drop around the circuit is zero, show that,

$$iR + L\frac{di}{dt} = E$$

Rearrange this equation to give $\dfrac{di}{dt} = -\dfrac{R}{L}\left(i - \dfrac{E}{R}\right)$

Integrate with respect to t and show that

$$\ln\left|i - \frac{E}{R}\right| = -\frac{Rt}{L} + C$$

Remember that $\left(i - \dfrac{E}{R}\right)$ may be negative, and hence the modulus signs must be used.

Using the initial condition that $i = 0$ when $t = 0$, show that

$$i = \frac{E}{R}\left(1 - e^{-Rt/L}\right)$$

Deduce that, when the switch S is closed, the current i rapidly increases to its ultimate value of E/R.

Exercise 13.3

1 A stone is thrown vertically downwards from the top of a tower with an initial speed of 2 m s^{-1}. Given that its acceleration is inversely proportional to its speed at any instant and that the acceleration is 5 m s^{-2} when the speed is 4 m s^{-1}, show that the equation of motion can be written as $\dfrac{dv}{dt} = \dfrac{20}{v}$. Find its speed after 6 s assuming that it has not reached the ground.

2 The number of bacteria in an experiment grow at a rate proportional to the number present at time t minutes. If the number of bacteria was 100 initially and is increasing at 10 per minute, find a differential equation to represent this situation and find the number of bacteria after 2 hours.

3 A long distance runner attempts to run 40 km at a speed proportional to $(40 - \frac{5}{4}x)$, where x km is his distance from the start at time t hours. If his initial speed is 12 km h^{-1}, find the time taken to complete (a) half the distance (b) 30 km. For what values of x is this a suitable model?

4 A liquid is placed in a room which is maintained at 15°C and cools at a rate proportional to the difference between its temperature and that of the room. If the initial temperature of the liquid is 80°C and it is then cooling at a rate of 1.3°C per minute, show that the rate of cooling at time t minutes is given by $\dfrac{d\theta}{dt} = -\dfrac{\theta}{50}$, where θ is the difference in temperature at time t. Hence find its temperature after 10 minutes.

5 In a chemical reaction, a substance X is formed from a substance Y such that the sum of the masses is a, where a is a constant. If the masses of X and Y at time t are x and y, respectively, and the rate at which X is increasing is proportional to the product of the two masses at that instant, show that $\dfrac{dx}{dt} = kx(a - x)$ where k is a constant.

If $x = \frac{1}{4}a$ when $t = 0$ and $x = \frac{1}{3}a$ when $t = \ln\frac{27}{8}$, show that $ak = \frac{1}{3}$ and find the value of t when $x = \frac{9}{10}a$.

6 A car is travelling at 90 km h^{-1} when the engine cuts out. The brakes are applied gently and the total retardation is proportional to the square of the speed, v m s^{-1}, at any instant. Write down a differential equation giving the speed of the car, t seconds after the engine cut out.

If the speed is observed to be 54 km h^{-1} after 10 seconds, show that $v = \dfrac{375}{t+15}$ and hence find the time that elapses before the speed is 18 km h^{-1}.

7 A radioactive substance decays so that the rate of change of its mass is proportional to the mass remaining at any instant. If x is the mass remaining at time t, show that $x = m_0 e^{-kt}$ where m_0 is the initial mass and k is a constant. If the initial mass is reduced to $\frac{3}{8}m_0$ in 40 days find the half life of the substance to the nearest day.

8 The volume of a spherical snowball decreases as it melts at a rate proportional to its volume at any instant. If the radius is 20 cm initially and is then changing at 2 cm per day, find how long it would take for **(a)** the volume **(b)** the radius, to be halved.

9 In a bimolecular reaction, one molecule of A combines with one molecule of B to form a molecule of C. If A and B each have a molecules present initially and the number of molecules of C present after time t is x, show that, assuming the rate of change of x is proportional to the product of the number of molecules of A and B that have not reacted, $\dfrac{dx}{dt} = k(a-x)^2$, where k is a constant.

Solve this equation and show that $t = \dfrac{x}{ka(a-x)}$. Deduce that for half the reaction to take place, $t = \dfrac{1}{ka}$.

Other first order differential equations

There are many other techniques for solving differential equations most of which are beyond the scope of the Advanced level syllabus. However, it is worth illustrating some of the possible approaches, since they may prove useful in some questions.

> WORKED EXAMPLE

Find the general solution of the equation $3xy^2 \dfrac{dy}{dx} + y^3 = \cos x.$

This is an equation in which the variables cannot be separated but the left-hand side can be recognized as the result of the differential of a product.

Since,
$$\frac{d}{dx}(xy^3) = x\frac{d}{dx}(y^3) + y^3\frac{d}{dx}(x) = 3xy^2\frac{dy}{dx} + y^3$$

we can write the equation as $\dfrac{d}{dx}(xy^3) = \cos x$

Integrating with respect to x gives $xy^3 = \sin x + C.$

When equations can be rewritten in this way they are called **exact differential equations**.

Investigation 7

Consider the first order equation $\dfrac{dy}{dx} + 2y = e^{-x}$ and see if you can rewrite it to give an **exact** equation.

In fact, you will find that this is not possible immediately, but we can make it exact by multiplying by e^{2x}.

Show that by multiplying by e^{2x} we obtain $\dfrac{d}{dx}(y\,e^{2x}) = e^x$.

Integrate this with respect to x and show that $y = e^{-x} + C\,e^{-2x}$.

Check by differentiation and substitution that this is the general solution of the original equation.

The term e^{2x} used as the multiplying factor is called the **integrating factor** and can be used when the equation is of the form,

$$\frac{dy}{dx} + Py = Q \qquad \text{where } P \text{ and } Q \text{ are functions of } x$$

Such equations are called **linear differential equations**.

It can be shown that the **integrating factor is** $e^{\int P\,dx}$.

So in Investigation 7, the integrating factor is $e^{\int 2\,dx} = e^{2x}$.

Investigation 8

Show that, by using the given integrating factor on the following equations, the stated general solutions are obtained.

(a) $\dfrac{dy}{dx} - 2y = e^x$; integrating factor $e^{\int -2\,dx} = e^{-2x}$

  solution $y = C\,e^{2x} - e^x$

(b) $\dfrac{dy}{dx} + 2xy = x$; integrating factor $e^{\int 2x\,dx} = e^{x^2}$

  solution $y = \tfrac{1}{2} + C\,e^{-x^2}$

(c) $\dfrac{dy}{dx} + y\cot x = \cos x$; integrating factor $e^{\int \cot x\,dx} = e^{\ln \sin x} = \sin x$

  solution $y = \tfrac{1}{2}\sin x + C\,\text{cosec}\,x$

WORKED EXAMPLE

By multiplying the equation $\dfrac{dy}{dx} + \dfrac{y}{x} = \dfrac{\ln x}{x^2}$ by x, find the general solution.

Multiplying the given equation by x gives,

$$x\frac{dy}{dx} + y = \frac{\ln x}{x}$$

Differential equations

Now since $\dfrac{d}{dx}(xy) = x\dfrac{dy}{dx} + y$ we have $\dfrac{d}{dx}(xy) = \dfrac{\ln x}{x}$

Integrating with respect to x,

$$xy = \int \frac{\ln x}{x}\,dx \qquad \text{............ (1)}$$

Integrating the right-hand side by parts,

let $u = \ln x \;\Rightarrow\; \dfrac{du}{dx} = \dfrac{1}{x}$ and $\dfrac{dv}{dx} = \dfrac{1}{x} \;\Rightarrow\; v = \ln x$

Using

$$\int u\frac{dv}{dx}dx = uv - \int v\frac{du}{dx}dx$$

$$\therefore \quad \int \frac{\ln x}{x}\,dx = (\ln x)^2 - \int \frac{1}{x}\ln x\,dx$$

$$\therefore \quad \int \frac{\ln x}{x}\,dx = \tfrac{1}{2}(\ln x)^2 + C$$

Substituting into Equation **(1)** gives $xy = \tfrac{1}{2}(\ln x)^2 + C$.

Exercise 13.4

1 Solve the differential equations by multiplying **(a)** by e^{2x} and **(b)** by x.

  (a) $\dfrac{dy}{dx} + 2y = e^{-2x}\cos x$ **(b)** $\dfrac{dy}{dx} + \dfrac{y}{x} = e^x$

2 Express the left-hand side of the following exact differential equations as a differential and hence, solve the equations.

  (a) $2xy\dfrac{dy}{dx} + y^2 = \sec^2 x$ **(b)** $x\dfrac{dy}{dx} + y = e^{-x}\sin x$

  (c) $xy^2 + x^2 y\dfrac{dy}{dx} = \sec^2 2x$ **(d)** $e^y + x e^y\dfrac{dy}{dx} = 1$

3 Use the integrating factor of x^{-4} to solve the equation $\dfrac{dy}{dx} - \dfrac{4y}{x} = 4x^3$.

4 Solve the equation $x\dfrac{dy}{dx} - y = x + 2$ by multiplying by $\dfrac{1}{x^2}$.

5 Find the particular solution of the following equations which satisfy the given conditions.

  (a) $x\dfrac{dy}{dx} - xy = e^x; \quad y = e$ when $x = 1$

  (b) $\dfrac{dy}{dx} - 4y\tan x = 12\cos x\sin x; \quad y = 0$ when $x = 0$

  (c) $\dfrac{dy}{dx} - \dfrac{y}{x} = x^2\cos x; \quad y = 0$ when $x = \pi$

  (d) $\dfrac{dx}{dt} - 3x = \sin 2t; \quad x = 0$ when $t = 0$

6 Use the substitution $z = \dfrac{1}{y}$ to show that the equation $\dfrac{dy}{dx} = 2y \tan x + y^2$ can be written as

$$\frac{dz}{dx} + 2z \tan x = -1.$$

By multiplying by $\sec^2 x$, find the general solution giving z in terms of x. Hence deduce an expression for y in terms of x.

7 If $y = vx$ show that $\dfrac{dy}{dx} = v + x\dfrac{dv}{dx}$. By using this substitution show that the equation $xy\dfrac{dy}{dx} = x^2 + 2y^2$ can be written as $x\dfrac{dv}{dx} = \dfrac{1+v^2}{v}$.

By separating the variables, solve this equation to find the particular solution for which $y = 0$ when $x = 1$.

Second order differential equations

Second order differential equations also arise from certain physical situations.

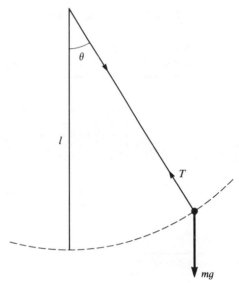

For example, the particle which oscillates to and fro in the vertical plane on the end of an inextensible string has, as its equation of motion,

$$\frac{d^2\theta}{dt^2} = -\frac{g}{l} \sin \theta$$

as seen on page 289.

In fact, if θ is small, $\sin \theta \simeq \theta$ and the equation becomes

$$\ddot{\theta} = -\frac{g}{l} \theta$$

where $\dfrac{d^2\theta}{dt^2}$ is written as $\ddot{\theta}$.

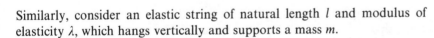

Similarly, consider an elastic string of natural length l and modulus of elasticity λ, which hangs vertically and supports a mass m.

Let the extension in the string be a in the equilibrium position. If the mass m is pulled down a further distance and released then, when the extension beyond the equilibrium position is x, the equation of motion is

$$mg - T = m\ddot{x}$$

Since, $\quad T = \dfrac{\lambda}{l}(a+x)$

and in equilibrium, $\qquad\qquad mg = \dfrac{\lambda}{l} a$

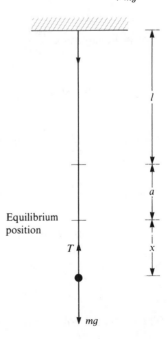

Equilibrium position

the equation becomes

$$mg - mg - \frac{\lambda}{l} x = m\ddot{x}$$

$$\Rightarrow \quad \ddot{x} = -\frac{\lambda}{ml} x$$

271

Differential equations

These two simple illustrations give equations of the second order which we can solve. It is possible to produce more complex second order equations by introducing, for example, resisting forces into oscillatory motion to damp out the vibrations (as in instruments or car shock absorbers).

Such an equation can be written $\dfrac{d^2x}{dt^2} + 2k\dfrac{dx}{dt} + \omega^2 x = 0$, but its solution is beyond the scope of this book.

Investigation 9

Find the general solution of the equation $\dfrac{d^2y}{dx^2} = 2x$.

(a) Firstly, show that by integrating with respect to x we obtain $\dfrac{dy}{dx} = x^2 + A$.

(b) Now integrate again with respect to x and show that $y = \frac{1}{3}x^3 + Ax + B$.

Notice that the general solution of a second order differential equation contains two arbitrary constants. Two conditions must be given if a particular solution is to be found.

Equations of the form $\dfrac{d^2y}{dx^2} = f(x)$ can be solved by direct integration.

WORKED EXAMPLE

Solve the equation $\dfrac{d^2y}{dx^2} = a\cos x$ given that $\dfrac{dy}{dx} = 2a$ and $y = a\pi$ when $x = \frac{1}{2}\pi$.

Integrate with respect to x, $\qquad \dfrac{dy}{dx} = a\sin x + A$

Substituting $\dfrac{dy}{dx} = 2a$ when $x = \frac{1}{2}\pi$ gives $\qquad 2a = a\sin\frac{1}{2}\pi + A \quad \Rightarrow \quad A = a$

Hence, $\qquad \dfrac{dy}{dx} = a\sin x + a$

Integrating again with respect to x gives $\qquad y = -a\cos x + ax + B$

Substituting $y = a\pi$ when $x = \frac{1}{2}\pi$ gives,

$$a\pi = -a\cos\tfrac{1}{2}\pi + \tfrac{1}{2}a\pi + B \quad \Rightarrow \quad B = \tfrac{1}{2}a\pi$$

The particular solution is $\quad y = -a\cos x + ax + \frac{1}{2}a\pi$.

Investigation 10

Use the substitution $p = \dfrac{dy}{dx}$ and differentiate with respect to x to show that $\dfrac{d^2y}{dx^2} = p\dfrac{dp}{dy}$.

Hence show that the equation $\dfrac{d^2y}{dx^2} = -\dfrac{16}{y^3}$ can be written $p\dfrac{dp}{dy} = -\dfrac{16}{y^3}$.

This is a first order equation which can be integrated with respect to y.

Show that $\frac{1}{2}p^2 = \dfrac{8}{y^2} + A$. If $\dfrac{dy}{dx} = 2$ when $y = 2$ show that $A = 0$.

Deduce that $y\dfrac{dy}{dx} = 4$ and, by integrating, show that $\frac{1}{2}y^2 = 4x + B$.

If $y = 2$ when $x = 0$, show that $B = 2$ and deduce that the particular solution of the equation is $y^2 = 4(2x + 1)$.

WORKED EXAMPLE

Solve the equation $\dfrac{d^2y}{dx^2} = -9y$ if $\dfrac{dy}{dx} = 0$ when $y = 1$ and $x = \frac{1}{3}\pi$.

Using the substitution $p = \dfrac{dy}{dx} \;\Rightarrow\; \dfrac{dp}{dx} = \dfrac{d^2y}{dx^2}$

Hence,
$$\frac{d^2y}{dx^2} = \frac{dp}{dx} = \frac{dp}{dy} \times \frac{dy}{dx} = p\frac{dp}{dy}$$

Thus, the equation can be written as $p\dfrac{dp}{dy} = -9y$

Integrating both sides with respect to y,
$$\int p\frac{dp}{dy}\,dy = \int -9y\,dy \;\Rightarrow\; \int p\,dp = \int -9y\,dy \;\Rightarrow\; \tfrac{1}{2}p^2 = -\tfrac{9}{2}y^2 + A$$

Since $\dfrac{dy}{dx} = p = 0$ when $y = 1 \;\Rightarrow\; A = \frac{9}{2}$

$$\therefore \qquad p^2 = 9 - 9y^2$$

$$\therefore \qquad p = \frac{dy}{dx} = \pm 3\sqrt{1 - y^2}$$

Rearranging gives
$$-\frac{1}{\sqrt{1 - y^2}}\frac{dy}{dx} = \pm 3$$

The negative sign introduced on the left-hand side simply makes the solution easier and is perfectly valid because of the $\pm$ sign on the right-hand side.

Differential equations

Integrating with respect to x gives

$$-\int \frac{1}{\sqrt{1-y^2}} \frac{dy}{dx} dx = \pm \int 3dx$$

$$\Rightarrow \quad -\int \frac{1}{\sqrt{1-y^2}} dy = \pm 3x + C$$

$$\Rightarrow \quad \cos^{-1} y = \pm 3x + C$$

or $\quad \cos^{-1} y = \pm(3x+C) \qquad$ by allowing the constant to be $\pm C$ as appropriate

Thus, $\qquad\qquad\qquad\qquad y = \cos[\pm(3x+C)] \quad \Rightarrow \quad y = \cos(3x+C)$

Since $y = 1$ when $x = \frac{1}{3}\pi \Rightarrow 1 = \cos(\pi+C) \Rightarrow \cos C = -1 \Rightarrow C = \pi.$

The particular solution is $\qquad y = \cos(3x+\pi)$

$$\Rightarrow \quad y = -\cos 3x$$

Equations of the form $\dfrac{d^2y}{dx^2} = f(y)$ can be solved by using the substitution $p = \dfrac{dy}{dx}$ from which we obtain

$$\frac{d^2y}{dx^2} = \frac{dp}{dx} = p\frac{dp}{dy}.$$

Investigation 11

Use the method of the Worked Example to find the general solution of the equation $\dfrac{d^2x}{dt^2} = -\omega^2 x$ where ω is a constant and $\dfrac{dx}{dt} = 0$ when $x = a$.

This equation is similar to the equation used in the previous Worked Example and represents simple oscillatory motion, such as the pendulum described on page 289 or the particle moving vertically on the end of an elastic string discussed on page 288. These are examples of simple harmonic motion.

Let $v = \dfrac{dx}{dt}$ and show that $\dfrac{d^2x}{dt^2} = v\dfrac{dv}{dx}$. Deduce that $v^2 = \omega^2(a^2 - x^2)$.

By writing v as $\dfrac{dx}{dt}$ show that $\dfrac{dx}{dt} = \pm\omega\sqrt{(a^2-x^2)}$.

By rearranging and integrating with respect to t, show that

$$\cos^{-1}\frac{x}{a} = \pm(\omega t + \alpha) \qquad \text{where } \alpha \text{ is a constant.}$$

Deduce that $\qquad\qquad\qquad\qquad x = a\cos(\omega t + \alpha)$

By writing $\alpha = -\frac{1}{2}\pi + \beta$ show that $\qquad\qquad x = a\sin(\omega t + \beta)$

By expanding $\sin(\omega t + \beta)$ show that if $a\cos\beta = A$ and $a\sin\beta = B$ then the solution can be written,

$$x = A\sin\omega t + B\cos\omega t$$

The general solution of the equation $\dfrac{d^2x}{dt^2} = -\omega^2 x$ can be written in the following forms

$$x = a \cos (\omega t + \alpha)$$
$$x = a \sin (\omega t + \beta)$$
$$x = A \sin \omega t + B \cos \omega t$$

Exercise 13.5

1 Solve the following differential equations.

(a) $\dfrac{d^2y}{dx^2} = e^{2x}$ (b) $\dfrac{d^2y}{dx^2} = \dfrac{1}{x^2}$ (c) $\dfrac{1}{x}\dfrac{d^2y}{dx^2} = e^x$ (d) $\dfrac{d^2y}{dx^2} = x \cos x$

2 Find the general solutions of the following differential equations.

(a) $\dfrac{d^2y}{dx^2} = -4y$ (b) $\dfrac{d^2y}{dx^2} + n^2y = 0$

3 Find the particular solutions of the following equations which satisfy the stated conditions.

(a) $\dfrac{d^2y}{dx^2} = \dfrac{1}{x}$; $y = 0$ and $\dfrac{dy}{dx} = 1$ when $x = 1$

(b) $\dfrac{d^2y}{dx^2} = 24x^2 - 5$; $y = 9$ when $x = 2$, and $y = 7$ when $x = 0$

(c) $\dfrac{d^2y}{dx^2} + 25y = 0$; $y = 2$ when $x = 0$, and $y = 3$ when $x = \tfrac{1}{2}\pi$

4 Use the substitution $p = \dfrac{dy}{dx}$ to show that $\dfrac{d^2y}{dx^2} = \dfrac{dp}{dx}$. Hence, solve the equations,

(a) $x\dfrac{d^2y}{dx^2} = 2\left(\dfrac{dy}{dx}\right)$ (b) $\dfrac{d^2y}{dx^2} = \left(\dfrac{2x}{1+x^2}\right)\dfrac{dy}{dx}$

(c) $(x+2)\dfrac{d^2y}{dx^2} - \dfrac{dy}{dx} = 1$ (d) $(1-x^2)\dfrac{d^2y}{dx^2} - x\dfrac{dy}{dx} = 0$

5 Use the substitutions $p = \dfrac{dy}{dx}$ and $\dfrac{d^2y}{dx^2} = p\dfrac{dp}{dy}$ to solve the following differential equations.

(a) $\dfrac{d^2y}{dx^2} = 2y\dfrac{dy}{dx}$ (b) $(1-y^2)\dfrac{d^2y}{dx^2} = -y\left(\dfrac{dy}{dx}\right)^2$

(c) $(y-1)\dfrac{d^2y}{dx^2} = \left(\dfrac{dy}{dx}\right)^2$ (d) $y\dfrac{d^2y}{dx^2} + \left(\dfrac{dy}{dx}\right)^3 = 0$

6 Show that $y = A e^{4x} + B e^{-4x}$ is the general solution of $\dfrac{d^2y}{dx^2} = 16y$.

7   If $y = v e^x$, where v is a function of x, show that

$$\frac{d^2y}{dx^2} = e^x \frac{d^2v}{dx^2} + 2 e^x \frac{dv}{dx} + v e^x$$

Use this to show that the equation $\dfrac{d^2y}{dx^2} = y$ can be written as $\dfrac{d^2v}{dx^2} + 2\dfrac{dv}{dx} = 0$. Hence solve the equation by direct integration.

8   Show that the general solution of the equation $\dfrac{d^2y}{dx^2} - n^2 y = 0$ is $y = A e^{nx} + B e^{-nx}$ where n, A and B are constants.

9   Show, by differentiation and substitution, that

(a) $y = A e^{2x} + B e^{-x}$ is a solution of $\dfrac{d^2y}{dx^2} - \dfrac{dy}{dx} - 2y = 0$.

(b) $y = (A + Bx) e^{-3x}$ is a solution of $\dfrac{d^2y}{dx^2} + 6\dfrac{dy}{dx} + 9y = 0$.

10   Verify that $y = A e^{-x} + B e^{-3x} + \frac{1}{10}(\sin x - 2 \cos x)$ is the general solution of the equation

$$\frac{d^2y}{dx^2} + 4\frac{dy}{dx} + 3y = \sin x$$

14 Variable acceleration

Introduction

If a body moves under the action of a constant force in a straight line, it will have a constant acceleration (by Newton's second law). Such motion was discussed in Chapters 11 and 15 of Book 1. However, if a body is subject to a **variable force**, this will cause the body to move with an **acceleration which varies**.

Consider a car pulling away from rest at a set of traffic lights with an acceleration which varies. Suppose the car reaches the maximum permitted speed in 10 seconds. A graph showing how the acceleration may vary with time is sketched below.

 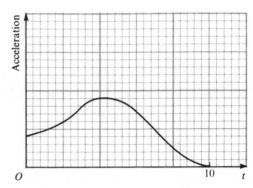

The graph shows a positive initial acceleration which increases for a few seconds and then decreases to zero as the speed increases to the maximum value.

Consider a wheel with a peg attached on which a piston rests. If the piston is able to move in a vertical cylinder, then it will move up and down as the wheel rotates. The graph below shows a sketch of the acceleration of the piston against time.

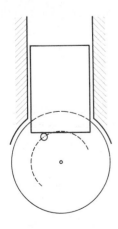 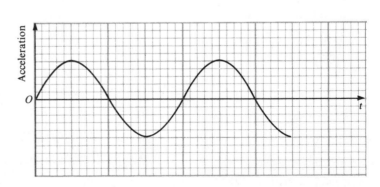

In this case the piston moves in an oscillatory motion. Its velocity has alternately positive and negative values between positions of instantaneous rest. Hence, the acceleration oscillates regularly taking positive and negative values. Clearly, the acceleration varies with time.

Variable acceleration

Acceleration as a function of time

Consider a parcel which is dropped from a hovering helicopter so that its acceleration at time t seconds is $10/t^2$ for $t \geq 1$. If it has a velocity of 4 m s^{-1} after one second, we can show that the velocity approaches a limiting value and draw graphs to show how the acceleration and velocity vary with time.

In this case we can formulate an equation of motion since the acceleration is given by $\dfrac{10}{t^2}$.

Acceleration $= \dfrac{dv}{dt}$ (the rate of change of the velocity)

and

$$\frac{dv}{dt} = \frac{10}{t^2}$$

This can be integrated with respect to t to give,

$$v = -\frac{10}{t} + C$$

Using the boundary conditions $v = 4$ when $t = 1$ gives,

$$4 = -\frac{10}{1} + C \quad \Rightarrow \quad C = 14$$

Hence the velocity at time t is given by $v = 14 - \dfrac{10}{t}$.

Clearly, as t increases, $\dfrac{10}{t}$ decreases to zero and hence the velocity approaches a limiting value of 14 m s^{-1}.

The graphs are shown below for $t \geq 1$.

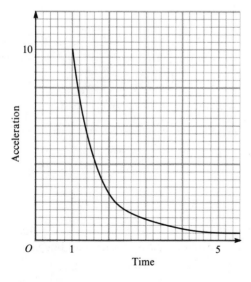

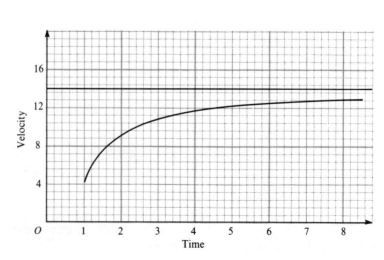

Investigation 1

A particle starts from rest and moves in a straight line with an acceleration of $3 \sin \pi t$ where t is the time in seconds. Find an expression for the velocity of the particle after t seconds and show that the displacement at this time is given by,

$$s = \frac{3}{\pi^2}(\pi t - \sin \pi t)$$

Draw graphs to show how the acceleration and the velocity vary with time.

Deduce that the acceleration is negative between $t = 1$ and $t = 2$ and that the velocity decreases from $\frac{6}{\pi}$ to zero in this interval.

Acceleration as a function of displacement

In the two examples discussed so far, the acceleration has been expressed as a function of time and it has been possible to integrate the expression for the acceleration with respect to time, directly, to obtain v in terms of t, i.e.

$$\text{If} \quad \frac{dv}{dt} = f(t) \quad \text{then} \quad v = \int f(t)\,dt + C$$

However, it is possible that the acceleration may be expressed as a function of the **displacement**. In this case an alternative form of the acceleration may be useful.

Using the chain rule for differentiation, where s is the displacement,

$$\frac{dv}{dt} = \frac{dv}{ds} \times \frac{ds}{dt} \quad \Rightarrow \quad \frac{dv}{dt} = \frac{dv}{ds} \times v$$

$$\text{or} \quad \boxed{\text{acceleration} = v\frac{dv}{ds}}$$

WORKED EXAMPLE

A particle of mass m is projected in a straight line in a medium which offers a total resistance to motion of $\dfrac{500m}{s^2}$, where s is the displacement from the point of projection, O. If the velocity of the particle is 20 m s^{-1} when it is 5 m from O, find its velocity at a point 20 m from O.

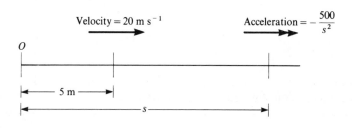

279

Variable acceleration

If v m s^{-1} is the velocity of the particle when it is s m from O then, by Newton's second law,

$$\text{Acceleration} = \frac{F}{m} = -\frac{500m}{s^2} \times \frac{1}{m} = -\frac{500}{s^2}$$

Since we wish to find a relationship between the velocity and the displacement, we use the form $v\dfrac{dv}{ds}$ for the acceleration, rather than $\dfrac{dv}{dt}$.

Hence,
$$v\frac{dv}{ds} = -\frac{500}{s^2}$$

To solve this differential equation we integrate both sides with respect to s (see Chapter 13 – Equations with separable variables).

$$\int v\frac{dv}{ds}\,ds = \int -\frac{500}{s^2}\,ds$$

$$\Rightarrow \quad \int v\,dv = -\int \frac{500}{s^2}\,ds \quad \Rightarrow \quad \tfrac{1}{2}v^2 = \frac{500}{s} + C$$

Using the boundary conditions, $v = 20$ when $s = 5$ gives,

$$\tfrac{1}{2}(20)^2 = \frac{500}{5} + C \quad \Rightarrow \quad C = 100$$

Thus,
$$v^2 = \frac{1000}{s} + 200$$

Now, when $s = 20$, $v^2 = \dfrac{1000}{20} + 200 \quad \Rightarrow \quad v^2 = 250$

The speed of the particle 20 m from O is $\sqrt{250} = 15.81$ m s^{-1}.

Investigation 2

A particle moves in a straight line such that its acceleration at time t is $-s$ where s is the displacement from a fixed point O. If the velocity is zero when $s = a$,   show that the equation of motion can be written as $v\dfrac{dv}{ds} = -s$.

By solving the equation, show that the velocity v, when the particle is at a displacement s from O, is given by $v^2 = a^2 - s^2$.

Can you describe the motion of this particle?

By writing $v = \dfrac{ds}{dt} = \sqrt{a^2 - s^2}$,   separate the variables and show that

$$\int \frac{1}{\sqrt{a^2 - s^2}}\,ds = \int dt$$

Hence solve the equation, given that $s = 0$ when $t = 0$. If you do not recognize the standard integral, use the substitution $s = a\sin\theta$.   You should find that $s = a\sin t$.

Exercise 14.1

1 A particle starts from rest and moves in a straight line with an acceleration of (t^2+t) m s^{-2}, where t is the time at any instant. Find its velocity and displacement at time t.

2 A particle moves in a straight line with an acceleration of $\dfrac{2}{1+t^2}$ where t is the time. If the particle has an initial velocity of 2 m s^{-1}, find its velocity after 5 seconds.

3 A particle of mass m moves in a straight line under the action of a force $-me^{t/2}$ N. If its initial speed is 6 m s^{-1} find when it first comes to rest.

4 A particle moves with an acceleration of $3s^2$ where s is the displacement from a point O after t seconds. If the particle has an initial speed of 4 m s^{-1}, show that the velocity after travelling 5 m is 16.31 m s^{-1}.

5 The acceleration of a particle moving in a straight line is $\left(3-\dfrac{s}{10}\right)$ m s^{-2}, where s is the displacement at time t. If the velocity of the particle is $5\sqrt{2}$ m s^{-1} when $s=10$ m, find an expression for the velocity at a displacement s. Hence, find the displacements from O when the particle is at rest.

6 A particle moves in a straight line with an acceleration of magnitude e^{-2x}, where x is the displacement from a fixed point O on the line at time t. If the initial velocity of the particle is 2 m s^{-1}, show that the velocity v at time t is given by $v^2=5-e^{-2x}$. Deduce that the velocity tends to a limiting value of $\sqrt{5}$ m s^{-1}.

7 A particle has a retardation of $4\sin^2 t$ m s^{-2} and moves in a straight line. If the velocity is 10 m s^{-1} initially, find the speed of the particle when $t=\frac{1}{2}\pi$ seconds. Find also the distance moved in the interval $0\leqslant t\leqslant\frac{1}{2}\pi$.

8 A particle moves from rest at O in a straight line with an acceleration of magnitude $\cos\omega x$, where x is the displacement from O at time t seconds. Show that the particle next comes to rest at a displacement $\dfrac{\pi}{\omega}$ from O.

9 A particle moves from rest at a point O in a straight line with an acceleration proportional to $(1+t)^{-1}$. If the velocity of the particle after 4 seconds is 10 m s^{-1}, show that the velocity after t seconds is given by $\dfrac{10\ln(1+t)}{\ln 5}$.

Acceleration as a function of velocity

If a racing car has an acceleration of

$$1-\frac{v^2}{2500}\ \text{m s}^{-2}$$

where v m s^{-1} is the speed of the car at a given time, t seconds, we can write the equation of motion as

$$\frac{dv}{dt}=1-\frac{v^2}{2500} \qquad \text{............ (1)}$$

or $\qquad v\dfrac{dv}{ds}=1-\dfrac{v^2}{2500} \qquad \text{............ (2)}$

Variable acceleration

Now, both of these equations can be solved but, clearly, produce results relating different variables. Equation **(1)** will give the velocity v in terms of the time t, but Equation **(2)** will relate the velocity v and the displacement s.

Thus, when the acceleration is given in terms of the velocity, it is necessary to decide which variables are required in the solution.

In general, $\dfrac{dv}{dt} = f(v)$ can be rearranged to give,

$$\frac{1}{f(v)}\frac{dv}{dt} = 1 \quad \Rightarrow \quad \int \frac{1}{f(v)}\,dv = \int dt$$

If, $v\dfrac{dv}{ds} = f(v)$ then,

$$\frac{v}{f(v)}\frac{dv}{ds} = 1 \quad \Rightarrow \quad \int \frac{v}{f(v)}\,dv = \int ds$$

The methods used to evaluate the integrals will depend upon the function given for $f(v)$. These methods were discussed in Chapter 2, but you will find the following two particularly useful:

$$\int \frac{f'(x)}{f(x)}\,dx = \ln k|f(x)| \qquad \text{and} \qquad \int \frac{dx}{a^2 + x^2} = \frac{1}{a}\tan^{-1}\left(\frac{x}{a}\right) + C$$

Investigation 3

Use Equation **(1)** from the example of the racing car at the beginning of this section, to show that,

$$\frac{2500}{2500 - v^2}\frac{dv}{dt} = 1$$

Use partial fractions to rearrange this equation and then integrate with respect to t.

Show that, $\quad \ln\left|\dfrac{50+v}{50-v}\right| = \dfrac{1}{25}t + C \quad$ where C is an arbitrary constant.

Using the initial condition that $v = 0$ when $t = 0$, deduce that the time, t, is given by,

$$t = 25\ln\left|\frac{50+v}{50-v}\right| \qquad \text{............ (3)}$$

Deduce that the time taken to reach half its maximum speed is $25\ln 3$ seconds.

Rearrange Equation **(3)** to show that

$$v = \frac{50(1 - e^{-t/25})}{(1 + e^{-t/25})}$$

Deduce that the maximum speed is 50 m s^{-1}.

Investigation 4

Using Equation **(2)** from the example of the racing car at the beginning of this section, show that,

$$s = \int \frac{2500v}{2500 - v^2}\, dv$$

Integrate the right-hand side of this result to show that, if $s = 0$ when $v = 0$, the displacement, s, is given by

$$s = 1250 \ln \left| \frac{2500}{2500 - v^2} \right|$$

Deduce that the distance travelled in reaching half its maximum speed from rest is $1250 \ln \frac{4}{3} = 359.6$ m.

The techniques discussed in this section and, in particular, in Investigations 3 and 4, can be applied specifically to the **vertical motion of a body under gravity in a resisting medium.**

WORKED EXAMPLE

A particle of mass m falls vertically under gravity from the top of a tower 100 m high. If the resistance to motion of the air is proportional to the speed of the particle at any instant, find an expression for the velocity after t seconds, assuming that the initial velocity is zero.

Since the resistance is proportional to the speed, we let its magnitude be mkv, where k is a constant. We use the constant of proportionality as mk rather than just k because it simplifies Newton's equation of motion.

The particle is subject to two forces: its weight and the resistance.

Hence, Newton's second law gives,

$$m \frac{dv}{dt} = mg - mkv$$

$$\Rightarrow \quad \frac{dv}{dt} = g - kv \qquad \text{............ (1)}$$

Rearranging the equation and integrating with respect to t gives,

$$\int \frac{1}{g - kv} \frac{dv}{dt}\, dt = \int dt \quad \Rightarrow \quad \int \frac{1}{g - kv}\, dv = \int dt \quad \Rightarrow \quad -\frac{1}{k} \ln |g - kv| = t + C$$

When $t = 0$, $v = 0$ and hence, by substitution $C = -\frac{1}{k} \ln g$

Variable acceleration

Thus,

$$t = \frac{1}{k}(\ln g - \ln|g - kv|)$$

$$\Rightarrow \quad t = \frac{1}{k}\ln\left|\frac{g}{g-kv}\right| \qquad \dots\dots\dots (2)$$

Rearranging

$$kt = \ln\left|\frac{g}{g-kv}\right| \quad \Rightarrow \quad e^{kt} = \frac{g}{g-kv}$$

Since $e^{kt} > 0$ for all finite values of t, it follows that

$$g - kv > 0 \quad \text{for all } t$$

Hence the modulus signs in Equation **(2)** could be omitted.

Now, since

$$e^{kt} = \frac{g}{g-kv} \quad \Rightarrow \quad (g-kv)e^{kt} = g \quad \Rightarrow \quad g\,e^{kt} - g = kv\,e^{kt}$$

Hence, $\quad g(1 - e^{-kt}) = kv$

So the velocity after t seconds is $\quad v = \frac{g}{k}(1 - e^{-kt}) \qquad \dots\dots\dots (3)$

Note that as t increases, e^{-kt} decreases, hence confirming that the limiting velocity is $\frac{g}{k}$.

Investigation 5

Use Equation **(3)** obtained as the result of the Worked Example and write it as

$$\frac{ds}{dt} = \frac{g}{k} - \frac{g}{k}e^{-kt}$$

Integrate this with respect to t and show that the displacement after t seconds is given by

$$s = \frac{g}{k^2}(kt + e^{-kt} - 1) \qquad \dots\dots\dots (4)$$

Now use the equation of motion **(1)** with the acceleration written in the form $v\frac{dv}{ds}$,

i.e. $\quad v\frac{dv}{ds} = g - kv \quad$ and show that,

$$\int \frac{v}{g-kv}\,dv = \int ds$$

By expressing $\frac{v}{g-kv}$ as $-\frac{1}{k}\left[1 - \frac{g}{g-kv}\right]$ integrate this equation and show that,

$$s = \frac{g}{k^2}\ln\left[\left(\frac{g}{g-kv}\right) - \frac{kv}{g}\right] \qquad \dots\dots\dots (5)$$

Verify that Equation **(5)** can also be obtained by substituting Equations **(2)** and **(3)** into Equation **(4)**.

WORKED EXAMPLE

A particle is projected vertically upwards with velocity u and experiences a resistance which varies as the square of its speed. Show that the time taken for the particle to reach its highest point is

$$\frac{1}{\sqrt{kg}} \tan^{-1} u \sqrt{\frac{k}{g}}$$

Find also the greatest height attained.

The particle is moving upwards subject to the force of gravity and the resistance.

The equation of motion is given by Newton's second law,

$$m\frac{dv}{dt} = -mg - mkv^2 \qquad \text{where } k \text{ is a constant}$$

$$\Rightarrow \quad \frac{dv}{dt} = -g - kv^2 \qquad\qquad \text{..........\ (1)}$$

Rearranging and integrating with respect to t gives,

$$\int \frac{1}{g+kv^2}\frac{dv}{dt}\,dt = -\int dt \quad \Rightarrow \quad \frac{1}{k}\int \frac{1}{\left(\dfrac{g}{k}+v^2\right)}\,dv = -t+C$$

The left-hand side is a standard integral giving the inverse tangent function   (see page 37).

$$\frac{1}{k}\sqrt{\frac{k}{g}}\tan^{-1} v\sqrt{\frac{k}{g}} = -t+C$$

When $t=0$, $v=u$ and hence, by substitution, $\quad C = \dfrac{1}{\sqrt{kg}}\tan^{-1} u\sqrt{\dfrac{k}{g}}$

Hence, $$t = \frac{1}{\sqrt{kg}}\left(\tan^{-1} u\sqrt{\frac{k}{g}} - \tan^{-1} v\sqrt{\frac{k}{g}}\right)$$

At the greatest height, $v=0$ and hence, $t = \dfrac{1}{\sqrt{kg}}\tan^{-1} u\sqrt{\dfrac{k}{g}}$, which represents the time taken to reach the greatest height.

To find the greatest height reached, we rewrite Equation (1) in the form,

$$v\frac{dv}{ds} = -g - kv^2$$

Rearranging we obtain $\quad \dfrac{v}{g+kv^2}\dfrac{dv}{ds} = -1$

Integrating with respect to s we have

$$\int \frac{v}{g+kv^2}\frac{dv}{ds}\,ds = -\int ds \quad \Rightarrow \quad \int \frac{v}{g+kv^2}\,dv = -s+C$$

Variable acceleration

The integral can be written in the form $\int \frac{f'(x)}{f(x)} dx = \ln |f(x)| + C.$

$$\therefore \quad \frac{1}{2k} \int \frac{2kv}{g + kv^2} dv = -s + C$$

$$\Rightarrow \quad \frac{1}{2k} \ln (g + kv^2) = -s + C$$

When $s = 0$, $v = u$, and hence, $C = \frac{1}{2k} \ln (g + ku^2)$

$$s = \frac{1}{2k} \ln (g + ku^2) - \frac{1}{2k} \ln (g + kv^2) = \frac{1}{2k} \ln \left(\frac{g + ku^2}{g + kv^2} \right)$$

At the maximum height, $v = 0$, and hence the greatest height reached by the particle is

$$\frac{1}{2k} \ln \left(\frac{g + ku^2}{g} \right)$$

Exercise 14.2

1 A particle moves in a straight line with an acceleration of $\frac{1}{v}$ m s^{-2}, where v is the velocity of the particle after t seconds. If the particle starts from rest, show that its velocity after t seconds is given by $v = \sqrt{2t}$. By writing v as $\frac{ds}{dt}$, find the displacement of the particle after 8 seconds.

2 A particle of unit mass moves in a straight line with a retardation of $(2 + v)$ m s^{-2} where v is the speed of the particle at time t seconds. If $v = 25$ when $t = 0$, find an expression for the speed of the particle in terms of t. Hence show that its speed after 2 seconds is 1.65 m s^{-1}.

3 A particle moves from rest in a straight line with an acceleration of $(16 - v^2)$ where v is the speed of the particle after t seconds. Show that

$$t = \frac{1}{8} \ln \left(\frac{4 + v}{4 - v} \right) \quad \text{and deduce that} \quad v = \frac{4(1 - e^{-8t})}{(1 + e^{-8t})}$$

4 Show that, if a particle has an acceleration of $(-100 - \frac{1}{4}v^2)$ m s^{-2} whilst moving in a straight line, the equation of motion can be written as $-4v \frac{dv}{ds} = 400 + v^2$. Solve this equation to find the distance travelled before it comes to rest if its initial speed is 20 m s^{-1}.

5 Show that the particle described in question 4 will come to rest in a time of $\frac{1}{5} \tan^{-1} 1 = 0.16$ seconds.

6 A body moves in a straight line with an acceleration of $(100 - v^2)$ m s^{-2}, where v is its speed at time t seconds. If the body increases its speed from u m s^{-1} to 8 m s^{-1} over a distance of 0.25 m, find the value of u.

7 A particle of mass m is projected vertically upwards with a velocity u in a medium which offers a resistance of kv per unit mass, where k is a constant. Show that the time taken to reach the greatest height is given by,

$$\frac{1}{k} \ln \left(1 + \frac{ku}{g} \right)$$

8 Show that the equation of motion of the particle in question **7** can also be written as $v\dfrac{dv}{ds} = -(g + kv)$. Solve this equation to find the greatest height reached.

9 A particle of mass m is projected vertically downwards from the top of a cliff with a speed u in a medium which offers a resistance to motion of kv per unit mass, where v is the speed of the particle at time t and k is a constant. Show that

$$\frac{g - kv}{g - ku} = e^{-kt}$$

and deduce that the velocity approaches the same limiting value as if it had fallen from rest.

10 A particle of mass m is projected vertically upwards with a speed u, where $u^2 = g/k$ in a medium which offers a resistance of kv^2 per unit mass, where v is the speed of the particle at time t, and k is a positive constant. If the particle comes to rest instantaneously at a height h after T seconds, show that

$$T^2 = \frac{\pi^2}{16kg}$$

Find a value for h in terms of k.

Simple harmonic motion

An important example of variable acceleration occurs in certain oscillatory motions or vibrations.

The motion of the balance wheel in a watch, the swinging of a pendulum or the vibrations of a mass on the end of a spring are all examples of oscillatory motion. This kind of motion occurs frequently and since the motion of the strings of many musical instruments oscillate when producing a note, the motion is called **simple harmonic motion** (abbreviated to **S.H.M.**).

In each of these motions the accelerations, displacements and velocities change periodically in both magnitude and direction. This can be clearly seen in the example of the mass, attached to an elastic spring, which is oscillating vertically about the equilibrium position O. Let A and B represent the upper and lower limits of the displacement from O.

Variable acceleration

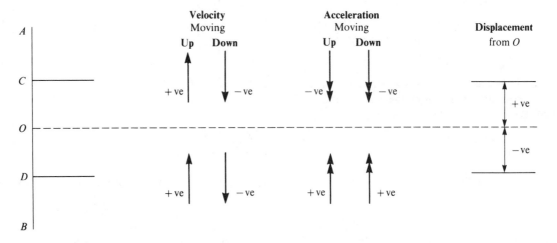

We can choose a sign convention so that, say, motion in an upwards direction is positive and that in the downwards direction is negative.

The figure above shows the sign associated with the velocities, accelerations and the displacements above and below the position of O.

Now, when the particle is at D, the total force acting on the mass is $T - mg$ towards O. However, when the particle is at C, the tension T is less than mg, and thus the effective force on the mass acts downwards towards O.

Let the elastic spring have a modulus of elasticity λ and natural length l.

If it has an extension a in the equilibrium position then at a point D, distant x below O, the total extension is $(a + x)$.

By Newton's second law,

$$mg - T = m\ddot{x}$$

$$\Rightarrow \quad mg - \frac{\lambda(x + a)}{l} = m\ddot{x}$$

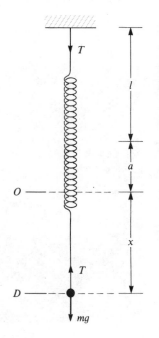

In the equilibrium position, $\quad T = \dfrac{\lambda a}{l} = mg$

Hence, $\qquad -\dfrac{\lambda x}{l} = m\ddot{x} \quad \Rightarrow \quad \ddot{x} = -\dfrac{\lambda x}{ml}$

Thus, the acceleration is proportional to the displacement from the equilibrium position. We use this relationship between acceleration and displacement to define **simple harmonic motion in a straight line**.

> A particle which moves in a straight line such that its acceleration is proportional to its displacement from a fixed point of the line and is always directed towards that point, is said to move with simple harmonic motion.

Using this definition, the simplest form for the equation of motion can be written $\ddot{x} = -kx$, where x is the displacement from the fixed point.

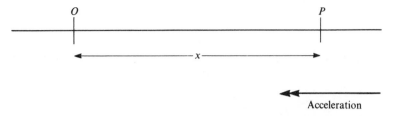

We usually let $k = \omega^2$ because $\omega^2 > 0$ and this simplifies some of the calculations involved.

Thus, the equation for S.H.M. is $\ddot{x} = -\omega^2 x$

The example and definition above refer to motion in a straight line but any motion which can be represented by such an equation is simple harmonic. So, the displacement x may be measured along a curve from a fixed point or be an angle between a given line on the body and a fixed line in space.

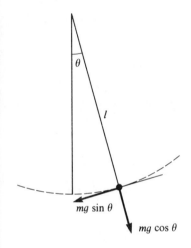

For example, in the case of a simple pendulum, the angle between the string and the vertical is θ.

The restoring force along the tangent to the path of the bob is $mg \sin \theta$ due to its weight component.

Now, the angular acceleration is $\dfrac{d^2\theta}{dt^2}$ i.e. $\ddot{\theta}$ and the linear acceleration of the bob is $l\ddot{\theta}$ along the direction of the tangent.

By Newton's second law,

$$ml\ddot{\theta} = -mg \sin \theta$$

$$\Rightarrow \quad \ddot{\theta} = -\frac{g}{l} \sin \theta$$

If θ is small, $\sin \theta \simeq \theta$ and hence for small oscillations $\ddot{\theta} = -\dfrac{g}{l}\theta$ and the motion is simple harmonic.

$$\ddot{\theta} = -\omega^2 \theta \quad \text{where} \quad \omega^2 = \frac{g}{l}$$

A particle which oscillates on a circular path such that its angular acceleration is proportional to its angular displacement from a fixed line and is directed towards that line, performs simple harmonic motion.

The equations of simple harmonic motion

It is the solution of the basic equation of motion $\ddot{x} = -\omega^2 x$ which uses the methods derived for variable acceleration.

Consider a particle P oscillating in simple harmonic motion in a straight line about a fixed point O.

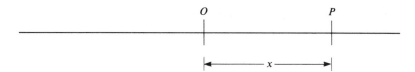

Let the acceleration of the particle be represented by $v\dfrac{dv}{dx}$ where v is the velocity at a general point P.

Hence, $$v\frac{dv}{dx} = -\omega^2 x$$

Integrating with respect to x gives

$$\int v\frac{dv}{dx}\,dx = -\int \omega^2 x\,dx$$

$$\Rightarrow \quad \int v\,dv = -\omega^2 \int x\,dx \qquad (\omega^2 \text{ is a constant})$$

$$\Rightarrow \quad \tfrac{1}{2}v^2 = -\tfrac{1}{2}\omega^2 x^2 + C$$

Now when $v = 0$, the displacement is a maximum say, $x = a$. The displacement a is called the **amplitude** of the motion.

Substituting gives $\quad 0 = -\tfrac{1}{2}\omega^2 a^2 + C \quad \Rightarrow \quad C = \tfrac{1}{2}\omega^2 a^2$

Thus $$\tfrac{1}{2}v^2 = \tfrac{1}{2}\omega^2 a^2 - \tfrac{1}{2}\omega^2 x^2$$

$$\boxed{\Rightarrow \quad v^2 = \omega^2(a^2 - x^2)}$$

This gives an expression for the square of the velocity of the particle at a displacement x from O.

By writing v as $\dfrac{dx}{dt}$ we can find an expression for x in terms of t.

$$\therefore \quad \frac{dx}{dt} = \pm\omega\sqrt{a^2 - x^2}$$

Rearranging and integrating with respect to t gives

$$\int \frac{1}{\sqrt{a^2 - x^2}}\,dx = \pm\omega\int dt$$

$$\Rightarrow \quad \sin^{-1}\frac{x}{a} = \pm(\omega t + \alpha) \qquad (\text{where } \alpha \text{ is a constant})$$

$$\Rightarrow \quad x = a\sin[\pm(\omega t + \alpha)]$$

Now, since $\sin(-\theta) = -\sin\theta$, this result can be written

$$x = \pm a\sin(\omega t + \alpha)$$

When $v = 0$, $x = \pm a$, so a can be positive or negative and thus there is no loss of generality if we write

$$\boxed{x = a\sin(\omega t + \alpha)}$$

This is the general solution of the differential equation $\ddot{x} = -\omega^2 x$.
    Compare the derivation in Investigation 11 of Chapter 13.

Particular solutions can be found by using the initial parameters.

(a) If $t=0$ when $x=0$ (i.e.   **time is measured** from the instant when the particle is at **the centre of the oscillation**).

Substituting into $x = a \sin (\omega t + \alpha)$ gives,

$$0 = \sin \alpha \quad \Rightarrow \quad \alpha = 0$$

Hence,
$$\boxed{x = a \sin \omega t}$$

(b) If $t=0$ when $x=a$ (i.e.   **time is measured** from the instant when the particle is at **the position of maximum displacement**).

Substituting into $x = a \sin (\omega t + \alpha)$ gives,

$$a = a \sin \alpha \quad \Rightarrow \quad \sin \alpha = 1 \quad \Rightarrow \quad \alpha = \tfrac{1}{2}\pi$$

Hence,
$$x = a \sin (\omega t + \tfrac{1}{2}\pi)$$

or
$$\boxed{x = a \cos \omega t}$$

We can show that the motion is **periodic**, as follows.

Since $x = a \sin (\omega t + \alpha)$ and velocity $= \dfrac{dx}{dt} = a\omega \cos (\omega t + \alpha)$

it follows that if ωt is increased to $\omega t + 2\pi$, the same values for the displacement x and the velocity will be obtained. Thus, the particle passes through the same points with the same velocities at times $t,\ t + \dfrac{2\pi}{\omega},\ t + \dfrac{4\pi}{\omega}, \ldots$

Hence, $\dfrac{2\pi}{\omega}$ is the time taken for one complete oscillation and this is called the **period** of the motion

$$\boxed{T = \frac{2\pi}{\omega}}$$

Investigation 6

Consider the equation $\ddot{\theta} = -n^2\theta$ and use the boundary conditions that when $t = 0,\ \theta = \phi$ and $\dfrac{d\theta}{dt} = 0$
to show that similar results can be obtained in the case of angular S.H.M., i.e.

$$\left(\frac{d\theta}{dt}\right)^2 = n^2(\phi^2 - \theta^2); \qquad \theta = \phi \cos nt; \qquad T = \frac{2\pi}{n}$$

Investigation 7

Consider a particle P which is describing a circular path of radius a with constant angular velocity ω. If O is the centre of the circle and N is the foot of the perpendicular from P to the diameter AA', show that the velocity of P in the direction parallel to AA' is $a\omega \sin \theta$, where θ is the angle that OP makes with the diameter.

(cont.)

291

Investigation 7—continued

Show also that the acceleration of P in a direction parallel to AA' is $a\omega^2 \cos\theta$.

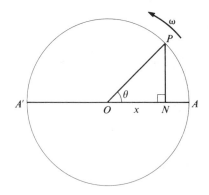

Deduce that if $ON = x$, then N has an acceleration $\omega^2 x$ towards O and hence N moves with S.H.M. along the diameter AA' as P moves round the circle with constant velocity.

Show that $a \sin\theta = \sqrt{a^2 - x^2}$ and deduce that the velocity of N is given by $v = \omega\sqrt{a^2 - x^2}$ along NO.

Also verify that the period is given by the time taken for P to complete one revolution at an angular velocity ω,

$$\text{i.e.} \qquad T = \frac{2\pi}{\omega}$$

> **WORKED EXAMPLES**

1 A particle is performing simple harmonic motion in a straight line with a period of 2 seconds. If the amplitude is 5 m, find the speed of the particle when it is 2 m from the centre of the oscillation.

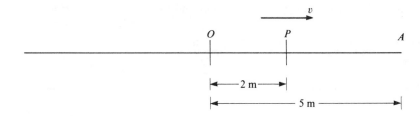

Let O be the centre of oscillation and v the speed of the particle at P, where $OP = 2$ m. Since OA is the amplitude, $OA = 5$ m.

Using the standard result for the velocity of the particle at a given point i.e. $v^2 = \omega^2(a^2 - x^2)$, it is clear that we need a value for ω.

Now, since the period $T = \dfrac{2\pi}{\omega}$ we have,

$$\frac{2\pi}{\omega} = 2 \quad \Rightarrow \quad \omega = \pi \text{ rad s}^{-1}$$

Hence,

$$v^2 = \pi^2(5^2 - 2^2) = 21\pi^2$$

Thus, the speed of the particle 2 m from O is $\sqrt{21\pi^2} = 14.4$ m s^{-1}.

2 A particle is performing simple harmonic motion in a straight line about a point O with an amplitude of 4 m and a period of $\frac{1}{3}\pi$ seconds. When the particle is at a point P, it has a velocity of 8 m s^{-1} away from O. Find the time that elapses before the particle reaches O again.

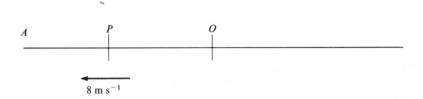

Let the amplitude of motion be $OA = a$. The equation of motion is given by $\ddot{x} = -\omega^2 x$ and the general solution is $x = a \sin(\omega t + \alpha)$.

However, by using the initial condition $x = a$ when $t = 0$, the particular solution becomes

$$x = a \cos \omega t \qquad \text{............ (1)}$$

This gives the time taken for the particle to cover the distance AP if $OP = x$.

Also, the period is $\frac{1}{3}\pi$ seconds and hence, since $T = \dfrac{2\pi}{\omega}$,

$$\tfrac{1}{3}\pi = \frac{2\pi}{\omega} \quad \Rightarrow \quad \omega = 6 \text{ rad s}^{-1}$$

To find the displacement from O when the velocity is 8 m s^{-1}, we use the result $v^2 = \omega^2(a^2 - x^2)$.

$$\therefore \qquad 8^2 = 6^2(4^2 - x^2) \quad \Rightarrow \quad 64 = 36(16 - x^2) \quad \Rightarrow \quad 16 = 9(16 - x^2) \quad \Rightarrow \quad \frac{16}{9} = 16 - x^2$$

$$\Rightarrow \quad x^2 = 16 - \frac{16}{9} \quad = 16\left(1 - \frac{1}{9}\right) \quad = 16 \times \frac{8}{9} = \frac{128}{9}$$

Thus, the displacement from O is $OP = x = \sqrt{\dfrac{128}{9}} = \dfrac{8\sqrt{2}}{3}$ m

Substituting into Equation **(1)** we have,

$$\frac{8\sqrt{2}}{3} = 4\cos 6t \quad \Rightarrow \quad \cos 6t = \frac{2\sqrt{2}}{3} \quad \Rightarrow \quad 6t = \cos^{-1}\frac{2\sqrt{2}}{3} = 0.34 \quad \Rightarrow \quad t = 0.057 \text{ s}$$

So the time for the particle to travel from P to A is 0.057 seconds.

Now, the time taken to move from A to O is $\frac{1}{4}$ of the period i.e. $\dfrac{\pi}{12}$ seconds.

The total time taken to move from P to O via A is $\left(0.057 + \dfrac{\pi}{12}\right)$ seconds.

$\therefore$ The time that elapses is 0.318 seconds.

Exercise 14.3

1 A particle moves in a straight line with simple harmonic motion. If the amplitude is 2 m and the period is $\frac{1}{2}\pi$ seconds, find the maximum speed of the particle.

2 A particle performs simple harmonic motion in a straight line with an amplitude of 1.2 m and a period of 5 seconds. Find the magnitude of its acceleration and velocity when it is **(a)** 1.2 m **(b)** 0.6 m **(c)** -0.3 m, from the centre of the oscillation.

3 A particle moves in a straight line in simple harmonic motion about a point O. Find the period of the oscillation if **(a)** the magnitude of its acceleration 2 m from O is 8 m s^{-2} and **(b)** its velocity 3 m from O is 5 m s^{-1}, and the amplitude is 7 m.

4 A particle moves in a straight line with simple harmonic motion between two points A and B. If $AB = 10$ m and the maximum acceleration is 20 m s^{-2}, find the time taken to travel **(a)** from A to B **(b)** from the centre of the motion to a point 2.5 m from B and **(c)** from A to a point 1 m from A.

5 A particle is performing simple harmonic motion in a straight line about a point O. If its speed at O is 15 m s^{-1} and its acceleration at a point 4 m from O is 10 m s^{-2}, find its amplitude and period.

6 A particle of mass m oscillates through a small angle at the end of a light string of length l which is fixed at its

other end. Show that the motion is approximately simple harmonic with period $T = 2\pi\sqrt{\dfrac{l}{g}}$.

Such a pendulum makes half an oscillation every second. Find the length of the string required.

7 A particle describes angular S.H.M. with a period of 2π seconds. If its maximum angular acceleration is $\dfrac{\pi}{3}$ rad s^{-2}, find the angular velocity of the particle when its angular displacement is half the greatest angular displacement.

8 A particle moving with S.H.M. in a straight line about a point O has a speed of 20 m s^{-1} when it is 5 m from O, and a speed of 10 m s^{-1} when it is 8 m from O. Find the amplitude and the period of the motion.

9 A particle is moving so that its displacement x, at time t, is given by $x = A \cos \omega t + B \sin \omega t$ where A and B are constants. Find expressions for $\dot{x}$ and $\ddot{x}$ and deduce that $x = A \cos \omega t + B \sin \omega t$ is a solution of $\ddot{x} = -\omega^2 x$.

10 A particle is performing S.H.M. in a straight line between two points A and B with a period of 10 seconds. If it takes the particle 2 seconds to travel from A to a point 1 m from A, find the amplitude and its speed 1 m from A.

11 A particle performs simple harmonic motion in a straight line so that the equation of motion is $\ddot{x} = -4x$, where x is the displacement of the particle from the mean position. Find x in terms of t, given that $x = 5$ and $\dot{x} = 0$ when $t = 0$. Find the time taken to move 3m from the position of maximum displacement.

15 Co-ordinate geometry

Introduction

In Investigations 1–5, $P(2, 3)$ is a point which satisfies various rules.
In each Investigation, you are asked to find other positions for P still satisfying the rule and to describe the complete locus, curve or path that P follows.
$F(2, 0)$ and $G(-2, 0)$ are fixed.

Investigation 1

P satisfies $PF + PG = 8$.

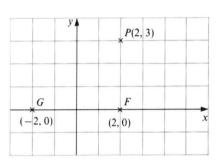

(a) Find more points which satisfy this rule.

(b) Find the x, y equation for the curve.

(c) What is its shape?

Investigation 2

P satisfies $PF = PD$, where D is the foot of the perpendicular from the line $x = -1$.

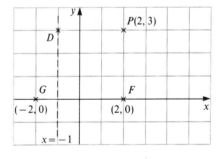

(a) Find more points satisfying the rule.

(b) Find the x, y equation for the locus.

(c) What is the shape of the curve?

Investigation 3

P satisfies $PF : PG = 3 : 5$.

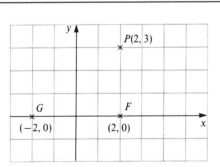

(a) Find more points which satisfy the rule.

(b) Find the x, y equation for the locus.

(c) What is the shape of the curve?

Investigation 4

$PF = \frac{1}{2}PE$, where E is on $x = 8$.

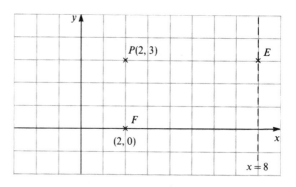

(a) Find more points P where $PF = \frac{1}{2}PE$, where PE is the distance from P to the line $x = 8$.

(b) Find the x, y equation of the curve.

(c) What is the shape of the curve?

Investigation 5

$PF = 2PD$, where D is the foot of the perpendicular from $x = \frac{1}{2}$.

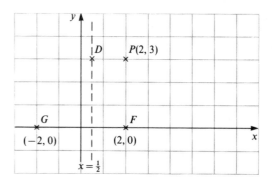

(a) Find more points P where $PF = 2PD$.

(b) For each point P work out $PG - PF$.

(c) Find the x, y equation for the curve.

(d) What is the shape of the curve?

Investigation 6

The shadow cast on the wall by my bedside lamp was a curve.

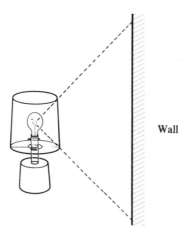

I picked up the lamp and the curve changed shape. How many different types of curve can you make?

Try rotating the lamp.

Do the same with a torchlight.

Investigation 7

Mark a cross, *C*, on a circular piece of paper (filter paper?) **not** at the centre of the circle, *O*. Fold over part of the paper so that a point on the circumference coincides with *C*. Rule in the fold line *L*. Do this repeatedly for points *A* around the circle and draw the lines *L*.

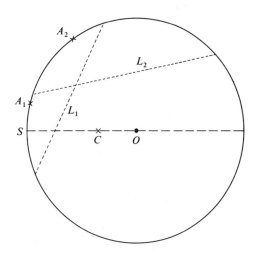

(a) What curve to the lines envelop?

(b) Where is the centre of the curve?

(c) Is the point *C* related to the curve?

(d) What happens as you vary the position of *C* along the line *CO*?

(e) What happens as *C* approaches *O* and as *C* approaches *S*?

Ellipse

Your answers to Investigations 1 and 4 should be the same, which highlights two properties (definitions?) of an ellipse.

Taking $PF = \frac{1}{2}PE$, we can derive the equation of the curve by letting *P* be a general point on the curve (x, y).

$$FP^2 = (x-2)^2 + y^2$$

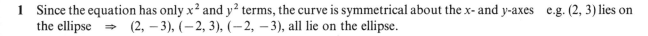

$PE =$ distance from *P* to the line $x = 8$.

$PE = 8 - x$, so $PF = \frac{1}{2}PE \Rightarrow$

$$\sqrt{(x-2)^2 + y^2} = \tfrac{1}{2}(8-x)$$

$$\Rightarrow \quad (x-2)^2 + y^2 = \frac{(8-x)^2}{4}$$

$$\Rightarrow \quad 4x^2 - 16x + 16 + 4y^2 = 64 - 16x + x^2 \quad \Rightarrow \quad 3x^2 + 4y^2 = 48 \quad \Rightarrow \quad \frac{x^2}{16} + \frac{y^2}{12} = 1$$

1 Since the equation has only x^2 and y^2 terms, the curve is symmetrical about the *x*- and *y*-axes   e.g. (2, 3) lies on the ellipse $\Rightarrow$ $(2, -3), (-2, 3), (-2, -3)$, all lie on the ellipse.

2 The curve cuts the *x*-axis at $(4, 0), (-4, 0)$; as $y = 0 \Rightarrow x^2 = 16 \Rightarrow x = 4$ or $x = -4$.

3 $x = 0 \Rightarrow y^2 = 12 \Rightarrow y = 2\sqrt{3} \Rightarrow$ the curve cuts the *y*-axis at $(0, 3.46)$ and $(0, -3.46)$.

$F(2, 0)$ is the **focus** of the ellipse associated with its **directrix**, $x = 8$.

Since the curve is symmetrical, the same ellipse can be defined with $G(-2, 0)$ as the **focus** and $x = -8$ as its associated **directrix**.

General equation of the ellipse

The most convenient form of the ellipse appears when the **focus** F is $(ae, 0)$ and the directrix is $x = a/e$ where $PF = ePD$ for all points P on the ellipse.

PD is the distance of P from the **line** $x = a/e$, i.e. PD is horizontal, and e is called the **eccentricity**.

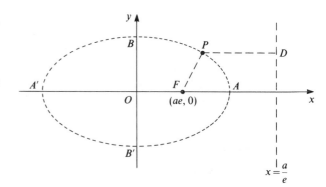

$P(x,y)$ and $PF = ePD$

$$\Rightarrow \quad \sqrt{(x-ae)^2 + y^2} = e\left(\frac{a}{e} - x\right)$$

Squaring $\quad \Rightarrow \quad (x-ae)^2 + y^2 = e^2\left(\frac{a}{e} - x\right)^2$

$$\Rightarrow \quad x^2 - 2aex + a^2e^2 + y^2 = a^2 - 2aex + e^2x^2 \quad \Rightarrow \quad x^2(1-e^2) + y^2 = a^2(1-e^2)$$

$$\Rightarrow \quad \frac{x^2}{a^2} + \frac{y^2}{a^2(1-e^2)} = 1 \quad \text{or} \quad \boxed{\frac{x^2}{a^2} + \frac{y^2}{b^2} = 1 \quad \text{where} \quad b^2 = a^2(1-e^2)}$$

$y = 0 \quad \Rightarrow \quad x = \pm a \quad \Rightarrow \quad A$ is $(a, 0)$, A' is $(-a, 0)$ and $AA' = 2a$, the **major axis**.

$x = 0 \quad \Rightarrow \quad y = \pm b \quad \Rightarrow \quad B$ is $(b, 0)$, B' is $(-b, 0)$ and $BB' = 2b$, the **minor axis**.

To produce an ellipse, $e < 1$. (You may have noticed in Investigation 5, when $e = 2$, that a **hyperbola** resulted.) e, the **eccentricity** of the ellipse, gives **its shape**. The larger e is, the narrower the ellipse. The smaller e is, the fatter the ellipse, i.e. approaching the shape of a circle.

Parametric form $\quad x = a\cos\alpha, \quad y = b\sin\alpha$

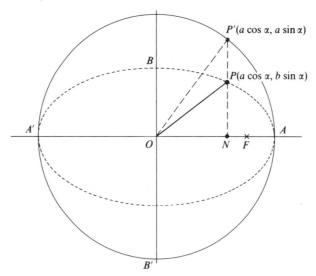

The point $P(a\cos\alpha, b\sin\alpha)$ lies on $\dfrac{x^2}{a^2} + \dfrac{y^2}{b^2} = 1$ whatever the value of α since,

$$\frac{a^2\cos^2\alpha}{a^2} + \frac{b^2\sin^2\alpha}{b^2} = \cos^2\alpha + \sin^2\alpha = 1$$

$\alpha = 0°$ gives $(a \cos \alpha, b \sin \alpha) = (a, 0)$ at A.

$\alpha = 90°$ gives $B(0, b)$; $\alpha = 180°$, at A'; $\alpha = 270°$, at B'.

However, $\angle PON \neq \alpha$.

$ON = a \cos \alpha \quad \Rightarrow \quad NP' = a \sin \alpha$,

where P' is on the (auxiliary) circle $x^2 + y^2 = a^2$.

$$\Rightarrow \quad \frac{NP'}{ON} = \tan \alpha \quad \Rightarrow \quad \angle P'ON = \alpha$$

Exercise 15.1

1 Starting from the relation $PF + FG = 8$ from Investigation 1, where F is $(2, 0)$ and G is $(-2, 0)$, derive the equation $\dfrac{x^2}{16} + \dfrac{y^2}{12} = 1$, for the ellipse.

The algebra is tricky and there is a 'best way' to proceed.

2 With the notation: $A(a, 0)$ lying on the ellipse, F the focus $(ae, 0)$ and $D(d, 0)$ where $x = d$ is the directrix, find the equations of the following ellipses.

(a) $A(4, 0)$; $F(3, 0)$; $D(6, 0)$ (b) $A(3, 0)$; $F(2, 0)$; $D(5, 0)$

(c) $A(3, 0)$; $F(2, 0)$; $D(6, 0)$

3 For the ellipse $3x^2 + 4y^2 = 48$, find the equation of (a) the tangent (b) the normal to the ellipse at the point $P(2, 3)$. (c) Find where the normal meets the ellipse again.

4 For Investigation 7, if the circle has equation $x^2 + y^2 = 4$ and the cross C is at $(2, 0)$, find the position of the resulting ellipse (if it is one) and its equation.

Find the focus, directrix and the eccentricity (value of e).

5 (a) Draw the ellipse which has centre $(0, 0)$ and passes through $(2, 0)$ and $(0, 1)$. Find its focus, directrix and eccentricity.

(b) Repeat (a) for the ellipse passing through $(3, 0)$ and $(0, 2)$.

(c) Repeat (a) for the ellipse passing through $(4, 0)$ and $(0, 3)$.

(d) What do you notice about the three ellipses? Can you prove it?

(e) Are the ellipses similar? Justify your answer.

6 Find the area of the ellipses (a) $3x^2 + 4y^2 = 48$ (b) $\dfrac{x^2}{a^2} + \dfrac{y^2}{b^2} = 1$.

Can you find their perimeters?

7 Find the equation of the tangent to the ellipse $\dfrac{x^2}{a^2} + \dfrac{y^2}{b^2} = 1$ at the points (a) (x', y') (b) $(a \cos \alpha, b \sin \alpha)$.

8 Prove that if TPS is a tangent to the ellipse $\dfrac{x^2}{a^2} + \dfrac{y^2}{b^2} = 1$ at P, then $\angle TPF = \angle SPG$, where F and G are the foci.

Hyperbola

From Investigation 5, the hyperbola is defined by $PF = 2 \times PD$ where D is on $x = \frac{1}{2}$.

$X(1,0)$, $P(2,3)$, $Q(3,\sqrt{24})$, $R(4,\sqrt{45})$, satisfy the condition.

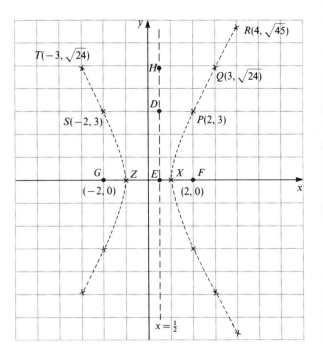

To fix Q, $x = 3 \Rightarrow QD = 2.5 \Rightarrow QF = 5$.

To fix R, $x = 4 \Rightarrow RD = 3.5 \Rightarrow RF = 7$.

If $P(x,y)$ is a general point on the curve,

$$FP^2 = (x-2)^2 + y^2 \quad \text{and} \quad PD = x - \tfrac{1}{2}$$

$$FP = 2PD \Rightarrow (x-2)^2 + y^2 = 4(x-\tfrac{1}{2})^2$$

$$\Rightarrow x^2 - 4x + 4 + y^2 = 4x^2 - 4x + 1$$

$$\Rightarrow y^2 - 3x^2 = -3 \quad \text{or} \quad 3x^2 - y^2 = 3$$

$$\Rightarrow x^2 - \frac{y^2}{3} = 1 \qquad \text{............ (1)}$$

In common with the ellipse, this equation involves only x^2 and y^2 terms, which implies that if $(2, 3)$ satisfies the equation, so do $(-2, 3)$, $(2, -3)$ and $(-2, -3)$, and the curve is symmetrical about both the x- and y-axes.

Check that $ZF = 2ZE$, $SF = 2SD$ and $TF = 2TH$, where E, D and H are the nearest points on the directrix, $x = \frac{1}{2}$. Both branches of the curve could be defined by the mirror image **focus**, $G(-2, 0)$, and its associated directrix, $x = -\frac{1}{2}$.

General equation of the hyperbola

As for the ellipse, we take the **focus**, F, as $(ae, 0)$ and the directrix, $x = a/e$.

(Remember, $e > 1$ this time.)

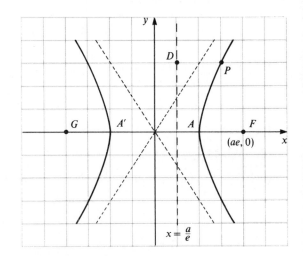

If $P(x,y)$ is a general point on the curve,

$$PF = ePD \Rightarrow FP^2 = e^2 PD^2$$

$$\Rightarrow (x - ae)^2 + y^2 = e^2 \left(x - \frac{a}{e}\right)^2$$

$$\Rightarrow x^2 - 2a\,ex + a^2 e^2 + y^2 = e^2 x^2 - 2a\,ex + a^2$$

$$\Rightarrow (e^2 - 1)x^2 - y^2 = a^2(e^2 - 1)$$

$$\Rightarrow \frac{x^2}{a^2} - \frac{y^2}{a^2(e^2-1)} = 1 \quad \text{or} \quad \boxed{\frac{x^2}{a^2} - \frac{y^2}{b^2} = 1 \quad \text{where} \quad b^2 = a^2(e^2-1)}$$

$y = 0 \Rightarrow x = a$ or $-a \Rightarrow A$ is $(a, 0)$, and A' is $(-a, 0)$.

$x = 0 \Rightarrow$ no values for $y \Rightarrow$ the curve does not cut the y-axis.

The hyperbola appears to approach asymptotes $y = \pm mx$. What is the value of m?

As x and y become large, from the equation $\dfrac{x^2}{a^2} - \dfrac{y^2}{b^2} = 1$, x^2 and y^2 are very large compared with 1.

$$\Rightarrow \quad \frac{x^2}{a^2} \simeq \frac{y^2}{b^2} \quad \Rightarrow \quad y \simeq \pm \frac{b}{a} x \quad \text{so,}$$

> the asymptotes have gradient $m = +\dfrac{b}{a}$ and $-\dfrac{b}{a}$.

For example, from Equation **(1)**, $x^2 - \dfrac{y^2}{3} = 1$, $\dfrac{b}{a} = \sqrt{3}$, so the asymptotes are inclined at $60°$ to the x-axis.

Parametric form for the hyperbola

$\dfrac{x^2}{a^2} - \dfrac{y^2}{b^2} = 1$ If the parametric form is $x = a f(\alpha)$ and $y = b g(\alpha)$, then we require $f^2 - g^2 = 1$, which suggests $f = \sec \alpha$ and $g = \tan \alpha$.

$$\Rightarrow \quad x = a \sec \alpha; \quad y = b \tan \alpha$$

For $0 \leqslant \alpha < 90°$ $\Rightarrow$ $1 \leqslant \sec \alpha < \infty$ and $0 \leqslant \tan \alpha < \infty$ which gives the part of the curve where $x > 0$ and $y > 0$. You should check the values of α which give points on the other branches of the curve.

Exercise 15.2

1 From the figure at the beginning of this section on the hyperbola, for the points X, P, Q and R, find PF and PG and discover a relation **similar** to the ellipse property $PF + PG = 8$.

2 Find the equations of **(a)** the tangent, **(b)** the normal, to the hyperbola $x^2 - \dfrac{y^2}{3} = 1$ at $P(2,3)$.

   Do either meet the hyperbola at another point, and if so, where?

3 Find the equation of the hyperbola with asymptotes $y = \sqrt{2}x$ and $y = -\sqrt{2}x$, passing through $(2, 0)$. Can you find any points on the hyperbola with integer coefficients?

4 Find the equation of the hyperbola passing through $(2, 0)$ with asymptotes $y = x$ and $y = -x$. Can you find any points on the hyperbola with integer coefficients?
   What is the angle between the asymptotes?

5 Rotate the hyperbola of question **4** through 90 degrees about the origin and find its equation.

6 Rotate the hyperbola of question **4** through $+45$ degrees about the origin and find its equation.

7 Find the equation of the hyperbola with asymptotes $y = x$ and $y = -x$ and focus $(2, 0)$.

8 Find the equation of **(a)** the tangent **(b)** the normal to the hyperbola $\dfrac{x^2}{a^2} - \dfrac{y^2}{b^2} = 1$ at the general point $(a \sec \alpha, b \tan \alpha)$.

Rectangular hyperbola

A hyperbola in which the asymptotes are perpendicular is called **rectangular**.

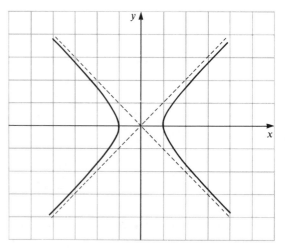

In these cases, the asymptotes are $y = x$ and $y = -x$, and with the normal notation, $b = a$, so

$$b^2 = a^2(e^2 - 1) \quad \Rightarrow \quad 1 = e^2 - 1$$

$$\Rightarrow \quad e^2 = 2 \quad \Rightarrow \quad e = \sqrt{2}$$

The equation of the hyperbola is now $x^2 - y^2 = 1$.

It is more convenient to rotate the curve through $+45$ degrees so that the asymptotes are now the axes and the axis of symmetry is $y = x$. To achieve this, we can apply the transformation matrix

$$\begin{pmatrix} 1 & -1 \\ 1 & 1 \end{pmatrix} \quad \text{which takes } (1, 0) \text{ to } (1, 1) \text{ and enlarges by a factor of } \sqrt{2}.$$

In general $\quad \begin{pmatrix} X \\ Y \end{pmatrix} = \begin{pmatrix} 1 & -1 \\ 1 & 1 \end{pmatrix} \begin{pmatrix} x \\ y \end{pmatrix} \quad$ so $\quad X = x - y \quad$ and $\quad Y = x + y.$

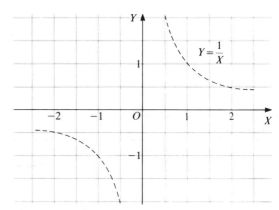

$y = x \quad \Rightarrow \quad X = 0,$

So the line $y = x$ becomes the new Y-axis, $\quad (X = 0).$

$x + y = 0 \quad \Rightarrow \quad Y = 0,$

So the line $y = -x$ becomes the new X-axis, $\quad (Y = 0).$

$x^2 - y^2 = 1 \quad \Rightarrow \quad (x - y)(x + y) = 1 \quad$ becomes

$$XY = 1 \quad \Rightarrow \quad Y = \frac{1}{X} \quad \text{on the new axes.}$$

Investigation 8

Find the focus and directrix of the hyperbolae **(a)** $XY = 1$ **(b)** $XY = c^2$.

We now dispense with the capital letter axes X and Y, and again call them x and y.

Investigation 9

Sketch the graphs of $xy = 1$, $xy = 2$, $xy = 3$, $xy = 4$, using the same axes.

What do you notice? Are the curves similar?

Parametric form

Since $xy = c^2$, we can use simply $x = ct$ and $y = c/t$ for parameters. You can now see the reason for using c^2 and not k or a or another constant.

So $t = 1$ gives $x = y = c$; $\quad t = 2$ gives the point $(2c, c/2)$; $\quad$ and $t = 1/2$ gives $(c/2, 2c)$.

$0 \leqslant t < \infty$ gives positive values for x and y and the part of the curve in the first quadrant, while $-\infty < t < 0$ gives x and y both negative.

Investigation 10

Sketch the graphs of $xy = -1$ and $xy = -2$ and compare them with $xy = 1$ and $xy = 2$. List the similarities and say which curves are odd, even or periodic.

WORKED EXAMPLE

For the hyperbola $xy = c^2$ **(a)** sketch the curve and find the equation of the tangent to the curve at the point **(b)** $(ct, c/t)$ **(c)** (x', y').

(a)

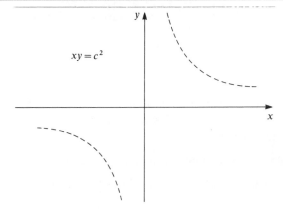

$xy = c^2$

(b) Remember, for this curve c is constant.

$$x = ct \Rightarrow \frac{dx}{dt} = c$$
$$y = \frac{c}{t} \Rightarrow \frac{dy}{dt} = \frac{-c}{t^2}$$
$$\frac{dy}{dx} = \frac{-c/t^2}{c} = \frac{-1}{t^2}$$

Tangent is $y - \frac{c}{t} = \frac{-1}{t^2}(x - ct)$

$\Rightarrow t^2 y - ct = -x + ct \Rightarrow t^2 y + x = 2ct$

(c) $xy = c^2 \Rightarrow x\frac{dy}{dx} + y = 0 \Rightarrow \frac{dy}{dx} = \frac{-y}{x} = \frac{-y'}{x'}$ at the point (x', y')

Tangent is $y - y' = \frac{-y'}{x'}(x - x') \Rightarrow x'y - x'y' = -xy' + x'y' \Rightarrow x'y + xy' = 2x'y'$

$x'y' = c^2 \Rightarrow$ Tangent is $x'y + xy' = 2c^2$ which can be modified to $\frac{y}{y'} + \frac{x}{x'} = 2$

Exercise 15.3

1 Find the closest point on each curve **(a)** $xy = 1$ **(b)** $xy = 2$ **(c)** $xy = 3$, to the origin. Are the curves similar? What transformation takes **(a)** to **(c)**?

2 For the curve $xy = 1$, identify the region between the curve, the x-axis and the lines $x = 1$ and $x = 2$. Transform this region using the matrix $\begin{pmatrix} 3 & 0 \\ 0 & 3 \end{pmatrix}$ and find the region's image.

What is the transformation and where have you seen this before?

3 Find the equation of **(a)** the tangent **(b)** the normal, to the hyperbola $xy = 3$ at the point $(1, 3)$. Find where each meets the curve again.

4 Find the equation of the normal to the rectangular hyperbola $xy = c^2$ at the point $P(ct, c/t)$. Find the co-ordinates of the point Q where it meets the curve again.

Find in terms of t the distance PQ and the value of t which minimizes PQ.

5 The tangent RPS to the hyperbola $xy = c^2$ at $P(ct, c/t)$ meets the x-axis at S and the y-axis at R. Find the area ORS (O is the origin).

6 For the points $T(ct, c/t)$ and $S(cs, c/s)$ on $xy = c^2$, find the equation of the chord ST (the line joining S and T). What happens if you put $s = t$?

7 (a) On graph paper draw the curve $xy = 4$ (both branches). Mark the point $S(2, 2)$ and join S to various points T on the other branch, marking the mid-points of all your lines ST. Describe the locus of the mid-points.

(b) Join $T(-2, -2)$ to various points S on the other branch and describe the locus of the mid-points of ST.

(c) What happens if each ST pair has gradient 1?

(d) What happens if each ST pair has gradient 2?

(e) What happens if you consider all possible ST pairs?

8 For the curve $xy = 4$, join the normals from various points on one branch to where they meet the other branch and find the locus of their mid-points.

9 From question **5**, find the centroid of $\triangle ORS$ in terms of t and sketch the locus as t varies. Find the xy equation for this locus.

10 The slipping ladder

(a) On axes marked 0 to 10 join $O(0, 0)$ to $L(0, 10)$ and imagine this to be a ladder. Pull the foot $F(0, 0)$ out to $(1, 0)$ and draw in the position of the ladder. Pull the ladder out further to $(2, 0)$ and draw in the new position of the ladder. Repeat this until the ladder is horizontal and you should have the envelope of a curve. Is it a rectangular hyperbola? Can you find its equation?

(b) On the falling ladder positions, mark in the mid-point each time. What is the locus and its equation?

(c) Mark in the positions of a point a quarter of the way up the ladder as the ladder falls. What is the locus and equation?

In each case, how do you complete the **complete** locus?

Conic sections

Producing the curves

From Investigation 6, with my torch (which throws out a **cone** of light) or my bedside lamp (which emits a **double cone** of light) I can

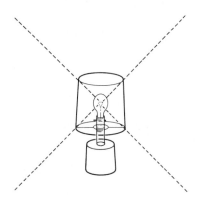

(a) shine the light directly at the wall and cast the shadow of a **circle**;

(b) angle the light slightly and produce an **ellipse**;

(c) angle the light more to produce a **parabola**(??) and a **hyperbola**.

You will need the lamp (double cone) to produce the two branches of the hyperbola, and you probably have realized that the angle at which you hold the light determines which shape you produce.

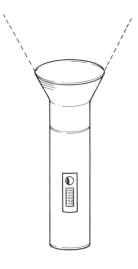

The shadow which traces the curve on the wall, is formed by the plane of the wall intersecting the cone of light, therefore producing various sections of the cone. It may now be easier to visualize the double cone being cut by planes which are inclined at different angles to the cone.

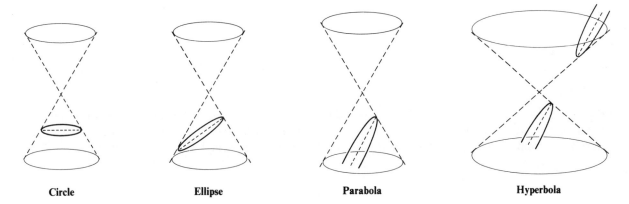

| Circle | Ellipse | Parabola | Hyperbola |

(a) If the cut is at right angles to the axis of the cone a **circle** results.

(b) An **ellipse** results if the cutting plane is angled to the cone's axis but is **not** parallel to the sloping edge (**generator**) of the cone.

(c) If the cutting plane **is** parallel to the sloping edge we get a **parabola**.

(d) If the cutting plane is steeper than the slant edge of the cone, it will cut the upper part of the cone also and produce the **two** branches of the **hyperbola**.

Can we **prove** all these statements or must we accept them? Are they **true**? Can you demonstrate the results with your torch or lamp?

Investigation 11

Make a cone out of plasticine or plastic and actually cut the cone to produce the four sections: a circle; an ellipse; a parabola and a hyperbola.
   (The author has seen a model made of wood, but perhaps you can use a large plastic funnel to chop up!)

Some Investigation results

Investigation 2 gave a **parabola**, $y^2 = 3(2x-1)$, from the definition $PF = PD$.

Investigation 4 gave an **ellipse**, from the definition $PF = \frac{1}{2}PD$.

Investigation 5 gave a **hyperbola**, from the definition $PF = 2PD$.

All three curves arise from the definition $PF = e \times PD$, where e is the eccentricity and the value of e determines the shape of the curve.

You may remember that, for an ellipse of width $2a$ (major axis) and height $2b$ (minor axis), $b^2 = a^2(1-e^2)$, where e is the eccentricity.

$b < a \ \Rightarrow \ e^2 < 1 \ \Rightarrow \ 0 < e < 1$ for the **ellipse**.

For the **circle**, $b = a$ (width = height) $\Rightarrow \ 1 = 1 - e^2 \Rightarrow \ e = 0$.

From Investigation 2 on the **parabola**, we can see that $PF = 1 \times PD$, so $e = 1$.

For the **hyperbola**, $e > 1$.

The various values for e determine whether our curve is a circle, parabola, ellipse or hyperbola.

Similarly, in our conic sections, the **angle of the intersecting plane** tells us which of the four curves we produce.

I wonder if there is a connection between e and the angle of the cutting plane?

Investigation 12

The diagram shows a plane cutting a cone to produce an ellipse. Two spheres touch the plane of the ellipse and the cone: the smaller sphere (above) at F; and the larger one (below) at G. A line on the surface of the cone (generator) from O touches the top sphere at A, the ellipse at P, and the bottom sphere at B.

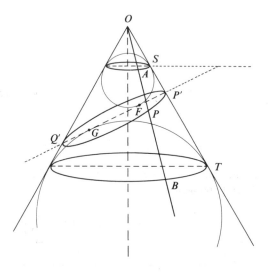

Can you show that $PF + PG = $ constant for all possible generators OB?

Hint Consider $PA + PB$ and $PF + PG$. Show that $AB = ST = P'Q'$, the major axis of the ellipse.

Solution

$PA = PF$ as both are tangents to the smaller sphere from the same external point P.

$PB = PG$ tangents from P to the larger sphere.

$$PF + PG = PA + PB = SP' + TP' = ST$$

and ST is the distance along any generator on the surface of the cone between the circles of contact of the cone with the two spheres, which is **constant** for these two spheres, and therefore this ellipse.

In Investigation 12, $PF + PG = k \Rightarrow F$ and G are the foci of this ellipse $\Rightarrow P'G = FQ'$.

$SP' = FP'$ (tangents from P' to the smaller sphere)
$GP' = TP'$ (tangents from P' to the larger sphere)

$ST = SP' + P'T = FP' + GP' = P'F + FQ' = P'Q' = 2a$, the major axis of the ellipse.

If $r = $ radius of the top sphere and C is the centre, then $\angle FCM = \alpha$, where α is the angle between the plane of the ellipse and the horizontal and CF is perpendicular to the line $Q'GMFP' \Rightarrow FM = r \tan \alpha$.

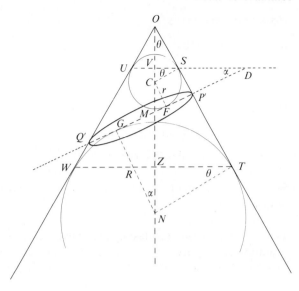

Similarly, if $R =$ radius of the bottom sphere and N is the centre, then $\quad GM = R \tan \alpha$.

$$\Rightarrow \quad FG = FM + GM = R \tan \alpha + r \tan \alpha$$

If e is the eccentricity of the ellipse,

$$FG = 2ae = (R + r) \tan \alpha$$

$$\Rightarrow \quad e = \frac{(R + r)}{2a} \tan \alpha$$

but perhaps R, r and a are related, possibly to α.

The diagram shows the vertical cross-section including the axis $P'Q' (= 2a)$ of the ellipse.

The intersection of the plane of the ellipse and the plane of the circle US is a line parallel to the minor axis of the ellipse and so, on this diagram, appears as a point D. We can show that this line is the directrix.

$$OS = r/\tan \theta \quad \text{and} \quad OT = R/\tan \theta$$

$$\Rightarrow \quad ST = OT - OS = R/\tan \theta - r/\tan \theta = 2a$$

$$\Rightarrow \quad R - r = 2a \tan \theta$$

where θ is $\frac{1}{2} \angle UOS$.

The vertical distance between the circles US and WT is $ST \cos \theta = 2a \cos \theta$ and also from the middle vertical line

$$VZ = VC + CM + MN - ZN = r \sin \theta + r/\cos \alpha + R/\cos \alpha - R \sin \theta$$

So, $\quad 2a \cos \theta = (r - R) \sin \theta + \dfrac{(R + r)}{\cos \alpha}$

and now we substitute $\quad R - r = 2a \tan \theta \quad$ and $\quad R + r = \dfrac{2ae}{\tan \alpha}$

$$\Rightarrow \quad 2a \cos \theta = -2a \tan \theta \sin \theta + \frac{2ae}{\tan \alpha \cos \alpha} \quad \Rightarrow \quad \cos \theta + \frac{\sin^2 \theta}{\cos \theta} = \frac{e}{\sin \alpha}$$

$$\Rightarrow \quad \frac{\cos^2 \theta + \sin^2 \theta}{\cos \theta} = \frac{e}{\sin \alpha} \quad \Rightarrow \quad e = \frac{\sin \alpha}{\cos \theta} \qquad \text{so } e \text{ is related to } \alpha, \text{ since } \theta \text{ is constant for this cone.}$$

$$\frac{P'F}{P'D} = \frac{P'S}{P'D} = \frac{\sin \alpha}{\sin (90 - \theta)} = \frac{\sin \alpha}{\cos \theta} = e \qquad \text{using the sine rule in } \triangle SDP', \text{ so we have shown that the line through } D \text{ is the directrix.}$$

We have also demonstrated the equivalence of the focus/directrix property and the constant sum property, i.e. $\quad FP + PG = 2a$.

Polar co-ordinates

A pilot or navigator uses polar co-ordinates rather than Cartesian co-ordinates to specify the distance (in a straight line) and the direction (given as a bearing) of one point from another.

Co-ordinate geometry

Polar co-ordinates

Relative to London:

Norwich is 115 miles in direction N 33° E, i.e. 33° to the East of North;

Doncaster is 170 miles in direction N 18° W;

Dover is 78 miles in direction S 65° E;

Plymouth is 216 miles in direction S 70° W.

Bearing is measured clockwise positive, starting from North.

The polar co-ordinates (**using bearings**), relative to London, are Norwich (115, 033°), Dover (78, 115°), Plymouth (216, 250°), and Doncaster (170, 342°).

Mathematical polar co-ordinates   (r, θ)

r represents the straight line distance and θ the angle measured anticlockwise positive from the positive x-axis (just as we do in trigonometry).

Taking London as the origin, the **mathematical polar co-ordinates** are,

  Norwich (115, 057°); Doncaster (170, 108°); Plymouth (216, 200°); Dover (78, 335°)

Notice that Dover could be (78, −25°) or (−78, 155°) or even (78, 335° + 360°) = (78, 695°).

Cartesian co-ordinates   (x, y)

$x = r\cos\theta$

$y = r\sin\theta$

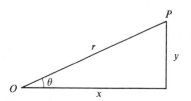

Polar co-ordinates   (r, θ)

$r = \sqrt{(x^2 + y^2)}$

where r is taken to be positive.

$\theta = \tan^{-1}\dfrac{y}{x}$

where θ is the angle for which $-180° < \theta \leqslant +180°$

and which satisfies the equations $\cos\theta = \dfrac{x}{r}$ and

$\sin\theta = \dfrac{y}{r}$.

Investigation 13

(a) Plot the points (1, 0), (2, 1), (3, 2), (4, 3) as Cartesian co-ordinates. What is its shape? What is its x,y equation?

(b) Plot the points (1, 0), (2, 1), (3, 2), (4, 3) as polar co-ordinates with the angle measured in right angles, i.e. (3, 2) is really (3, 180°). What is the r,θ equation? Plot more points, such as (1.5, 0.5). What is the locus?

Investigation 14

Find points which satisfy $r = \theta/10$, where θ is measured in degrees, by drawing up a table of values,

θ	0°	30°	60°	90°	120°
r	0	3	6	9	12

Plot the points and describe the locus.

Investigation 15

Draw up a table of values for θ and $\cos \theta$ and draw the curve $r = \cos \theta$.

θ	0°	30°	60°	90°	120°
r	1	0.87	0.5	0	−0.5

What happens if you ignore the negative values of r, i.e. when $90° < \theta < 270°$? What happens if you do not ignore them?

Investigation 16

Draw the locus $r = \sec \theta$ for $0° \leqslant \theta \leqslant 360°$.

What is the shape and can you derive the x,y equation?

Investigation 17

Plot **(a)** $r = \cos 2\theta$ **(b)** $r = \cos 3\theta$ **(c)** $r = \cos 4\theta$.

Can you derive the x,y equations?

Investigation 18

For polar co-ordinates (r, θ), plot the curves $r = \dfrac{2}{1 + e \cos \theta}$, when **(a)** $e = 1$ **(b)** $e = 2$ **(c)** $e = \frac{1}{2}$.

It may help to have a computer graph plotting program (or a calculator one), or to program your calculator to give the r values for any θ input.

WORKED EXAMPLE

(a) Sketch the curve whose polar equation is $r = 2(1 + \cos\theta)$.

(b) Find the area enclosed.

(c) Find the length of the curve. (This topic is usually only found in Further Mathematics syllabi, and so can be omitted, but it is interesting and instructive.)

(a) Table of values

θ	0°	30°	60°	90°	120°	150°	180°	210°	240°	270°	300°	330°	360°
$\cos\theta$	1	0.87	0.5	0	−0.5	−0.87	−1	−0.87	−0.5	0	0.5	0.87	1
$1 + \cos\theta$	2	1.87	1.5	1	0.5	0.13	0	0.13	0.5	1	1.5	1.87	2
$r = 2(1 + \cos\theta)$	4	3.73	3	2	1	0.27	0	0.27	1	2	3	3.73	4

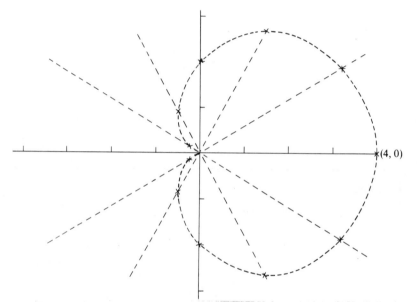

The curve, a **cardioid**, is symmetrical about the x-axis since between 180° and 360°, cos θ repeats the values it takes from 0° to 180°.

At $\theta = 180°$, does the locus approach the origin to form a point (cusp)? How can we tell?

Can we evaluate the value of the gradient at $\theta° = 180°$? i.e. find $\dfrac{dy}{dx}$.

$y = r\sin\theta = 2(1 + \cos\theta)\sin\theta$

$x = r\cos\theta = 2(1 + \cos\theta)\cos\theta$

$$\frac{dy}{dx} = \frac{dy/d\theta}{dx/d\theta} = \frac{2\cos\theta(1 + \cos\theta) - 2\sin^2\theta}{-2\sin\theta(1 + \cos\theta) - 2\sin\theta\cos\theta} \overset{c}{=} \frac{\cos 2\theta + \cos\theta}{-\sin 2\theta - \sin\theta} = \frac{2\cos 3\theta/2 \cos\theta/2}{-2\sin 3\theta/2 \cos\theta/2} = -\cot\left(\frac{3\theta}{2}\right)$$

$\qquad\qquad$ differentiating as a product $\qquad\qquad$ simplifying $\qquad\qquad$ factor formulae

$\theta = 180° \;\Rightarrow\; \dfrac{dy}{dx} = 0 \;\Rightarrow\;$ a cusp, pointing in towards the origin.

This analysis may appear difficult, but it is satisfying to check your result.

(b) Area in polar co-ordinates

For two neighbouring points, (r, θ) and $(r + \delta r, \theta + \delta\theta)$, OPP' is approximately a sector of area $\frac{1}{2}r^2\delta\theta$.

So, $\qquad\qquad$ Area $OAB = \displaystyle\lim_{\delta\theta \to 0}\sum_{\theta=\alpha}^{\theta=\beta} \tfrac{1}{2}r^2\,\delta\theta = \int_{\alpha}^{\beta} \tfrac{1}{2}r^2\,d\theta$

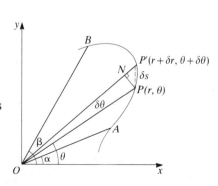

Area of the cardioid

$$= 2 \int \tfrac{1}{2} r^2 \, d\theta = \int_{\theta=0}^{\theta=\pi} 4(1+\cos\theta)^2 \, d\theta$$

$$= \int_0^\pi (4 + 8\cos\theta + 4\cos^2\theta) \, d\theta = \int_0^\pi (4 + 8\cos\theta + 2(1+\cos 2\theta)) \, d\theta = \left[6\theta + 8\sin\theta + \sin 2\theta \right]_0^\pi = 6\pi$$

From the diagram in part **(a)**, the area of the cardioid would appear to lie between a circle radius 2 (area 4π) and a circle radius 2.5 (area 6.25π), which it does.

(c) Length of curve

By Pythagoras in $\triangle P'PN$ where $\angle PNP' = 90°$, $P'N \simeq \delta r$ and $PN = r\sin\delta\theta \simeq r\,\delta\theta$, $PP' = \sqrt{(r\delta\theta)^2 + (\delta r)^2} \simeq \delta s$

$$\Rightarrow \quad \delta s \simeq \sqrt{r^2\delta\theta^2 + \delta r^2} \Rightarrow \quad \frac{\delta s}{\delta\theta} \simeq \sqrt{r^2 + \left(\frac{\delta r}{\delta\theta}\right)^2} \quad \Rightarrow \quad \frac{ds}{d\theta} = \sqrt{r^2 + \left(\frac{dr}{d\theta}\right)^2} \quad \Rightarrow \quad s = \int \sqrt{\left[r^2 + \left(\frac{dr}{d\theta}\right)^2 \right]}\, d\theta$$

$$\text{Length of cardioid} = \int_0^{2\pi} \sqrt{4(1+\cos\theta)^2 + (-2\sin\theta)^2} \, d\theta = 2\int_0^\pi 2\sqrt{1 + 2\cos\theta + \cos^2\theta + \sin^2\theta}\, d\theta$$

$$= 4\int_0^\pi \sqrt{2(1+\cos\theta)}\, d\theta = 4\int_0^\pi \sqrt{2 \times 2\cos^2\frac{\theta}{2}}\, d\theta = 8\int_0^\pi \cos\frac{\theta}{2}\, d\theta = \left[16\sin\frac{\theta}{2} \right]_0^\pi = 16$$

The length of the cardioid would appear to be close to the circumference of a circle of radius 2.5, i.e. area $5\pi = 15.7$, and it is wise to build checks like this into your working to make sure your answer is reasonable.

Such exercises in finding area and length of curve provide good examples of integration, using all your knowledge of formulae, and manipulation of trigonometrical functions.

The use of **polar co-ordinates** enables many interesting curves to be expressed in simple terms and drawn easily. Can you design a computer program to plot curves using polar co-ordinates? Perhaps the double-valued nature of many of the 'functions' gives too much difficulty, but the symmetry of the curves emphasizes the properties of periodic trigonometrical functions and provides an aesthetic satisfaction in our subject of algebraic and numerical intricacy.

Exercise 15.4

1 $x = 2$ and $y = 3$ define straight lines when using Cartesian co-ordinates. What is the locus of **(a)** $r = 2$
   (b) $\theta = 3$ radians?

2 What is the difference in appearance of the graphs **(a)** $r = \cos\theta$ and $r = \sin\theta$ **(b)** $r = \cos 2\theta$ and $r = \sin 2\theta$ **(c)** $r = \cos\theta$ and $r = |\cos\theta|$ **(d)** $r = |\cos\theta|$ and $r = \cos^2\theta$ **(e)** $r = \cos^2\theta$ and $r = \cos^3\theta$ **(f)** $r = \sin 2\theta$ and $r = \sin 3\theta$?

3 Using $x = r\cos\theta$ and $y = r\sin\theta$, find the polar equations and state the shape of each curve.

   (a) $y = 4x$ **(b)** $y^2 = 4x$ **(c)** $x^2 + y^2 = 16$ **(d)** $x^2 - y^2 = 9$ **(e)** $x^2 + y^2 + 4y = 0$

   (f) $xy = 2$ **(g)** $y = x(x-2)$ **(h)** $6y = 9 - x^2$ **(i)** $y^2 = x^2 - x^4$

(4) Find the Cartesian equation of the following polar curves, stating the shape.

   (a) $r = 4\cos\theta$ **(b)** $r = 2/\cos\theta$ **(c)** $r(1+\cos\theta) = 3$ **(d)** $r = 3$

   (e) $r^2\sin\theta\cos\theta = 9$ **(f)** $r^2 = 4(\cos^2\theta - \sin^2\theta)$ **(g)** $r(\tfrac{1}{2} + \cos\theta) = 3$

---

16 Numerical methods

Introduction

In this chapter we shall be considering some numerical ways to find areas under curves and to solve equations in cases when our algebraic methods do not work.

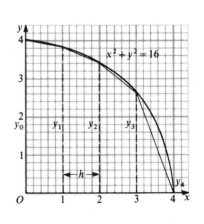

Investigation 1

Find the area of the quadrant of the circle $x^2 + y^2 = 16$.

(a) Divide the quadrant into 4 strips of width $h = 1$ and sum the areas of each to find the total area. Treat each strip like a trapezium by cutting off the top curved part and finding the heights y_0, y_1, y_2 etc. using the equation of the circle.

$$y_0 = 4; \quad y_1 = \sqrt{16 - x_1^{\,2}} = \sqrt{15} = 3.873;$$

$$y_2 = \sqrt{12} = 3.464$$

First trapezium area $= \dfrac{(y_0 + y_1)h}{2}$.

(cont.)

Investigation 1—continued

(b) By a trapezium method with 8 strips (divide into 8 strips of width 0.5).

(c) Use 4 strips, width $h = 1$, the y values and the area, formula S_4.

For $x = 0$, $y_0 = 4$

$x = 1$, $y_1 = \sqrt{15} = 3.873$

$x = 2$, $y_2 = \sqrt{12} = 3.464$

$x = 3$, $y_3 = \sqrt{7} = 2.646$

$x = 4$, $y_4 = 0$ $\qquad S_4$, Area $= \dfrac{h}{3}(y_0 + y_4 + 4(y_1 + y_3) + 2y_2)$

(d) Use 8 strips width $h = 0.5$ and the S_8 formula below.

For $x = 0$, $y_0 = 4$

$x = 0.5$, $y_1 = \sqrt{16 - 0.25} = \sqrt{15.75} = 3.969$

$x = 1$, $y_2 = \sqrt{16 - 1} = \sqrt{15} = 3.873$

$x = 1.5$, $y_3 = \sqrt{16 - 2.25} = \sqrt{13.75} = 3.708$

. . . .

$x = 4$, $y_8 = 0$ $\qquad S_8$, Area $= \dfrac{h}{3}\left[y_0 + y_8 + 4(y_1 + y_3 + y_5 + y_7) + 2(y_2 + y_4 + y_6) \right]$

(e) Using $A = \frac{1}{4}\pi r^2$ for a quadrant of a circle.

(f) By integration, Area $= \displaystyle\int y\,dx = \int_0^4 \sqrt{(16 - x^2)}\,dx$

(g) By equating each of your answers for **(a)**, **(b)**, **(c)** and **(d)** with **(e)**, find which of the approximate numerical methods **(a)** to **(d)** gives the best value for π and is therefore the most accurate?

Investigation 2

To find the area of the ellipse $\dfrac{x^2}{16} + \dfrac{y^2}{9} = 1$.

(a) Use the trapezium method with 4 strips (width $h = 1$) on a quadrant of the ellipse.

(b) Use the trapezium method with 8 strips (width $h = 0.5$).

(c) Use the S_4 formula with 4 strips $(h = 1)$.

(d) Use the S_8 formula with 8 strips $(h = 0.5)$.

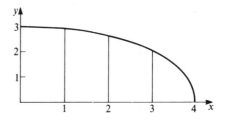

(e) Integrate $\displaystyle\int y\,dx = \int_0^4 \dfrac{\sqrt{(144 - 9x^2)}}{4}\,dx$.

(cont.)

Investigation 2—continued

(f) Compare your answers with the areas of a circle radius 3 inside the ellipse and the circle radius 4 outside the ellipse. Can you suggest a formula for the area of the ellipse?

What is the area of the ellipse $\dfrac{x^2}{a^2} + \dfrac{y^2}{b^2} = 1$?

(g) Find out the formula for the area of an ellipse and see which of **(a)** to **(d)** gives the best approximation.

Trapezium rule

The trapezium rule gives an approximate value for the area under a curve, by dividing that area into strips (each of width h) and approximating each strip to a trapezium.

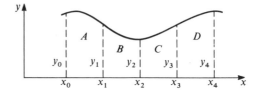

Area between x_0 and x_4

$= \text{area } A + \text{area } B + \text{area } C + \text{area } D$

$= \tfrac{1}{2}(y_0 + y_1)h + \tfrac{1}{2}(y_1 + y_2)h + \tfrac{1}{2}(y_2 + y_3)h + \tfrac{1}{2}(y_3 + y_4)h$

$= \dfrac{h}{2}[y_0 + 2(y_1 + y_2 + y_3) + y_4]$

WORKED EXAMPLES

1 Find the area of that part of the ellipse $\dfrac{x^2}{16} + \dfrac{y^2}{9} = 1$ in the positive quadrant, using the trapezium rule with 8 strips.

$$\frac{x^2}{16} + \frac{y^2}{9} = 1 \;\Rightarrow\; \frac{y^2}{9} = 1 - \frac{x^2}{16} \;\Rightarrow\; y^2 = \frac{9}{16}(16 - x^2) \;\Rightarrow\; y = \frac{3}{4}\sqrt{(16 - x^2)}$$

Sketch the graph between $x_0 = 0$ and $x_8 = 4$ strip width $h = 0.5$.

x	$y = \tfrac{3}{4}\sqrt{16 - x^2}$
0	$y_0 = 3$
0.5	$y_1 = 2.976$
1	$y_2 = 2.905$
1.5	$y_3 = 2.781$
2	$y_4 = 2.598$
2.5	$y_5 = 2.342$
3	$y_6 = 1.984$
3.5	$y_7 = 1.452$
4	$y_8 = 0$

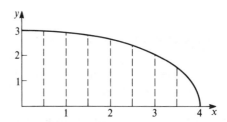

$$\text{Area} = \frac{h}{2}\left[y_0 + 2(y_1 + y_2 + y_3 + \ldots + y_7) + y_8 \right]$$

$$= 0.25\,[3 + 2(17.038) + 0] = 9.269$$

Using 4 strips gives $\qquad \text{Area} = 0.5\,[3 + 2(7.487) + 0] = 8.987$

Clearly from the diagram, taking more strips gives a better value for the area since more of the actual area is enclosed by the trapezia.

2 Using the trapezium rule, find the area of the quadrant of the circle $x^2+y^2=16$ in the first quadrant using **(a)** 4 strips **(b)** 8 strips **(c)** $\frac{1}{4}\pi r^2$ and estimate a value for π.

(a)

x	$y=\sqrt{16-x^2}$
0	$y_0=4$
1	$y_1=3.873$
2	$y_2=3.464$
3	$y_3=2.646$
4	$y_4=0$

Area $=\frac{1}{2}[4+2(9.983)+0]$

$=11.983$

(c) Area $=\frac{1}{4}\pi r^2=4\pi=12.566$

Using **(b)** and comparing with **(c)**, $4\pi=12.359$

$\Rightarrow\quad \pi\simeq 3.0898$ (a poor estimate!)

(b)

x	$y=\sqrt{16-x^2}$
0	$y_0=4$
0.5	$y_1=3.969$
1	$y_2=3.873$
1.5	$y_3=3.708$
2	$y_4=3.464$
2.5	$y_5=3.122$
3	$y_6=2.646$
3.5	$y_7=1.936$
4	$y_8=0$

Area $=\frac{1}{4}[4+2(22.718)]$

$=12.359$

Note that, with the circle being a convex curve, each trapezium excludes some of the area under the curve, so the trapezium rule, in such cases, will always under-estimate the true value.

Simpson's rule

A more refined method joins neighbouring strips together, for example A and B, and finds the area under a parabola drawn through P, Q and R.

Consider A and B, where $x_0=x_1-h$ and $x_2=x_1+h$.

If the parabola through P, Q and R is $y=ax^2+bx+c$

then $\quad y_0=a(x_1-h)^2+b(x_1-h)+c$

$\qquad y_1=ax_1^2\qquad +bx_1\qquad +c$

$\qquad y_2=a(x_1+h)^2+b(x_1+h)+c$

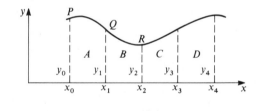

$$\text{Area } A+B=\int_{x_0}^{x_2}(ax^2+bx+c)\,dx=\left[\frac{ax^3}{3}+\frac{bx^2}{2}+cx\right]_{x_1-h}^{x_1+h}$$

$$=\frac{a}{3}\left[(x_1+h)^3-(x_1-h)^3\right]+\frac{b}{2}\left[(x_1+h)^2-(x_1-h)^2\right]+c\left[(x_1+h)-(x_1-h)\right]$$

$$=\frac{a}{3}(6x_1^2h+2h^3)+\frac{b}{2}(4x_1h)+2ch$$

$$=\frac{h}{3}(6ax_1^2+2ah^2+6bx_1+6c)$$

$$=\frac{h}{3}\left[(2ax_1^2+2ah^2+2bx_1+2c)+(4ax_1^2+4bx_1+4c)\right]$$

$$=\frac{h}{3}\left[(y_0+y_2)+4y_1\right]\qquad \text{using the equations above for } y_0,\ y_1,\ y_2.$$

Similarly, the area of $C+D = \dfrac{h}{3}(y_2+4y_3+y_4)$,   so

Area for 4 strips (5 ordinates y_0 to y_4) is $\dfrac{h}{3}(y_0+4y_1+2y_2+4y_3+y_4)$.

For 8 strips, **Simpson's rule** is $A = \dfrac{h}{3}\left[y_0+4(y_1+y_3+y_5+y_7)+2(y_2+y_4+y_6)+y_8 \right]$

Some people like to remember $A = \dfrac{h}{3}(\text{ends}+4\times\text{odds}+2\times\text{evens})$

> For Simpson's rule, you **must** use an **even** number of **strips** and therefore you **must** use an **odd** number of **ordinates** (y values).

WORKED EXAMPLE

Using Simpson's rule, find the area of the positive ellipse quadrant for $\dfrac{x^2}{16}+\dfrac{y^2}{9}=1$ using **(a)** 8 strips **(b)** 4 strips **(c)** formula $\dfrac{\pi ab}{4}$ and compare them also with the trapezium rule results.

x	$y=\tfrac{3}{4}\sqrt{16-x^2}$		
0	$y_0 = 3$		
0.5		$y_1 = 2.976$	
1			$y_2 = 2.905$
1.5		$y_3 = 2.781$	
2			$y_4 = 2.598$
2.5		$y_5 = 2.342$	
3			$y_6 = 1.984$
3.5		$y_7 = 1.452$	
4	$y_8 = 0$		
	—		
	3	9.551	7.487

(a) $A = \dfrac{0.5}{3}(3+(4\times 9.551)+(2\times 7.487))$

$\qquad = \dfrac{1}{6}(3+38.204+14.974)$

$\qquad = 9.363$

(b) Re-number $y_0 = 3$; $y_1 = 2.905$; $y_2 = 2.598$

$\qquad\qquad y_3 = 1.984$; $y_4 = 0$

$A\ (4\ \text{strips}) = \dfrac{1}{3}\left[3+4(2.905+1.984)+2\times 2.598 \right]$

$\qquad = 9.251$

(c) Formula: Area $=\pi ab$, where $2a=$ major axis and $2b=$ minor axis, is best derived by stretching the circle $x^2+y^2=b^2$ in the x-direction, using the transformation $\begin{pmatrix} 1 & 0 \\ 0 & a/b \end{pmatrix}$ so that the circle area πb^2 is multiplied by a/b to give,

$$\text{Area of Ellipse} = \pi b^2 \times a/b = \pi ab$$

For $\dfrac{x^2}{16}+\dfrac{b^2}{9}=1$, $a=4$ and $b=3$ $\Rightarrow$ Quadrant Area $=\dfrac{\pi\times 4\times 3}{4}=3\pi=9.425$

Trapezium rule with 4 strips gave $T_4 = 8.987$; $T_8 = 9.269$.

Simpson's rule with 4 strips gave $S_4 = 9.251$; $S_8 = 9.363$.   Formula gave 9.425.

As you might expect, in both cases the more strips you take, the more accurate is your estimate for the area under the curve, and Simpson's rule appears to be more accurate than the trapezium rule.

Exercise 16.1

T_n means using the trapezium rule with n strips, i.e. $n+1$ ordinates (y_0 to y_n).
S_n means using Simpson's rule with n strips (n even), i.e. $n+1$ ordinates (y_0 to y_n)

1 Evaluate S_4 and S_8 for the circle quadrant for $x^2 + y^2 = 16$ and compare the results with T_4 and T_8. Which gives the best result? Use this to estimate a value for π.

2 Find the area under the graph of $y = 1/x$ from $x = 1$ to $x = 2$ using   **(a)** T_5 **(b)** T_{10} **(c)** S_{10}.
   (d) Check your answers by algebraic integration.

3 Repeat question **2** for values $x = 2$ to $x = 4$.   Explain your result.

4 Use   **(a)** T_6 **(b)** S_6 to estimate the area under the graph of $y = \sin x$ for $x = 0$ to $x = \pi/2$.
   (c) Compare your results with the true value by integration.

5 Repeat question **4** for $y = \sqrt{\sin x}$.

6 Repeat question **4** for $y = \sin^2 x$.

7 Use   **(a)** T_4 **(b)** S_4 to estimate the value of $\displaystyle\int_0^1 \frac{1}{1+x^2}\,dx$.   **(c)** Evaluate the integral algebraically and compare the results.

8 Repeat question **7** for $\displaystyle\int_0^1 \frac{1}{\sqrt{1-x^2}}\,dx$ but for   **(a)** T_6 **(b)** S_6.

9 The result $\displaystyle\int_0^\infty \frac{1}{\sqrt{2\pi}} e^{-x^2/2}\,dx = 0.5$ is used in statistics for the Normal Distribution. Evaluate   **(a)** T_8
   and   **(b)** S_8 for $\displaystyle\int_0^4 e^{-x^2/2}\,dx$ and compare your values with $\dfrac{\sqrt{2\pi}}{2}\left(=\sqrt{\dfrac{\pi}{2}}\right)$.

Polynomial equations

Investigation 3

Solve the following equations:

(a) $x^2 = 4$ **(b)** $x^3 = 8$ **(c)** $x^4 = 16$ **(d)** $x^3 = 6$ **(e)** $x^3 = 6x$ **(f)** $x^3 = 6x - 5$
(g) $x^3 = 6x^2 - 6x - 5$ **(h)** $x^4 = 6x^2 - 5$ **(i)** $x^4 = 6x - 5$

The first few equations look simple enough, but take care not to miss (overlook) any solutions.

(a) $x^2 = 4 \Rightarrow x = +2$ or $x = -2$.

(b) $x^3 = 8 \Rightarrow x = 2$, but $x^3 - 8 = (x-2)(x^2+2x+4) = 0 \Rightarrow x = 2$ or $x^2 + 2x + 4 = 0$

$\qquad x^2 + 2x + 4 = 0 \Rightarrow (x+1)^2 + 3 = 0 \Rightarrow (x+1)^2 = -3 \Rightarrow x = -1 \pm \sqrt{-3}$

$\qquad \Rightarrow x = -1 + \sqrt{3}i$ or $x = -1 - \sqrt{3}i$.

In addition to $x = 2$, we have two complex roots $x = -1 + \sqrt{3}i$ and $x = -1 - \sqrt{3}i$.

(cont.)

Investigation 3—continued

(c) $x^4 = 16 \Rightarrow x^2 = +4 \text{ or } x^2 = -4 \Rightarrow x = +2 \text{ or } x = -2 \text{ or } x = +2i \text{ or } x = -2i$

Two real roots and two complex roots.

(d) $x^3 = 6 \Rightarrow x = \sqrt[3]{6}$ but $x^3 - 6 = (x - \alpha)(x^2 + \alpha x + \alpha^2)$ where $\alpha = \sqrt[3]{6}$.

$x = \alpha$ or $x^2 + \alpha x + \alpha^2 = 0 \Rightarrow x = \dfrac{-\alpha \pm \sqrt{\alpha^2 - 4\alpha^2}}{2} = \dfrac{-\alpha + \sqrt{3}\alpha i}{2}$ or $\dfrac{-\alpha - \sqrt{3}\alpha i}{2}$

One real root or two very complicated complex roots.

(e) $x^3 = 6x \Rightarrow x^3 - 6x = 0 \Rightarrow x(x^2 - 6) = 0 \Rightarrow x = 0 \text{ or } x = +\sqrt{6} \text{ or } x = -\sqrt{6}$

Three real roots.

(f) $x^3 = 6x - 5 \Rightarrow x^3 - 6x + 5 = 0$. The factor theorem gives $x = 1$ as an 'obvious' root.

$x^3 - 6x + 5 = (x - 1)(x^2 + x - 5) = 0 \Rightarrow x = 1 \text{ or } x^2 + x - 5 = 0$

$\Rightarrow x = 1$ or $x = \dfrac{-1 + \sqrt{21}}{2}$ or $x = \dfrac{-1 - \sqrt{21}}{2}$

Again we have 3 real roots. Does every cubic equation have three roots?

(b) and **(d)** gave 1 real and 2 complex roots, but could we have 2 real and 1 complex?

(g) $x^3 = 6x^2 - 6x - 5 \Rightarrow f(x) = x^3 - 6x^2 + 6x + 5 = 0$, but I cannot spot an easy factor(!)

Table of values

x	-1	0	1	2	3	4	5
$f(x)$	-8	5	6	1	-4	-3	10

The table of values identifies three roots p, q, r, where,

$$-1 < p < 0; \qquad 2 < q < 3; \qquad 4 < r < 5$$

and this is confirmed by the graph of $f(x)$.

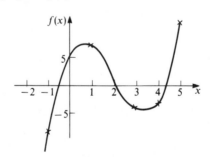

How can we calculate p, q and r more accurately?

(h) $x^4 = 6x^2 - 5 \Rightarrow x^4 - 6x^2 + 5 = 0 \Rightarrow (x^2 - 1)(x^2 - 5) = 0$

$\Rightarrow x = 1, \; x = -1, \; x = +\sqrt{5} \text{ or } x = -\sqrt{5}$

And we have four real roots.

(cont.)

Investigation 3—continued

(i) $x^4 = 6x - 5 \Rightarrow x^4 - 6x + 5 = 0 \Rightarrow (x-1)(x^3 + x^2 + x - 5) = 0$

$\Rightarrow x = 1$ or $x^3 + x^2 + x - 5 = 0$

Table of values for $g(x) = x^3 + x^2 + x - 5$

x	-2	-1	0	1	2	3
$g(x)$	-11	-6	-5	-2	9	34

so a root lies between 1 and 2.

The gradient of $g(x)$ is $g'(x) = 3x^2 + 2x + 1 = 3(x + \frac{1}{3})^2 + \frac{2}{3} > 0$ for all x, so that with a positive gradient $g(x)$ can only cut the x-axis at one point.

How can we identify this second real root accurately? i.e. the real root of $x^3 + x^2 + x - 5 = 0$. We shall look at this in detail in the next section.

Methods to solve $x^3 + x^2 + x - 5 = 0$

We know, from Investigation 3, that the one real root, $x = p$, lies between $x = 1$ and $x = 2$.

Interval bisection

If $g(x) = x^3 + x^2 + x - 5$ then $g(1) = -2$ and $g(2) = 9$.

$g(1.5) = (1.5)^3 + (1.5)^2 + 1.5 - 5 = 2.125 \Rightarrow 1 < p < 1.5$

$g(1.25) = 1.953 + 1.563 + 1.25 - 5 = -0.2345$

$\Rightarrow 1.25 < p < 1.5$

We might now guess that the root is closer to 1.25 than 1.5, and find $g(1.3)$ instead of $g(1.375)$.

$g(1.3) = 2.197 + 1.69 + 1.3 - 5 = 0.187$ so $1.25 < p < 1.3$

This method seems a long, laborious way to get to a solution which, so far, appears to be close to $p = 1.28$.

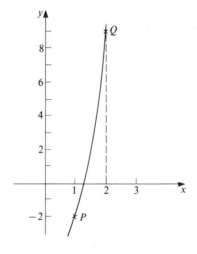

Linear interpolation

From the graph of $g(x)$, P is $(1, -2)$ and Q is $(2, 9)$.

Join PQ to meet the x-axis at R and, by similar triangles,

$\dfrac{SR}{RT} = \dfrac{SP}{QT}$ or $\dfrac{SR}{ST} = \dfrac{SP}{PS + TQ} \Rightarrow \dfrac{SR}{1} = \dfrac{2}{11} \simeq 0.2$

$\Rightarrow OR = 1.2$

$g(1.2) = 1.728 + 1.44 + 1.2 - 5 = -0.632$

$\Rightarrow U = (1.2, -0.632)$

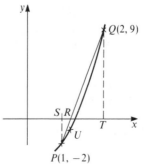

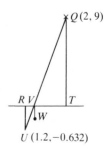

Join UQ meeting the x-axis at V and similar triangles RUV, VTQ

$\Rightarrow \dfrac{RV}{RT} = \dfrac{RU}{RU + TQ} \Rightarrow \dfrac{RV}{0.8} = \dfrac{0.632}{9.632} \Rightarrow RV = 0.0525$ So V is where $x = 1.2525$.

$g(1.2525) = 1.9649 + 1.5688 + 1.2525 - 5 = -0.2139$,   and we are getting closer to the root.

A better solution is at X, where $\dfrac{VX}{VT} = \dfrac{VW}{VW + TQ}$ $\Rightarrow$ $\dfrac{VX}{2 - 1.2525} = \dfrac{0.2139}{9 + 0.2139}$ $\Rightarrow$ $VX = 0.017\,35$.

This better solution is $1.2525 + 0.0174 = 1.2699 \simeq 1.27$.

Check $g(1.27) = 2.0484 + 1.6129 + 1.27 - 5 = -0.0687$ and hence X is $(1.27, -0.0687)$. It may be quicker and more accurate, now, to work with our values for $x = 1.2525$ and $x = 1.27$.

An even better solution is at Z, where

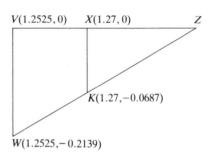

$\dfrac{XZ}{ZV} = \dfrac{XK}{VW}$ $\Rightarrow$ $XZ = \dfrac{0.0687ZV}{0.2139} = 0.3212(0.0175 + XZ)$

$0.6788\,XZ = 0.3212 \times 0.0175$ $\Rightarrow$ $XZ = 0.008\,28$

Now our solution is $x = 1.278\,28$ after 4 applications of **linear interpolation**.

Is there a quicker way?

Newton–Raphson

Draw the tangent at $Q(2,9)$ to meet the x-axis at A.

$g'(2) = 3(2)^2 + 4 + 1 = 17$ $\Rightarrow$ $\dfrac{QT}{TA} = 17$ $\Rightarrow$ $AT = \dfrac{9}{17} = 0.53$ $\Rightarrow$ $A(1.47, 0)$

Draw AB vertically to meet $g(x)$ at $B(1.47, 1.81)$.

Draw tangent at B to meet the x-axis at C.

$g'(1.47) = 10.42$ $\Rightarrow$ $\dfrac{AB}{AC} = 10.42$ $\Rightarrow$ $AC = \dfrac{1.81}{10.42} = 0.174$

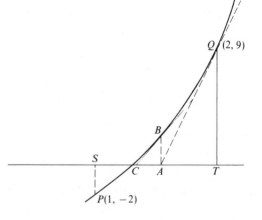

At C, $x = 1.47 - 0.174 = 1.296$.

Draw CD vertically to meet $g(x)$ at $D(1.296, 0.1524)$.

$g'(1.296) = 8.631$ $\Rightarrow$ a better solution at E, where

$$OE = 1.296 - CE = 1.296 - \dfrac{0.1524}{8.631} = 1.278$$

Check $g(1.278) = -0.0014$,   a little more accurate than linear interpolation and the advantage is that we can derive a general formula for this process.

In general,   if x_1 is the first approximation $(x = 2$,   above) then a better approximation x_2 is given by

$$\dfrac{g(x_1)}{x_1 - x_2} = g'(x_1) \quad \Rightarrow \quad x_1 - x_2 = \dfrac{g(x_1)}{g'(x_1)} \quad \Rightarrow \quad \boxed{x_2 = x_1 - \dfrac{g(x_1)}{g'(x_1)}}$$

This formula, the **Newton–Raphson Formula**, can be applied several times to obtain a closer approximation to your solution.

For $g(x) = x^3 + x^2 + x - 5 = 0$, we put $x_1 = 2 \Rightarrow x_2 = 1.47 \Rightarrow x_3 = 1.296 \Rightarrow x_4 = 1.278$

One more application gives, $x_5 = 1.278 - \dfrac{(1.278^3 + 1.278^2 + 1.278 - 5)}{3 \times 1.278^2 + 2 \times 1.278 + 1.278} = 1.278 + \dfrac{0.001\,38}{8.733\,85}$

So $x_5 = 1.278 + 0.000\,16 = 1.278\,16$.

Note $x_5 - x_4 = 0.000\,16$, which indicates that your solution is correct to 2 decimal places, as $0.000\,16$ may only affect the third decimal place.

If the gradient of $g'(x_1)$ is negative, then $g'(x_1) = \dfrac{-g(x_1)}{x_2 - x_1} \Rightarrow x_2 - x_1 = \dfrac{-g(x_1)}{g'(x_1)}$, leading to the same formula.

WORKED EXAMPLES

1 Evaluate the roots of $x^3 - 6x^2 + 6x + 5 = 0$ correct to 2 decimal places. (See Investigation 3 **(g)**.)

The graph of $f(x) = x^3 - 6x^2 + 6x + 5$ and its table of values appear in Investigation 3 **(g)**. The three real roots are p, q and r, where

$$-1 < p < 0; \quad 2 < q < 3; \quad 4 < r < 5$$

To find p: $f(-1) = -8$ and $f(0) = 5$, so by linear interpolation take $x_1 = \dfrac{-5}{13} \simeq -0.385$.

By the Newton–Raphson method, $f'(x) = 3x^2 - 12x + 6$.

$x_2 = -0.385 - \dfrac{1.744}{11.065} = -0.385 - 0.158 = -0.543$.

I programmed $x_2 = x_1 - \dfrac{(x_1{}^3 - 6x_1{}^2 + 6x_1 + 5)}{3x_1{}^2 - 12x_1 + 6}$ into my calculator to give,

$x_2 = -0.529\,02$, $x_3 = -0.528\,918$, $x_4 = -0.528\,918$,

$$\text{so} \quad p = -0.528\,918 \text{ to 6 d.p.}$$

$p = -0.53$ to 2 d.p.

To find q: $2 < q < 3$.

Starting at $x_1 = 2 \Rightarrow x_2 = 2.16\dot{6} \Rightarrow x_3 = 2.167\,40 \Rightarrow x_4 = 2.167\,45$

so $q = 2.17$ to 2 d.p.

Starting at $x_1 = 3 \Rightarrow x_2 = 1.6\dot{6} \Rightarrow x_3 = 2.1895 \Rightarrow x_4 = 2.167\,40 \Rightarrow x_5 = 2.167\,45$ and it takes a little more time to converge to q, but not much.

To find r: $4 < r < 5$.

Starting at $x_1 = 4 \Rightarrow x_2 = 4.5 \Rightarrow x_3 = 4.372\,55 \Rightarrow x_4 = 4.361\,55 \Rightarrow x_5 = 4.361\,47$.

$r = 4.36$ to 2 d.p.

2 Locate the roots of $2x^3 - x^2 - 6x + 3 = 0$ and find the largest root correct to 2 decimal places, using the Newton–Raphson process.

$f(x) = 2x^3 - x^2 - 6x + 3$

Table of values:

x	-2	-1	0	1	2	3
$f(x)$	-5	6	3	-2	3	30

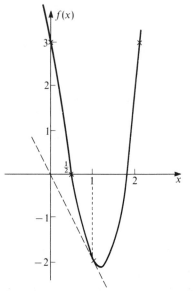

The roots lie between -2 and -1; 0 and 1; 1 and 2.

The largest root is α where $1 < \alpha < 2$.

Newton–Raphson formula gives $x_2 = x_1 - \dfrac{(2x_1^3 - x_1^2 - 6x_1 + 3)}{6x_1^2 - 2x_1 - 6}$

Using $x_1 = 1$ $\Rightarrow$ $x_2 = 1 - \dfrac{(-2)}{(-2)} = 0,$

$x_2 = 0$ $\Rightarrow$ $x_3 = 0 - \dfrac{3}{-6} = \dfrac{1}{2}$

$x_3 = \dfrac{1}{2}$ $\Rightarrow$ $x_4 = \dfrac{1}{2} - \dfrac{(1/4) - (1/4) - 3 + 3}{(3/2) - 1 - 6} = \dfrac{1}{2}$

$f\left(\dfrac{1}{2}\right) = 0$ $\Rightarrow$ $x = \dfrac{1}{2}$ is an exact root, but it is not the root we require, as $1 < \alpha < 2$.

Try, $x_1 = 2$ $\Rightarrow$ $x_2 = 2 - \dfrac{3}{14} = 1.786$ $\Rightarrow$ $x_3 = 1.786 - \dfrac{0.488}{9.567} = 1.786 - 0.051 = 1.735.$

$x_3 = 1.735$ $\Rightarrow$ $x_4 = 1.735 - \dfrac{0.025\,26}{8.591} = 1.735 - 0.0029 = 1.7321 = 1.73$ (to 2 d.p.)

The graph shows why starting at $x = 1$, where $f'(1) = -2$, led to $x_2 = 0$ and the tangent at $(0, 3)$ being -6, gave $x_3 = \dfrac{1}{2}$, the exact solution.

Care must be taken to ensure that the first approximation will lead to the correct solution, in this case by starting at $x = 2$.

In fact, the original equation can be solved algebraically.

$f(1/2) = 0$ $\Rightarrow$ $2x - 1$ is a factor and $f(x) = x^2(2x - 1) - 3(2x - 1) = (2x - 1)(x^2 - 3)$

Possible factors are $(2x + 1)$, $(x + 1)$, $(x + 3)$, and $(2x + 3)$, so perhaps we should have tried this, although the question asked for the Newton–Raphson process to be applied.

$f(x) = 0$ $\Rightarrow$ $x = \dfrac{1}{2}$ or $x^2 = 3$ $\Rightarrow$ $x = \dfrac{1}{2}$; $x = -\sqrt{3}$ or $x = +\sqrt{3} = 1.7321$, the root required.

Exercise 16.2

Solve the following equations using an appropriate method.

1 $x^3 = 27$ **2** $x^2 = -4$ **3** $x^3 = 3x - 2$ **4** $x^3 = 3x^2 - 2$

5 $x^3 = 3x^2 - 2x$ **6** $x^3 = 3x^2 - 3x + 1$ **7** $x^3 = 3x^2 - 3x + 2$ **8** $x^4 = 5x^2 - 4$

9 $x^4 = 5x - 4$ **10** $x^5 = 3x^2 - 2x$ **11** $x^4 = 1$ **12** $x^3 = 1$ **13** $x^5 = 1$

Investigation 4

If two circles of radii 10 cm overlap to give 3 equal areas, what is the distance between their centres?

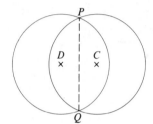

This is equivalent to a chord PQ dividing a circle into areas in the ratio $1:3$.

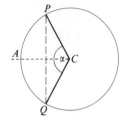

If the angle $PCQ = \alpha$, then the area of sector $CPAQ = \frac{1}{2}r^2\alpha$ and the area of $\triangle PCQ = \frac{1}{2}r^2 \sin \alpha$.

Area of minor segment is $\frac{1}{2}r^2\alpha - \frac{1}{2}r^2 \sin \alpha = \frac{1}{4}$ area of whole $= \frac{1}{4}\pi r^2$

$$\Rightarrow \quad \alpha - \sin \alpha = \frac{\pi}{2} \quad \text{or} \quad \alpha = \sin \alpha + \frac{\pi}{2} \quad \text{remembering } \alpha \text{ is in radians}$$

Once we have solved this equation, $CD = 2 \times 10 \cos \dfrac{\alpha}{2}$.

But how do we solve $\alpha = \sin \alpha + \dfrac{\pi}{2}$ $\left(\text{or} \quad x = \sin x + \dfrac{\pi}{2} \right)$?

Here are several methods to try.

A Graphical solution Draw the graphs of $y = \sin x$ and $y = x - \dfrac{\pi}{2}$ (x in radians) and find where they intersect.

B Graphical continuation Draw the graph of $y = \sin x - x + \dfrac{\pi}{2}$ between suitable values and find where it crosses the x-axis.

C Spiral method Start with $x_1 = 2$ and find $y_1 = \sin x_1 + \dfrac{\pi}{2}$

put $x_2 = y_1$ and find $y_2 = \sin x_2 + \dfrac{\pi}{2}$

put $x_3 = y_2$ and find $y_3 = \sin x_3 + \dfrac{\pi}{2}$

What is happening?

Why is this called the Spiral method? **Hint** Draw $y = x$ and $y = \sin x + \dfrac{\pi}{2}$.

(cont.)

Investigation 4—continued

D From your graph of $y = \sin x - x + \dfrac{\pi}{2}$, choose convenient values $x = a$ and $x = b$ either side of the solution. Find y_a when $x = a$ and y_b when $x = b$. Join $A(a, y_a)$ and $B(b, y_b)$ and find x_3, where AB meets the x-axis. Is x_3 a solution to our equation?

Use $B(b, y_b)$ and $C(x_3, y_3)$ to repeat this process. (You can do this by calculation.)

E Use the Newton-Raphson formula to solve the equation $\sin x - x + \dfrac{\pi}{2} = 0$.

To solve the equation $\sin x = x - \dfrac{\pi}{2}$

Note that, in Investigation 4, x is in radians, since sector area, $S = \frac{1}{2} r^2 \alpha$, depends on α being in radians.

1 First I drew $y = \sin x$ and $y = x - \dfrac{\pi}{2}$ and noticed that they crossed at approximately $x = 2.32$ radians.

2 I thought I could enlarge the graphs between $x = 2.2$ and $x = 2.4$ for a more accurate answer.

3 I also thought I would draw

$$y = \sin x - x + \dfrac{\pi}{2}$$

to see where this graph cuts the x-axis.

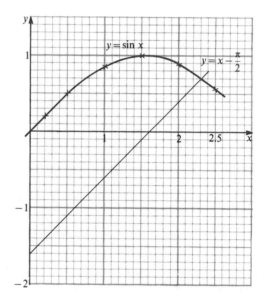

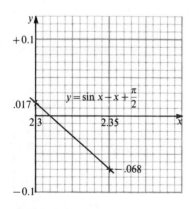

Table of values

x (radians)	2.2	2.25	2.3	2.35	2.4
$\sin x$	0.808	0.778	0.746	0.711	0.675
$\sin x - x + \dfrac{\pi}{2}$	0.179	0.099	0.017	−0.068	−0.154

From the table of values, the solution lies between 2.30 and 2.35. Joining the point (2.3, 0.017) to (2.35, −0.068) with a straight line, indicates that the solution lies close to $x = 2.31$ radians.

Can we do better?

The **Interval bisection** method would appear long-winded and less accurate in this case, but by **Linear interpolation** (using similar triangles), the next approximation divides 2.3 and 2.35 in the ratio $17 : 68 = 1 : 4$ i.e. $x = 2.31$.

$$x = 2.31 \quad \Rightarrow \quad y = \sin 2.31 - 2.31 + \dfrac{\pi}{2} = 0.000\,198\,4$$

which indicates that $x = 2.31$ is a very accurate solution.

In fact, $x = 2.311 \quad \Rightarrow \quad y = -0.001\,87$, so $x = 2.310$ is the solution correct to 3 d.p.

Newton–Raphson's formula gives $x_2 = x_1 - \dfrac{(\sin x_1 - x_1 + \pi/2)}{\cos x_1 - 1}$.

$$x_1 = 2 \quad \Rightarrow \quad x_2 = 2 - \frac{0.480\,09}{-1.4161} = 2 + 0.3390 = 2.3390$$

$$x_2 = 2.3390 \quad \Rightarrow \quad x_3 = 2.3390 - \frac{(-0.049\,04)}{(-1.694\,84)} = 2.3390 - 0.028\,93 = 2.310\,065$$

So two applications of Newton–Raphson give $x = 2.3101$.

The **Spiral method** is as follows for the same equation, in the form $x = \sin x + \pi/2$.

$x_1 = 2 \quad \Rightarrow \quad y_1 = \sin 2 + \pi/2 = 2.480\,093\,7$

$x_2 = 2.480\,093\,7 \quad \Rightarrow \quad y_2 = \sin 2.480\,093\,7 + \pi/2 = 2.185\,096\,6$

$x_3 = 2.185\,096\,6 \quad \Rightarrow \quad y_3 = 2.387\,973\,3$

$x_4 = 2.387\,973\,3 \quad \Rightarrow \quad y_4 = 2.255\,078\,9$

$x_5 = 2.255\,078\,9 \quad \Rightarrow \quad y_5 = 2.345\,669\,1$

$x_6 = 2.345\,669\,1 \quad \Rightarrow \quad y_6 = 2.285\,306\,4$

$x_7 = 2.285\,306\,4 \quad \Rightarrow \quad y_7 = 2.326\,210\,7$

$\Rightarrow \quad y_8 = 2.298\,783\,8$

$\Rightarrow \quad y_9 = 2.317\,311\,3$

$\Rightarrow \quad y_{10} = 2.304\,856\,3$

$\Rightarrow \quad y_{11} = 2.313\,257\,1$

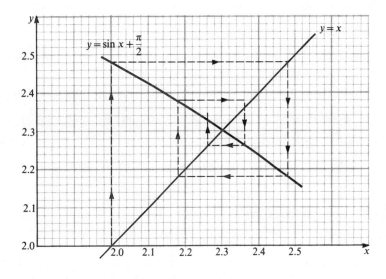

and this sequence is taking a long time to approach its limit. The diagram shows why. The spiral is 'long-winded' because the gradient of $y = \sin x + \pi/2$ needs to be less for this process to converge quickly.

For the Spiral method to be effective, the equation must be put in the form $x = f(x)$ and the gradient of $f(x)$ must be less than 1. (The smaller the gradient the faster the process converges.)

If the gradient of $f(x)$ is greater than 1, the spiral moves outwards and the sequence diverges, as you can see in the diagram.

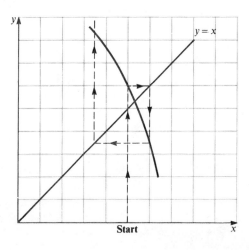

325

Investigation 5

Draw the graph of $y=x$ on co-ordinate axes with equal scales on both axes so that $y=x$ appears as a line at 45 degrees to the x-axis. Mark both scales from 0 to 6, i.e. $0<x<6$ and $0<y<6$.

(a) On the same diagram, draw the line through (1,4) and (5,2) to represent $f(x)$, as though you were solving $x=f(x)$ by the Spiral method. Start at $x=1$ and construct the spiral. Where do you finish?

(b) On another diagram repeat the process using $f(x)=8-2x$, starting at $x=2$.

(c) What happens if you use $f(x)=6-x$?

Investigation 6

Imagine you are using the Newton–Raphson process to solve $x^3-4x=0$. What happens if you start at $x=1$?

Exercise 16.3

1 For the equation $x^3+x-1=0$,

(a) locate limits for the positive root, and find a better value for this root by,

(b) Linear interpolation, starting with $x_1=0$ and $x_2=1$;

(c) The Newton–Raphson method, starting with $x=0$ and with $x=1$;

(d) The Spiral method, writing $x=1-x^3$ and starting with $x=0.5$. What happens if you start with $x=0$ or $x=1$?

2 Solve the equation $x=\cos x$, **(a)** with a sketch graph **(b)** by Linear interpolation **(c)** using Newton–Raphson **(d)** by the Spiral method for $x=g(x)$.

Solve the following equations by **(a)** locating an interval in which the root lies **(b)** Linear interpolation **(c)** Newton–Raphson **(d)** the Spiral method (if appropriate).

3 $2x=\tan x$ **4** $x^2=\sin x$ **5** $x^2=\cos x$ **6** $\dfrac{1}{x}=\ln x$ **7** $e^x=\dfrac{1}{x}$ **8** $x^3=2x-1$

9 $x^3=2x^2-1$ **10** $e^x=3x$ **11** $2^x=\dfrac{1}{x}$ **12** $2^x=\dfrac{1}{x^2}$

17 Inequalities

Introduction

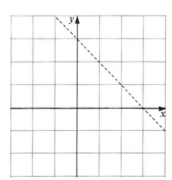

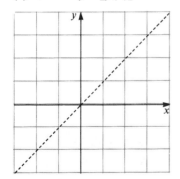

Investigation 1

Identify the following graphs, equations, inequalities and regions.

1 (a) dotted line $y = f(x)$?
 (b) region below line
 (c) region above line
 (d) draw $y = [f(x)]^2$

2 (a) dotted line equation?
 (b) region below line
 (c) region above line

(cont.)

Investigation 1—continued

3 **(a)** dotted line equation?

(b) region 'inside' curve

(c) region 'outside' curve

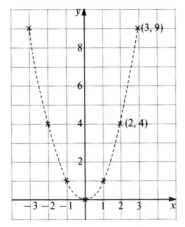

4 **(a)** dotted line equation?

(b) region inside

(c) region outside

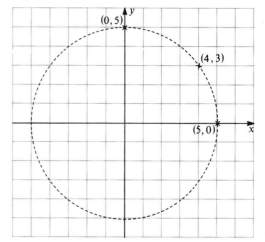

5 unshaded region?

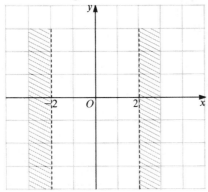

6 unshaded region?

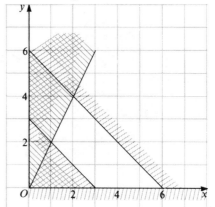

7 **(a)** dotted lines (together) $y = f(x)$

(b) region above

(c) draw $y = [f(x)]^2$

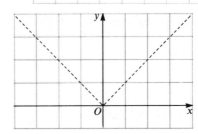

(a) dotted line $y = A(x)$

(b) dotted line $y = B(x)$

(c) x values for $A > B$

(d) x values for $A < B$

(e) draw $y = A(x) + B(x)$

(f) draw $y = [A(x)]^2$ and $y = [B(x)]^2$

(g) x values for $[A(x)]^2 > [B(x)]^2$

(h) x values for $[A(x)]^2 < [B(x)]^2$

8

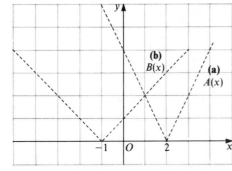

What do you notice about your answers to **(c)** and **(g)** and also to **(d)** and **(f)**?

Investigation 2

(a) On the same axes draw **(i)** $y = \sin x$ **(ii)** $y = |\sin x|$ **(iii)** $y = \sin^2 x$ **(iv)** $y = |\sin x|^2$

(b) On the same axes draw **(i)** $y = \cos x$ **(ii)** $y = |\cos x|$ **(iii)** $y = \cos^2 x$ **(iv)** $y = |\cos x|^2$

Investigation 3 (These you may find difficult!)

Describe the following regions using inequalities.

(a) inside the square

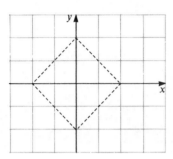

(b) inside the curves

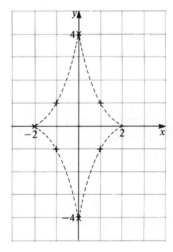

(c) inside the curves, which are quadrants with centres $(4, 4)$, $(4, -4)$, $(-4, 4)$, $(-4, -4)$

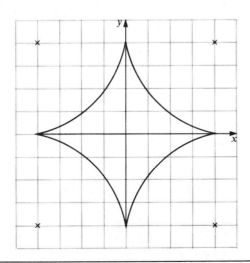

(d) the inside region

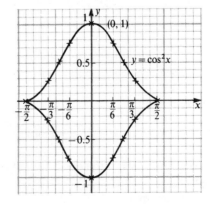

329

Inequalities and equations

WORKED EXAMPLES

1 Solve the inequalities **(a)** $2x-3>6$ **(b)** $3-2x>6$.

These inequalities are one-dimensional (linear), having only one unknown x, and can be solved most easily algebraically.

(a) $2x-3>6$

 Add 3 to both sides $2x>9$

 Divide both sides by 2 $\Rightarrow$ $x>4.5$

(b) $3-2x>6$

 Add $2x$ $3>6+2x$

 Subtract 6 $-3>2x$

 Divide by 2 $\Rightarrow$ $-1.5>x$ or $x<-1.5$

 Alternatively: $3-2x>6$

 Subtract 3 $-2x>3$

 Divide by -2 $\Rightarrow$ $x<-1.5$

> When dividing (or multiplying) an inequality by a negative number, the inequality sign has to change.

2 Find the largest and smallest values of $2x+y$ within the restrictions $y\geqslant0$, $y\leqslant2x$ and $3\leqslant x+y\leqslant6$.

The restrictions (inequalities) are best represented in the x,y diagram.

For $y\geqslant0$ we are restricted to values of y greater than or equal to 0, i.e. the region above $y=0$ (the x-axis).

For $y\leqslant2x$, draw $y=2x$ which divides the x-y plane into two regions $y>2x$ and $y<2x$.

The point (2,2) satisfies $y<2x$, so the region required for $y\leqslant2x$ includes (2,2) i.e. the region to the right of $y=2x$ and **includes** the line.

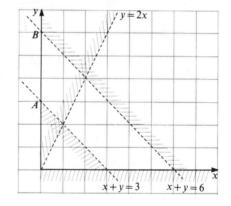

The line A represents $x+y=3$ and the line B represents $x+y=6$. For $3\leqslant x+y\leqslant6$ we require the region between the lines A and B **including** both lines.

The inequalities restrict us to the quadrilateral region left **unshaded.**

$2x+y=4$ represents a straight line through the points $(2,0)$ and $(0,4)$.

$2x+y=8$ represents a straight line through the points $(4,0)$ and $(0,8)$.

$2x+y=k$ represents a straight line through the points $(\frac{1}{2}k,0)$ and $(0,k)$.

The lines $2x+y=k$ all have the same gradient, (-2), and as k increases these lines move further away from the origin.

The smallest value of $2x+y$ satisfying all the inequalities is given by the smallest value of k for a line within the unshaded quadrilateral, i.e. the line nearest the origin, which is the line passing through $(1, 2)$.

For this line, $k=2x+y=2+2=4$, so the smallest value of $2x+y$ is 4.

The largest value of $2x+y$ is the k value for the line passing through $(6,0)$.

For $(6,0)$, $2x+y=12$; so 12 is the largest value of $2x+y$.

(You may have met problems like this in **Linear Programming** in your GCSE course.)

Exercise 17.1

Solve the following linear inequalities.

1 $3x+7<16$ **2** $3(x+2)>6$ **3** $6(x+4)<3(x-7)$ **4** $\dfrac{5}{x}>10$ **5** $\dfrac{2}{x}+6>14$

6 $\dfrac{1}{x}+\dfrac{2}{x}>3$ **7** $x^2+4>(x+4)^2$ **8** $x^2<9$ **9** $x^2>16$ **10** $(x+3)^2>16$

11 Find **(i)** the smallest and **(ii)** the largest value, of $x+2y$, within the following restrictions.
   (a) $|x|\leqslant 3$ and $|y|\leqslant 2$ **(b)** $x+y\geqslant 3$; $y\leqslant 3x$; $2x+y\leqslant 10$
   (c) $|x|+|y|\leqslant 4$ **(d)** $x^2+y^2\leqslant 25$

12 Repeat question **11** for the smallest and largest values of $x+y$.

13 On an Argand diagram, describe the region satisfied by the following inequalities for complex numbers.
   (a) $|z|<3$ **(b)** $\arg z=\dfrac{\pi}{4}$ **(c)** $|z|=3$ and $\arg z=\dfrac{\pi}{4}$

   (d) $2<|z|<3$ **(e)** $\dfrac{\pi}{4}<\arg z<\dfrac{\pi}{2}$ **(f)** $|z-1|=|z+1|$

14 What region is represented by **(a)** $y>x^2$ **(b)** $y+x^2<2$ **(c)** $y>x^2$ and $y+x^2<2$? For **(c)**, find the area of the region.

15 For $x>0$, find the area of the region $x^3<y<x^2$.

16 Find the areas inside the curves of Investigation 3. (**(b)** is formed by parts of parabolae and **(d)** by $\cos^2 x$.)

17 Find the area between the circle centre $(0,0)$ and the rectangular hyperbola (with asymptotes x- and y-axes), both curves passing through the points $(1, 2)$ and $(2, 1)$.

Investigation 4

If I throw a ball vertically upwards into the air from a height of 1.5 m above the ground with a speed of $10 \, \text{m s}^{-1}$,

(a) how long will it be in the air?   (Take $g = 10 \, \text{m s}^{-2}$.)

(b) how long will it be above the level of the bedroom window (4 m above ground)?

(c) between what times will it be above the roof (8 m above the ground)?

(a) Using $s = ut + \frac{1}{2}at^2$ gives the height above the level of projection, h, as

$$h = 10t - \tfrac{1}{2} \times 10t^2 = 10t - 5t^2$$

When I catch the ball (at the same height as throwing it), $h = 0$, so

$$10t - 5t^2 = 0 \quad \Rightarrow \quad 5t(2-t) = 0 \quad \Rightarrow \quad t = 0 \quad \text{or} \quad t = 2$$

and the ball is in the air for 2 seconds.

If I let the ball hit the ground (where $h = -1.5$), then $-1.5 = 10t - 5t^2$

$$\Rightarrow \quad 5t^2 - 10t - 1.5 = 0 \quad \Rightarrow \quad t = \frac{10 \pm \sqrt{100 + 4 \times 5 \times 1.5}}{10} = \frac{10 \pm \sqrt{130}}{10} = 2.14 \text{ s}$$

discounting the negative root.

(b) Above $h = 4$ we require $10t - 5t^2 > 2.5 \quad \Rightarrow \quad 5t^2 - 10t + 2.5 < 0$

Solving $5t^2 - 10t + 2.5 = 0 \quad \Rightarrow \quad 2t^2 - 4t + 1 = 0$

$$\Rightarrow \quad t = \frac{4 \pm \sqrt{16 - 4 \times 2 \times 1}}{4} = \frac{4 \pm \sqrt{8}}{4} = \frac{2 \pm \sqrt{2}}{2} \quad \Rightarrow \quad t_1 = \frac{2 - \sqrt{2}}{2} \quad \text{and} \quad t_2 = \frac{2 + \sqrt{2}}{2}$$

So the difference in time, $t_2 - t_1 = \dfrac{2\sqrt{2}}{2} = \sqrt{2} = 1.41$ s

or from the original equation, $2t^2 - 4t + 1 = 0$, we require the difference in time $t_2 - t_1$ (the time the ball is above the window) and, using the sum and product of the roots,

$$t_2 - t_1 = \sqrt{(t_2 + t_1)^2 - 4t_1 t_2} = \sqrt{\left(\frac{4}{2}\right)^2 - 4 \times \frac{1}{2}} = \sqrt{4 - 2} = \sqrt{2} = 1.41 \text{ s}$$

(c) Above the roof (8 m) $\quad \Rightarrow \quad h > 6.5 \quad \Rightarrow \quad 10t - 5t^2 > 6.5 \quad \Rightarrow \quad 5t^2 - 10t + 6.5 < 0$

The critical times are given by $5t^2 - 10t + 6.5 = 0$

$$\Rightarrow \quad t = \frac{10 \pm \sqrt{100 - 4 \times 5 \times 6.5}}{10} \quad \Rightarrow \quad \textbf{no solutions}, \text{ so the ball is never above the roof.}$$

Its greatest height is $\dfrac{10^2}{2g} = \dfrac{100}{20} = 5$ m above the point of projection.

WORKED EXAMPLES

1 Solve the inequality $3x^2 + 10x - 8 > 0$.

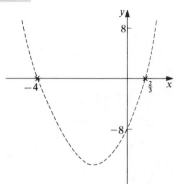

The graph of $y = 3x^2 + 10x - 8$ is a parabola cutting the x-axis at points given by

$$3x^2 + 10x - 8 = 0 \qquad \ldots\ldots\ldots\ldots \text{ (1)}$$

As the parabola is U-shaped ($+$ve, x^2 coefficient), we require the x values 'outside' the roots of Equation **(1)**.

Factorizing $\Rightarrow (3x - 2)(x + 4) = 0 \Rightarrow x = \frac{2}{3}$ or $x = -4$

$$3x^2 + 10x - 8 > 0 \Rightarrow x > \tfrac{2}{3} \text{ or } x < -4$$

2 Find the range of values of m for which the line $y = mx$ intersects **(a)** the circle $(x - 3)^2 + y^2 = 4$ **(b)** the parabola $y^2 = x - 4$

(a) $y = mx$ intersects the circle when

$$(x - 3)^2 + (mx)^2 = 4$$

$$\Rightarrow x^2 + m^2x^2 - 6x + 5 = 0$$

If $y = mx$ is a tangent, this equation has equal roots

$$\Rightarrow 36 = 4 \times 5(1 + m^2) \Rightarrow m^2 = \frac{4}{5}$$

$$\Rightarrow m = \pm\frac{2}{\sqrt{5}}$$

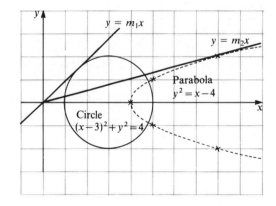

m_1 is the gradient of the tangent $y = m_1x$ to the circle.

So m_1 lies between $\pm\dfrac{2}{\sqrt{5}}$

i.e. $-\dfrac{2}{\sqrt{5}} \leqslant m_1 \leqslant +\dfrac{2}{\sqrt{5}}$.

(b) By a similar method, substituting $y = mx$ in $y^2 = x - 4 \Rightarrow m^2x^2 - x + 4 = 0$

Roots equal $\Rightarrow 1 = 4m^2 \times 4 \Rightarrow m^2 = \dfrac{1}{16} \Rightarrow m = \pm\dfrac{1}{4} \Rightarrow -\dfrac{1}{4} \leqslant m_2 \leqslant +\dfrac{1}{4}$

Inequalities

Solve $\dfrac{x+1}{2x-3} > \dfrac{1}{x-3}$

The temptation is to cross-multiply, but we do not know whether $2x-3$ and $x-3$ are positive or negative, and multiplying by a negative number changes the sign of the inequality.

$$\frac{x+1}{2x-3} > \frac{1}{x-3} \quad \Rightarrow \quad \frac{x+1}{2x-3} - \frac{1}{x-3} > 0 \quad \Rightarrow \quad \frac{(x+1)(x-3)-(2x-3)}{(2x-3)(x-3)} > 0$$

$$\Rightarrow \quad \frac{x^2-4x}{(2x-3)(x-3)} > 0 \quad \Rightarrow \quad \frac{x(x-4)}{(2x-3)(x-3)} > 0$$

The critical values are $x=0$, $x=1.5$, $x=3$, $x=4$.

Examine the signs of the brackets: for $x<0$ 　　　LHS $= \dfrac{(-)\times(-)}{(-)\times(-)} > 0$

for $0<x<1.5$　LHS $= \dfrac{(+)\times(-)}{(-)\times(-)} < 0$

for $1.5<x<3$　LHS $= \dfrac{(+)\times(-)}{(+)\times(-)} > 0$

for $3<x<4$　　LHS $= \dfrac{(+)\times(-)}{(+)\times(+)} < 0$

for $x>4$ 　　　LHS $= \dfrac{(+)\times(+)}{(+)\times(+)} > 0$

The solution is $x<0$　or　$1.5<x<3$　or　$x>4$

Exercise 17.2

Solve the inequalities.

1 $(x-3)(x-4)>0$ 　　**2** $(x-3)(x-4)>6$ 　　**3** $(x-3)(x-4)>6x$

4 $x(x-2)^2 \geqslant 0$ 　　**5** $x(x-2)^2 \geqslant 3$ 　　**6** $x(x-2)^2 \geqslant 1$

7 $\dfrac{1}{x-2} < 1$ 　　**8** $\dfrac{x}{x+2} \geqslant 3$ 　　**9** $4-x > \dfrac{2}{x}$ 　　**10** $\dfrac{x}{x+1} \geqslant \dfrac{x+1}{x+2}$

11 $\dfrac{1-x}{x-2} \geqslant \dfrac{3}{x}$ 　　**12** $\dfrac{2}{x+1} > \dfrac{x+1}{2}$ 　　**13** $0<x+1<4$

14 $0 \leqslant \dfrac{3}{x+1} \leqslant 4$ 　　**15** $0 \leqslant \dfrac{x+2}{2x+1} \leqslant 2$ 　　**16** $1<|x|<2$

17 Find the range of values of x for which $x^2 > 2x$.

18 Find the range of values of x for which $x^3 > 3x$.

19 Find the range of values of k for which $x^2-(k+1)x+k=0$ has real roots.

20 If a square has side x metres, for what values of x is its area (in m^2) greater than its perimeter (in metres)?

21 **(a)** Repeat question **20** for a circle radius r, i.e. area > circumference.
Can you do the same for **(b)** a rectangle **(c)** an equilateral triangle **(d)** a rhombus?

22 For a cube of side x, for what values of x are **(a)** volume > surface area **(b)** volume > side **(c)** surface area > side, numerically speaking?

23 Find the values of m for which **(a)** $x^2+mx+m=0$ has real roots **(b)** x^2+mx+m is always positive.

24 If $p>0$ and $q>0$, prove that $\dfrac{p}{q}+\dfrac{q}{p}\geqslant 2$.

25 Prove that the arithmetic mean of two unequal positive numbers, $\dfrac{(a+b)}{2}$, is greater than their geometric mean $(\sqrt{ab})$.
Can you extend this to 3 numbers, i.e. prove $\dfrac{(a+b+c)}{3}>\sqrt[3]{abc}$?

26 For which values of x are **(a)** $ax>bx$, given $a>b>0$ **(b)** $2^x>2x$ **(c)** $3^x>3x$?

The modulus function

In Book 1 (page 301) we met the modulus function $y=|x|$ which is a V-shaped graph and is always positive. Investigations 1(7) and 1(8) were designed to revise and develop some of the properties of this type of function.

In the diagram below, the line FGC represents $y=x+1$, but to draw $y=|x+1|$, the negative part of $y=x+1$, (i.e. FG) is reflected in the x-axis to give DG. So the dashed V-shaped graph DGC represents the function $y=|x+1|$

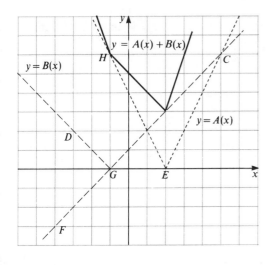

To draw the graph of $y=|2x-4|$.

Values of x	-2	-1	0	1	2	3	4
$2x-4$	-8	-6	-4	-2	0	2	4
$y=|2x-4|$	8	6	4	2	0	2	4

All the y values are positive so the graph of $y=|2x-4|$ lies above the x-axis and is also V-shaped with a steeper gradient than $y=|x|$.

Remember when drawing $y=|2x-4|$ to first draw $y=2x-4$, which passes through $(2,0)$ and reflect the negative portion in the x-axis to give EH.

From the graph we can see that $y=|x+1|$ and $y=|2x-4|$ intersect at $(1,2)$ and $(5,6)$.
The solution of $|x+1|=|2x-4|$ is $x=1$ or $x=5$.

The figure also shows the graph of $y=A(x)+B(x)$, i.e. $y=|x+1|+|2x-4|$, which was investigated on page 328.

Inequalities

$|x+1| > |2x-4|$ in between these values i.e. when $1 < x < 5$. .... **(1)**

$|x+1| < |2x-4|$ outside these values i.e. when $x < 1$ or $x > 5$. .... **(2)**

Note You must **not** write this result as $5 < x < 1$ as this implies that $5 < 1$.

Squaring

If we draw $y = |x+1|^2$ we obtain $y = (x+1)^2$, which is easier to deal with algebraically as the modulus signs have disappeared.

From the graphs we can see that $y = (x+1)^2$ and $y = (2x-4)^2$ intersect at $(1,4)$ and $(5,36)$ so the solution of $(x+1)^2 = (2x-4)^2$ is $x = 1$ or $x = 5$.

The solution of $|x+1| = |2x-4|$ is $x = 1$ or $x = 5$.

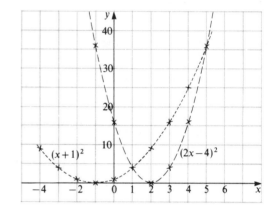

Algebraic solution

For an algebraic solution to $|x+1| = |2x-4|$ we can square both sides to get

$(x+1)^2 = (2x-4)^2 \Rightarrow x^2 + 2x + 1 = 4x^2 - 16x + 16$

$\Rightarrow \quad 3x^2 - 18x + 15 = 0$

$\Rightarrow \quad x^2 - 6x + 5 = (x-1)(x-5) = 0$

$\Rightarrow \quad x = 1$ or $x = 5$.

Similarly, squaring the inequality **(1)**, $|x+1| > |2x-4| \Rightarrow (x+1)^2 > (2x-4)^2 \Rightarrow 3x^2 - 18x + 15 < 0$

$\Rightarrow \quad x^2 - 6x + 5 = (x-1)(x-5) < 0.$

To solve this, think of $y = (x-1)(x-5)$, which is a parabola cutting the x-axis at $x = 1$ and $x = 5$ and is negative between $x = 1$ and $x = 5$.

So the solution is $1 < x < 5$.

This can be deduced mathematically from the inequalities,

$(x-1)(x-5) < 0 \Rightarrow x-1 > 0$ and $x-5 < 0 \Rightarrow x > 1$ and $x < 5$ .... **(1)**

or $x-1 < 0$ and $x-5 > 0 \Rightarrow x < 1$ and $x > 5$ .... **(2)**

No value of x can satisfy **(2)** and hence the only solution is **(1)**.

So $1 < x < 5$.

Again squaring, the inequality **(2)**, $|x+1| < |2x-4| \Rightarrow x^2 - 6x + 5 = (x-1)(x-5) > 0.$

To obtain the solution think again of the U-shaped parabola $y = (x-1)(x-5)$ which is positive (above the x-axis) 'outside' the values $x = 1$ and $x = 5$.

So the solution is $x < 1$ or $x > 5$.

WORKED EXAMPLES

Solve **(a)** $2|x+1|<|2x-4|$ **(b)** $|x+1|>|x-1|$ **(c)** $|x+1|+|x-1|=2$.

(a) Squaring both sides $\Rightarrow$ $4(x+1)^2<(2x-4)^2$ $\Rightarrow$ $4x^2+8x+4<4x^2-16x+16$ $\Rightarrow$ $24x<12$ $\Rightarrow$ $x<0.5$

The graph of $y=2|x+1|$ has double the gradient of $y=|x+1|$ and will only intersect $y=|2x-4|$ at one point, where $x=0.5$.

(b) $|x+1|>|x-1|$ $\Rightarrow$ $(x+1)^2>(x-1)^2$ $\Rightarrow$ $x^2+2x+1>x^2-2x+1$ $\Rightarrow$ $4x>0$ $\Rightarrow$ $x>0$

(c) $|x+1|+|x-1|=2$ can be squared on both sides but it is easier to write it as $|x+1|=2-|x-1|$ before squaring.

Then,
$$(x+1)^2=(2-|x-1|)^2$$
$$\Rightarrow x^2+2x+1=4-4|x-1|+x^2-2x+1$$
$$\Rightarrow 4|x-1|=4-4x$$
$$\Rightarrow |x-1|=1-x$$
$$\Rightarrow (x-1)^2=(1-x)^2 \quad \text{(always true??)}$$

What has happened? Look at the graph.

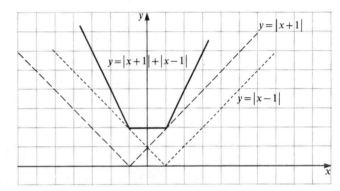

The graph of $y=|x+1|+|x-1|$ has three sections and takes the value 2 for any x value between $x=-1$ and $x=+1$.

Did we square both sides too many times? Clearly our algebraic procedures did not produce the correct answer so we must not always blindly follow them but keep our eyes on what is happening graphically with these unusual functions.

Exercise 17.3

Solve the inequalities and equations

1 $|x|>4$ **2** $|x+1|>4$ **3** $(x+1)^2>4$ **4** $|x+1|+|x-1|=4$

5 $2|x+3|=5$ **6** $|2x+1|=5$ **7** $|2x+1|<5$ **8** $|2x|+1>5$

9 $|x-1|\leqslant|2x-4|$ **10** $|x-1|\geqslant2|x-4|$ **11** $|x-1|\geqslant2|x-2|$ **12** $|x|=2-|x|$

13 $|x+1|=-|x-1|$ **14** $|x|+|y|=2$ and $y=|x|$ **15** $|x|+|y|=2$ and $y=|x+1|$

16 $|\sin x|=\sin^2 x$ **17** $|\sin x|=|\cos x|$ **18** $\sin^2 x=\cos^2 x$

19 $3|x-1|>2|x+2|$ **20** $5|2x-3|<4|x-5|$ **21** $2|3x-1|\leqslant3|2x-1|$

22 $\dfrac{|x+1|}{|x+2|}\geqslant4$ **23** Prove $|x+1|+|x-1|\geqslant2$

Examination Techniques

A driving test consists of a single examination taken at the end of a period of tuition and training and similarly nearly every Advanced Level mathematics course is assessed by two or three formal examination papers taken at the end of the course. Although you practise every manoeuvre you may need to perform in the driving test, you cannot anticipate everything that may happen and, if you do not perform well on the day, you fail the test. How do you prepare best for your driving test? And are there any similarities in preparing for your Advanced Level mathematics examinations?

Driving test		Maths examination
lessons from instructor	=	lessons from teacher
learn the basic techniques and the Highway Code	=	learn necessary formulae and facts
practise manoeuvres	=	practise solving problems
look, signal, manoeuvre	=	read, analyse, write solution
more driving time	=	more studying time
experience enables you to handle difficult situations with confidence	=	experience of questions enables you to cope with difficult problems under examination pressure
practise is essential	=	practise is essential

This may all appear obvious, but the more thorough your preparation the better your chance will be to overcome nerves and boost your self-confidence.

Preparation

Much of your preparation will take place during your course when certain facts and processes must become second nature.

1   Learn each topic and process thoroughly. During examinations you will need to recall facts quickly and accurately.

2   Learn any formulae you can. This will help when you have to solve problems under the pressure of examination conditions.

3   Scrutinize the syllabus to make sure you have studied all the required topics.

4   Familiarize yourself with the notations your examination board uses e.g. notations for functions and vectors.

5   Practise writing solutions to as many past paper questions as you can.
    My university professor said, 'You only do mathematics by solving problems.' The more questions you have done the less an examiner can surprise you.

6   Practise doing questions under examination conditions i.e. against the clock. Remember that you will be required to concentrate for two to three hours in the actual examination.

7   Measure your rate of working. If you are a slow worker, seek advice on ways to speed up. If you work quickly, make sure you do not make mistakes and that your work is not untidy.

8 Develop a neat style of presentation, so that an examiner can follow your methods. Remember that marks are awarded for methods used as well as for correct answers. Here is a quote from the instructions to a JMB paper:

'Careless and untidy work will be penalised.'

9 Read and understand the instructions at the beginning of the paper about which questions, and how many, you must answer.

Style of questions

In addition to different types of question, each examination board will have different styles of papers and the lay-out may be different for each syllabus.

Multiple-choice questions

Most multiple-choice questions require that you select the **one** correct answer out of the **five** suggested. This is straightforward, but sometimes boards adopt a much more complicated selection method which will need scrutiny and practice.

If your board includes multiple-choice papers it is essential to have practised answering this type of question before the real examination, because the multiple-choice paper may contain 30 questions to be answered in 75 minutes, i.e. a change of topic every two and a half minutes.

Short questions

Short questions are perhaps the easiest as they contain one simple idea and do not require expertise from several different parts of the syllabus. Students prefer this type of question which allows the examiner to cover wide sections of the syllabus with many shorter questions. Students must be aware of the number of marks allotted to each question and budget their time accordingly. It may be more profitable to spend more time on the longer questions.

Longer questions

Longer questions explore one particular topic in depth or combine topics from different parts of the syllabus. They may well get progressively more difficult towards the end, and for this reason you may gain more marks for answering one **whole** question rather than two **half** questions.

Choice of questions

In a **multiple-choice paper**, usually you must attempt all the questions and there is no penalty for guessing, so you can fill in the answer sheet at the end of the paper if you cannot answer all the questions.

A **short question paper** usually gives you little choice so the best practice is to work through the paper in the order of the questions, omitting those questions you cannot readily answer. Some students like to try questions on their favourite topics first to give their self-confidence a boost. Some papers indicate the number of marks allotted to each question or part of a question so students can plan their strategy.

With **long questions** where, for instance, six out of eight questions have to be answered, it is wise to read the whole paper first, deciding which questions to answer and in which order to tackle them. Some students put their watches on their desks to make sure they do not spend an undue amount of time on one particular question.

Many times we, as tutors, hear of students who did not leave enough time for each question, or only attempted three questions when they should have done four. Of course it is a different matter if you cannot do the questions, but correct planning and disciplined technique can help enormously. Very often it pays to return to a particular question when you may appreciate a different slant by looking at the problem from a new angle.

IF YOU HAVE TIME AT THE END OF A PAPER, CHECK YOUR WORKING. It is so easy to make an arithmetical slip or press the wrong buttons on your calculator.

We advise our students to make a habit of building in checks to their working so that they say to themselves, 'Is this a reasonable answer?'.

Past paper examination questions

The next section of this book contains a selection of questions taken from the past examination papers from several different examining boards. They have been grouped in topics, but we would advise students to obtain the past papers for their own particular syllabus, to check that the style and format of the examination has not changed significantly from previous years. You can check this by looking at an up-to-date syllabus, published by each examination board.

There are books of past examination papers (with answers) and also books of written answers to typical examination questions.

Past Paper Questions

Examination Boards abbreviated below

(JMB) Joint Matriculation Board (GCE)
(LON) University of London (GCE Exam)
(NI)   Northern Ireland (GCE)

(O & C SMP) Oxford and Cambridge Schools Examination
             Board, School Mathematics Project
(WJEC)   Welsh Joint Education Committee (GCE)

General Questions

1 The equation $ax^2 + bx + c = 0$ has roots p and q. Express $(p+1)(q+1)$ in terms of a, b and c. (LON)

2 Give counter-examples to show that the following statements are false.

   (i)   The sum of two unequal, positive, irrational numbers is irrational.

   (ii) The product of two unequal, positive, irrational numbers is irrational. (JMB)

3 **(a)** Obtain partial fractions for the expression $\dfrac{(x+1)(4x-1)}{(2-x)(x^2+3)}$.

   (b) Points P, T are the extremities of a diameter of the circle whose centre is O and whose radius is r. An arc of a second circle, centre P and radius $r/2$ is drawn to cut the first circle at R and S. Show that the area of the crescent RTS is r^2. (NI)

4 You are given that positive variables x and y are related by the equation $2 \ln y = c - \ln x$ where c is constant and that $y = 3$ when $x = 4$. Find in as simple a form as possible (and not involving logarithms) an expression for y in terms of x. (O & C SMP)

5 **(i)** Solve the simultaneous equations

$$\log_2 x - \log_4 y = 4$$
$$\log_2 (x - 2y) = 5$$

   (ii) Prove by induction, or otherwise, that

$$\frac{1}{1.2} + \frac{1}{2.3} + \frac{1}{3.4} + \cdots + \frac{1}{n(n+1)} = \frac{n}{n+1}$$

(LON)

Inequalities

1 Find the set of real values of x for which $|x-2| > 2|x+1|$ (LON)

2 Find in each case the set of values of x for which

   (a) $\dfrac{1}{3-x} < \dfrac{1}{x-2}$ **(b)** $|3+2x| \leqslant |4-x|$

(LON)

341

3 Find the set of values of x for which $x^2 - 4x + 3 > 0$. (LON)

4 Find the set of values of x for which $3x^2 > 5x + 2$. (O & C SMP)

5 Find the set of values of x for which $\dfrac{2x + 3}{3x + 4} < 1$.

(LON)

6 Find a positive number k and an acute angle α (in degrees) such that

$$\sin x + \sqrt{3} \cos x = k \sin (x + \alpha)$$

Use your answer to find the set of values of x in the interval $0 < x < 360$ such that $\sin x + \sqrt{3} \cos x > 1$.
(O & C SMP)

Numerical Methods

1 Using the substitution $x = \sin \theta$, or otherwise, find

$$\int_{-\frac{1}{2}}^{\frac{1}{2}} \frac{1}{\sqrt{(1 - x^2)}} \, dx$$

Use Simpson's rule with three ordinates to calculate an approximate value of this integral. Deduce that

$$\pi \approx 2\left(1 + \frac{1}{\sqrt{3}}\right)$$

(O & C SMP)

2 A train starts from rest and the pull P of the locomotive in newtons after travelling a distance of s metres is given by the table

s	0	10	20	30	40
P	8400	9100	9600	9900	10 000

Estimate, using Simpson's rule, the work in joules done in hauling the train through the first 40 metres.
(LON)

3 Use the trapezium rule with 5 strips of equal width to find approximately the value of $\displaystyle\int_0^1 10^x \, dx$. (LON)

4 Show that the equation $\sin x - \ln x = 0$ has a root lying between $x = 2$ and $x = 3$.
Given that this root lies between $a/10$ and $(a + 1)/10$, where a is an integer, find the value of a.
Estimate the value of the root to 3 significant figures. (LON)

5 Show that the equation $f(x) = 0$, where $f(x) \equiv x^3 + x^2 - 2x - 1$, has a root in each of the intervals $x < -1$, $-1 < x < 0$, $x > 1$.
Use the Newton–Raphson procedure, with initial value 1, to find two further approximations to the positive root of $f(x) = 0$, giving your final answer to 2 decimal places. (LON)

Co-ordinate Geometry

1 Referred to O as origin, A is the point $(6, 3, -2)$ and B is the point $(8, 1, -4)$. Write down the lengths of OA and OB and determine (as a fraction) the value of $\cos AOB$. Find also the area of the triangle AOB, leaving your answer in surd form. (JMB)

2 Find the equation of the tangent to the curve $xy(x + y) + 16 = 0$ at the point on the curve where the gradient is -1. (JMB)

3   Show that the point of intersection of the tangents at the points $(ap^2, 2ap)$ and $(aq^2, 2aq)$ on the curve $y^2 = 4ax$ is $[apq, a(p+q)]$.

    The line PQ passes through the point $(a, 0)$. Show that the tangents at P and Q are perpendicular and their point of intersection lies on a fixed line, whatever the values of p and q. Show further that the locus of the midpoint of PQ is $y^2 = 2a(x-a)$. (WJEC)

4   Use the focus-directrix property to obtain, in any form, the equation of the parabola which has the point $(4, 1)$ as focus and the line $y = 2$ as directrix. (LON)

5   Calculate the values of $\dfrac{6}{2 + \cos 2\theta}$ when $\theta = 45°$ and $\theta = 60°$.

    Write down the greatest and least values of this expression as θ varies. For what values of θ (between $0°$ and $360°$ inclusive) does the expression take these values?

    Without further calculation, sketch the complete curve whose polar equation is

    $$r = \frac{6}{2 + \cos 2\theta}$$

    (O & C SMP)

6   (a)   Write down the co-ordinates of the mid-point M of the line joining $A(0, 1)$ and $B(6, 5)$.

    (b)   Show that the line $3x + 2y - 15 = 0$ passes through M and is perpendicular to AB.

    (c)   Calculate the co-ordinates of the centre of the circle which passes through A, B and the origin O. (LON)

Functions

1   Given that f and g are two functions, each with period π, show by means of a counter-example that $f \cdot g$ does not necessarily have period π. (JMB)

2   Given that $P(x) = x^3 - 6x^2 + 9x + a$ for all real x, find the values of x for which $P'(x) = 0$. Determine the values of the constant a such that the equation $P(x) = 0$ has a repeated root.

    By sketching graphs of $y = P(x)$ for these values of a, or otherwise, find the set of values of a for which $P(x) = 0$ has only one real root.

    Given now that $a = -5$, and that 4 is an approximation to the root of the equation $P(x) = 0$, use Newton's method once to obtain another approximation to the root, giving your answer correct to one decimal place. (JMB)

3   The function f is defined by $f: x \to \dfrac{x+3}{x-1}$, $x \in \mathbb{R}$, $x \neq 1$.

    Find   (a)   the range of f (b) $ff(x)$ (c) $f^{-1}(x)$. (LON)

4   State for each of the following functions whether it is odd, even, periodic, or none of these, justifying your answers.

    (i) $x \sin x$ (ii) $\sin x + \cos x$ (iii) $x + \cos x$ (O & C SMP)

5   (a)   State the domains and ranges of the functions f and g defined by $f(x) = \sin x$, $g(x) = \sqrt{(1 - 4x^2)}$.
        Find the composite function $f \circ g$ and state its domain and range. Explain why $g \circ f$ cannot be formed.

    (b)   Investigate the function f defined by $f(x) = 3x^5 - 10x^3 + 10$ for local maxima, minima and points of inflection. Sketch the graph of the function, showing clearly the stationary points and any points of inflection. (WJEC)

    [Note that $f \circ g$ means 'f acting on g' or 'apply g first, followed by f'.]

Differentiation

1 Differentiate with respect to x **(a)** $\dfrac{e^{-x}}{x}$ **(b)** $\ln(1+\sin x)$.

(LON)

2 The volume of a spherical balloon is increasing at a constant rate $C\,\text{m}^3\,\text{s}^{-1}$. Find, in $\text{m}\,\text{s}^{-1}$, the rate of increase of the radius of the balloon when the surface area of the balloon is $S\,\text{m}^2$. (LON)

3 Differentiate with respect to x,   **(a)** $x^3 \ln x$ **(b)** $\dfrac{1+\cos x}{x}$

(LON)

4 Show that the curve $y = \dfrac{3}{(2x+1)(1-x)}$ has only one turning point. Find the co-ordinates of this point and determine its nature. Sketch the curve.

Find the area of the region enclosed by the curve and the line $y = 3$. (JMB)

5 If $f(x) = \sqrt{(1+x^2)}$, find $f'(x)$.

Prove that, if $g(x) = \dfrac{x}{\sqrt{(1+x^2)}}$, then $g'(x) = \dfrac{1}{\{\sqrt{(1+x^2)}\}^3}$.

One of the three graphs **A**, **B**, **C** shown here is the graph of $y = g(x)$. Say which it is, and for each of the others give **one** reason why it could not be.

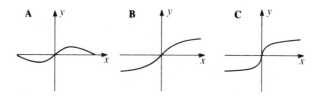

(O & C SMP)

Integration

1 Evaluate the integrals   **(a)** $\displaystyle\int_0^{3/4} x\sqrt{(1+x^2)}\,dx$ **(b)** $\displaystyle\int_0^{\pi/4} \tan^2 x\,dx$ **(c)** $\displaystyle\int_0^{\pi} x\sin x\,dx$

   (d) $\displaystyle\int_3^4 \dfrac{1}{x^2-3x+2}\,dx$.

Express your answer to **(d)** as a natural logarithm. (LON)

2 Use integration by parts to find the indefinite integral $\displaystyle\int x^4 \ln x\,dx$. (O & C SMP)

3 Find by integration the volume of the sphere formed by rotating the circle $x^2+y^2=a^2$ through π radians about the x-axis. (LON)

4 A point moves in a plane so that, at time t, its position referred to rectangular axes OX, OY is given by $x = 4\cos^3 t$, $y = 4\sin^3 t$, $0 \leqslant t \leqslant \tfrac{1}{2}\pi$.

Find its speed at time t. Show that $\dfrac{dy}{dx} = -\tan t$ and that, if the tangent to its path meets the axes in P and Q, then PQ is of constant length. Sketch the path. *(cont.)*

4 *cont.*
Show that the area A enclosed by the path and the axes is given by

$$A = \int_0^{\pi/2} 48 \sin^4 t \cos^2 t \, dt$$

Either by first using the identities $\quad 2 \sin t \cos t = \sin 2t \quad$ and $\quad 2 \sin^2 t = 1 - \cos 2t \quad$ to show that $16 \sin^4 t \cos^2 t = 1 - \cos 4t - 2 \sin^2 2t \cos 2t, \quad$ or otherwise, find the value of A. (O & C SMP)

5 Tabulate, to 3 decimal places, the function $f(x)$, where $f(x) \equiv \sqrt{(1 + x^2)}$, for values of x from 0 to 0.8 at intervals of 0.1. Use these values to estimate I, to 3 decimal places, where $\quad I = \int_0^{0.8} f(x) \, dx, \quad$ by Simpson's rule.

Obtain a second estimate of the value of I, also to 3 decimal places, by using the first 3 terms in the binomial expansion of $\sqrt{(1 + x^2)}$. (LON)

Trigonometry

1 Determine the period of the function

$$f : x \longrightarrow \sin^2\left(\frac{x}{3}\right), \quad x \in \mathbb{R}$$ (JMB)

2 Find all values of θ between 0 and 360 which satisfy the equation $\quad \cos (60 + \theta)° + 2 \sin (30 + \theta)° = 0$. (LON)

3 Express $3 \cos 2t + 4 \sin 2t$ in the form $r \cos (2t - \alpha)$, where $r > 0$ and $0 < \alpha < 90°$, giving the value of α to the nearest degree. (LON)

4 **(i)** Find the general solution of the equation $\sin 2\theta = \cos \theta$.

(ii) The points A, B, C lie on a circle with centre O.
Given that $AB = 11$ m, $BC = 13$ m, $CA = 20$ m, find the angles AOB, BOC, COA to the nearest tenth of a degree and the radius of the circle to the nearest tenth of a metre. (LON)

5 Find all the angles θ between $0°$ and $360°$ such that $2 \cos 2\theta = 1$. (O & C SMP)

6 Use the formula $\quad 2 \cos \theta \cos \phi = \cos (\theta + \phi) + \cos (\theta - \phi) \quad$ to write the product $\quad 2 \cos 3x \cos 2x \quad$ in an alternative form.
Hence show that the product $4 \cos 3x \cos 2x \cos x$ can be written in the alternative form $\cos \alpha x + \cos \beta x + \cos \gamma x + 1$, where α, β and γ are numbers to be determined. (O & C SMP)

Series

1 Expand $(1 - x)^{-\frac{1}{2}}$ in a series of ascending powers of x as far as the term in x^3. (LON)

2 An infinite geometric series with first term 2 converges to the sum 3. Find the fourth term in the series. (LON)

3 The second and fourth terms of an arithmetic progression are -3 and 7 respectively. Find the hundredth term. (LON)

4 Obtain the first two non-zero terms in the expansion of $\tan x$ as a series of ascending powers of x. (LON)

5 Given that
$$f(x) = \frac{(1+nx)^{1/n}}{(1+mx)^{1/m}}$$

where m and n are non-zero constants, and that x is so small that terms in x^3 and higher powers may be neglected, prove that $f(x) = 1 + \frac{1}{2}(m-n)x^2$. (LON)

6 Prove that $1 + \frac{1}{2} + \frac{1}{4} + \cdots + \frac{1}{2^n} = 2 - \frac{1}{2^n}$.

Use this finite summation to explain carefully what you mean when you say that 'the sum of the infinite series

$$1 + \frac{1}{2} + \frac{1}{4} + \cdots + \frac{1}{2^n} + \cdots \quad \text{is 2.'}$$

(O & C SMP)

7 **(i)** If $f(n) = \dfrac{1}{n(n+1)}$, prove that $f(n) - f(n+1) = \dfrac{2}{n(n+1)(n+2)}$.

Hence, or otherwise, find the sum of the series

$$\frac{1}{1.2.3} + \frac{1}{2.3.4} + \cdots + \frac{1}{n(n+1)(n+2)}$$

(ii) Find the set of real values of x for which the geometric series

$$\frac{x+1}{x^2} + \frac{1}{x} + \frac{1}{x+1} + \cdots \quad (x \neq 0 \text{ or } -1)$$

is convergent and state the sum to infinity when this sum exists. (LON)

Differential Equations

1 Solve the differential equation $\dfrac{dy}{dx} = 2\dfrac{y}{x}$, $x > 0$, given that $y = 4$ when $x = 1$. Hence express y in terms of x. (LON)

2 Express $\dfrac{x}{(x+1)(x+2)}$ in partial fractions.

Solve the differential equation $(x+1)(x+2)\dfrac{dy}{dx} = x(y+1)$ for $x > -1$, given that $y = \frac{1}{2}$ when $x = 1$. Express your answer in the form $y = f(x)$. (LON)

3 Solve the differential equation $2x(x+1)y\dfrac{dy}{dx} = y^2 + 1$.

Find the integral curve which passes through the point $(-2, 1)$. Sketch this curve and state the equations of its asymptotes. (LON)

4 A tank initially contains 200 litres of pure water. Starting at time $t = 0$, brine containing $\frac{1}{4}$ kg of dissolved salt per litre of water flows into the tank at the rate of 12 litres per minute. The mixture is kept uniform by stirring and the well stirred mixture simultaneously flows out of the tank at the same rate.

(i) If x denotes the amount of salt in the tank at any time t, show that salt is leaving the tank at the rate of $\dfrac{3x}{50}$ kilogrammes per minute.

(ii) Find the rate at which salt enters the tank and hence obtain the differential equation for the rate of change of x with time.

(iii) Find the amount of salt dissolved in the mixture in the tank after 25 minutes and show that there is a limit to the amount of salt which can be dissolved in the water in the tank. (NI)

5   A water tank has the shape of an open rectangular box of length 1 m, width 0.5 m and height 0.5 m. Water may be drained from the tank through a tap at the bottom of the tank, and it is known that, when the tap is open, water leaves at a rate of 100 h litres per minute, where h m is the depth of water in the tank. When the tap is open, water is also fed into the tank at a constant rate of 50 litres per minute and no water is fed into the tank when the tap is closed. Show that, t minutes after the tap has been opened, the variable h satisfies the differential equation $10\dfrac{dh}{dt} = 1 - 2h.$

On a particular occasion the tap was opened when $h = 0.25$ and closed when $h = 0.375$. Show that the tap was opened for $5 \ln 2$ minutes.   (LON)

6   Given that k is a positive constant, find the general solution of the differential equation $\dfrac{d\theta}{dt} = k(1000 - \theta)$. (Your solution should involve a new arbitrary constant.)

A block initially at $40°C$ is put into an oven at time $t = 0$. The oven is kept at a constant temperature of $1000°C$. At any subsequent instant the rate of increase of the temperature θ of the block is proportional to the difference in temperature that exists between the block and the oven. If after 1 minute the temperature of the block has risen to $160°C$, show that at time t minutes $(t > 0)$ the block has temperature θ given by $\theta = 1000 - 960(\frac{7}{8})^t$.   (O & C SMP)

Permutations and Combinations

1   In how many ways can a committee of 3 women and 3 men be chosen from 5 women and 7 men? In how many ways can the committee be chosen if Mrs A refuses to serve when Mr B is a member?   (NI)

2   How many 4 digit numbers can be formed using the digits $1, 2, 3, 4, 5, 6, 7$ if each digit can appear at most once in any number? How many of these numbers are **(i)** even, **(ii)** greater than 7600, **(iii)** even or greater than 7600?   (NI)

3   **(i)**   Find the number of different permutations of four letters that can be made from the letters of the word AMALGAM. Find also the number of these permutations which contain exactly one pair of letters the same.
    (ii)   Use the binomial expansion to find the value of $\sqrt{4.04}$ to six places of decimals.   (LON)

4   A cricket club has 20 members, of whom 5 are good bowlers and 6 others are good batsmen. How many different teams of eleven players can be selected containing exactly 2 of these bowlers and exactly 4 of these batsmen?   (LON)

5   A student must answer exactly 7 out of 10 questions in an examination. Given that she must answer **at least** 3 of the first 5 questions, determine the number of ways in which she may select the 7 questions.   (LON)

6   Calculate the number of ways in which six people can form

   (a)   a queue (i.e. a single file) of six people,
   (b)   a queue of two people and another queue of four people,
   (c)   a group of two people and another group of four people,
   (d)   first, second and third pairs,
   (e)   three pairs.

Assume that the order within a group or a pair is **not** significant.   (LON)

Complex Numbers

1   Given that $z_1 = 2 + i$ and $z_2 = 3 + 4i$, find the modulus and the tangent of the argument of each of
   (a) $z_1 z_2^*$ **(b)** z_1 / z_2.   (LON)

2   Given that $z = \sqrt{3} + i$, find the value of $\arg(z^7)$ which lies between $-\pi$ and $+\pi$.   (LON)

3   Given that $\dfrac{1}{x+iy} + \dfrac{1}{1+2i} = 1$ where x, y are real, find x and y.   (LON)

4   Find the complex numbers z and w satisfying the simultaneous equations

$$iz + (1+2i)w = 1$$
$$(2+i)z + (2-i)w = -3 \qquad \text{(JMB)}$$

5   **(i)**   Given that $z = \cos\theta + i\sin\theta$, where $z \neq -1$, show that

$$\frac{2}{1+z} = 1 - i\tan\tfrac{1}{2}\theta$$

 (ii)   One root of the equation $z^3 + z^2 + 4z + \lambda = 0$, where λ is a real number, is $1 - 3i$. Find the other roots and the value of λ.   (LON)

6   Show by substitution that $1 + 2j$ is a solution of the equation $x^3 + x + 10 = 0$.

   Write down the other complex solution. Hence, or otherwise, express the left-hand side of the equation as the product of a quadratic factor and a linear factor, both with real coefficients.   (O & C SMP)

7   **(a)**   Find the real numbers a and b such that $\dfrac{a}{1+2i} + \dfrac{b}{3+2i} = \dfrac{5+6i}{8i-1}$.

 (b)   If $z_1 = \dfrac{1}{\sqrt{2}}(1+i)$ and $z_2 = \tfrac{1}{2}(1+i\sqrt{3})$,   find the modulus and argument of $z_1{}^2 + iz_2$.

 (c)   Find the Cartesian equation for the locus of points represented in the Argand diagram by complex numbers satisfying the relation $|z-1| = 2|z-i|$. Find a complex number α such that $\arg(z-\alpha) = \tfrac{1}{2}\pi$ is the equation of a tangent to the locus.   (WJEC)

Probability

1   Two fair dice are thrown. Calculate the probability that the product of the two scores will be

 (a)   a multiple of 5,
 (b)   a multiple of 3,
 (c)   a multiple of 3 or a multiple of 5 or a multiple of both.   (LON)

2   **(i)**   Two cards are drawn without replacement from ten cards which are numbered from 1 to 10. Find the probability that
 (a)   the numbers on both cards are even,
 (b)   the number on one card is odd and the number on the other card is even.
 (c)   the sum of the numbers on the two cards exceeds 4.

 (ii)   Events A and C are independent. Probabilities relating to events A, B and C are as follows:

$$P(A) = 1/5, \quad P(B) = 1/6, \quad P(A \cap C) = 1/20, \quad P(B \cup C) = 3/8$$

   Evaluate $P(C)$ and show that events B and C are independent.   (LON)

3   Four cards are to be dealt at random, without replacement, from a pack of ten cards, of which two are red and eight are black. Find the most probable number of red cards that will be dealt.   (JMB)

4   A child throws two fair dice. If the numbers showing are unequal, he adds them together to get his final score. On the other hand, if the numbers showing are equal, he throws two more fair dice and adds all four numbers showing to get his final score. Calculate the probabilities that his final score is   **(a)** 2   **(b)** 3   **(c)** 4   **(d)** 5   **(e)** 6   **(f)** more than 6.   (LON)

5 (i) Events A and B are independent. The probability of A occurring is $\frac{1}{3}$ and the probability of B occuring is $\frac{1}{4}$. Find the probability of

    (a) neither event occurring,

    (b) one and only one of the two events occurring.

(ii) A bag contains 10 balls, of which 3 are red, 3 are blue and 4 are white. Three balls are to be drawn one at a time, at random and without replacement, from the bag. Find the probability that

    (a) the first 2 balls drawn will be of different colours,

    (b) all 3 balls drawn will be of the same colour,

    (c) exactly 2 of the balls drawn will be of the same colour. (LON)

6 (a) The three events A, B and C have respective probabilities $\frac{2}{5}$, $\frac{1}{3}$ and $\frac{1}{2}$. Given that A and B are mutually exclusive, $P(A \cap C) = \frac{1}{5}$ and $P(B \cap C) = \frac{1}{4}$,

    (i) show that only two of the three events are independent,

    (ii) evaluate $P(C|B)$ and $P(A' \cap C')$.

(b) When Alec, Bert and Chris play a particular game their respective probabilities of winning are 0.3, 0.1 and 0.6, independently for each game played. They agree to play a series of up to five games, the winner of the series (if any) to be the first player to win three games. Given that Bert wins the first two games of the series show that

    (i) Bert is just over 10 times more likely than Alec to win the series,

    (ii) there is a slightly better than even chance that there will be a winner in the series. (WJEC)

Impulse and Impact

1 A particle of mass m moving with speed $2u$ is overtaken by and coalesces with a particle of mass $2m$ moving with speed $3u$ in the same straight line. Calculate the loss in kinetic energy due to the collision. (LON)

2 Two identical smooth spheres, S and T, moving in opposite directions with speeds u, $3u$ respectively, collide directly. The sphere T is reduced to rest. Find the coefficient of restitution between the spheres. (LON)

3 Two masses, one 3 kg and the other 2 kg, are connected by a light inextensible string which passes over a fixed smooth pulley and the system is released from rest. Find the acceleration of the system and show that the tension in the string is 23.52 N.

After falling 2 metres the heavier mass strikes a fixed horizontal plane and is brought to rest. If the lighter mass does not reach the pulley, show that after $\frac{4}{7}$ seconds the 3 kg mass is again in motion. Find the speed with which it begins to ascend and the impulsive tension in the string. (WJEC)

4 (i) A sphere of mass $3m$ is moving with speed $2u$ when it collides with another sphere, of the same radius but of mass m, which is moving in the opposite direction with speed u. The coefficient of restitution between the spheres is $\frac{1}{3}$. Calculate

    (a) the speed of each sphere immediately after impact,

    (b) the magnitude of the impulse received by each sphere on impact.

(ii) A pump, working at an effective rate of 41 kW, raises 80 kg of water per second from a depth of 20 m. Calculate the speed with which the water is delivered.
(Take g as 10 m s^{-2}.) (LON)

5 Three perfectly elastic smooth spheres A, B and C, of equal radii but of masses $3m$, $2m$ and $1m$ respectively, lie at rest on a horizontal plane with their centres in a straight line. Sphere A is projected with speed u to collide directly with sphere B which then collides directly with sphere C. Find the speed of each sphere after each of these collisions. Explain why there are only two collisions between the spheres. (LON)

Centre of Mass

1 Three particles of masses 3, 4 and 5 have position vectors $2\mathbf{i}-3\mathbf{j}+3\mathbf{k}$, $5\mathbf{i}-3\mathbf{j}-4\mathbf{k}$, and $2\mathbf{i}-3\mathbf{j}-\mathbf{k}$, respectively, where $\mathbf{i, j, k}$ are mutually perpendicular unit vectors. Find the position vector of the centre of mass of this system.

   A second system of particles having total mass 8 and position vector of its centre of mass $\frac{1}{2}(\mathbf{i}+4\mathbf{j}+3\mathbf{k})$ is added to the three particles. Find the position vector of the centre of mass of the combined systems.   (NI)

2 Find the co-ordinates of the centroid of a uniform lamina bounded by the curve $y^2=x^3$ where $y>0$, the x-axis and the line $x=4$. This lamina is suspended freely from the origin O. Find, to the nearest degree, the inclination to the vertical of the x-axis.   (LON)

3 Show by integration that the centre of mass of a uniform triangular lamina PQR is at a distance $\frac{1}{3}h$ from QR, where h is the length of the altitude PS.

   A uniform piece of cardboard is in the form of a rectangle $ABCD$, in which $AB=10a$, $BC=6a$, and E is the point in AB such that $AE=4a$. The cardboard is folded along CE so that the edge CB lies along the edge CD to form a trapezium shaped lamina $AECD$ and the triangular part CEB is of double thickness. Find the distance of the centre of mass of the lamina $AECD$ from   (a) AD (b) AE.   Show that, when this lamina is freely suspended from the vertex A, the edge AD makes an angle $\tan^{-1}(11/9)$ with the vertical.

   The lamina $AECD$, which is of weight W, is next suspended by two vertical strings attached at A and D and is at rest with AD horizontal. Find the tension in each string.   (LON)

4 Prove, by integration, that the centre of mass of a uniform solid right circular cone, of vertical height h and base radius r, is at a distance $3h/4$ from the vertex of the cone.

   Such a cone is joined to a uniform solid right circular cylinder, of height h, base radius r and made of the same material so that the plane base of the cone coincides with a plane face of the cylinder. Find the distance of the centre of mass of the composite solid from the centre of the base of the cone.

   When the composite solid hangs in equilibrium from a point A on the circumference of the base of the cone, the line joining A to the vertex of the cone is horizontal. Show that $4r=h\sqrt{5}$.   (LON)

5 The centre of mass of a thin uniform semicircular lamina of radius a lies on the axis of symmetry, distance $\dfrac{4a}{3\pi}$ from the mid-point, O, of the straight edge. A semicircular segment of radius b (where $b<a$), the mid-point of whose straight edge is also O, is removed from the lamina. What is the position of the centre of mass of the remnant?   (NI)

6 Prove that the centre of gravity of a uniform circular arc subtending an angle 2θ at the centre of a circle of radius a is $(a\sin\theta)/\theta$ from the centre. Deduce that the centre of gravity of a uniform sector bounded by that arc and the radii to its extremities is $(\frac{2}{3}a\sin\theta)/\theta$ from the centre.

   Show also that the centre of gravity of a segment of a circular lamina cut off by a chord subtending a right angle at the centre of the circle is $\frac{2}{3}\sqrt{2}a/(\pi-2)$ from the centre.   (WJEC)

Work, Energy and Power

1 A truck has an engine which develops 24 kW at the truck's maximum speed, on a level road, of 25 m s^{-1}. Calculate the total resistance to motion at this speed.   (LON)

2 An elastic string, of natural length 2 m and modulus of elasticity 6 N, is stretched on a horizontal table by applying a force of magnitude 2 N. Find the extension of the string. Find also the work done in stretching the string from its natural length.   (LON)

3 A car of mass 100 kg moves with its engine shut off down a slope of inclination α, where $\sin\alpha=1/20$, at a steady speed of 15 m s^{-1}. Find the resistance, in newtons, to the motion of the car. Calculate the power delivered by the engine when the car ascends the same inclination at the same steady speed, assuming that resistance to motion is unchanged. (Take g as 10 m s^{-2}.)   (LON)

4 A car of mass 1500 kg tows a caravan of mass 500 kg. The car and caravan move on a straight level road with the engine of the car exerting a constant pull of 4100 N. Given that there are frictional resistances of 800 N on the car and 300 N on the caravan, find the magnitude of

(a) the acceleration of the car,

(b) the tension in the tow bar between the car and the caravan.

The car and caravan then go up a straight road which is inclined at an angle $\sin^{-1}(1/8)$ to the horizontal, the frictional resistances on the car and caravan remaining unchanged. The speed of the car and caravan increases from 10 m s^{-1} to 20 m s^{-1} in 16 seconds with the engine of the car now exerting a constant pull of P newtons. Find the value of P and the rate at which the engine is working when the speed is 15 m s^{-1}.

When the speed is 20 m s^{-1}, the tow-bar breaks. Find, to the nearest 10 m, the further distance travelled by the caravan before it comes momentarily to rest. (Take g as 10 m s^{-2}.) (LON)

5 A pump delivers 3 m^3 of water per minute through a nozzle in a pipe of cross-sectional area 0.002 m^2. Find the average speed of the water in m s^{-1}, and deduce that the water is given 15 625 J of kinetic energy per second.

Given that the pump also raises the water through a height of 10 m, find, in kW, the effective power of the pump.

Find, in N, the magnitude of the force which the water would exert on a flat surface perpendicular to the jet at the nozzle, assuming that the water is reduced to rest immediately on impact. (1 m^3 of water has mass 10^3 kg.) (Take g as 10 m s^{-2}.) (LON)

6

A dog of mass m dives off the stern of a punt of mass M, which was previously at rest. Immediately after jumping off, the dog has a horizontal speed of V (as seen by an observer on the bank). Calculate the speed with which the punt begins to move.

Prove that the total kinetic energy of the dog and the punt is then $\dfrac{m(M+m)V^2}{2M}$. (O & C SMP)

Relative Velocity

1 A cyclist is travelling along a straight road at 20 kilometres per hour in a direction $\theta°$ west of south where $\tan\theta = \dfrac{3}{4}$. A tractor is travelling at 22 kilometres per hour due west along a second straight road running from east to west. Both the cycle and the tractor are approaching an intersection of the two roads. When the cyclist passes the intersection the tractor is approaching it and is 5 kilometres away. Determine the time after the cyclist passes the intersection when the cyclist and tractor are nearest to each other and find their distance apart at that time. (NI)

2 A ship is steaming at 15 knots due east, while the wind speed is 20 knots from due north. Find the magnitude and the direction, to the nearest degree, of the wind velocity relative to the ship. Find also the course, between east and south, along which the ship would have to steer at 16 knots for the wind velocity relative to the ship to be at right angles to the course of the ship. Obtain the magnitude of the velocity of the wind relative to the ship in this case. (LON)

3 A river flows at a constant speed of 5 m s⁻¹ between straight parallel banks which are 240 m apart. A boat crosses the river, travelling relative to the water at a constant speed of 12 m s⁻¹. A man cycles at a constant speed of 4 m s⁻¹ along the edge of one bank of the river in the direction opposite to the direction of flow of the river. At the instant when the boat leaves a point O on the opposite bank, the cyclist is 80 m downstream of O. The boat is steered relative to the water in a direction perpendicular to the banks. Taking **i** and **j** to be perpendicular horizontal unit vectors downstream and across the river from O respectively, express, in terms of **i** and **j**, the velocities and the position vectors relative to O of the boat and the cyclist t seconds after the boat leaves O. Hence, or otherwise, calculate the time when the distance between the boat and the cyclist is least, giving this least distance.

If, instead, the boat were to be steered so that it crosses the river from O to a point on the other bank directly opposite to O, show that this crossing would take approximately 22 seconds. (LON)

4 The vectors **i** and **j** are unit vectors in the directions East and North respectively. A man who walks with constant speed 6 km h⁻¹ due North and to him the wind appears to have a velocity $u_1(\sqrt{3}\mathbf{i} - 3\mathbf{j})$ km h⁻¹. Without changing speed the man alters his course so that he is walking in the direction of the vector $\left(-\dfrac{\sqrt{3}}{2}\mathbf{i} + \dfrac{1}{2}\mathbf{j}\right)$ and the velocity of the wind now appears to him to be $u_2\mathbf{i}$ km h⁻¹. Find u_1 and u_2. Find also the actual velocity of the wind. (LON)

5 At a given instant two ships X and Y are at a distance 10 km apart with Y due East of X. X moves with a constant speed of 9 km h⁻¹ in the direction N $(\tan^{-1}\frac{12}{5})$ E and Y moves with a constant speed of 12 km h⁻¹ in the direction N $(\tan^{-1}\frac{5}{12})$ W. Show that, relative to Y, X moves with speed 15 km h⁻¹ in a direction E $(\tan^{-1}\frac{33}{56})$ S.

Find the minimum value of the distance between the two ships. How long is it before this occurs? (N.B. $33^2 + 56^2 = 65^2$.) (WJEC)

Vectors

1 The vectors **a** and **b** are such that $|\mathbf{a}| = 3$ and $|\mathbf{b}| = 5$. Calculate the magnitude of **a** + **b** given that the angle between **a** and **b** is $\pi/3$. (LON)

2 Vectors **i** and **j** are at right angles to each other and are of unit length. Express $\mathbf{v} = 4\mathbf{i} + 7\mathbf{j}$ in terms of the vectors **a** and **b** where $\mathbf{a} = -\mathbf{i} + 2\mathbf{j}$ and $\mathbf{b} = -4\mathbf{i} - 2\mathbf{j}$.

Verify that $\mathbf{c} = -3\mathbf{i} + \mathbf{j}$ is equally inclined to **a** and **b** and find the resolute of **v** in the direction of **c**. (WJEC)

3 Find a vector equation for the plane passing through the points A, B, C with position vectors $(\mathbf{i} - \mathbf{k})$, $(2\mathbf{i} + \mathbf{j} + 3\mathbf{k})$, $(3\mathbf{i} + 2\mathbf{j} + \mathbf{k})$ respectively. By finding the scalar product $\overrightarrow{AC} \cdot \overrightarrow{BC}$, or otherwise, show that angle ACB is a right angle. (LON)

4 With respect to an origin O the points A and B have position vectors $3a\mathbf{i} - 2a\mathbf{k}$ and $a\mathbf{i} + 2a\mathbf{j} + a\mathbf{k}$ respectively, where $a > 0$. Find

(a) a vector equation of the line AB,
(b) Cartesian equations of the line AB,
(c) the cosine of the acute angle between the line OA and the line AB. (LON)

5 Prove that the two lines with equations

$$\mathbf{r} = \begin{pmatrix} 0 \\ 2 \\ -3 \end{pmatrix} + s \begin{pmatrix} 1 \\ -1 \\ -1 \end{pmatrix} \quad \text{and} \quad \mathbf{r} = \begin{pmatrix} -1 \\ 6 \\ -1 \end{pmatrix} + t \begin{pmatrix} 2 \\ 1 \\ -1 \end{pmatrix}$$

have a point in common. Find, in parametric form, an equation for the plane containing the two lines. (O & C SMP)

6 The position vectors of the vertices of a triangle ABC referred to a given origin are $\mathbf{a}, \mathbf{b}, \mathbf{c}$. P is a point on AB such that $AP/PB = \frac{1}{2}$, Q is a point on AC such that $AQ/QC = 2$, and R is a point on PQ such that $PR/RQ = 2$. Prove that the position vector of R is $\frac{4}{9}\mathbf{a} + \frac{1}{9}\mathbf{b} + \frac{4}{9}\mathbf{c}$.

Prove that R lies on the median BM of the triangle ABC, and state the value of BR/RM. (O & C SMP)

7 The position vectors, with respect to a fixed origin, of the points L, M and N are given by $\mathbf{l}$, $\mathbf{m}$ and $\mathbf{n}$ respectively, where

$$\mathbf{l} = a(\mathbf{i}+\mathbf{j}+\mathbf{k}), \quad \mathbf{m} = a(2\mathbf{i}+\mathbf{j}), \quad \mathbf{n} = a(\mathbf{j}+4\mathbf{k})$$

and a is a non-zero constant. Show that the unit vector $\mathbf{j}$ is perpendicular to the plane of the triangle LMN.

Find a vector perpendicular to both $\mathbf{j}$ and $(\mathbf{m}-\mathbf{n})$, and hence, or otherwise, obtain a vector equation of that perpendicular bisector of MN which lies in the plane LMN.

Verify that the point K with position vector $a(5\mathbf{i}+\mathbf{j}+4\mathbf{k})$ lies on this bisector and show that K is equidistant from L, M and N. (LON)

Projectiles

1 A projectile is fired from a point O. The speed of the projectile when at its greatest height h above O is $\sqrt{(2/5)}$ times its speed when at height $h/2$ above O. Show that the initial angle which the velocity of the projectile makes with the horizontal is $\pi/3$. (LON)

2 A stone thrown upwards from the top of a vertical cliff 56 m high falls into the sea 4 seconds later, 32 m from the foot of the cliff. Find the speed and direction of projection. (The stone moves in a vertical plane perpendicular to the cliff.)

A second stone is thrown at the same time, in the same vertical plane, at the same speed and at the same angle to the horizontal, but downwards. Find how long it will take to reach the sea and the distance between the points of entry of the stones into the water. (Take g to be 10 m/s^2.) (LON)

3 A particle is projected with speed u and at an angle of elevation α from a point O. Show that, at time t after projection, the position vector of the particle relative to O is $\mathbf{r}$, where $\mathbf{r} = (u\cos\alpha)t\,\mathbf{i} + [(u\sin\alpha)t - \frac{1}{2}gt^2]\mathbf{j}$ and $\mathbf{i}$ and $\mathbf{j}$ are unit vectors directed horizontally and vertically upwards respectively.

Given that $u = 40$ m s^{-1} and that the particle strikes a target A on the same horizontal level as O, where $OA = 60$ m, find the least possible time, to the nearest tenth of a second, that elapses before the particle hits the target A. (Take g as 10 m s^{-2}.) (LON)

4 A particle is projected with velocity u at an elevation α. Show that its horizontal range through the point of projection is $u^2\sin 2\alpha/g$.

When projected at an elevation of $\tan^{-1}\frac{3}{4}$ a projectile falls 20 metres short of a target in a horizontal plane through the point of projection. When the elevation is $\tan^{-1}1$ the projectile overshoots the target by 30 metres. Show that the target is at a horizontal distance of 1220 metres from the point of projection and find the correct elevations so that the projectile hits the target. (WJEC)

5 A particle is projected with speed u at an angle of elevation α from a point A on a level plane. At time t, the particle passes through a point P which is vertically above a point Q on the plane. On passing through P the particle is moving in a direction which is at an angle of elevation θ. Given that the angle $PAQ = \beta$, express $\tan\beta$ and $\tan\theta$ in terms of u, α, g and t. Hence, or otherwise, show that $2\tan\beta = \tan\alpha + \tan\theta$.

A batsman hits a ball from a point on level ground so that the ball starts to move with speed u m s^{-1} at an angle of elevation $\tan^{-1}(4/5)$. The ball is caught by a fielder at a height of 2 m above the ground when it is moving downwards at an angle $\tan^{-1}(2/3)$ to the horizontal. Calculate the distance between the batsman and the fielder. (LON)

Circular Motion

1   A particle P, moving on the smooth inside surface of a fixed spherical bowl, of radius r, describes a horizontal circle at depth $r/2$ below the centre of the bowl. Show that P takes time $\pi\sqrt{(2r/g)}$ to complete one revolution of its circular path.   (LON)

2   A particle of mass m is connected by an inextensible light string of length l to a fixed point on a smooth horizontal table. The string breaks when subject to a tension whose magnitude exceeds mg. Find the maximum number of revolutions per second that the particle can make without breaking the string.   (LON)

3   Two particles A and B each of mass m are connected by a light inextensible string which passes through a hole in a smooth horizontal table, A being on the table, B hanging underneath. How many revolutions per second in a circle of $\frac{1}{2}$ m radius would A have to make on the table in order that B remain at rest?   (NI)

4   A light elastic string AB of natural length a and modulus of elasticity mg is joined to a light inextensible string BC of length a. The ends A and C of the string are fastened to two fixed points with A vertically above C and $AC = a$. A particle of mass m is fixed at B and rotates with speed v in a horizontal circle. Show that, if AB makes an angle $\frac{1}{6}\pi$ with the downward vertical, $v^2 = \frac{1}{2}\sqrt{3ag}$, and find the tension in BC.   (WJEC)

5   (a)   An engine governor consists of two equal small spheres A and B, each of mass m, connected by two rods smoothly hinged at a fixed point O and by two other rods to a small sleeve C of mass $2m$ which can slide on a smooth vertical shaft OO'. The rods are light and of the same length. If the system rotates uniformly about OO' with C a constant depth h below O, show that the period of revolution is $2\pi\sqrt{(h/6g)}$.

   (b)   A particle, attached to a fixed point O by a light inextensible string of length l, moves in a vertical plane through O with the string taut throughout the motion. Show that the speed u of the particle when it is vertically below O cannot be between $\sqrt{(2gl)}$ and $\sqrt{(5gl)}$.

      Suppose that the particle is released from rest from a point which is at a distance l from O and on the same horizontal level as O, and that at a depth x ($<l$) vertically below O a thin peg P has been fixed with its axis horizontal and perpendicular to the plane in which motion takes place. If, during that motion, the particle completes a vertical circle about P, show that x must be at least $(3/5)l$.   (WJEC)

Motion of a Particle

1   The position vector of a particle P at time t is given by $\mathbf{r} = t\sin t\,\mathbf{i} + t\cos t\,\mathbf{j}$ where $\mathbf{i}$ and $\mathbf{j}$ are constant orthogonal unit vectors. Calculate the velocity and acceleration vectors of P, and show that its speed is $\sqrt{(1+t^2)}$.   (WJEC)

2   (a)   State Newton's laws of motion.

   (b)   Derive, from first principles, the constant acceleration formula $v^2 = u^2 + 2as$, where a particle has changed its speed from u to v in travelling a distance s under a constant acceleration a.

   (c)   A particle moves along a horizontal straight line, with a constant acceleration. The particle leaves point A on the line travelling towards the right with speed 4 m s^{-1} and 6 s later reaches point B (to the left of A). At point B the particle has a speed of 8 m s^{-1} and is *travelling towards the left*.

      (i)   What is the magnitude and direction of the acceleration?
      (ii) What is the total distance travelled by the particle?
      (iii) What is the distance between the points A and B?   (NI)

3 A particle of mass $2m$ is on a plane inclined at an angle $\tan^{-1}\frac{3}{4}$ to the horizontal. The particle is attached to one end of a light inextensible string. This string runs parallel to a line of greatest slope of the plane, passes over a small smooth pulley at the top of the plane and then hangs vertically carrying a particle of mass $3m$ at its other end. The system is released from rest with the string taut. Find the acceleration of each particle and the tension in the string when the particles are moving freely, given that

(a) the plane is smooth,
(b) the plane is rough and the coefficient of friction between the particle and the plane is $\frac{1}{4}$.   (LON)

4 A particle of mass m moves along the x-axis under the action of a force F which is directed along the positive x-axis and at time t is given by $F = mk\sin\omega t$, where ω and k are positive constants. At time $t = 0$ the particle is projected from the origin O with speed u along the positive x-axis. Find the acceleration, velocity and displacement from O at time t.
            (LON)

5 **(i)** The brakes of a train, which is travelling at $108\ \text{km h}^{-1}$, are applied as the train passes point A. The brakes produce a constant retardation of magnitude $3f\ \text{m s}^{-2}$ until the speed of the train is reduced to $36\ \text{km h}^{-1}$. The train travels at this speed for a distance and is then uniformly accelerated at $f\ \text{m s}^{-2}$ until it again reaches a speed of $108\ \text{km h}^{-1}$ as it passes point B. The time taken by the train in travelling from A to B, a distance of $4\ \text{km}$, is 4 minutes. Sketch the speed/time graph for this motion and hence calculate

(a) the value of f,
(b) the distance travelled at $36\ \text{km h}^{-1}$.

(ii) A particle moves in a straight line with variable acceleration $\dfrac{k}{1+v}\ \text{m s}^{-2}$, where k is a constant and $v\ \text{m s}^{-1}$ is the speed of the particle when it has travelled a distance x m. Find the distance moved by the particle as its speed increases from 0 to $u\ \text{m s}^{-1}$.   (LON)

6 Show that the acceleration dv/dt of a particle moving in a straight line can be written, in terms of its velocity v and its displacement x from a point of the line, in the form $v\,dv/dx$.

At a distance x km from the centre of the earth the gravitational acceleration in km/s^2 is given by the formula c/x^2 where $c = 4 \times 10^5$. If a lunar vehicle $10\,000$ km from the centre of the earth is moving directly away from it at a speed of 10 km/s, at what distance will its speed be half that value?   (O & C SMP)

7 A particle moving on a straight line with speed v experiences a retardation of magnitude $be^{v/u}$, where b and u are constants. Given that the particle is travelling with speed u at time $t = 0$, show that the time t_1 for the speed to decrease to $\frac{1}{2}u$ is given by $bt_1 = u(e^{-\frac{1}{2}} - e^{-1})$. Find the further time t_2 for the particle to come to rest. Deduce that $t_2/t_1 = e^{\frac{1}{2}}$.

Find, in terms of b and u, an expression for the distance travelled in decelerating from speed u to rest.
            (LON)

8 A swimmer crosses a straight river of width 40 metres which has a uniform current. In crossing the river as quickly as possible he takes 90 seconds and is carried 30 metres downstream to a point P. Show that the swimmer, starting from P, can swim back in a straight line to his original starting point.

Deduce that the time t in *minutes* that it would take him to do so satisfies $28t^2 - 108t - 225 = 0$. Verify that this equation has a root $75/14$.

Find the angle between the line of the river bank and the direction in which he would have to head.
            (WJEC)

9 A particle P moves in a straight line and experiences a retardation of $0.01\ v^3\ \text{m s}^{-2}$, where $v\ \text{m s}^{-1}$ is the speed of P. Given that P passes through a point O with speed $10\ \text{m s}^{-1}$, show that, when it is a distance 10 m from O, its speed is $5\ \text{m s}^{-1}$.

Find the time taken for the speed of P to be reduced from $10\ \text{m s}^{-1}$ to $5\ \text{m s}^{-1}$.   (LON)

10 **(i)** A particle starts with speed 20 m s^{-1} and moves in a straight line. The particle is subjected to a resistance which produces a retardation which is initially 5 m s^{-2} and which increases uniformly with the distance moved, having a value of 11 m s^{-2} when the particle has moved a distance of 12 m. Given that the particle has speed v m s^{-1} when it has moved a distance of x m, show that, while the particle is in motion,

$$v\frac{dv}{dx} = -(5+\tfrac{1}{2}x)$$

Hence, or otherwise, calculate the distance moved by the particle in coming to rest.

(ii) At time t seconds, where $t>0$, the position vector of particle A with respect to a fixed origin O is $(\mathbf{i}t-\mathbf{j}\cos t+\mathbf{k}\sin t)$ m and the position vector of particle B with respect to O is $(\mathbf{i}+\mathbf{j}e^{-t}-\mathbf{k}e^{-t})$ m. Find the least value of t for which the velocities of A and B are perpendicular and show that, at this instant, the accelerations of the particles are parallel. (LON)

Statics

1 A uniform ladder of weight W and length $2l$ rests with its upper end against a smooth vertical wall. Its lower end stands on rough horizontal ground, and the coefficient of friction between the ladder and the ground is $\frac{1}{2}$. The ladder is in limiting equilibrium. Find the angle the ladder makes with the horizontal. Find also the reaction at the wall.

A man of weight W climbs the ladder. Find how far up it he can go before the ladder will slip.

Find also how far up the ladder he can go when a load of weight W is placed on the foot of the ladder. (LON)

2 A uniform triangular lamina ABC has mass m and each side is of length $2a$. The lamina rests in a fixed vertical plane with A on a rough horizontal table and the side AC is vertical. Equilibrium is maintained by a force of magnitude T which acts along the side BC. Show that $T=mg/3$. Find the magnitude of the reaction at A.

Show also that equilibrium is possible only when $\mu\geqslant\sqrt{3}/5$, where μ is the coefficient of friction between the lamina and the table. (LON)

3 Two equal light rods AB, BC of length $2a$, rigidly joined at right angles to each other at B, are placed astride a fixed rough horizontal circular cylinder of radius a so that they are equally inclined to the vertical, and the plane formed by them is perpendicular to the axis of the cylinder. Weights w and W $(W>w)$ are hung from A and C respectively, W being adjusted so that slipping is about to take place. Find the normal reactions between the cylinder and the rods and prove that $W/w=(1+\mu+\mu^2)/(1-\mu+\mu^2)$, where μ is the coefficient of friction between the cylinder and the rods.

What is the condition on μ for such equilibrium to be possible? (WJEC)

4 A straight uniform rod $CDEF$, of length $6a$ and mass M, is suspended from a fixed point A by two light inextensible strings AD and AE, each of length $2a$, where D and E are the points of trisection of the rod. A particle of mass m is attached to the rod at a point distant x from the end C and the system hangs in equilibrium. Given that the tension in the string AD is twice the tension in the string AE, show that the rod is inclined to the horizontal at an angle θ, where $\tan\theta=1/(3\sqrt{3})$. Hence find, in terms of $M, m, a,$ and g, **(a)** the tensions in the strings **(b)** the distance x. (LON)

5 **(i)** A uniform rod PQ, of mass m and length $4a$, is smoothly pivoted to a fixed point at its end P, and the end Q is attached by a light spring, obeying Hooke's Law and of unstretched length $4a$, to a fixed point S at a distance $3a$ vertically above P. If the rod is in equilibrium in a horizontal position, find the tension in the spring and show that the force required to double its length is $10\,mg/3$.

(ii) The ends A and E of a light inextensible string $ABCDE$, where $AB=DE$ and $BC=CD$, are attached to two fixed points on the same horizontal level. Particles of weight $W, 3W, W$ are attached to B, C, D respectively and hang in equilibrium with AB, BC making acute angles θ, ϕ respectively with the horizontal. Calculate, in terms of W, θ, ϕ, the tensions in the parts AB and BC of the string and show that $5\tan\phi=3\tan\theta$. (LON)

Coplanar Forces

1   Forces of magnitude 2N, 4N and 4N act in the sense indicated by the letters along the sides *AB, BC, CA* respectively of an equilateral triangle *ABC* of side 2 m. Find the magnitude and direction of their resultant and the point where its line of action cuts *AC* produced.

   This system of forces is to be reduced to equilibrium by the addition of a couple in the plane *ABC* and a force which acts through *A*. Find the magnitude and direction of the force and the magnitude and sense of the couple.

(LON)

2

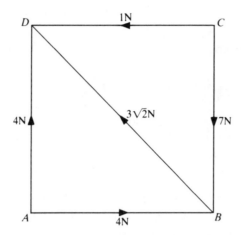

Forces of magnitude 4, 7, 1, 4 and $3\sqrt{2}$ newtons act along the sides *AB, CB, CD, AD* and the diagonal *BD* respectively of a square *ABCD* of side 0.1 m, as shown in the diagram. Show that the force system is equivalent to a couple and find the moment of this couple.

(LON)

3   In a rectangular *ABCD*,   *AB* = *DC* = 4 m   and   *AD* = *BC* = 3 m.   Forces of magnitude 3, 5, 4, 6, *P*, *Q* newtons act along *AB, BC, CD, DA, BD, AC* respectively in the directions indicated by the order of the letters. Show that this system of forces cannot be in equilibrium.

   (i)   If the system reduces to a couple, show that $Q = 7P$.

   (ii)   If the system reduces to a single resultant force acting through *B*, show that $Q = 15$.   Given also $P = 10$, find the magnitude of the resultant force.

(LON)

4   A uniform hexagonal lamina *ABCDEF* has weight *W*. The lamina is smoothly pivoted at *B*. It is kept in equilibrium in a vertical plane with *AB* horizontal and the vertices *C, D, E* and *F* above *AB* by a force of magnitude *P* acting at *D* in the direction *AD*. Find *P* and the magnitude and direction of the force exerted on the lamina by the pivot.

   The force at *D* is reversed in direction whilst retaining its magnitude. Find the weight which must be attached at *C* to keep the lamina in this position of equilibrium.

(LON)

Answers

The authors take sole responsibility for the accuracy of the answers given.

The University of Cambridge Local Examinations Syndicate bears no responsibility for the example answers to questions taken from its part question papers which are contained in this publication.

The University of London School Examinations Board accepts no responsibility whatsoever for the accuracy or method of working in the answers given.

Revision Exercises A

Algebra

1 5 or -6 **2** (a) 4×4 (b) 10×2.5; $\dfrac{50}{9} \times \dfrac{25}{8}$; $\dfrac{29}{2} \times \dfrac{58}{25}$

3 (a) $x = \dfrac{4-7y}{5y-3}$ (b) $x = \dfrac{3y+4}{5y+7}$ (c) $x = \sqrt{(y-2)} - 1$

4 (a) 6 (b) -3 (c) $\log_2 6 = 2.585$ (d) 0 (e) No solution (f) 3

5 (a) ± 2 (b) ± 64 (c) ± 4 (d) ± 32 (e) ± 8 (f) $\pm\frac{1}{4}$

6 (a) 3 (b) 3 (c) $\sqrt{6} = 2.245$ (d) $\log_2 6 = 2.585$

7 (a) 3 (b) $2^8 = 256$ (c) $\frac{1}{3}$ (d) 64

8 (a) 2 (b) 2 (c) ± 4 (d) ± 2

9 (a) 1 (b) No solution (c) $\log_3 4 = 1.262$

10 (a) $2^4 = 16$ (b) $\sqrt[4]{2}$ (c) 2 or $\sqrt[3]{2}$

11 (a) $x = 1$ (b) 9 or $3\sqrt{3}$

12 (a) $\dfrac{1}{x-1} - \dfrac{1}{x+1}$ (b) $\dfrac{1}{x-1} + \dfrac{1}{x+1}$ (c) $2 + \dfrac{1}{x-1} - \dfrac{1}{x+1}$ (d) $\dfrac{2}{x} - \dfrac{2}{x+1}$ (e) $\dfrac{2}{x-1} - \dfrac{2}{x}$

(f) $\dfrac{1}{x+1} + \dfrac{1}{x-1} - \dfrac{2}{x}$

13

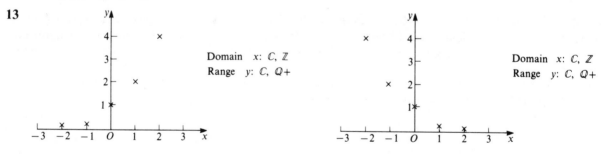

Domain x: C, $\mathbb{Z}$
Range y: C, $\mathbb{Q}+$

Domain x: C, $\mathbb{Z}$
Range y: C, $\mathbb{Q}+$

Functions

1
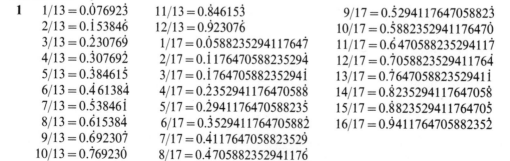

$1/13 = 0.\dot{0}7692\dot{3}$	$11/13 = 0.\dot{8}4615\dot{3}$	$9/17 = 0.\dot{5}294117647058 82\dot{3}$
$2/13 = 0.\dot{1}5384\dot{6}$	$12/13 = 0.\dot{9}2307\dot{6}$	$10/17 = 0.\dot{5}882352941176 47\dot{0}$
$3/13 = 0.\dot{2}3076\dot{9}$	$1/17 = 0.\dot{0}588235294117 64\dot{7}$	$11/17 = 0.\dot{6}470588235294 11\dot{7}$
$4/13 = 0.\dot{3}0769\dot{2}$	$2/17 = 0.\dot{1}176470588235 29\dot{4}$	$12/17 = 0.\dot{7}058823529411 76\dot{4}$
$5/13 = 0.\dot{3}8461\dot{5}$	$3/17 = 0.\dot{1}764705882352 94\dot{1}$	$13/17 = 0.\dot{7}647058823529 41\dot{1}$
$6/13 = 0.\dot{4}6138\dot{4}$	$4/17 = 0.\dot{2}352941176470 58\dot{8}$	$14/17 = 0.\dot{8}235294117647 05\dot{8}$
$7/13 = 0.\dot{5}3846\dot{1}$	$5/17 = 0.\dot{2}941176470588 23\dot{5}$	$15/17 = 0.\dot{8}823529411764 70\dot{5}$
$8/13 = 0.\dot{6}1538\dot{4}$	$6/17 = 0.\dot{3}529411764705 88\dot{2}$	$16/17 = 0.\dot{9}411764705882 35\dot{2}$
$9/13 = 0.\dot{6}9230\dot{7}$	$7/17 = 0.\dot{4}117647058823 52\dot{9}$	
$10/13 = 0.\dot{7}6923\dot{0}$	$8/17 = 0.\dot{4}705882352941 17\dot{6}$	

2 2; 2.59374246; 2.704813829; 2.716923932; 2.718145927; 2.718268237; 2.718280469; limit $e = 2.718281828 \ldots$.

3 $a_2 = 2.5$; $a_3 = 2.\dot{6}$; $a_4 = 2.708\dot{3}$; $a_5 = 2.71\dot{6}$; $a_6 = 2.7180\dot{5}$; $a_7 = 2.718253969 \ldots$; $a_{10} \simeq e$ and $\lim\limits_{n \to \infty} a_n = e$

4 $(x + \frac{1}{2})^2 = x^2 + x + \frac{1}{4}$ by expansion

5 (a) $x \longrightarrow 2x + 1$ (b) $x \longrightarrow 4 - x$ (c) $x \longrightarrow 2^{x-1}$ (d) $x \longrightarrow x^2 + 1$ (e) $x \longrightarrow x$th prime or $x \longrightarrow \dfrac{x^2 - x + 4}{2}$

6 (a) $x \longrightarrow \dfrac{x + 1}{2}$ (b) $x \longrightarrow 4 - x$ (c) $x \longrightarrow 1 + \log_2 y$ (d) $x \longrightarrow \sqrt{x - 1}$ (e) prime $\longrightarrow$ its number or $x \longrightarrow 1/2 + \sqrt{2y - 15/4}$

7 (a) $AB(x) = 2(4 - x) - 1 = 7 - 2x$ (b) $BA(x) = 4 - (2x - 1) = 5 - 2x$ (c) $AC(x) = 2 \times 2^{x-1} - 1 = 2^x - 1$ (d) $CA(x) = 2^{2x-1}$ (e) $AD(x) = 2(x^2 + 1) - 1 = 2x^2 + 1$ (f) $DA(x) = (2x - 1)^2 + 1$ (g) $CD(x) = 2^{x^2}$

8 (a) $fg(x) = 6x - 3$ (b) $gf(x) = 6x + 1$ (c) $ff(x) = 4x + 3$ (d) $gg(x) = 9x - 8$ (e) $f^{-1}(x) = \dfrac{x - 1}{2}$ (f) $g^{-1}(x) = \dfrac{x + 2}{3}$ (g) $f^{-1}g^{-1}(x) = \dfrac{x - 1}{6}$ (h) $g^{-1}f^{-1}(x) = \dfrac{x + 3}{6}$

9

	I	f	g	fg	gf	h
I	I	f	g	fg	gf	h
f	f	I	fg	g	h	gf
g	g	gf	I	h	f	fg
fg	fg	h	f	gf	I	g
gf	gf	g	h	I	fg	f
h	h	fg	gf	f	g	I

$$h = fgf = \frac{x}{x - 1} = gfg$$

10 The set of six functions is closed under the operation of combining functions as the combination of any two gives another member of the set, i.e. no further (different) functions are introduced.

11 (a) $f(gh) = 6x^2 - 5 = (fg)h$
(b) $f(hg) = (fh)g = 18x^2 - 24x + 7$
(c) The above results are true in general for all functions.

12 Only fg and gf.

13 2 and 3; (a) $6x^2 - 5x + 1 = 0$ (b) $x^2 - 13x + 36 = 0$

14 (a) $3x^2 + 5x + 1 = 0$ (b) $x^2 - 19x + 9 = 0$

Co-ordinates

2 In our tower block of 4 floors, the first floor rooms number 11, 12, 13, 14, 15, 16, the second floor rooms are 21, 22, 23, 24, 25, 26 and the third floor 31, 32, 33, 34, 35, 36, the first number referring to the floor on which the room is situated.

3 $y = 2x + 1$ (a) $y = -2x + 1$ (b) $y = -2x - 1$ (c) $2y = x - 1$
(d) $2y = x + 1$ (e) $2y = -x + 1$ (f) $y = 2x - 1$ (g) $2y = -x - 1$

4 **(a)** $(3,4)$ **(b)** $(-1,2)$ **(c)** $(1,-2)$ **(d)** $(-3,-4)$ **(e)** $90°$
(f) $45°$ **(g)** 2.5 sq units **(h)** $2y=x+5$

5 **(a)** $4y=3x$ **(b)** $5y+3x=18$ **(c)** $G=(\frac{8}{3},2)$ **(d)** G satisfies $y+6x=18$
(e) Q is $(2,0)$ **(f)** R is $(\frac{10}{3},4)$ **(g)** G satisfies $y=3x-6$ **(h)** $1:2$

6 **(a)** $x\geqslant0;$ $2y\geqslant x;$ $x+2y\leqslant8$ **(b)** $2y<x;$ $y>0;$ $x+2y<8$

(c) 8 sq units **(d)** 8 sq units **(e)** $\dfrac{8\sqrt{5}}{5}=3.58$ **(f)** 0.8

7 **(a)** $y=2x-1$ **(b)** $y=-2x+5$ **(c)** W is $(1.5,2)$ **(d)** $WO=WA=WB=2.5$ and W is the
circumcentre of $\triangle OAB$. **(e)** $(4,-3)$ **(f)** $B=(4,2)$

Differentiation

1 **(a)** $10x-4$ **(b)** $4x^3-\dfrac{6}{x^3}$ **(c)** $2x+3$ **(d)** $3x^2+8x+3$

2 **(a)** $5(x-1)(x+1)(x+4)^2$ **(b)** $\dfrac{7}{(3-x)^2}$ **(c)** $\dfrac{\cos x}{(1+x)}-\dfrac{\sin x}{(1+x)^2}$
(d) $x\sec^2 x+\tan x$ **(e)** $5(8x-1)(4x^2-x)^4$ **(f)** $-4\cos^3 x\sin x$

3 **(a)** $y=-\frac{1}{2}$ **(b)** $7x+y+2=0$ **5** Min $(1,\frac{1}{3})$; Max $(-1,3)$

6 **(b)** $\dfrac{y\sin x-2x\sin y}{\cos x+x^2\cos y}$ **7** $3x-2y-3=0$ **8** $\dfrac{dy}{dx}=\dfrac{2t-1}{3t^2}$; Min $(\frac{1}{8},-\frac{1}{4})$ **10** $\dfrac{1}{12\pi}$ cm s^{-1}

12 $y=2x+11;$ $27y=54x-58$ **13** 6 cm **14** 9%

Integration

1 **(a)** $\frac{1}{3}x^3+x^2+3x+C$ **(b)** $-\frac{2}{3}x^{-3}+C$ **(c)** $\frac{1}{5}x^5-2x^3+9x+C$ **(d)** $\frac{1}{2}x^2+4x+\dfrac{2}{x}+C$

2 **(a)** $\frac{1}{2}x^2-5x+C$ **(b)** $\frac{1}{2}x^2+\frac{3}{2}x-10x+C$ **(c)** $\frac{2}{3}x^{\frac{3}{2}}+C$ **(d)** $\frac{1}{2}x^2-2x^{\frac{1}{2}}+C$

3 **(a)** $\frac{1}{4}x^4+\dfrac{1}{2x^2}+C$ **(b)** $\frac{3}{4}x^{\frac{4}{3}}+\frac{3}{2}x^{\frac{2}{3}}+C$ **(c)** $\frac{1}{3}x^3-x-\dfrac{1}{x}+C$ **(d)** $\frac{2}{5}x^{\frac{5}{2}}+4x^{\frac{1}{2}}+C$

(e) $\frac{2}{7}x^{\frac{7}{2}}-\frac{2}{3}x^{\frac{3}{2}}+C$

4 $y=x^3-5x+3$ **5** $y=\tan x+\sin x+3$ **6** $v=6-4\cos t$

7 **(a)** $1\frac{5}{6}$ **(b)** $7\frac{11}{15}$ **(c)** $\frac{1}{3}$ **(d)** $2\frac{1}{4}$ **8** $1\frac{5}{6}$ **9** $2\frac{2}{3}$ **10** $4\frac{1}{2}$ **11** 5.21 **12** 34.13π

13 6π **14** $\dfrac{\pi}{30}$ **15** **(a)** $\dfrac{\pi}{6}$ **(b)** $\dfrac{\pi}{2}$ **(c)** $3\sin^{-1}\left(\dfrac{x}{2}\right)+C$

Trigonometry

1 **(a)** $AC=t^2+1$ **(b)** $PR=s^2+t^2$

	AB	BC	AC
$t=2$	3	4	5
$t=3$	8	6	10
$t=4$	15	8	17
$t=5$	24	10	26
$t=6$	35	12	37

		PQ	QR	PR
$s=2$	$t=1$	3	4	5
$s=3$	$t=2$	5	12	13
$s=3$	$t=1$	8	6	10
$s=4$	$t=3$	7	24	25
$s=4$	$t=2$	12	16	20
$s=4$	$t=1$	15	8	17

Certain values of t and combinations of s and t give Pythagorean Triples, or multiples of them. Will both systems give them all?

2 **(a)** 12 and 13 **(b)** $8a$ by $3\sqrt{3}a$ where a is the hexagon side.
(c) $4H$ and $3H$ where $H=$ area of hexagon.

(d) The first looks tidier the second more compact, but more difficult to count? **(e)** 16 in the first box, 18 in the second. **(f)** To pack bottles (cylinders) the second method will require a larger box. Milk cartons seem to be appearing in cuboid packs but have you seen the triangular pyramid cartons?

3 **(a)** $\frac{4}{5}$ **(b)** $\frac{3}{4}$ **(c)** $\frac{5}{13}$ **(d)** $\frac{5}{12}$ **(e)** $\frac{56}{65}$ **(f)** $\frac{33}{65}$ **(g)** $\frac{16}{65}$ **(h)** $\frac{63}{65}$ **(i)** $\frac{24}{25}$ **(j)** $\frac{7}{25}$ **(k)** $\frac{24}{7}$ **(l)** $\frac{120}{169}$ **(m)** $\frac{119}{169}$ **(n)** $\frac{120}{119}$ **(o)** 45 degrees **(p)** 22.6 degrees

4 104.5°, 46.5°, 29° **5** 28.1°, 61.9°, 90°

6 $C = 32.4°$, $A = 107.6°$, $BC = 8.92$ cm **7** **(a)** $\tan x$ **(b)** $\sec x$ **(c)** $\sin x$

8 **(a)** 199.5°, 340.5° **(b)** 90°, 41.8°, 138.2° **(c)** 109.5°, 250.5°

9

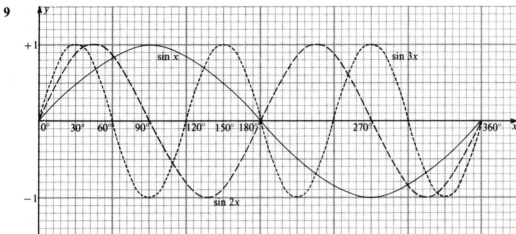

10

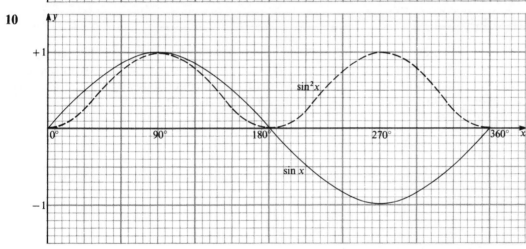

11

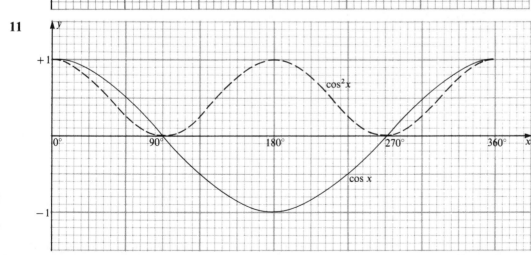

12

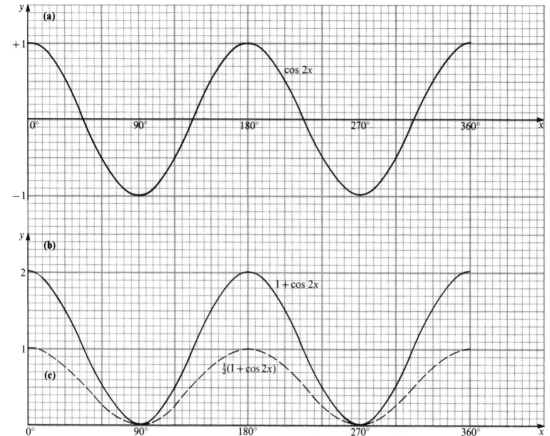

13 (a) 15.9 m **(b)** 218.8 m **(c)** 18.8 m **(d)** 37.6 m **(e)** Inside, other racers in view: outside, the bends are not so tight. **(f)** 39.8 m by 89.8 m. **(g)** To accommodate 60 m dash, usually in centre of course. **(h)** To prevent disorientation of balance which occurs when continuously flying in a circle; the straights give an opportunity to regain balance when flying on a level course.

14 (a) The width (diameter) of the 50p piece is constant and equal to the length from one vertex to the opposite side.

$d = 30 \Rightarrow r = 16.65;$ $d = 21 \Rightarrow r = 11.66;$ $r = 16 \Rightarrow d = 28.8$

(b) $d(10p) = 15.22$ $d(5p) = 8.66$

(c)

	50p	20p	10p	5p	
Diameter	30	21	15.22	8.66	Volumes
Radius	16.65	11.66	8	5	(and weights)
Area	695.2	340.9	175.7	52.9	depend on
Ratio of areas	4	2	1	0.3	thickness
	13.1	6.4	3.32	1	

So, $4:2:1:0.3$ compared with 10p; $13.1:6.44:3.32:1$ compared with 5p.

(d) Seven-sided by design, but curved for slot machines.

Chapter 1

Investigation 1 **(a)** $\dfrac{dy}{dx} \propto 2^x$ **(b)** Both $y = 2^x$ and $y = 3^x$ pass through $(0, 1)$.

For $x > 0$, $3^x > 2^x$; for $x < 0$, $3^x < 2^x$.

$y = a^x$ passes through $(0, 1)$ and its gradient is proportional to a^x.

Investigation 2   **(a)** and **(b)** $\dfrac{dy}{dx} \propto \dfrac{1}{x}$ or $x\dfrac{dy}{dx} = $ constant.

Exercise 1.1

1   **(a)** $2^x \ln 2$ **(b)** $3^x \ln 3$ **(c)** $4^x \ln 4$ **(d)** e^x **(f)** $2e^{2x}$

2   **(a)** $2^x (2^{x^2}) \ln 2$ **(b)** $3x^2 (3^{x^3}) \ln 3$ **(c)** $4x^3 e^{x^4}$ **(d)** $4e^{4x}$ **(e)** $4e^{4x}$

3   **(a)** $10e^{2x}$ **(b)** $3e^{3x}$ **(c)** $3e^{3x+2}$ **(d)** $\dfrac{1}{2\sqrt{x}} e^{\sqrt{x}}$ **(e)** $-\dfrac{e^{1/x}}{x^2}$

4   **(a)** $-e^{-x}$ **(b)** e^{-x} **(c)** $e^x - e^{-x}$ **(d)** $e^x + e^{-x}$

5   **(a)** $\cos x \, e^{\sin x}$ **(b)** $-\sin x \, e^{\cos x}$ **(c)** $\sec^2 x \, e^{\tan x}$ **(d)** $\sec x \tan x \, e^{\sec x}$

6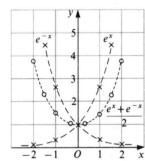

7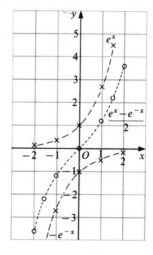

8

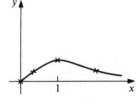
(a)   max $x = 1$

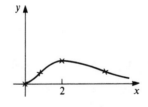
(b)   max $x = 2$

(c)   max $x = n$

9   **(a)** $(x+1)e^x$ **(b)** $e^x(x+2)$ **(c)** $e^x(x+n)$ **(d)** $e^{2x}(x+1)^2$

10   **(a)** $c'(x) = s(x);$ $s'(x) = c(x)$ **(b)** $c^2 - s^2 = 1$

     (c) c, see **6**; s, see **7**   **(d)** $c(-x) = c(x);$ $s(-x) = -s(x)$ **(e)** $c(a-b)$ **(f)** $s(a+b)$

Exercise 1.2

1   **(a)** $\dfrac{1}{x+1}$ **(b)** $\dfrac{5}{x+4}$ **(c)** $\tan x$ **(d)** $\dfrac{2\ln x}{x}$

2   **(a)** $\dfrac{3}{3x} = \dfrac{1}{x}$ **(b)** $\dfrac{3x^2}{x^3} = \dfrac{3}{x}$ **(c)** $\dfrac{2x}{x^2-1}$ **(d)** $-\dfrac{\csc^2 x}{\cot x} = \dfrac{-1}{\sin x \cos x} = -2\csc 2x$

3   Same answers as for **2**.   **(c)** $\ln(x^2-1) = \ln(x-1) + \ln(x+1) \to \dfrac{1}{x-1} + \dfrac{1}{x+1}$

4 (a) $\dfrac{2}{x}$ (b) $\dfrac{2x}{x^2+1}$ (c) $\dfrac{2}{x+1}$ (d) $\dfrac{3}{x+1}$ (e) $\dfrac{n}{x+1}$

5 (a) $\dfrac{3}{x}$ (b) $\dfrac{4}{x}$ (c) $\dfrac{3x^2+4x^3}{x^3+x^4}$ (d) $\dfrac{7}{x}$ (e) $\dfrac{3}{x}-\dfrac{4}{x}=-\dfrac{1}{x}$ (f) 0

6 (a) $\dfrac{1}{2x}$ (b) $\dfrac{1}{2(x+1)}$ (c) $\dfrac{x}{x^2+1}$ (d) $\dfrac{1}{2(x+1)(x+2)}=\dfrac{1}{2(x+1)}-\dfrac{1}{2(x+2)}$

7 (a) $-\dfrac{1}{x}$ (b) $-\dfrac{1}{x+1}$ (c) $-\dfrac{2}{x}$ (d) $-\dfrac{3}{x+1}$

8 (a) $\cot x$ (b) $-\tan x$ (c) $\tan x$ (d) $-\cot x$

9 (a) $\dfrac{2\cos 2x}{\sin 2x}=2\cot 2x$ (b) $-3\tan x$ (c) $3x^2\cot x^3$ (d) $\dfrac{\cos x-\sin x}{\cos x+\sin x}$

10 (a) $1+\ln x$ (b) $2x\ln x+x$ (c) $3x^2\ln x+x^2$ (d) $x^{n-1}+(n-1)x^{n-1}\ln x$

11 (a) $\sec x$ (b) $-\dfrac{\operatorname{cosec} x\cot x+\operatorname{cosec}^2 x}{\operatorname{cosec} x+\cot x}=-\operatorname{cosec} x$

12 (a) 1 (b) 1 (c) $e^x\ln x+\dfrac{e^x}{x}$ (d) $\dfrac{1}{x}+1$

Exercise 1.3

1 (a) $e^{x+2}+C$ (b) $e^4-e^3=34.5$ **2** (a) e^x+2x+C (b) $e^2-e+2=6.67$

3 (a) $\dfrac{e^{2x}}{2}+C$ (b) $\tfrac{1}{2}(e^4-e^2)=23.6$ **4** (a) $e^{x-2}+C$ (b) $e^0-e^{-1}=0.632$

5 (a) $-\dfrac{e^{-2x}}{2}+C$ (b) $-\dfrac{e^{-4}}{2}+\dfrac{e^{-2}}{2}=0.0585$ **6** (a) $\dfrac{e^{3x}}{3}+C$ (b) $\dfrac{e^9}{3}-\dfrac{e^3}{3}=2694$

7 (a) $e^x+\dfrac{e^{2x}}{2}+C$ (b) $e^3+\dfrac{e^6}{2}-e-\dfrac{e^2}{2}=215.4$ **8** (a) $\dfrac{e^{3x}}{3}+C$ (b) 2694

9 (a) $\dfrac{e^{x^3}}{3}+C$ (b) $\dfrac{e^{27}}{3}-\dfrac{e}{3}=1.77\times10^{11}$ **10** (a) $\dfrac{e^{nx}}{n}+C$ (b) $\dfrac{e^{3n}-e^n}{n}$

11 (a) $e^{\sin x}+C$ (b) $e^{1/\sqrt2}-1$ **12** (a) $-e^{\cos x}+C$ (b) $e-e^{1/\sqrt2}$

13 (a) $e^{\tan x}+C$ (b) $e-1$ **14** (a) $-e^{1/x}+C$ (b) $1-e^{1/4}=-0.284$

15 (a) $2e^{\sqrt x}+C$ (b) 9.34 **16** (a) $42e^{x/7}+C$ (b) 25.9

17 (a) $x+2e^x+\dfrac{e^{2x}}{2}+C$ (b) 1594 **18** (a) $\dfrac{x^2}{2}+C$ (b) 7.5

19 (a) $\dfrac{x^2}{2}+C$ (b) 7.5 **20** (a) $x\ln x-x+C$ (b) 2.545 **21** (a) $\ln|x+2|+C$
(b) 0.6931

22 (a) $x-\ln|x+1|+C$ (b) 2.084 **23** (a) $6\ln|x|+C$ (b) $\ln 4=8.318$

24 (a) $\tfrac{1}{2}\ln|2x-1|+C$ (b) 0.973 **25** (a) $\tfrac{1}{2}\ln|x^2+2|+C$ (b) 0.896

26 (a) $-\tfrac{1}{2}\ln|1-2x|+C$ (b) -0.255 **27** (a) $\tfrac{1}{2}\ln|x^2-1|+C$ (b) 0.49

28 (a) $\tfrac{1}{2}x^2-x+\ln|x+1|+C$ (b) 1.788 **29** (a) $\tfrac{1}{2}\ln\left|\dfrac{x-1}{x+1}\right|+C$ (b) 0.203

30 (a) $-\ln\cos x+C$ (b) 0.347 **31** (a) $\ln\sin x+C$ (b) 0.549

32 (a) $\ln(\sec x+\tan x)+C$ (b) 0.436

Exercise 1.4

1 (a) 4 or $\sqrt{2}$ (b) 9 or $\sqrt{3}$ (c) 2

2 (a) 1.585 (b) 1.262 (c) 1.387 (d) 1

3 (a) 0 or 0.631 (b) 0.631 or 1 **4** $x=2,\ y=8$ or $x=8,\ y=2$

5 (a) $\dfrac{\sec^2 x}{\tan x} = 2\operatorname{cosec} 2x$ (b) $a^x \ln a$ (c) $\dfrac{\log_{10} e}{x} = \dfrac{0.434}{x}$ (d) $\dfrac{3}{x}$ (e) $\dfrac{3}{x}(\ln x)^2$

6 (a) 3 (b) 4 (c) 2.5

7 $(1/e, 0.692) = (0.37, 0.69)$

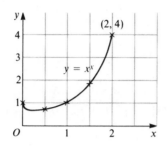

8 (a) $\frac{1}{2}\ln\dfrac{y}{3}$ (b) $1+\ln(y-6)$ (c) $\sqrt{\dfrac{y}{e^3}}$

10 (a) 1.792 (b) -1.099 (c) 1.195

Chapter 2

Exercise 2.1

1 $\frac{1}{2}\sin 2x + C$ **2** $\frac{1}{3}(x-3)^3 + C$ **3** $-\cot x + C$ **4** $-\frac{1}{5}\cos(5x-2)+C$ **5** $x^4 + \frac{4}{3}x^3 + \frac{1}{2}x^2 + C$

6 $\frac{1}{8}(x^2-1)^4 + C$ **7** $\frac{1}{4}\sin^4 x + C$ **8** $-\frac{2}{3}\cos^{\frac{3}{2}}x + C$ **9** $-\frac{1}{2}(x^2-5)^{-1}+C$

10 $-(x^2+x-3)^{-1}+C$ **11** $-\frac{1}{2}\cos 2x + C$ **12** $\sin^2 x + A = -\cos^2 x + B = -\frac{1}{2}\cos 2x + C$

13 $\frac{1}{2}(x^4-4x)^{\frac{1}{2}}+C$ **14** $\frac{1}{7}x^7 + \frac{3}{5}x^5 + x^3 + x + C$ **15** $\frac{1}{2}\tan^2 x + C$

16 $-\frac{1}{2}\cos(x^2)+C$ **17** $2\sin\sqrt{x}+C$ **18** $-\frac{1}{4}\cot^4 x + C$ **19** $\frac{1}{10}(x^2-6x+1)^5+C$

20 $\frac{1}{3}\sec^3 x + C$ **21** $\dfrac{-1}{2(2x+3)}+C$ **22** $\dfrac{1}{3}\left(1 - \dfrac{\sqrt{2}}{4}\right)$ **23** $\frac{76}{15}$

24 $\frac{1}{6}$ **25** $\sqrt{11}-1 = 2.32$ **26** $\dfrac{\sqrt{2}}{12}$ **27** $\sqrt{3}$

Exercise 2.2

1 (a) 0.077 (b) 0.589 **2** (a) $\dfrac{\pi}{4}$ (b) $\dfrac{\pi}{4}$

3 (a) $-\frac{1}{4}\cos 2x - \frac{1}{2}\cos x + C$ (b) $\frac{1}{16}\sin 8x + \frac{1}{4}\sin 2x + C$

4 (a) $\dfrac{5x}{16} + \dfrac{1}{8}\sin 4x + \dfrac{3}{128}\sin 8x - \dfrac{1}{96}\sin^3 4x + C$ (b) $-4\cos\left(\dfrac{x}{4}\right) - \dfrac{4}{3}\cos^3\left(\dfrac{x}{4}\right) + C$

5 0.589

Answers

Exercise 2.3

1 $\frac{1}{7}(x-3)^7+C$ **2** $\frac{1}{15}(3x+1)^5+C$ **3** $-\frac{1}{16}(1-4x)^4+C$ **4** $\frac{1}{21}(2+x)^6(3x-1)+C$

5 $-\frac{1}{8}(3-x^2)^4+C$ **6** $\frac{2}{15}(3x-2)(x+1)^{\frac{3}{2}}+C$ **7** $\frac{1}{5}(x+1)(2x-3)^{\frac{3}{2}}+C$ **8** $\frac{2}{3}(x^2+9)^{\frac{3}{2}}+C$

9 $\frac{1}{5}\sin^5 x+C$ **10** $\frac{1}{5}\tan^5 x+C$ **11** $-\frac{1}{8}\cos^8 x+C$ **12** $\frac{1}{56}(x-2)^6(7x^2+20x+16)+C$

13 $\frac{1}{3}(x+7)\sqrt{2x-1}+C$ **14** $\frac{2}{15}(3x-7)(x-4)^{\frac{3}{2}}+C$

15 $-\frac{1}{30}(5x+1)(1-x)^5+C$ **16** $-\frac{1}{12}(2-x^2)^6+C$ **17** $\frac{1}{6}(3+4x)^{\frac{3}{2}}+C$

18 $\frac{4}{3}(x-4)\sqrt{x+2}+C$ **19** $\ln|x-3|-\dfrac{8}{(x-3)}+C$ **20** $\frac{4}{9}(x^3+1)^{\frac{3}{2}}+C$ **21** $2\sqrt{\sec x}+C$

22 $\frac{2}{3}(\sin x)^{\frac{3}{2}}+C$ **23** $2\sin\sqrt{x}+C$ **24** $\frac{2}{5}\sqrt{\cos x}\,(\cos^2 x-5)+C$

Exercise 2.4

1 $\frac{1}{2}\tan^{-1}\left(\dfrac{x}{2}\right)+C$ **2** $\sin^{-1}\left(\dfrac{x}{2}\right)+C$ **3** $8\sin^{-1}\left(\dfrac{x}{4}\right)+\frac{1}{2}x\sqrt{16-x^2}+C$

4 $\frac{1}{3}\sin^{-1}\left(\dfrac{3x}{2}\right)+C$ **5** $3\sin^{-1}x+C$ **6** $\frac{25}{2}\sin^{-1}\left(\dfrac{x}{5}\right)+\frac{1}{2}x\sqrt{25-x^2}+C$

7 $\frac{1}{12}\tan^{-1}\left(\dfrac{3x}{4}\right)+C$ **8** $\tan^{-1}(x-2)+C$ **9** $\frac{1}{3}\tan^{-1}\left(\dfrac{x-3}{3}\right)+C$ **10** $\dfrac{1}{\sqrt{2}}\tan^{-1}\left(\dfrac{x+1}{\sqrt{2}}\right)+C$

11 $\frac{1}{5}\tan^{-1}\left(\dfrac{2x}{5}\right)+C$ **13 (a)** $\frac{1}{8}\tan^{-1}\left(\dfrac{2x+1}{4}\right)+C$ **(b)** $\sin^{-1}x-x\sqrt{1-x^2}+C$

Exercise 2.5

1 $\dfrac{\pi}{6}$ **2** $\dfrac{5\pi}{3}$ **3** π **4** $\dfrac{\pi}{12}$ **5** $\dfrac{\pi}{48}$ **6** $\dfrac{\pi}{6}$ **7** 4.3 **8** 359.47 **9** $\frac{7}{36}$

10 $\frac{1}{2}(2-\sqrt{3})=0.134$ **11** $-\frac{1}{20}$ **12** 6.7 **14** $\frac{64}{5}$ **15** $\dfrac{3\pi}{4}-2=0.36$ **16** $\frac{1}{21}$ **17** 12π

Exercise 2.6

1 $\sin x-x\cos x+C$ **2** $\frac{1}{4}e^{2x}(2x-1)+C$ **3** $\frac{1}{3}x\sin 3x+\frac{1}{9}\cos 3x+C$ **4** $-e^{-x}(x+1)+C$

5 $x\sin^{-1}x+\sqrt{1-x^2}+C$ **6** $\frac{1}{2}(x^2+1)\tan^{-1}x-\frac{1}{2}x+C$ **7** $x^2\sin x+2x\cos x-2\sin x+C$

8 $e^x(x^3-3x^2+6x-6)+C$ **9** $\dfrac{x^3}{3}\ln|x|-\dfrac{x^3}{9}+C$ **10** $\frac{1}{2}x^2(\ln x)^2-\frac{1}{2}x^2\ln x+\frac{1}{4}x^2+C$

11 $x\tan x+\ln|\cos x|-\frac{1}{2}x^2+C$ **12** $x\ln|2x|-x+C$ **13** $\frac{1}{2}e^x(\sin x-\cos x)+C$

14 $\frac{1}{2}e^{-x}(\sin x-\cos x)+C$ **15** $\frac{1}{4}e^{2x}(\sin 2x-\cos 2x)+C$ **16** $-\dfrac{1}{x}(1+\ln|x|)+C$

17 $x\tan x+\ln|\cos x|+C$ **18** $\dfrac{x}{n}\sin nx+\dfrac{1}{n^2}\cos nx+C$ **19** 1

20 $2(e^2-1)$ **21** $\frac{1}{8}(\pi-2)$ **22** $\frac{1}{27}(5e^3-2)$ **23** $\dfrac{2^{n+1}\ln 2}{(n+1)}-\dfrac{2^{n+1}}{(n+1)^2}+\dfrac{1}{(n+1)^2}$

24 $2\ln 4-2$ **25** $\dfrac{\pi^2}{16}-\dfrac{1}{4}$ **26** $-\frac{1}{5}(e^{\pi/2}+1)$ **27** 0.658 **28** $\frac{1}{3}e^{x^3}(x^3-1)+C$ **29** 0.636 sq units

30 $\frac{1}{5}(e^{2\pi}+1)$ sq units **31** $\dfrac{e^{-2N}}{13}(3\sin 3N-2\cos 3N)+\frac{2}{13}$

Exercise 2.7

1 $\frac{2}{5}\tan^{-1}\left(\frac{x}{5}\right)$ **2** $\ln\left|\dfrac{2+\tan(x/2)}{2-\tan(x/2)}\right|+C$ **3** $\frac{1}{2}\tan^{-1}\left|\frac{1}{2}\tan(x/2)\right|+C$ **4** $\dfrac{-2}{(1+\tan(x/2))}+C$

5 $\frac{1}{5}\ln\left|\dfrac{2\tan(x/2)-1}{\tan(x/2)+2}\right|+C$ **6** $\ln\left(\dfrac{2\sqrt{3}}{\sqrt{3}+1}\right)$ **9** **(a)** $\frac{1}{4}\ln\left|\dfrac{1+\tan 2x}{1-\tan 2x}\right|+C$ **(b)** $\tan^{-1}\left(\frac{1}{2}\tan(x/4)\right)+C$

11 **(a)** $\frac{1}{2}\tan^{-1}(2\tan x)+C$ **(b)** $\dfrac{1}{\sqrt{2}}\tan^{-1}\dfrac{\tan x}{\sqrt{2}}+C$

13 **(a)** $\frac{1}{4}\ln\left(\dfrac{2+\sqrt{3}}{2-\sqrt{3}}\right)$ **(b)** $\frac{1}{3}\ln 5$ **(c)** $\sqrt{2}\tan^{-1}\left(\dfrac{1}{\sqrt{2}}\right)-\dfrac{\pi}{4}$

Exercise 2.8

1 **(a)** $\frac{1}{3}(x^2+3)^{\frac{3}{2}}+C$ **(b)** $\frac{1}{2}\sqrt{(x^4-4x^2-2)}+C$ **2** **(a)** π **(b)** $\frac{9}{16}$

3 **(a)** $2-3\ln 2$ **(b)** 2 **4** $\frac{1}{2}\ln\left|\dfrac{3+x}{3-x}\right|+C$ **5** **(a)** $\frac{1}{4}-3\ln 2$ **(b)** $13\frac{11}{15}=13.73$

6 **(a)** $-\frac{1}{2}\cot 2x+C$ **(b)** $\sec x+C$ **7** **(a)** $-\frac{1}{5}\cos^5 x+C$ **(b)** $\frac{1}{5}\sec^5 x+C$

8 **(a)** $\frac{4}{189}(9x+1)(3x-2)^6+C$ **(b)** $-\frac{1}{4}\ln\left|\dfrac{1+\sqrt{1-4x^2}}{1-\sqrt{1-4x^2}}\right|+C$ **9** **(a)** 1 **(b)** $\frac{1}{3}\ln 2$

10 **(a)** $\dfrac{\pi}{18}$ **(b)** $\dfrac{\pi}{3\sqrt{3}}$ **14** $\dfrac{3\pi^2}{8}$ cubic units   **15** $\frac{1}{4}(1-5e^{-4})$ sq units

16 **(a)** $\dfrac{e^{2x}}{13}(3\sin 3x+2\cos 3x)+C$ **(b)** $\dfrac{e^x}{10}(5-2\sin 2x-\cos 2x)+C$

Chapter 3

Investigation 1

(a) $\dfrac{3}{8}=\dfrac{0}{2}+\dfrac{1}{4}+\dfrac{1}{8}=\dfrac{1}{2}+\dfrac{0}{4}-\dfrac{1}{8}=\dfrac{1}{2}+\dfrac{1}{4}-\dfrac{3}{8}=\dfrac{1}{2}+\dfrac{3}{4}-\dfrac{7}{8}$

i.e. four solutions: $(0,1,1)$; $(1,0,-1)$ or $(1,1,-3)$ or $(1,3,-7)$ using proper fractions, but the first is the simplest.

(b) $(1,1)$ **(c)** $(1,0,1)$; $(1,1,-1)$; $(0,1,3)$; $(0,3,-1)$; $(1,3,-5)$ **(d)** $(1,2,0)$; $(3,2,-1)$; $(-1,2,1)$

(e) $(0,1,1,0)$; $(0,-1,1,1)$; $(5,-1,0,0)$; $(1,1,-1,1)$ **(f)** $(1,2)$

(g) $\dfrac{3}{8}+\dfrac{3}{4}+\dfrac{1}{2}$ **(h)** $\dfrac{1}{16}+\dfrac{1}{8}+\dfrac{1}{4}+\dfrac{1}{2}$ **(i)** $\dfrac{1}{2}+\dfrac{1}{9}+\dfrac{-1}{3}$ or $\dfrac{1}{18}+\dfrac{2}{9}$ or $\dfrac{-1}{18}+\dfrac{1}{3}$ or $\dfrac{1}{2}+\dfrac{-2}{9}$ **(j)** $\dfrac{17}{17}+\dfrac{2}{17}=1+\dfrac{2}{17}$

Exercise 3.1

1 **(a)** $\dfrac{1}{x-1}-\dfrac{1}{x+1}$ **(b)** $\dfrac{1}{x-1}+\dfrac{1}{x+1}$ **(c)** $2+\dfrac{1}{x-1}-\dfrac{1}{x+1}$ **(d)** $2x+\dfrac{1}{x-1}+\dfrac{1}{x+1}$

2 **(a)** $\dfrac{1}{6(x-2)}-\dfrac{x+2}{6(x^2+2)}$ **(b)** $\dfrac{1}{3(x-1)}-\dfrac{x-2}{3(x^2+2)}$ **(c)** $\dfrac{9}{10(x-3)}+\dfrac{x+3}{10(x^2+1)}$ **(d)** $1+\dfrac{1}{2(x-1)}+\dfrac{x-1}{2(x^2+1)}$

3 **(a)** $\dfrac{4}{9(x-2)} - \dfrac{4}{9(x+1)} - \dfrac{4}{3(x+1)^2}$ **(b)** $\dfrac{5}{9(x-1)} - \dfrac{5}{9(x+2)} + \dfrac{7}{3(x+2)^2}$

(c) $\dfrac{19}{8(x-3)} + \dfrac{13}{8(x+1)} - \dfrac{3}{2(x+1)^2}$ **(d)** $4 + \dfrac{2}{3(x-1)} - \dfrac{38}{3(x+2)} + \dfrac{12}{(x+2)^2}$

4 **(a)** $\dfrac{2}{(x+2)^2} - \dfrac{3}{(x+2)^3}$ **(b)** $\dfrac{5}{7(x-2)} - \dfrac{5x-4}{7(x^2+3)}$ **(c)** $\dfrac{1}{x-1} - \dfrac{x-1}{x^2-2}$ **(d)** $\dfrac{8x-13}{x^2-3} - \dfrac{8}{x+2}$

5 **(a)** $\dfrac{3}{2x^2} - \dfrac{3}{4x} + \dfrac{3}{4(x+2)}$ **(b)** $\dfrac{3}{x-1} - \dfrac{3}{x}$ **(c)** $\dfrac{13}{4(x-2)} - \dfrac{1}{2x^2} - \dfrac{1}{4x}$

(d) $3 - \dfrac{1}{x} + \dfrac{1}{x^2} - \dfrac{2}{x+1}$

Exercise 3.2

1 **(a)** $\ln(x-1) - \ln(x+1) = \ln\left(\dfrac{|x-1|}{|x+1|}\right)$ **(b)** $\ln(x^2-1)$ **(c)** $2x + \ln\left(\dfrac{|x-1|}{|x+1|}\right)$ **(d)** $x^2 + \ln|x^2-1|$

2 **(a)** $\dfrac{1}{6}\ln|x-2| - \dfrac{1}{12}\ln|x^2+2| - \dfrac{1}{6}\int\dfrac{1}{1+\frac{1}{2}x^2}\,dx = \dfrac{1}{6}\ln(x-2) - \dfrac{1}{12}\ln(x^2+2) - \dfrac{\sqrt{2}}{6}\tan^{-1}\left(\dfrac{x}{\sqrt{2}}\right)$

(b) $\dfrac{1}{3}\ln|x-1| - \dfrac{1}{6}\ln|x^2+2| + \dfrac{\sqrt{2}}{3}\tan^{-1}\dfrac{x}{\sqrt{2}}$; $I_2^3 = \dfrac{1}{3}\ln 2 - \dfrac{1}{6}\ln\dfrac{11}{6} + \dfrac{\sqrt{2}}{3}\left(\tan^{-1}\dfrac{3}{\sqrt{2}} - \tan^{-1}\sqrt{2}\right) = 0.2125$

(c) $\dfrac{9}{10}\ln|x-3| + \dfrac{1}{20}\ln|x^2+1| + \dfrac{3}{10}\tan^{-1}x$ **(d)** $x + \dfrac{1}{2}\ln|x-1| + \dfrac{1}{4}\ln|x^2+1| - \dfrac{1}{2}\tan^{-1}x$

3 **(a)** $\dfrac{4}{9}\ln\left(\dfrac{|x-2|}{|x+1|}\right) + \dfrac{4}{3(x+1)}$ **(b)** $\dfrac{5}{9}\ln\left(\dfrac{|x-1|}{|x+2|}\right) - \dfrac{7}{3(x+2)}$; $I_2^3 = 0.3778$

(c) $\dfrac{19}{8}\ln|x-3| + \dfrac{13}{8}\ln|x+1| + \dfrac{3}{2(x+1)}$; $I_4^5 = 1.8924$

(d) $4x + \dfrac{2}{3}\ln|x-1| - \dfrac{38}{3}\ln|x+2| - \dfrac{12}{x+2}$; $I_2^3 = 2.2356$

4 **(a)** $\dfrac{-2}{x+2} + \dfrac{3}{2(x+2)^2}$; $I_0^1 = \dfrac{1}{8}$ **(b)** $\dfrac{5}{7}\ln(x-2) - \dfrac{5}{14}\ln(x^2+3) + \dfrac{4\sqrt{3}}{21}\tan^{-1}\dfrac{x}{\sqrt{3}}$

(c) $\ln(x-1) - \dfrac{1}{2}\ln(x^2-2) + \dfrac{1}{2\sqrt{2}}\ln\left(\dfrac{|x-\sqrt{2}|}{|x+\sqrt{2}|}\right)$; $I_2^3 = 0.3281$

(d) $4\ln(x^2-3) - \dfrac{13\sqrt{3}}{6}\ln\left(\dfrac{|x-\sqrt{3}|}{|x+\sqrt{3}|}\right) - 8\ln(x+2)$

5 **(a)** $-\dfrac{3}{2x} - \dfrac{3}{4}\ln x + \dfrac{3}{4}\ln(x+2)$; $I_1^2 = 0.4459$ **(b)** $3\ln(x-1) - 3\ln x = 3\ln\left(\dfrac{x-1}{x}\right)$; $I_2^3 = 0.8630$

(c) $3.25\ln(x-2) + \dfrac{1}{2x} - 0.25\ln x$; $I_3^4 = 2.1391$ **(d)** $3x - \ln x - \dfrac{1}{x} - 2\ln(x+1)$; $I_1^2 = 1.996$

Exercise 3.3

1 **(a)** $2[1-(2/3)^n]$ **(b)** 2 **(c)** $\dfrac{9}{22}$ **(d)** $\dfrac{9}{22}$ **(2) (a)** $\dfrac{n}{2n+1}$ **(b)** $\dfrac{3n^2+5n}{4(n+1)(n+2)}$ **(c)** $\dfrac{3}{4}$

3 (a) $\dfrac{n}{4(n+1)}$ **(b)** $\dfrac{n}{4(n+1)}$ **(c)** $\dfrac{1}{4}$ **4 (a) and (b)** $\dfrac{6n^2+5n}{4(n+1)(2n+1)}$ **(c)** $\dfrac{3}{4}$

5 (a) $\dfrac{n}{2n+1}$ **(b)** $\dfrac{2n}{2n+1}$ **6 (a)** $\dfrac{1}{2}\left(\dfrac{3}{2}-\dfrac{1}{n+1}-\dfrac{1}{n+2}\right)$ **(b)** $\dfrac{3}{4}-\dfrac{1}{2n}-\dfrac{1}{2(n+1)}$

7 (a) $n(n+1)[2n(n+1)-1]$ **(b)** $\dfrac{1}{2}\left(\dfrac{1}{2}-\dfrac{1}{n+1}+\dfrac{1}{n+2}\right)$

Exercise 3.4

1 $2+2x^2+2x^4+2x^6+\ldots$ **2** $3-3x+9x^2-15x^3+\ldots$ **3** $1+2x+2x^2+2x^3+\ldots$

4 $2+x+2x^2+x^3+\ldots$ **5** $\dfrac{5}{2}x+\dfrac{15}{4}x^2+\dfrac{65}{8}x^3$

Revision Exercise B

Vectors

1 $\mathbf{PR}=\mathbf{p}+\mathbf{q};\quad \mathbf{SR}=\mathbf{p}+\mathbf{q}-\mathbf{r};\quad \mathbf{QS}=\mathbf{r}-\mathbf{p}$

2 $\mathbf{AD}=\mathbf{a}+\mathbf{b}-\mathbf{c};\quad \mathbf{OD}=\tfrac{1}{2}(\mathbf{a}+\mathbf{b}-\mathbf{c});\quad \mathbf{OB}=\tfrac{1}{2}(\mathbf{a}-\mathbf{b}+\mathbf{c})$

3 $1.52\,\text{m}$, N $51.17°$ E **4** 10.24 N at $12.4°$ to AB

5 $\mathbf{XY}=\tfrac{2}{15}(2\mathbf{b}+3\mathbf{c}-5\mathbf{a})$ **6 (a)** $\sqrt{2}$ **(b)** $\sqrt{14}$ **(c)** 2 **(d)** $\sqrt{10}$

7 $\sqrt{45};\quad 2\mathbf{i}-4\mathbf{j}+5\mathbf{k}$ **8** $\sqrt{14};\quad \sqrt{50};\quad \dfrac{1}{\sqrt{90}}(5\mathbf{i}+\mathbf{j}+8\mathbf{k})$

9 (a) $-5\mathbf{i}+\tfrac{5}{2}\mathbf{j}$ **(b)** $-3\mathbf{i}+\tfrac{9}{2}\mathbf{j}$

10 $\mathbf{AB}=\mathbf{i}-\mathbf{j}+5\mathbf{k};\quad \mathbf{BC}=\mathbf{i}-\mathbf{j}-5\mathbf{k};\quad \mathbf{AC}=2\mathbf{i}-2\mathbf{j}$

11 $0\mathbf{i}+0\mathbf{j};\quad \mathbf{i}+2\mathbf{j};\quad 4\mathbf{i}+4\mathbf{j};\quad 9\mathbf{i}+6\mathbf{j};\quad 16\mathbf{i}+8\mathbf{j};\quad \mathbf{v}=2t\mathbf{i}+2\mathbf{j}$

12 $3\mathbf{i}-4\mathbf{j}+11\mathbf{k};\quad \mathbf{a}=2\mathbf{i}+12\mathbf{k}$ **13** $\mathbf{v}=3\mathbf{i}+(4-gt)\mathbf{j};\quad \mathbf{r}=3t\mathbf{i}+(4t-\tfrac{1}{2}gt^2)\mathbf{j}$

14 $\mathbf{a}=3\mathbf{i}-\mathbf{j}+\mathbf{k};\quad \mathbf{v}=12\mathbf{i}-4\mathbf{j}+4\mathbf{k};\quad \mathbf{r}=19\mathbf{i}-9\mathbf{j}+7\mathbf{k}$

15 $\mathbf{v}=-2\sin 2t\,\mathbf{i}+2\cos 2t\,\mathbf{j}+\mathbf{k};\quad \mathbf{a}=-4\cos 2t\,\mathbf{i}-4\sin 2t\,\mathbf{j}$

$|\mathbf{a}|=4\sqrt{\cos^2 2t+\sin^2 2t}=4$

16 $3\mathbf{i}-3\mathbf{j}$ **17** $\mathbf{i}+2\mathbf{j}$ **18** $\mathbf{v}_r=16\mathbf{j};\quad \mathbf{v}_c=-24\cos 45°\,\mathbf{i}-24\sin 45°\,\mathbf{j}$

velocity of wind relative to runner $=a\mathbf{i}+(b-16)\mathbf{j}$

velocity of wind relative to cyclist $=(a-12\sqrt{2})\mathbf{i}+(b-12\sqrt{2})\mathbf{j}$

actual velocity of wind $=12\sqrt{2}\mathbf{i}+16\mathbf{j}$

19 $3\,\text{s}$

Chapter 4

Exercise 4.1

1 (a) $\mathbf{r}=\lambda\mathbf{i}+(1-2\lambda)\mathbf{j}+(2+2\lambda)\mathbf{k}$ **(b)** $\mathbf{r}=(-2-4\lambda)\mathbf{i}+(3+\lambda)\mathbf{j}+(1-5\lambda)\mathbf{k}$

(c) $\mathbf{r}=(3+3\lambda)\mathbf{i}+\tfrac{1}{2}\mathbf{j}-\lambda\mathbf{k}$ **(d)** $\mathbf{r}=\lambda(2\mathbf{i}+3\mathbf{j}-\mathbf{k})$

2 (a) $\mathbf{r}=\mathbf{i}-2\mathbf{j}-3\mathbf{k}+\lambda(-\mathbf{i}+4\mathbf{k})$ **(b)** $\mathbf{r}=3\mathbf{i}+\mathbf{j}+4\mathbf{k}+\lambda(3\mathbf{i}-\mathbf{j}+5\mathbf{k})$

(c) $\mathbf{r}=-2\mathbf{i}-3\mathbf{j}+\mathbf{k}+\lambda(\mathbf{i}+7\mathbf{j})$

3 (a) yes **(b)** yes **(c)** yes **(d)** no **4** $\mathbf{r}=(2+3\lambda)\mathbf{i}+(-1+\lambda)\mathbf{j}+(1-\lambda)\mathbf{k}$

5 **(a)** $\mathbf{i}-\mathbf{j}-2\mathbf{k}$ **(b)** $(2\frac{1}{2},\frac{1}{2},0)$ **(c)** $(3,0,-1)$

7 AB: $\mathbf{r}=3\mathbf{a}+\mathbf{b}+\lambda(-\mathbf{a}+2\mathbf{b})$; AC: $\mathbf{r}=3\mathbf{a}+\mathbf{b}+\mu(2\mathbf{a}-4\mathbf{b})$

8 $\mathbf{r}=\frac{1}{3}(2\mathbf{c}+\mathbf{b})$; $\mathbf{r}=\frac{1}{3}(2\mathbf{c}+\mathbf{b}-3\mathbf{a})$; $\mathbf{r}=(1-\lambda)\mathbf{a}+\frac{1}{3}\lambda\mathbf{b}+\frac{2}{3}\lambda\mathbf{c}$ **9** $\mathbf{r}=(3-\lambda)\mathbf{i}-2\lambda\mathbf{j}-(2-\lambda)\mathbf{k}$

Investigation 7   **(a)** $x=1$ and $y=z-1$

                        (b) $y=-1$ and $x+2z=4$

Investigation 8   **(a)** $x=3$ and $y=1$ **(b)** $x=-1$ and $z=3$

Investigation 9   **(b)** $(1,0,2)$ PT: $\dfrac{x-1}{4}=\dfrac{y}{1}=\dfrac{z-2}{-3}$

Exercise 4.2

1   **(a)** $\dfrac{x-2}{1}=\dfrac{y+1}{-2}=\dfrac{z-3}{-1}$ **(b)** $\dfrac{x-1}{-1}=\dfrac{y}{1}=\dfrac{z-4}{3}$

     (c) $\dfrac{x+1}{2}=\dfrac{2y-6}{-1}=\dfrac{2z-4}{-3}$ **(d)** $4x=3z$ and $y=2$ **(e)** $\dfrac{x}{1}=\dfrac{y}{1}=\dfrac{z}{-2}$

2   **(a)** $\mathbf{r}=(4\lambda+1)\mathbf{i}+(3\lambda+2)\mathbf{j}+(2\lambda+5)\mathbf{k}$ **(b)** $\mathbf{r}=(3-\lambda)\mathbf{i}+5\lambda\mathbf{j}+(4-2\lambda)\mathbf{k}$

     (c) $\mathbf{r}=\frac{1}{2}(5+3\lambda)\mathbf{i}+\frac{1}{3}(1+2\lambda)\mathbf{j}+\lambda\mathbf{k}$ **(d)** $\mathbf{r}=4\mathbf{i}+3\lambda\mathbf{j}+(\lambda-2)\mathbf{k}$

3   **(a)** $\dfrac{x-1}{1}=\dfrac{y-2}{-1}=\dfrac{z-3}{1}$ **(b)** $\dfrac{2x}{1}=\dfrac{3y}{1}=\dfrac{z-6}{-1}$ **(c)** $\dfrac{x+1}{2}=\dfrac{4(y-4)}{-1}=\dfrac{z+3}{2}$

4   **(a)** $\dfrac{x}{1}=\dfrac{y}{-3}=\dfrac{z-1}{1}$ **(b)** $\dfrac{x-4}{1}=\dfrac{y-6}{5}=\dfrac{z+2}{-4}$ **(c)** $x=3$ and $y=-1$

5   **(a)**   yes   **(b)**   yes   **(c)**   no   **(d)**   yes

6   **(a)** $\mathbf{r}=\mathbf{i}+\mathbf{j}+\mathbf{k}+\lambda(\mathbf{i}-2\mathbf{j}-4\mathbf{k})$; $\dfrac{x-1}{1}=\dfrac{y-1}{-2}=\dfrac{z-1}{-4}$

     (b) $\mathbf{r}=3\mathbf{i}-2\mathbf{j}-\mathbf{k}+\lambda(\mathbf{i}-2\mathbf{j}-4\mathbf{k})$; $\dfrac{x-3}{1}=\dfrac{y+2}{-2}=\dfrac{z+1}{-4}$

7 $D(0,-7,0)$; AC: $\dfrac{x-1}{-1}=\dfrac{y-2}{6}=\dfrac{z-3}{2}$

                 BD: $\dfrac{x-3}{3}=\dfrac{y-5}{12}=\dfrac{z-4}{4}$

8 $(2,-1,1)$ **9** $(-1,4,2)$ **10** -2

11 AC: $\dfrac{x-1}{1}=\dfrac{y-2}{-5}=\dfrac{z-3}{2}$

     BD: $\dfrac{x-3}{1}=\dfrac{y-4}{3}=\dfrac{z-7}{2}$;   intersect at $(\frac{3}{2},-\frac{1}{2},4)$

12 $\dfrac{x-1}{5}=\dfrac{z-2}{-3}$ and $y=1$; $\dfrac{x-1}{3}=\dfrac{y-1}{-4}=\dfrac{z-2}{5}$

13 $\frac{1}{3}(7\mathbf{b}-4\mathbf{a})$ **14** $t=2$; $(3,4,-3)$ **15** $(8,12,-3)$

16 $\mathbf{r}=(8+7\lambda)\mathbf{i}+(1+2\lambda)\mathbf{j}-(1+3\lambda)\mathbf{k}$

     $\mathbf{r}=(7+3\mu)\mathbf{i}-(3+\mu)\mathbf{j}+(12+5\mu)\mathbf{k}$

     $(1,-1,2)$; $\mathbf{F}_1+\mathbf{F}_2=10\mathbf{i}+\mathbf{j}+2\mathbf{k}$; $\dfrac{x-1}{10}=\dfrac{y+1}{1}=\dfrac{z-2}{2}$

17 $F_1+F_2=7i-11j+14k;$ $\dfrac{x+1}{7}=\dfrac{y-2}{-11}=\dfrac{z+15}{14}$

Exercise 4.3

1 **(a)** 5   **(b)** -6 **(c)** 31   **(d)** -5

2 $-\dfrac{9}{5\sqrt{10}};$ -9

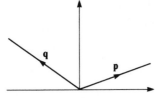

3 **(a)** 0   **(b)** 9   **(c)** -4 **(d)** 9   **4** **(a)** -2 **(b)** 0   **(c)** 8   **(d)** 2

5 **(a)** -3 **(b)** -12 **(c)** $6\sqrt{2}$ **(d)** -4.5 **(e)** $12\cos 75°=3.106$

Investigation 11 $AP=a+b;$ $PB=a-b;$ $AP.PB=a^2-b^2=0$

Exercise 4.4

1 **(a)** 10   **(b)** -8 **(c)** -3 **2** **(a)** 87.7°   **(b)** 90°   **(c)** 63.02°

3 **(a)** -1 **(b)** 12   **4** **(a)** $a^2+b^2+2(a.b)$ **(b)** p^2-q^2

7 **(a)** $a^2-3b^2-2(a.b)$ **(b)** $a^2-b^2+c^2$ **8** -2

9 **(a)** -2 **(b)** 15   **(c)** 13   **(d)** 34   **10** $\widehat{A}=35.3°;$ $\widehat{B}=15.8°;$

   $\widehat{C}=128.9°$

12 **(a)** 2   **(b)** $\frac{11}{5}$ **(c)** $-\frac{17}{7}$ **13** **(a)** 51.9°   **(b)** 84.5°

15 $\widehat{B}=60°;$ area $=6.06$ **17** $\dfrac{x-7}{-4}=\dfrac{y+4}{3}=\dfrac{2z-9}{-1}$

Exercise 4.5

1 **(a)** $r\cdot(i+4j-k)=18$ **(b)** $r\cdot(-2i+3j+k)=14$

2 **(a)** 0   **(b)** $\dfrac{4}{\sqrt{11}}$

3 **(a)** $r.(i-3j+4k)=-1$ **(b)** $r.(2i-8j+3k)=-17$ **(c)** $r.(6j-2k)=10$

4 **(a)** $3x+9y+z=17$ **(b)** $29x-11y+3z=106$ **(c)** $x+y+z=10$

5 **(a)** $\dfrac{4}{\sqrt{11}}$ **(b)** $\dfrac{5}{\sqrt{29}}$ **(c)** $\dfrac{2}{\sqrt{38}}$ **(d)** $\dfrac{1}{\sqrt{2}}$ **6** **(a)** $\frac{1}{3}\sqrt{2}$ **(b)** $\dfrac{3}{\sqrt{14}}$

7 **(a)** $r=3a+\lambda(a-b)+\mu(2a+b)$ **(b)** $r=4a-b+\lambda(3a+2b)+\mu(3a-2b)$

   (c) $r=5a+\lambda(c-3a)+\mu(b-2a)$

8 **(a)** $7x+4y+5z=12$ **(b)** $x+3y+5z=12$ **(c)** $5x-4y-z=7$

9 **(a)** $2x+y-z=3$ **(b)** $x+5y+z=1$ **(c)** $4x-y=13$

10 **(a)** $r=i-j+3k+\lambda(i-j+k)+\mu(2i-j-k)$

   (b) $r=3i+2j+3k+\lambda(i+2j)+\mu(2j+3k)$

11 $r=\frac{1}{2}(-i+12j-5k)+\lambda(i+j+k)+\mu(i+2j-k)$ **12** $(-\frac{2}{3},\frac{5}{3},\frac{8}{3})$

13 $r.(i+5j+3k)=0$ **14** $r.(i+2j+2k)=5$

Answers

Exercise 4.6

1 **(a)** 26.9° **(b)** 18.8° **(c)** 82.5° **(2)** 38° **3** **(a)** 0.5976 **(b)** 0.0825

4 **(a)** 28.3° **(b)** 31.9° **(c)** 45.6° **(d)** 80.7°

5 **(a)** 0.8642 **(b)** 0.9901 **6** **(a)** $\frac{19}{7}$ **(b)** $\frac{4}{3}$

7 **(a)** $\mathbf{r} = -\mathbf{k} + \lambda(3\mathbf{i} - \mathbf{j} + 5\mathbf{k})$ **(b)** $\mathbf{r} = -\mathbf{j} + 2\mathbf{k} + \lambda(4\mathbf{i} + \mathbf{j} - 5\mathbf{k})$

(c) $\mathbf{r} = \frac{11}{3}\mathbf{j} + \frac{7}{3}\mathbf{k} + \lambda(6\mathbf{i} - \mathbf{j} - 5\mathbf{k})$ **(d)** $\mathbf{r} = -\frac{19}{5}\mathbf{j} - \frac{32}{5}\mathbf{k} + \lambda(5\mathbf{i} + 7\mathbf{j} + 11\mathbf{k})$

8 $\mathbf{r} = \frac{19}{7}\mathbf{i} + \frac{22}{7}\mathbf{j} - \mathbf{k}$

9 $\mathbf{r} = 2\mathbf{i} + \mathbf{k} + \lambda(2\mathbf{i} - \mathbf{j} + 3\mathbf{k})$; $(\frac{11}{7}, \frac{3}{14}, \frac{5}{14})$; $\dfrac{x-2}{1} = \dfrac{y}{1} = \dfrac{z-1}{-1}$

Chapter 5

Investigation 3 With a weight at one end, the supports should be at that end and in the middle. With equal weights at each end, the supports should be one-quarter and three-quarters of the way along the light.

Investigation 5 For 2 coins and 1 coin; one-third of the way along the ruler. For 3 coins and 1 coin; one-quarter of the way along. For x and y: $y/(x+y)$ of the way along. Supporting force = sum of weight of coins.

Exercise 5.1

1 **(a)** 0.6 m from A **(b)** 0.48 m from A **(c)** 0.5$\dot{7}$ m from A

2 **(a)** 50 N downwards, 0.6 m from A. **(b)** 50 N downwards, 0.48 m from A.

(c) 90 N upwards, 0.5$\dot{7}$ m from A.

3 **(a)** 80 N upwards at C where $AC = 0.625$ m; $R = 80$ N downwards at C.

(b) 30 N downwards at D where $BD = \frac{1}{6}$ m; $R = 30$ N upwards at D.

(c) 90 N upwards at E where $AE = 0.51$ m; $R = 90$ N upwards at E.

4 **(a)** $M(A) = 10$ N m clockwise; $M(B) = 0$; $M(C) = 5$ N m clockwise; $R = 10$ N down, at B.

(b) $M(A) = 17$ N m anticlockwise; $M(B) = 27$ N m anticlockwise; $M(C) = 22$ N m anticlockwise $R = 10$ N m downwards at X, where $AX = 1.7$ m.

(c) $M(A) = M(B) = M(C) = 5$ N m clockwise; $R = 0$ ($\Rightarrow$ couple).

Exercise 5.2

1 **(a)** 50 N up at A; $M = 30$ N m clockwise **(b)** 50 N up at A; $M = 24$ N m clockwise

(c) 90 N up at A; $M = 52$ N m clockwise

3 **(a)** 80 N up at A; $M = 50$ N m clockwise **(b)** 30 N down at A; $M = 25$ N m anticlockwise

(c) 90 N up at A; $M = 46$ N m clockwise

4 **(a)** 10 N up at A; $M = 10$ N m clockwise **(b)** 10 N up at A; $M = 17$ N m anticlockwise

(c) 0, couple already; $M = 5$ N m clockwise

Exercise 5.3

1 (a) $R = 2i + 2j$ (b) $M(O) = 25$ N m anticlockwise; $M(A) = 15$ N m anticlockwise; $M(B) = 25$ N m anticlockwise; $M(C) = 35$ N m anticlockwise   (c) $y = x - 12.5$

2 (a) $R = (2 - 1.5\sqrt{2})i + (2 - 1.5\sqrt{2})j$

  (b) $M(O) = 25$ N m anticlockwise; $M(A) = 15 + 7.5\sqrt{2}$ N m anticlockwise; $M(B) = 25$ N m anticlockwise; $M(C) = 35 - 7.5\sqrt{2}$ N m anticlockwise   (c) $y = x + 100 + 75\sqrt{2}$

3 (a) $R = 0$;   (b) $M(O) = M(A) = M(B) = M(C) = 25$ N m anticlockwise;

  (c)   Couple, moment $= 25$ N m anticlockwise

4 (a) $R = 4i$ (b) $M(O) = M(A) = 15$ N m anticlockwise; $M(B) = M(C) = 35$ N m anticlockwise;

  (c) $y = -3.75$

5 (a) $R = 3i - 2j$ (b) $M(O) = 20$ N m anticlockwise; $M(A) = 30$ N m anticlockwise; $M(B) = 45$ N m anticlockwise; $M(C) = 35$ N m anticlockwise   (c) $2x + 3y + 30 = 0$

6   $3i + 3\sqrt{3}j$,   origin A, with AB as x-axis, AE as y-axis; $R = 6$ N $\| BC$ through $(7a, 0)$.

7   $R = 0$, couple, moment $21\sqrt{3}a$, where $2a =$ side of hexagon.

8   $R = 0$, couple, moment $6\sqrt{3}a$, where $2a =$ side of hexagon.

9   $R = 0$, couple, moment $4a + 2\sqrt{3}a$, where $2a =$ side of hexagon.   10 $R = 6$ N along BC.

11 (a) $P = 1, Q = 2 \Rightarrow$ Resultant $= 0$ but a couple is formed and the forces cannot be in equilibrium.   (b) $P = 1, Q = 2$

  (c) $P = -4, Q = 2$; R along $XY \| OA$, where $OX = 0.4$.

  (d) $P = 1, Q = -3$; R along $ST \| OC$, where $OS = 0.6$.

12 (a) $R = 0$; F, G, H act along the sides of a $\triangle$ $\Rightarrow$ couple   (b) A is $2i + 3j$

  (c)   1 anticlockwise   (d)   8 clockwise   (e)   2 anticlockwise   (f)   5 clockwise   (g)   0

Exercise 5.4

1 (a) 100 N; 200 N; $\alpha = 45°$ (b) $\mu = \frac{1}{8}$ 2 $\alpha = \arctan 0.75 = 36.9°$

3   $T = 60\sqrt{5} = 134$ N; $R = 60\sqrt{5} = 134$ N at $48.2°$ above horizontal.

(T minimum when $\alpha = 45°$ when rope attached $1.5\sqrt{2} = 2.1$ m from base which is not possible; the rope is already in the position for least value of T.)

4 (a) $B = C = 25$ N   (b) $B = 37.5$ N, $C = 12.5$ N

  (c) $B = 12.5$ N, $C = 37.5$ N,   (d) $B = 10$ N, $C = 40$ N

5   144.6 N; Reaction $= 431$ N   6   Reaction $= \dfrac{\sqrt{7}}{3} mg$

7   Coefficient of friction $= 36/77$. Greatest force $= 72\, Mg/125$ at A towards bar.

Chapter 6

Investigation 1   (a) $-3.45, 1.45$ (b) $0, -4$ (c) $-\frac{1}{2}$ twice   (d) $-0.78, 1.28$

Investigation 4   (i) 29   (ii) 25   (a) $3 - i$ (b) $-6 - 2i$ (c) $7 + 4i$ (d) $-5 + 3i$

Answers

Exercise 6.1

1 (a) $8i$ (b) -1 (c) 1 (d) i (e) $3\sqrt{2}\,i$

2 (a) $x=\pm 3$ (b) $x=\pm 3i$ (c) $x=\tfrac{1}{2}(-1\pm i\sqrt{3})$

 (d) $x=1\pm i\sqrt{2}$ (e) $x=\tfrac{1}{4}(-5\pm 7i)$ (f) $x=2\pm i\sqrt{3}$

3 (a) $1+i$ (b) $2-i$ (c) $3-2i$ (d) $-4+3i$ (e) $-5i$

4 (a) $2+5i$ (b) $5-4i$ (c) $2+2i$ (d) $-6+6i$

5 (a) $8-i$ (b) $14-5i$ (c) $12+i$ (d) 25 (e) $6+10i$ (f) $5-12i$

6 (a) $-i$ (b) $-i$ (c) -1 (d) $\tfrac{1}{2}$

7 (a) $-2i$ (b) $2+11i$ (c) $5+12i$ (d) $(x^2-y^2)+(2xy)i$

8 (a) $\tfrac{1}{10}(3-i)$ (b) $1+i$ (c) $\tfrac{1}{7}(\sqrt{3}+2i)$ (d) $\tfrac{1}{29}(15+6i)$

9 (a) i (b) $\tfrac{1}{2}(1-i)$ (c) $\tfrac{1}{13}(1-5i)$ (d) $\tfrac{1}{3}(-6-13i)$

10 $(a+ib)+(a-ib)=a$ which is real

11 (a) $2+i$: $-2-i$ (b) $3-2i$ $-3+2i$ (c) $3-i$; $3+i$ (d) $2+2i$; $-2-2i$

12 $a=-\tfrac{3}{5}$, $b=\tfrac{8}{5}$ **14** $z^2=-3+4i$; $1+2i$, $-1-2i$

Investigation 7 (b) $\tfrac{1}{2}(-1\pm i\sqrt{3})$

Exercise 6.2

1 (a) $|2i|=2$

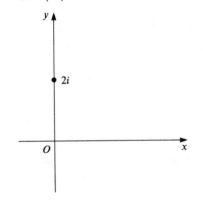

(b) $|1+i|=\sqrt{2}$

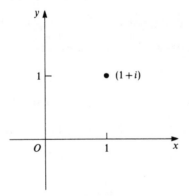

(c) $|2-4i|=2\sqrt{5}$

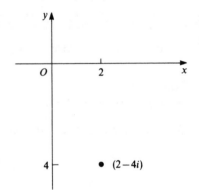

(d) $|-1+3i|=\sqrt{10}$

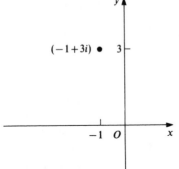

(e) $|-2-i|=\sqrt{5}$

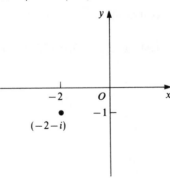

2 **(a)** $3i$ **(b)** 1 **(c)** $2+i$ **(d)** $-3+2i$ **(e)** $2-5i$

3 **(a)** $\sqrt{2};\ -\dfrac{\pi}{4}$ **(b)** $5;\ \tan^{-1}\left(\dfrac{4}{3}\right)$ **(c)** $5;\ \tan^{-1}\left(\dfrac{4}{3}\right)-\pi$

    (d) $\sqrt{29};\ -\tan^{-1}\left(\dfrac{5}{2}\right)$ **(e)** $\sqrt{13};\ \pi-\tan^{-1}\left(\dfrac{2}{3}\right)$

4 **(a)** $\sqrt{2};\ -\dfrac{3\pi}{4}$ **(b)** $2;\ \dfrac{\pi}{6}$ **(c)** $\sqrt{17};\ -\tan^{-1}4$ **(d)** $5;\ \tan^{-1}\left(\dfrac{3}{4}\right)-\pi$ **(e)** $2;\ \dfrac{2\pi}{3}$

5 **(b)** $z^2=2i$

    (c) $\dfrac{1}{z}=\tfrac{1}{2}(1-i)$

    (d) $\dfrac{1}{z^2}=-\tfrac{1}{2}i$

6 **(b)** $\bar{z}=3-4i$

    (c) $iz=-4+3i$

    (d) $i^2z=-3-4i$

    (e) $i\bar{z}=4+3i$

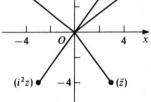

7

(c) $z_1+z_2=3+4i$

(d) $z_1-z_2=1-2i$

(e) $z_1-2z_2=-5i$

8 **(a)** 1 **(b)** $-5+2i$ **(c)** $-5+5i$ **(d)** $\tfrac{1}{10}(-7+i)$ **(e)** $\tfrac{1}{5}(-7-i)$

9 **(a)** $\sqrt{13};\ \pi-\tan^{-1}\left(\tfrac{3}{2}\right)$ **(b)** $\sqrt{17};\ \tan^{-1}\tfrac{1}{4}$ **(c)** $\sqrt{17};\ -(\pi-\tan^{-1}\tfrac{1}{4})$

    (d) $5\sqrt{2};\ -\dfrac{3\pi}{4}$ **(e)** $\tfrac{1}{2}\sqrt{2};\ \tan^{-1}7-\pi$

10 **(a)** $z_2=3k+2ki;\ \ k\geqslant0$ **(b)** $z_2=3k+2ki;\ \ -1\leqslant k\leqslant0$

11 **(a)** $\sqrt{5}$ **(b)** $\sqrt{58}$ **(c)** $\sqrt{40}$ **(d)** $\sqrt{85}$

Answers

12

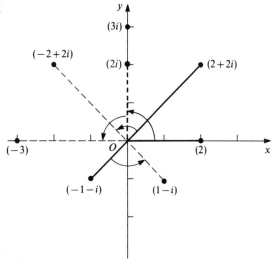

It produces a rotation of the radius vector through 90° about the origin in an anticlockwise sense.

13

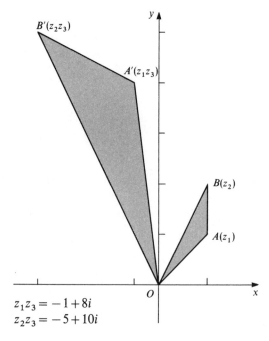

$z_1 z_3 = -1 + 8i$

$z_2 z_3 = -5 + 10i$

Exercise 6.3

1 (a) $3(\cos 0 + i \sin 0)$ (b) $2\left(\cos \dfrac{\pi}{2} + i \sin \dfrac{\pi}{2}\right)$ (c) $\sqrt{2}\left(\cos \dfrac{\pi}{4} + i \sin \dfrac{\pi}{4}\right)$ (d) $2\left(\cos \dfrac{\pi}{6} + i \sin \dfrac{\pi}{6}\right)$

   (e) $\sqrt{5}\,[\cos(-63.4°) + i \sin(-63.4°)]$ (f) $2\sqrt{2}\,[\cos(-\tfrac{3}{4}\pi) + i \sin(-\tfrac{3}{4}\pi)]$

   (g) $5(\cos 53.1° + i \sin 53.1°)$ (h) $\sqrt{13}(\cos 123.7° + i \sin 123.7°)$ (i) $13(\cos 112.6° + i \sin 112.6°)$

2 (a) $1 + i\sqrt{3}$ (b) $2\sqrt{3} + 2i$ (c) $-\dfrac{5}{2} + \dfrac{5\sqrt{3}}{2}i$ (d) $\dfrac{3}{\sqrt{2}}(1 - i)$ (e) $\sqrt{2} - \sqrt{2}i$

3 (a) $2;\ \dfrac{\pi}{12}$ (b) $\dfrac{3}{2};\ \dfrac{5\pi}{12}$ (c) $1;\ -2\theta$ (d) $\dfrac{1}{2};\ -\dfrac{2\pi}{3}$

4 (a) 1 (b) -1 (c) $-2\sqrt{3} + 2i$ (d) $\tfrac{1}{2}(\sqrt{3} - i)$

5 (a) $\cos \dfrac{\pi}{12} + i \sin \dfrac{\pi}{12};\ \ \cos \dfrac{\pi}{12} - i \sin \dfrac{\pi}{12}$

   (b) $\dfrac{1}{2}\left(\cos \dfrac{\pi}{4} + i \sin \dfrac{\pi}{4}\right) = \dfrac{\sqrt{2}}{4}(1 + i);\ \ 2\left(\cos \dfrac{\pi}{4} - i \sin \dfrac{\pi}{4}\right) = \sqrt{2}(1 - i)$ (c) $-2\sqrt{2};\ -\dfrac{\sqrt{2}}{4}$

6 (a) $1;\ -\theta$ (b) $1;\ -\theta$ (c) $1;\ 2\theta$

7 (a) $\sqrt{2 + 2\cos\theta} = 2\cos\tfrac{1}{2}\theta;\ \ \tfrac{1}{2}\theta$ (b) $\sqrt{2 - 2\cos\theta} = 2\sin\tfrac{1}{2}\theta;\ \ \tfrac{1}{2}(\pi + \theta)$ (c) $\operatorname{cosec}\tfrac{1}{2}\theta;\ \ \tfrac{1}{2}(\theta - \pi)$

8

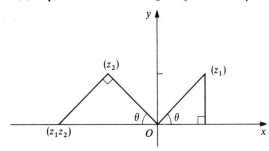

$z_1 = \sqrt{2}\left(\cos \dfrac{\pi}{4} + i \sin \dfrac{\pi}{4}\right)$

$z_2 = \sqrt{2}\left(\cos \dfrac{3\pi}{4} + i \sin \dfrac{3\pi}{4}\right)$

$z_1 z_2 = 2\,(\cos\pi + i \sin\pi) = -2$

9 (a) $2 + 2\sqrt{3}\,i$ (b) $8i$ (c) $-16\sqrt{3} + 16i$

Investigation 12

(a)

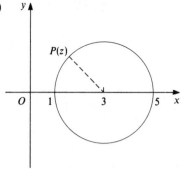

$$(x-3)^2 + y^2 = 4$$

(b)

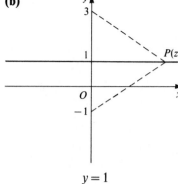

$$y = 1$$

(c)

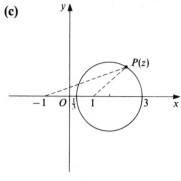

$$(x - \tfrac{5}{3})^2 + y^2 = \tfrac{16}{9}$$

Exercise 6.4

1 (a)

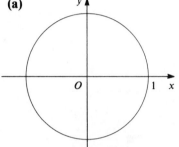

$$x^2 + y^2 = 1$$

(b)

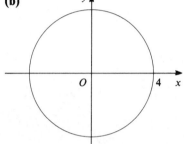

$$x^2 + y^2 = 16$$

(c)

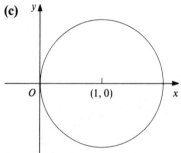

$$(x-1)^2 + y^2 = 1$$

(d)

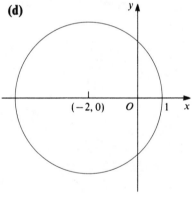

$$(x+2)^2 + y^2 = 9$$

(e)

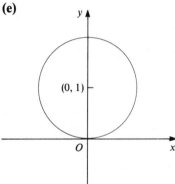

$$x^2 + (y-1)^2 = 1$$

(f)

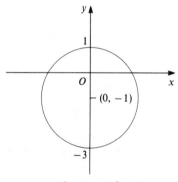

$$x^2 + (y+1)^2 = 4$$

(g)

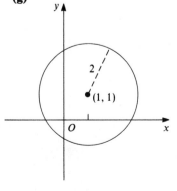

$$(x-1)^2 + (y-1)^2 = 4$$

(h)

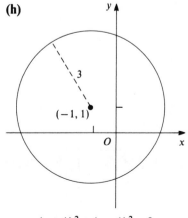

$$(x+1)^2 + (y-1)^2 = 9$$

(i)

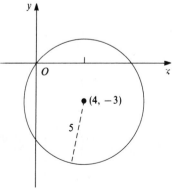

$$(x-4)^2 + (y+3)^2 = 25$$

2 (a)

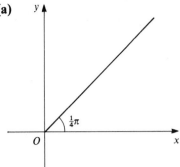

(b)

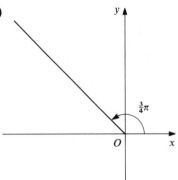

(c)

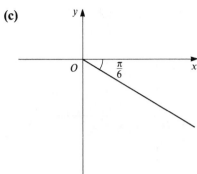

(d)

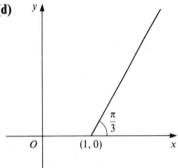

(e)

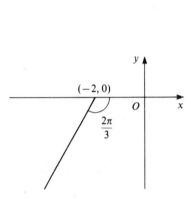

(f)

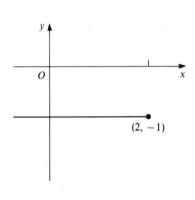

(g)

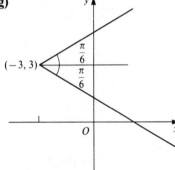

3 (a)

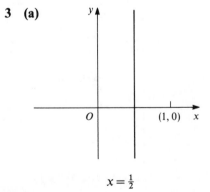

$x = \frac{1}{2}$

(b)

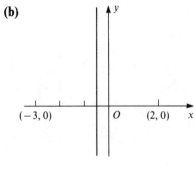

$x = -\frac{1}{2}$

(c)

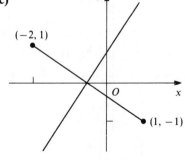

$4y = 6x + 3$

(d)

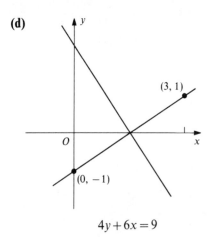

$$4y + 6x = 9$$

4 (a)   **(b)**   **(c)**

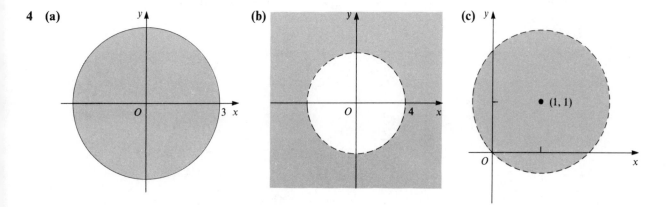

(d)   **(e)**

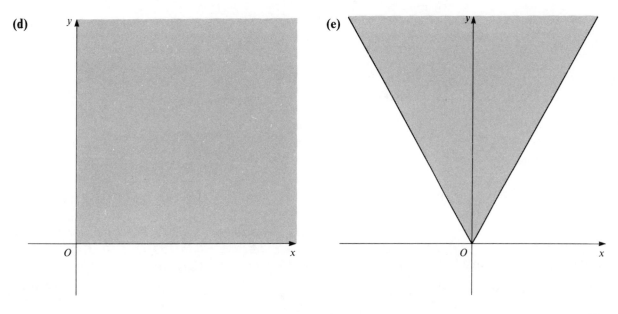

4 **(f)**

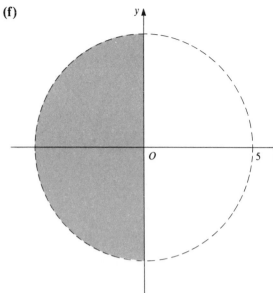

(g)

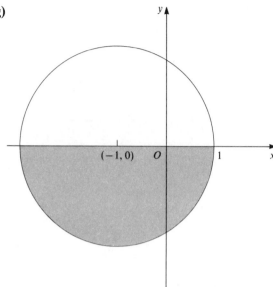

5 **(a)** $2x - 4y = 3$ **(b)** $3x^2 + 3y^2 - 25x + 50 = 0$ **(c)** $x^2 + y^2 - 2y - 15 = 0$ **(d)** $3x^2 + 3y^2 - 22x + 35 = 0$

7 $3 + \sqrt{5}; \quad 3 - \sqrt{5}$ **8** $\sqrt{2} + 1; \quad \sqrt{2} - 1$

9 **(a)**

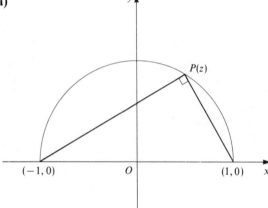

(b)

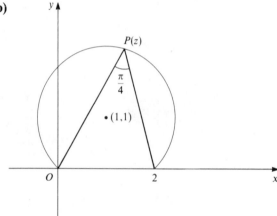

Exercise 6.5

1 **(a)** $\cos 3\pi + i \sin 3\pi = -1$ **(b)** $\cos \tfrac{1}{2}\pi + i \sin \tfrac{1}{2}\pi = i$

   (c) $\cos(-2\pi) + i \sin(-2\pi) = 1$ **(d)** $\cos(-\tfrac{1}{2}\pi) + i \sin(-\tfrac{1}{2}\pi) = -i$

2 **(a)** $-4 - 4i$ **(b)** $128(-1 + \sqrt{3}\,i)$ **(c)** $16(\sqrt{3} - i)$

3 **(a)** 1 **(b)** $-\dfrac{1}{2} + i\dfrac{\sqrt{3}}{2}$ **4** **(a)** $\dfrac{2\tan\theta}{1 - \tan^2\theta}$ **(b)** $\dfrac{5\tan\theta - 10\tan^3\theta + \tan^5\theta}{1 - \tan^2\theta + 5\tan^4\theta}$

7 $32\cos^6\theta - 48\cos^4\theta + 18\cos^2\theta - 1$

Exercise 6.6

1 **(a)** $\tfrac{1}{8}(\cos 4\theta + 4\cos 2\theta + 3)$ **(b)** $\tfrac{1}{64}(\cos 7\theta + 7\cos 5\theta + 21\cos 3\theta + 35\cos\theta)$

2 **(a)** $\tfrac{1}{4}(3\sin\theta - \sin 3\theta)$ **(b)** $\tfrac{1}{8}(\cos 4\theta - 4\cos 2\theta + 3)$

5 (a) 3π (b) 0.53 **6** $\dfrac{5\sqrt{2}}{6} - \dfrac{2}{3} = 0.51$ **7** $1, i, -1, -i$

8 (a) 1, $\cos\frac{2}{3}\pi + i\sin\frac{2}{3}\pi = \frac{1}{2}(-1 + i\sqrt{3})$, $\cos\dfrac{4\pi}{3} + i\sin\dfrac{4\pi}{3} = \frac{1}{2}(-1 - i\sqrt{3})$

    (b) -1, $\frac{1}{2}(1 + i\sqrt{3})$, $\frac{1}{2}(1 - i\sqrt{3})$

9 $\cos\dfrac{(2r+1)\pi}{5} + i\sin\dfrac{(2r+1)\pi}{5}$ for $r = 0, 1, 2, 3, 4$

10 $\cos\dfrac{(2r+1)\pi}{4} + i\sin\dfrac{(2r+1)\pi}{4}$ for $r = 0, 1, 2, 3$

11 $4 + 4i = 4\sqrt{2}\left[\cos\dfrac{(8r+1)\pi}{4} + i\sin\dfrac{(8r+1)\pi}{4}\right]$

Revision Exercise C

Forces

1 (a) T and $3g$ (b) T and $5g$ (c) $f = 0.25g$ m s^{-2}; $T = 3.75g$ N   **2** $AB = 10$; $8g$ N   **3** $AB = 6$

4 The coefficient of friction between the plastic cover and the table is greater than that between the paper and the table, i.e. the paper is smoother than the plastic cover.

5 Coefficient of friction $= \tan\alpha$, where α is the angle at which the book begins to slide. Tilt the surface until sliding starts.

6 $\mathbf{R} = 5\mathbf{i} + 3\mathbf{j}$ (Magnitude $\sqrt{34}$ at angle $31°$ to the x-axis.)

7 $\mathbf{R} = 45.5\mathbf{i} + 20.3\mathbf{j}$ (Magnitude 49.8 at $24°$ to Ox.)

8 $P = R = \dfrac{7}{\sqrt{3}} = 4.04$ N; $Q = \sqrt{\dfrac{19}{3}} = 2.52$ N; $\alpha = \arctan\dfrac{3\sqrt{3}}{7} = 36.6°$

Kinematics

1 $t = 1$ and $t = 2$;   at rest: $t = 0.5$ $(s = \frac{9}{16})$; $t = 1.62$ $(s = -1)$

    First second, $s = 1.125$;   second second, $s = 3.24$;   third second, $s = 24$.

2 $t = \dfrac{3\pi}{4} + k\pi$ $(k = 0, 1, 2\dots)$.   At rest when $t = \dfrac{\pi}{4} + k\pi$ at $s = +\sqrt{2}$ and $-\sqrt{2}$.

    First second, $s = 0.447$;   second second, $s = 0.887$;   third second, $s = 1.342$.

3 (a) (i) 1.58 s   (ii) 1.77 s   (b) (i) 9.8 m s^{-1} (ii) 11.7 m s^{-1}

4 $t = 0$; $\mathbf{v} = 20\mathbf{i}$.   Motion under gravity starting with horizontal velocity.

5 Motion in a circle with constant speed $= 4$. Velocity along tangent (speed $= 4$). Acceleration towards centre $(= 4)$.

6 Motion in an ellipse $\left(\dfrac{x^2}{16} + \dfrac{y^2}{9} = 1\right)$.

    Velocity along tangent $(v = -4\sin t\,\mathbf{i} + 4\cos t\,\mathbf{j})$.   Acceleration towards centre $(a = -4\cos t\,\mathbf{i} - 4\sin t\,\mathbf{j})$.

7 $v_0 = 1.5$, $v_1 = 2.5$, $v_2 = 3.5$

8 (a) $v_1 = 8\mathbf{i} - 4\mathbf{j}$; $\mathbf{r}_1 = 8\mathbf{i} + \mathbf{j}$ (b) $v_2 = 8\mathbf{i} - 14\mathbf{j}$; $\mathbf{r}_2 = 16\mathbf{i} - 8\mathbf{j}$

    (c) $v_3 = 8\mathbf{i} - 24\mathbf{j}$; $\mathbf{r}_3 = 24\mathbf{i} - 27\mathbf{j}$ (d) $y = \dfrac{3x}{4} - \dfrac{5x^2}{64}$

Newton's laws

1 (a) $\dfrac{1}{\sqrt{3}}$ **(b)** $2g$ N **(c)** $g\,\mathrm{m\,s^{-2}}$ **2 (a)** 17.32 N **(b)** 9.8 N at 30° to the horizontal

3 354 N **4 (a)** 236 N **(b)** 136 N **5** $4\,\mathrm{m\,s^{-2}}$; $26\,\mathrm{m\,s^{-1}}$

6 25 kN **7** $0.9\,\mathrm{ms^{-2}}$; 1920 N **8** 3.75 km

9 $\dfrac{5g}{12}\,\mathrm{m\,s^{-2}}$; $\sqrt{\dfrac{5g}{6}}\,\mathrm{m\,s^{-1}}$; $\dfrac{1}{3}\sqrt{\dfrac{40}{3g}}\,\mathrm{s}$ **10** $3.13\,\mathrm{m\,s^{-2}}$

Chapter 7

Investigation 2 **(b)** centroid

Investigation 4 m_1 at (a,b); m_2 at (c,d); m_3 at (e,f): C of M $\left(\dfrac{m_1a+m_2c+m_3e}{m_1+m_2+m_3},\ \dfrac{m_1b+m_2d+m_3f}{m_1+m_2+m_3}\right)$

Exercise 7.1

1 (a) C **(b)** D **(c)** D **(d)** C **(e)** C **(f)** 2.6 m from A towards F

2 (a) B **(b)** C **(c)** F **(d)** Mid-point of BC

3 (a) 3.5 **(b)** 8 **(c)** 10.5 **(d)** 6 **4** C, C, F; A, E, F; B, E, E

5 (a) B, C, D, E; A, C, C, F **(b)** C, D, E, F **(c)** $2m$ at B, m at C, $3m$ at C, $4m$ at F

6 (a) $(0.75, 2.25)$ **(b)** $(2.4, 2.4)$ **(c)** $(5, 3)$ **(d)** $(3, 1.8)$ **(e)** $(5, 3)$ **(f)** $(3.9, 3.6)$

(g) $(3.5, \frac{10}{3})$ **(h)** $(\frac{21}{8}, 2.5)$

7 (a) $(5, 2)$ **(b)** $(4, 4)$ **(c)** $(2.4, 3.6)$ **(d)** $(3.6, 4.1)$

Exercise 7.2

1 $(4, 3)$ **2** $(4, 4)$ **3** $(5, 4)$ **4** $(3, 4)$ **5** $(3, 4)$ **6** $(3, 3)$ **7** $(3, 3)$ **8** $(3, 3)$

9 $(3, 3)$ **10** $(\frac{11}{3}, \frac{10}{3})$ **11** $(\frac{11}{3}, \frac{4}{3})$ **12** $(\frac{19}{7}, \frac{19}{7})$ **13** $(\frac{7}{3}, 3)$ **14** $(\frac{7}{2}, \frac{7}{2})$ **15** $(3, 3)$

16 $(\frac{11}{3}, 3)$ **17** $(\frac{7}{2}, 3)$ **18** $(\frac{7}{2}, \frac{7}{2})$

19

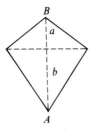

Centre of mass is on line of symmetry AB, distance $\dfrac{a+2b}{3}$ from A or $\dfrac{b-a}{3}$ from the intersection of the diagonals.

20

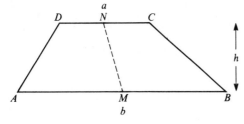

Centre of mass is on NM, the line joining the mid-points N and M of the parallel sides, at a distance

$\dfrac{(2a+b)h}{3(a+b)}$ from AB or $\dfrac{(a+2b)h}{3(a+b)}$ from CD.

Exercise 7.3

1 On line of symmetry Ox; $\quad X = \dfrac{2r \sin q}{3q}$ **2** On line of symmetry, distance $\dfrac{4\sqrt{2}\,r}{3\pi}$ from centre.

3 (a) On line of symmetry, distance $\dfrac{4\sqrt{2}\,r}{9\pi}$ from centre.

(b) On line of symmetry, $X = Y = \dfrac{4r}{9\pi}$ (c) Same as (b)

4 $\dfrac{2r \sin^3 q}{3(q - \sin q \cos q)}$ from centre of circle. **5** $\dfrac{4r(1 - \sin^3 q)}{3(\pi - 2q + \sin 2q)}$ from centre.

6 $\dfrac{2r}{(\pi + 2)}$ from centre of semicircle. **7** $\dfrac{r}{2 + \sqrt{2}}$ from centre of DE.

8 $\dfrac{(2 \sin q + \cos q)r}{2 + 2q}$ from O along OB. **9** $\dfrac{3r}{4}$ from O along OB.

10 On OQ, distance $\dfrac{2\sqrt{2}\,r}{3(4 - \pi)}$ from O. **11** $\dfrac{2r}{3(2 + \pi)}$ from O along OB; $\quad \sqrt{2}\,r$

Investigation 8 The combined C of M is as low as possible, i.e. as far from the apex as possible. My pen was made of plastic (hollow) with a metal weight at the end opposite the apex.

Exercise 7.4

1 (a) $\dfrac{(2\sqrt{5} - 3)}{3(\sqrt{5} + 2)} r$ from centre of base (into cone). (b) $\dfrac{r}{16}$ from centre of base (into cone).

(c) $\dfrac{(3 - \sqrt{2})}{3(2 + \sqrt{2})} r$ from centre (into hemisphere). (d) $\dfrac{r}{6}$ from centre (into hemisphere).

(c) and (d) will stand upright from the position where the line of symmetry is horizontal, but not from lying on the edge of the cone. (a) and (b) will stand still wherever you place them.

2 $\dfrac{3(2h^2 - r^2)}{4(h + 2r)}$ from centre of hemisphere (into cylinder); $\quad h = \dfrac{r}{\sqrt{2}}$

3 (a) $18.4°$ (b) $14°$ (c) $20.6°$ (d) $26.6°$

4 When filled it might not right itself; bad design, unstable.

5 $\left(\dfrac{3b}{5}, 0 \right)$ **6** $\left(\dfrac{2b}{3}, 0 \right)$ **7** $X = \dfrac{3(a^4 - b^4)}{8(a^3 - b^3)}$ $\left(\text{Factorize, cancel } (a - b), \text{ put } b = a, \text{ to get } \dfrac{a}{2} \right)$

8 $\dfrac{(H^4 - h^4)}{4(H^3 - h^3)}$ $\left(\text{Factorize, cancel } (H - h), \text{ put } h = H \text{ to get } \dfrac{H}{3} \right)$

9 (a) $\frac{5}{12}$ of the height from the base.

(b) On the line of symmetry, $\frac{8}{15}$ of a square from the centre of the middle square of the vertical line of 5 squares.

(c) $X = \frac{13}{11}$ of a square from the centre of the LHS vertical line of squares.

$Y = \frac{14}{11}$ of a square down from the centre of the top row of squares.

The best method is to consider each square to be a point mass at the centre of each square.

10 $\bar{X} = \bar{Y} = \dfrac{24a^3 - x^3}{24a^2 - 3x^2}$, origin A; $\quad x = 1.1a$

Chapter 8

Exercise 8.1

1 35.28 kJ   **2** **(a)** 8 J **(b)** 0 **(c)** 0   **3** 9.8 J

4 8 J; 2 J; 10 J   **5** 300 kJ   **6** 10.83 J   **7** 60 J

8 1690.5 kJ   **9** **(a)** −3 J; −14 J; 8 J   **(b)** −1 J; 25 J; −20 J   **(c)** −9 J; 4 J   **10** 14 kg

Investigation 4

t	2	4	6	8	10	12
v	3	6	9	12	15	18
Power	5.7	11.4	17.1	22.8	28.5	34.2

Maximum speed $= 50$ m s^{-1}

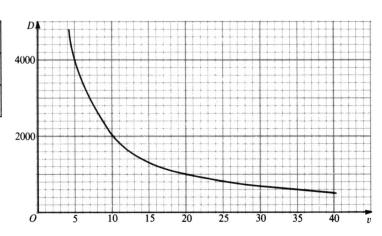

Exercise 8.2

1 1500 kW   **2** 30 m s^{-1} **3** 735 kW   **4** $\mathbf{v} = \frac{1}{2}t\,\mathbf{i} + (t+1)\mathbf{j} + \mathbf{k}$; $(2 + \frac{5}{2}t)$ W

5 $\mathbf{v} = (4t+2)\mathbf{i} + \mathbf{j} + t^2\mathbf{k}$; $(2t^3 + 16t + 8)$ W   **6** 69.2 N   **7** 17.96 kW   **8** 36.25 m s^{-1}; 0.065 m s^{-2}

9 28.2 kW   **10** 21.67 kW; 38.3 km h^{-1} **11** 20 m s^{-1} **12** 0.17 m s^{-2}; 10.3 m s^{-1}; 21.3 m s^{-1}

Exercise 8.3

1

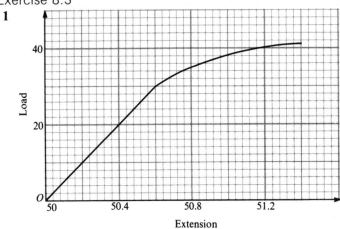

2 4.32 N   **3** 3.98 m   **4** 2 N   **5** 1.2 m   **6** 1.5 J

7 150 J   **8** 0.33 m   **9** 1.35 m   **10** $\dfrac{mg \tan \theta}{2(2 - \sin \theta)}$ **11** 1.99 m

Exercise 8.4

1 2.4 m; 7.2 J   **2** 2.45 J   **3** **(a)** 2 J **(b)** 1050 J   **4** 16 N   **5** $\dfrac{3mg}{2}$ J   **6** $\dfrac{mgl}{12}$; $a = \dfrac{4l}{3}$

Chapter 9

Exercise 9.1

3 14.7 J **4** 100 J **5** 4.13 J **6** 42.88 J **7** (a) 254.8 J (b) 294 J **8** 20.1 J

9 (a) $3g$ J; $-3g$ J (b) $6g$ J; 0 **10** 39.64 J

Exercise 9.2

1 4901.9 J **2** 7.96 m s^{-1}; 316.6 J; 316.6 W **3** 10.96 kW **4** 4.28 kW **5** 5.38 kW

Exercise 9.3

1 31.3 m s^{-1} **2** 17.15 m s^{-1} **3** 0.82 m **4** 7.67 m s^{-1}

5 6.57 m s^{-1} **6** (a) 12.56 m s^{-1} (b) 10.05 m **8** (a) 4.78 m s^{-1} (b) 2.34 m

10 3.46 m s^{-1}

Chapter 10

Exercise 10.1

1 30 N s **2** (a) $4\sqrt{5}$ N s (b) $2\sqrt{26}$ N s **3** 3 N s **4** 4000 N

5 8 N **6** 7.5 N s in direction of final velocity

7 (a) $\frac{9}{4}\mathbf{i}+6\mathbf{j}$ (b) $\frac{1}{2}(1-e^{-2})\mathbf{i}+\frac{1}{2}\mathbf{j}+\mathbf{k}$ (c) $\dfrac{1}{\sqrt{2}}\mathbf{i}+(2-\sqrt{2})\mathbf{j}+(\frac{1}{2}\ln 2)\mathbf{k}$

8 0.15 N s **9** $\frac{1}{4}(7\mathbf{i}+14\mathbf{j}+7\mathbf{k})$ **10** (a) 7.81 N s (b) 7 N s

11 2.937 N s; 58.7 m s^{-1} **12** 10.12 N s

Exercise 10.2

1 45.24 N **2** 3750 N **3** 1572 N **4** 21.2 N **5** 400 N **6** 12 m s^{-1} **7** 1 N

Exercise 10.3

1 $\frac{16}{5}$ m s^{-1} **2** $\frac{8}{3}$ m s^{-1} **3** $\frac{9}{14}$ m s^{-1} in direction of A

4 2.24 m s^{-1} **5** 150 N **6** 0.25 kg **8** 6.4 m s^{-1}; 1073 k N **9** $\frac{2}{5}$ m s^{-1}; 48.6 J

10 (a) $\frac{25}{16}$ m s^{-1} (b) 1.35 m s^{-1}

11 (a) $\frac{1}{2}$ m s^{-1} (b) $9\frac{1}{2}$ m s^{-1} **12** $\frac{28}{3}$ m s^{-1} **13** $\frac{1}{4}mu^2$ **14** $\dfrac{u^2}{32g}$

Exercise 10.4

1 20 m s^{-1} **2** $\sqrt{\dfrac{16}{5g}}$ s **3** $\frac{1}{3}\sqrt{g}$ m s^{-1}; $1\frac{1}{3}$ m

4 6.67 m s^{-1} at 58° to BA produced; 3.54 m s^{-1} along BA; 10.61 N s

5 5 m s^{-1}; $\sqrt{13}$ m s^{-1}; 2 m s^{-1} **6** $\sqrt{\dfrac{g}{8}}$ m s^{-1}

7 Tensions: $AB \dfrac{11I}{35}$; $BC, \dfrac{4I}{35}$; Speeds: $\dfrac{11I}{35m}$; $\dfrac{2\sqrt{31}\,I}{35m}$; $\dfrac{4I}{35m}$

Exercise 10.5

1 $1\frac{1}{8}$ m **2** $1\frac{1}{3}$ s **3** $2\frac{1}{2}$ m s^{-1} and 3 m s^{-1} in original direction of motion.

4 $\frac{4}{15}$ m s^{-1} and $\frac{5}{3}$ m s^{-1} in original direction of 5 kg mass.

6 9.375 J **7** 3 m s^{-1}; $\frac{2}{3}$ kg **8** $e = \frac{1}{4}$; 2:3

10 4 m s^{-1}; $e = \frac{1}{2}$ **11** A, $\frac{4}{5}$ m s^{-1}; B, $\frac{8}{5}$ m s^{-1}; C, $\frac{32}{5}$ m s^{-1}

13 $\frac{2}{9}\sqrt{ga}$; $\frac{14}{9}\sqrt{ga}$ **14** A, $\dfrac{13u}{64}$; B, $\dfrac{15u}{64}$; C, $\dfrac{9u}{16}$

Revision Exercises D

Permutations and combinations

1 (a) 24 **(b)** 120 **(c)** 36 **(d)** 51 **(e)** 72 **2 (a)** 7! **(b)** $\dfrac{17!}{12!}$ **(c)** $\dfrac{9!}{5!4!}$

3 (a) $9 \times 7!$ **(b)** $180 \times 12!$ **(c)** $4 \times \dfrac{6!}{3!}$ **(d)** $33 \times 8!$ **(e)** $n^2(n-2)!$ **(f)** $\dfrac{9!}{6!3!}$

4 40 320 **5** 720 **6** 7! **7** 840 **8** 360

9 (a) 120 **(b)** 2520 **(c)** 10 080 **(d)** $\dfrac{10!}{3!3!2!} = 50\,400$

10 (a) $2 \times 4! = 48$ **(b)** 72 **11** 240 **12** 42 **13** 40 319 **14** 72 **15 (a)** 180

(b) 648 **17 (a)** 56 **(b)** 210 **(c)** 792 **18** 4368 **19** 126 **20** 3150

21 700 **22** 79; **24** **25** 200

Series

1 (a) 1, 2, 4; 2^{14} **(b)** $\dfrac{1}{4}, \dfrac{2}{9}, \dfrac{3}{16}$; $\dfrac{15}{256}$ **(c)** $1, \dfrac{9}{10}, \dfrac{8}{9}$; $\dfrac{128}{135}$

2 (a) 5, 16, 60, 236 **(b)** 100, 50, 25, $12\frac{1}{2}$ **(c)** 1, 3, 12, 60

3 (a) $u_n \rightarrow 0$ **(b)** $u_n \rightarrow 2$ **(c)** $u_n \rightarrow 5$ **(d)** $u_n \rightarrow 4$ **(e)** u_n oscillates finitely between 2 and 4

4 (a) $7 + 14 + 23 + 34 + 47$ **(b)** $\frac{1}{2} + \frac{2}{3} + \frac{3}{4} + \frac{4}{5}$ **(c)** $3 - 7 + 15 - 31 + 63 - 127 + 255$

5 (a) $\displaystyle\sum_{r=1}^{26} (2r - 1)$ **(b)** $\displaystyle\sum_{r=1}^{n} (r+1)r^2$ **(c)** $\displaystyle\sum_{r=1}^{14} (r+1)(r+4)$

6 152; 670 **7** 30 **8** 11, 17, 23 **9** -41; 7 **10** $20(-\frac{1}{2})^{17}$; $13\frac{9}{32}$

11 $5\frac{109}{256}$ **12** 3; $3^{10} - 1$ **13** -4; 205 **14** 10; 8 **15** 12

16 (a) $\frac{4}{9}$ **(b)** $\frac{32}{99}$ **(c)** $\frac{19}{37}$ **17** $-1 < x < 5$

18 (a) $1 + 6x + 15x^2 + 20x^3 + 15x^4 + 6x^5 + x^6$

(b) $32 + 80x + 80x^2 + 40x^3 + 10x^4 + x^5$

(c) $1 - 8x^2 + 28x^4 - 56x^6 + 70x^8 - 56x^{10} + 28x^{12} - 8x^{14} + x^{16}$

19 $1 - 11x + 55x^2 - 165x^3 + 330x^4$; 0.757

20 (a) $1 + 4x + 12x^2 + 32x^3$ **(b)** $1 + 2x - 2x^2 + 4x^3$ **(c)** $\dfrac{1}{2} - \dfrac{x}{4} + \dfrac{x^2}{8} - \dfrac{x^3}{16}$

21 (a) $3 - 2x + 2x^2 - 2x^3$ **(b)** $\dfrac{1}{8}\left(1 + \dfrac{x}{2} - \dfrac{x^2}{2} + \dfrac{x^3}{4}\right)$ **(c)** $2 - 6x + \dfrac{85}{4}x^2 - \dfrac{175}{2}x^3$

22 $1-5x+15x^2-45x^3$; $-\frac{1}{3}<x<\frac{1}{3}$

23 **(a)** $\frac{1}{6}(n+1)(n+2)(2n+3)$ **(b)** $n^2(2n^2-1)$ **(c)** $\frac{2}{3}n(4n^2+12n+11)$ **(d)** $\frac{1}{4}n(n+1)(n^2+9n+20)$

24 **(a)** $\frac{3}{4}-\frac{2n+3}{2(n+1)(n+2)}$; $\frac{3}{4}$ **(b)** $\frac{23}{144}-\frac{1}{12}\left(\frac{3}{n+1}-\frac{1}{n+2}-\frac{1}{n+3}-\frac{1}{n+4}\right)$; $\frac{23}{144}$

25 $\frac{r}{(r+1)!}$; $1-\frac{1}{(n+1)!}$

Chapter 11

Exercise 11.1

1 **(a)** $\frac{1}{6}$ **(b)** $\frac{1}{2}$ **(c)** $\frac{1}{2}$ (remember 1 is not a prime number)   **(d)** $\frac{1}{3}$ **(e)** 0

2 **(a)** $\frac{1}{2}$ **(b)** $\frac{1}{4}$ **(c)** $\frac{3}{13}$ **(d)** $\frac{1}{13}$ **(e)** $\frac{1}{26}$ **(f)** $\frac{16}{52}$ **(g)** $\frac{1}{52}$ **(h)** 0   **(i)** $\frac{1}{52}$

3 **(a)** $\frac{1}{100}$ **(b)** $\frac{99}{100}\times\frac{1}{99}=\frac{1}{100}$ **(c)** $\frac{99}{100}\times\frac{98}{99}\times\frac{1}{98}=\frac{1}{100}$

4 **(a)** $\frac{1}{3}$ (I think, according to the latest statistics)

5 **(a)** $\frac{59}{557}$ **(b)** $\frac{23}{557}$ **(c)** $\frac{24}{557}$

6 **(a)** 14 ones;   14 twos;   14 sixes;   **(b)** 7, 9   **(c)** 'Ones' occur in probability numerical descriptions and 'sixes' occur because of the dice example. But why the 'twos'?

Exercise 11.2

1 **(a)** $\frac{1}{8}$ **(b)** $\frac{1}{8}$ **(c)** $\frac{3}{8}$ **(d)** $\frac{7}{8}$

2 Not all equally likely; there are two ways of getting one of each.

3 **(a)** $\frac{5}{32}$ **(b)** $\frac{5}{16}$ **(c)** $\frac{1}{2}$

4 **(a)** $\frac{3}{8}$ **(b)** $\frac{5}{16}$ **5** **(a)** $\frac{1}{2}$ **(b)** $\frac{3}{8}$ **(c)** $\frac{5}{16}$ **(d)** $\frac{35}{128}$ **(e)** 0

6 **(a)** $\frac{1}{3}$ **(b)** $\frac{2}{9}$ **(c)** $\frac{4}{27}$ **(d)** $\frac{1}{3}\left(\frac{2}{3}\right)^{n-1}$ **7** $\frac{9}{16}$

8 **(a)** 0 (You need to get 12 questions correct.)   **(b)** and **(c)**   Use the Binomial approxiation to the Normal Distribution i.e. ask a statistics teacher (!)

9 0.0173   **(a)** Yes   **(b)** It does not seem so.

10 0.0708   **11** **(a)** $\frac{1}{4}$ **(b)** $\frac{49}{64}$ **(c)** $\frac{123}{128}$

12 **(a)** 1   **(b)** 1   **(c)** $1-p$(all in different weeks) $=1-\frac{51}{52}\times\frac{50}{52}\times\frac{49}{52}\cdots\frac{34}{52}\times\frac{33}{52}\simeq0.995$

Exercise 11.3

1 (a) $\dfrac{1}{4}$ (b) $\dfrac{1}{2}$ (c) $\dfrac{1}{13}$ (d) $\dfrac{3}{13}$ (e) $\dfrac{1}{4}$ (f) 0 (g) $\dfrac{1}{2}$ (h) $\dfrac{1}{52}$ (i) $\dfrac{4}{13}$ (j) $\dfrac{11}{26}$

(k) $\dfrac{3}{52}$ (l) $\dfrac{1}{26}$ (m) $\dfrac{7}{13}$ (n) $\dfrac{3}{26}$ (o) $\dfrac{8}{13}$ (p) $\dfrac{1}{13}$ (q) $\dfrac{3}{13}$ (r) $\dfrac{1}{3}$ (s) 1 (t) $\dfrac{1}{2}$

(u) 1 (v) $\dfrac{3}{13}$ (w) $\dfrac{1}{2}$ (x) $\dfrac{1}{4}$ (y) $\dfrac{1}{13}$ (z) $\dfrac{1}{2}$

2 (a) $\dfrac{25}{204}$ (b) $\dfrac{1}{26}$ (c) $\dfrac{13}{102}$ (d) $\dfrac{11}{663}$ (e) $\dfrac{470}{663}$ (f) $\dfrac{193}{663}$ (g) $\dfrac{77}{204}$ (h) $\dfrac{13}{102}$ (i) $\dfrac{179}{204}$

(j) $\dfrac{32}{51}$ (k) $\dfrac{19}{51}$ (l) $\dfrac{25}{102}$ (m) $\dfrac{1}{3}$ (n) $\dfrac{1}{3}$ (o) 0 (p) $\dfrac{1}{2}$

3 (a) $\dfrac{1}{8}$ (b) $\dfrac{1}{26}$ (c) $\dfrac{1}{8}$ (d) $\dfrac{3}{169}$ (e) $\dfrac{120}{169}$ (f) $\dfrac{49}{169}$ (g) $\dfrac{3}{8}$

(h) $\dfrac{1}{8}$ (i) $\dfrac{7}{8}$ (j) $\dfrac{5}{8}$ (k) $\dfrac{3}{8}$ (l) $\dfrac{1}{4}$ (m) $\dfrac{1}{2}$ (n) $\dfrac{1}{13}$ (o) $\dfrac{1}{13}$ (p) $\dfrac{1}{2}$

4 (a) $\dfrac{3}{104}$ (b) $\dfrac{25}{884}$ (c) $\dfrac{5}{1326}$ (d) $\dfrac{11}{1326}$ (e) $\dfrac{11}{1326}$ (f) $\dfrac{55}{1326}$

5 (a) $\dfrac{3}{52}$ (b) $\dfrac{1}{52}$ (c) $\dfrac{1}{26}$ (d) $\dfrac{1}{52}$ (e) $\dfrac{8}{13}$ (f) $\dfrac{8}{13}$

(g) $\dfrac{11}{26}$ (h) $\dfrac{8}{13}$ (i) $\dfrac{7}{13}$ (j) $\dfrac{1}{4}$ (k) $\dfrac{7}{13}$ (l) $\dfrac{2}{13}$

6 (a) $\dfrac{1}{10}$ (b) 0 (c) $\dfrac{1}{10}$ (d) $\dfrac{1}{10}$ (e) $\dfrac{3}{10}$ (f) $\dfrac{1}{5}$ (g) $\dfrac{3}{5}$ (h) $\dfrac{9}{10}$

7 (a) $\dfrac{1}{100}$ (b) $\dfrac{1}{100}$ (c) $\dfrac{1}{100}$

8 (a) $\dfrac{1}{6}$ (b) $\dfrac{5}{6}$ (c) $\dfrac{1}{3}$ (d) 0 (e) 1 (f) 0 (g) $\dfrac{1}{6}$ (h) $\dfrac{2}{3}$

(i) $\dfrac{1}{3}$ (j) 1 (k) 0 (l) 0 (m) $\dfrac{1}{3}$ (n) $\dfrac{2}{3}$ (o) $\dfrac{1}{18}$ (p) $\dfrac{1}{36}$

(q) $\dfrac{1}{6}$ (r) $\dfrac{1}{6}$ (s) $\dfrac{1}{6}$ (t) $\dfrac{1}{18}$ (u) $\dfrac{11}{36}$ (v) $\dfrac{1}{6}$ (w) $\dfrac{1}{6}$ (x) $\dfrac{1}{12}$ (y) $\dfrac{1}{36}$ (z) $\dfrac{1}{3}$

9 (a) $\dfrac{9}{20}$ (b) $\dfrac{1}{4}$ (c) $\dfrac{1}{12}$ (d) $\dfrac{2}{15}$ (e) $\dfrac{1}{3}$

Exercise 11.4

1 (a) $\dfrac{1}{3}$, $\dfrac{2}{3}$; No (b) $\dfrac{1}{3}$, $\dfrac{2}{3}$; No (c) $\dfrac{2}{3}$, $\dfrac{1}{3}$; No (d) $\dfrac{2}{3}$, $\dfrac{1}{3}$; No (e) No

2 (a) Yes (b) No (c) Yes (d) Yes (e) No (f) Yes

3 (a) $\dfrac{1}{7}$ (b) $\dfrac{2}{7}$ (c) $\dfrac{4}{7}$; No

4 (a) $\dfrac{25}{102}$ (b) $\dfrac{25}{102}$ (c) $\dfrac{26}{51}$ (d) $\dfrac{1}{17}$ (e) $\dfrac{1}{221}$ (f) $\dfrac{4}{663}$ (ace followed by king); $\dfrac{8}{663}$ (any order)

(g) $\dfrac{32}{663}$ (h) $\dfrac{2}{663}$

5 **(a)** $p(P \cap C) = \dfrac{1}{16}$ **(b)** $p(P \cup C) = \dfrac{7}{8}$ **(c)** $p(P \cap \bar{C}) = \dfrac{9}{16}$ **(d)** $p(\bar{P} \cap C) = \dfrac{1}{4}$

(e) $p(\bar{P} \cap \bar{C}) = \dfrac{1}{8}$ **(f)** $p(P|C) = \dfrac{1}{5}$ **(g)** $p(C|P) = \dfrac{1}{10}$ **(h)** $p(C|\bar{P}) = \dfrac{2}{3}$

(i) $p(P|\bar{C}) = \dfrac{9}{11}$ **(j)** $p(\bar{P} \cap \bar{C}) = \dfrac{1}{8}$ **(k)** $p(P) = \dfrac{5}{8}$ **(l)** $p(C) = \dfrac{5}{16}$ **(m)** No

Exercise 11.5

1 Expected winnings $= \dfrac{1}{6} \times £6 - \dfrac{5}{6} \times £1 = \dfrac{£1}{6}$ at each turn, so the player wins in the long run. £5 for a six would make the game fair. **2** £5

3 You have an even chance of doubling your stake each time; it depends on the option of when each player can call a halt to the game.

4 A very involved analysis; the possibilities are enormous. In one throw of 5 dice: $p(\text{no fives}) = 0.402 = p(\text{one five})$; $p(2 \text{ fives}) = 0.161$; $p(3 \text{ fives}) = 0.002$; $p(4 \text{ fives}) = 0.003$; $p(5 \text{ fives}) = 0.0001$.
It is a wonder any good scores are made.

5 **(a)** $1 - p(\text{all different}) = 1 - 0.507 = 0.493$, if the cards are dealt successively to the first player.

(b) $\dfrac{^{13}C_4}{52 \times 51 \times 50 \times 49} = 0.000\,11$ i.e. one deal in ten thousand.

(c) 0.397, if dealt to the first player. **(d)** Very low.

6 Very low; there are 45 winning lines out of $^{50}C_8, = 536\,878\,650$ lines.
536 million lines at $\frac{1}{8}$p per line costs £671 000 stake to pay to win. But how much would you win ??

7 $\dfrac{1}{2}$ **8.** No; 50p; No; Yes

9 Ask the Chancellor of the Exchequer.

Chapter 12

Exercise 12.1

1 $y = 1 - \frac{1}{2}x^2 + \frac{1}{24}x^4$ lies very close to $y = \cos x$. **2** $x + \frac{1}{3}x^3$

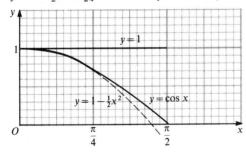

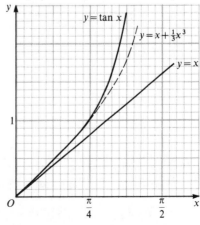

Answers

3 $1+\frac{1}{2}x^2+\frac{5}{24}x^4$

4 (a) $4x-\frac{32}{3}x^3$ (b) $x^2-\frac{1}{3}x^4$ (c) $1+2x+2x^2+\frac{4}{3}x^3+\frac{2}{3}x^4$ (d) $x+\frac{1}{6}x^3$ (e) $x-\frac{1}{3}x^3$

(f) $1+nx+\frac{n(n-1)}{2!}x^2+\frac{n(n-1)(n-2)}{3!}x^3+\frac{n(n-1)(n-2)(n-3)}{4!}x^4$

(g) $-\frac{1}{2}x^2-\frac{1}{12}x^4$ (h) $-x-\frac{1}{2}x^2-\frac{1}{3}x^3-\frac{1}{4}x^4$ (i) $x-\frac{1}{2}x^2+\frac{1}{6}x^3-\frac{1}{12}x^4$

5 (a) $x+x^2+\frac{1}{3}x^3-\frac{1}{30}x^5$ (b) $1+x+\frac{1}{2}x^2-\frac{1}{8}x^4$

Exercise 12.2

1 (a) $1-x+\frac{1}{2}x^2-\frac{1}{6}x^3$; $\frac{(-1)^{r-1}x^r}{r!}$ (b) $1+3x+\frac{9}{2}x^2+\frac{9}{2}x^3$; $\frac{3^r x^r}{r!}$

(c) $1+x^2+\frac{1}{2}x^4+\frac{1}{6}x^6$; $\frac{x^{2r}}{r!}$ (d) $1+\frac{1}{2}x+\frac{1}{8}x^2+\frac{1}{48}x^3$; $\frac{x^r}{2^r(r!)}$

(e) $e(1+x+\frac{1}{2}x^2+\frac{1}{6}x^3)$; $e\left(\frac{x^r}{r!}\right)$ (f) $1-\frac{1}{2}x^2-\frac{1}{3}x^3-\frac{1}{8}x^4$; $\frac{(1-r)x^r}{r!}$

2 (a) $2+4x^2+\frac{4}{3}x^4$ (b) $1+\frac{3}{2}x+\frac{13}{8}x^2$ (c) $1-x-\frac{1}{2}x^2$

(d) $1+8x+32x^2$ (e) $e(1-x^2+\frac{1}{2}x^4)$ (f) $1-7x^2+\frac{38}{3}x^3$

3 (a) e^{4x} (b) $e^{-\frac{1}{2}x}$ **4** $(1+2x)e^x$ **5** $\frac{1}{2}(e^x-e^{-x})$

6 (a) $\frac{1}{2}$ (b) 1 **7** (a) e^{-x} (b) e^{x^2} (c) $(1+x)e^x$ (d) $2xe^x$

9 (a) $\frac{2x^{2r}}{(2r)!}$ (b) $\frac{2x^r}{r!}$ for odd powers; $\left[\frac{2}{(2r)!}+\frac{1}{r!}\right]x^{2r}$ for even powers (c) $\frac{(r-1)^2}{r!}x^r$

Exercise 12.3

1 (a) $-2x-2x^2-\frac{8}{3}x^3$ (b) $\frac{1}{3}x-\frac{1}{18}x^2+\frac{1}{81}x^3$ (c) $\ln 3-\frac{1}{3}x-\frac{1}{18}x^2$

2 (a) $-x^2-\frac{1}{2}x^4-\frac{1}{3}x^6-\frac{1}{4}x^8$; $-\frac{2x^r}{r}$ (b) $\ln 4-\frac{1}{8}x-\frac{1}{128}x^2-\frac{1}{1536}x^3$; $-\frac{x^r}{(8^r)r}$

(c) $\ln 4+\frac{5}{4}x+\frac{7}{32}x^2+\frac{91}{192}x^3$; $\left[1-(-\frac{3}{4})^r\right]\frac{x^r}{r}$

3 (a) $\ln 4-\frac{5}{4}x-\frac{17}{32}x^2$; $-1\leqslant x<1$ (b) $x+\frac{1}{2}x^2-\frac{1}{6}x^3$; $-1<x\leqslant 1$

(c) $1+\frac{1}{2}x+\frac{11}{8}x^2$; $-1<x<1$ (d) $-\frac{1}{2}x-\frac{5}{8}x^2-\frac{5}{12}x^3$; $-2\leqslant x<2$

(e) $\frac{3}{2}x-\frac{3}{4}x^2+\frac{15}{4}x^3$; $-\frac{1}{3}<x\leqslant\frac{1}{3}$ (f) $-2x+x^2-\frac{2}{3}x^3$; $-1\leqslant x<1$

6 0.5108 **7** $x+\frac{1}{2}x^2-\frac{2}{3}x^3+\frac{1}{4}x^4+\frac{1}{5}x^5-\frac{1}{3}x^6$

Exercise 12.4

7 $f(k+1)-f(k)=(3^{k+1}+1)-(3^k+1)=3^k(3-1)=2(3^k)$

8 $f(k+1)-f(k)=(5^{k+1}+4(k+1)+3)-(5^k+4k+3)=4(5^k+1)$

9 $f(k+1)-f(k)=(4^{k+1}+3^{k+1}+2)-(4^k+3^k+2)=3(4^k)+2(3^k)=3[4^k+2(3^{k-1})]$

17 $[(k-1)^3+k^3+(k+1)^3]-[(k-2)^3+(k-1)^3+k^3]=9(k^2-k+1)$

Chapter 13

Exercise 13.1

5   **(a)** $y = x^2 - x + C$ **(b)** $y = \sin x + C$

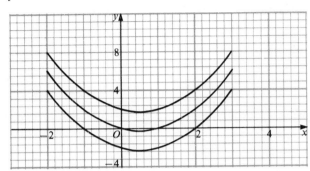

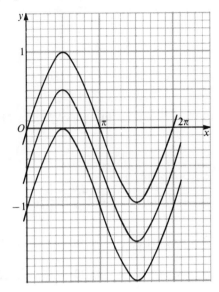

     (c) $y = \tfrac{1}{2} e^{2x} + C$

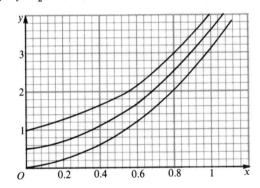

6   **(a)** $y = x^3 - 3x^2 + x + C$ **(b)** $y = 2\sin\tfrac{1}{2}x - 2\cos\tfrac{1}{2}x + C$ **(c)** $y = \tfrac{1}{3}(x^2 + 1)^{\frac{3}{2}} + C$ **(d)** $y = \tan x + C$

7   **(a)** $y = \tfrac{1}{2}x^2 - \ln|x| + C$ **(b)** $y = -2e^{-x} - \tfrac{1}{2}e^{2x} + C$ **(c)** $y = Ax + B - \tfrac{2}{9}\sin 3x$

     (d) $y = \tfrac{1}{2}e^{2x} - e^x + Ax + B$

8   **(a)** $x\dfrac{dy}{dx} - y = 0$ **(b)** $x\dfrac{dy}{dx} + y = 0$ **(c)** $\dfrac{dy}{dx} = y$

     (d) $\dfrac{dy}{dx} - 3y = 0$ **(e)** $x\dfrac{dy}{dx} - 2y = x(1 - \ln x)$

9   **(a)** $x = -\dfrac{1}{y} + C$ **(b)** $y = \tan(x + C)$ **(c)** $\tan y = x + C$

     (d) $e^y = x + C$ **(e)** $\ln|y + 1| = x + C$ **(f)** $y = \dfrac{e^{2x} - 1}{e^{2x} + 1}$

10   **(a)** $y = \tfrac{4}{3}x^3 - 2x^2 + x + \tfrac{5}{3}$ **(b)** $y = \sin^{-1} x$ **(c)** $y = 2e^{4x}$ **(d)** $e^{-2y} = 3 - 2x$

11 $p = \dfrac{1}{k}e^x - 1; \quad y = \dfrac{1}{k}e^x - x + C$

Exercise 13.2

1 $y^3 = kx$ **2** $y(\frac{1}{2}x^2 + C) = -1$ **3** $y = ke^{-\frac{1}{x}}$ **4** $\sin y = e^x + C$

5 $\tan y = 4x + C$ **6** $y = k(1-x)^{-3}$ **7** $kx = \dfrac{1}{(2 - \sin y)}$ **8** $y = k\left(\dfrac{x-1}{x+1}\right)$ **9** $e^y = \frac{1}{2}x^2 + C$

10 $y(1+x) = k\,e^x$ **11** $y(1 - x^2) = k$ **12** $\frac{1}{2}y^2 = -e^{-x}(1+x) + C$ **13** $y^2 = (\ln x)^2 + C$

14 $e^{\frac{1}{y}} = k\cot^2 x$ **15** $2y^3 = 3(x^2 + 2x + 10)$

16 $y = e^{-\cos x}$ **17** $y = \tan\left(x^2 + \dfrac{\pi}{4}\right)$ **18** $\ln\left|\frac{1}{2}y\right| = x - \sin x$

19 $y = \dfrac{x-1}{x+1}$ **20** $y = e^{x^2 - 4}$ **21** $v = \ln(51 - \frac{1}{2}t^2)$

Exercise 13.3

1 26.9 m s^{-1} **2** $\dfrac{dN}{dt} = \dfrac{N}{10};$ $N = 100e^{12}$

3 $t = \frac{8}{3}\ln\left|\dfrac{32}{32-x}\right|;$ **(a)** 2.62 hours   **(b)** 7.39 hours; $x < 32$

4 $68.2°\text{C}$ **5** $3\ln 27$ **6** 60 seconds   **7** 59 days   **8** **(a)** 2.3 days   **(b)** 6.9 days

Exercise 13.4

1 **(a)** $y = e^{-2x}(\sin x + C)$ **(b)** $xy = xe^x - e^x + C$

2 **(a)** $xy^2 = \tan x + C$ **(b)** $2xy = -e^{-x}(\cos x + \sin x) + C$

   (c) $x^2 y^2 = \tan 2x + C$ **(d)** $e^y = 1 + \dfrac{C}{x}$

3 $y = 4x^4 \ln k|x|$ **4** $y = x\ln|x| - 2 + Cx$

5 **(a)** $ye^{-x} = \ln e|x|$ **(b)** $y = 2\sec^4 x - 2\cos^2 x$

   (c) $y = x^2 \sin x + x\cos x + x$ **(d)** $\frac{1}{13}(2e^{3t} - 2\cos 2t - 3\sin 2t)$

6 $z = \dfrac{C - \tan x}{\sec^2 x};$ $y = \dfrac{\sec^2 x}{C - \tan x}$ **7** $x^2 + y^2 = x^4$

Exercise 13.5

1 **(a)** $y = \frac{1}{4}e^{2x} + Ax + B$ **(b)** $y = Ax - \ln|x| + B$

   (c) $y = e^x(x-2) + Ax + B$ **(d)** $y = -x\cos x + 2\sin x + Ax + B$

2 **(a)** $y = A\cos(2x + \alpha)$ **(b)** $y = A\cos(nx + \alpha)$

3 **(a)** $y = x\ln|x|$ **(b)** $y = 2x^4 - \frac{5}{2}x^2 - 10x + 7$ **(c)** $y = 3\sin 5x + 2\cos 5x$

4 **(a)** $y = \frac{1}{3}Ax^3 + B$ **(b)** $y = Ax(1 + \frac{1}{3}x^2) + B$

   (c) $y = A(\frac{1}{2}x^2 + 2x) - x + B$ **(d)** $y = A\sin^{-1} x + B$

5 **(a)** $x = \dfrac{1}{A}\tan^{-1}\left(\dfrac{y}{A}\right) + B$ **(b)** $\frac{1}{2}\sin^{-1} y + \frac{1}{2}y\sqrt{1 - y^2} = Ax + B$

   (c) $\ln|y-1| = Ax + B$ **(d)** $x = y\ln|y| + Ay + B$ **7** $y = Ae^x - Be^{-x}$

Chapter 14

Exercise 14.1

1 $v = \frac{1}{3}t^3 + \frac{1}{2}t^2$; $s = \frac{1}{12}t^4 + \frac{1}{6}t^3$ **2** $4.75\,\text{m s}^{-1}$ **3** $2.77\,\text{s}$ **5** $v = \sqrt{6s - \frac{1}{10}s^2}$; $0\,\text{m}$ and $60\,\text{m}$

7 $10 - \pi\,\text{m s}^{-1}$; $14.24\,\text{m}$

Exercise 14.2

1 $21\frac{1}{3}\,\text{m}$ **2** $27e^{-t} - 2$ **4** $2\ln 2\,\text{m}$ **6** 6.375 **8** $\frac{u}{k} + \frac{g}{k^2}\ln\left(\frac{g}{g+ku}\right)$ **10** $\frac{1}{2k}\ln 2$

Exercise 14.3

1 $8\,\text{m s}^{-1}$

2 **(a)** $1.89\,\text{m s}^{-2}$; $0\,\text{m s}^{-1}$ **(b)** $0.95\,\text{m s}^{-2}$; $1.31\,\text{m s}^{-1}$ **(c)** $0.47\,\text{m s}^{-2}$; $1.46\,\text{m s}^{-1}$

3 **(a)** $\pi\,\text{s}$ **(b)** $7.95\,\text{s}$ **4** **(a)** $\frac{1}{2}\pi\,\text{s}$ **(b)** $\frac{1}{12}\pi\,\text{s}$ **(c)** $0.32\,\text{s}$

5 $9.49\,\text{m}$; $3.97\,\text{s}$ **6** $0.993\,\text{m}$ **7** $\frac{\pi\sqrt{3}}{6}\,\text{rad s}^{-1}$ **8** $8.775\,\text{m}$; $2.265\,\text{s}$

9 $\dot{x} = \omega(-A\sin\omega t + B\cos\omega t)$; $\ddot{x} = -\omega^2(A\cos\omega t + B\sin\omega t) = -\omega^2 x$

10 $1.45\,\text{m}$; $0.86\,\text{m s}^{-1}$ **11** $x = 5\cos 2t$; $0.58\,\text{s}$

Chapter 15

Investigation 2   Parabola: $y^2 = 3(2x - 1)$

Investigation 3   Circle: $2x^2 + 2y^2 - 17x + 8 = 0$

Investigation 6   Circle, parabola, ellipse and hyperbola.

Exercise 15.1

1 $\sqrt{(x-2)^2 + y^2} = 8 - \sqrt{(x+2)^2 + y^2}$ and square both sides. Simplify and square both sides again.

2 **(a)** $\dfrac{(x-2)^2}{4} + \dfrac{y^2}{3} = 1$ **(b)** $\dfrac{(x-1)^2}{4} + \dfrac{y^2}{3} = 1$ **(c)** $\dfrac{(x-1.5)^2}{2.25} + \dfrac{y^2}{2} = 1$

3 **(a)** $x + 2y = 8$ **(b)** $y = 2x - 1$ **(c)** $\left(-\frac{22}{19}, -\frac{63}{19}\right)$

4 Equation $\dfrac{(x-0.5)^2}{1} + \dfrac{y^2}{0.75} = 1$;   focus $(1,0)$;   directrix $x = 2.5$; $e = 0.5$

5 **(a)** $e = \dfrac{\sqrt{3}}{2}$; $F = (\sqrt{3}, 0)$; $x = \dfrac{4}{\sqrt{3}}$ **(b)** $e = \dfrac{\sqrt{5}}{3}$; $F = (\sqrt{5}, 0)$; $x = \dfrac{9}{\sqrt{5}}$

(c) $e = \dfrac{\sqrt{7}}{4}$; $F = (\sqrt{7}, 0)$; $x = \dfrac{16}{\sqrt{7}}$

6 **(a)** 12π **(b)** πab **7** **(a)** $\dfrac{xx'}{a^2} + \dfrac{yy'}{b^2} = 1$ **(b)** $\dfrac{x\cos\alpha}{a} + \dfrac{y\sin\alpha}{b} = 1$

8 If normal at P meets x-axis at X, show that $GP:PF = GX:XF$, which implies that PX bisects $\angle GPF$. $\angle SPX = 90° = \angle XPT$ $\Rightarrow$ $\angle SPG = \angle FPT$.

Answers

Exercise 15.2

1 $PF=3$, $QF=5$, $RF=7$, $PG=5$, $QG=7$, $RG=9$;   In general, $PG-PF=2$.

2 (a) $y=2x-1$ **(b)** $x+2y=8$;   Normal meets the curve again at $(-\frac{38}{11}, \frac{63}{11})$

3 $x^2-\dfrac{y^2}{2}=4$; $(\pm 6, \pm 8)$ **4** $x^2-y^2=4$;   Only $(\pm 2, 0)$; $90°$

5 $y^2-x^2=4$ **6** $xy=2$ **7** $x^2-y^2=2$

8 (a) Tangent: $\dfrac{x\sec\alpha}{a}-\dfrac{y\tan\alpha}{b}=1$ **(b)** Normal: $\dfrac{by}{\tan\alpha}+\dfrac{ax}{\sec\alpha}=a^2+b^2$

Investigation 8   **(a)** Focus $(\sqrt{2}, \sqrt{2})$;   Directrix $x+y=\sqrt{2}$
(b) Focus $(c\sqrt{2}, c\sqrt{2})$;   Directrix $x+y=c\sqrt{2}$

Investigation 9   For $xy=k$, the larger k is, the further away from the origin the curves lie.   Yes, the curves are similar.

Investigation 10 $xy=-k$ is the reflection of $xy=k$ in the y-axis (or the x-axis), or a rotation of $\pm 90°$ about $(0,0)$.   All the curves are **odd**.

Exercise 15.3

1 (a) $(1,1)$ **(b)** $(\sqrt{2},\sqrt{2})$ **(c)** $(\sqrt{3},\sqrt{3})$;   Yes;   Enlargement scale factor $\sqrt{3}$ from the origin.

2 Region under the curve through $(1,1)$, $(2, \frac{1}{2})$.
Region under $xy=1$ between $x=3$ and $x=6$.

Two-way stretch, scale factor 3 in x-direction and $\dfrac{1}{3}$ in y-direction, preserving area.

Proof of $\displaystyle\int_1^2 \frac{1}{x}\,dx = \int_3^6 \frac{1}{x}\,dx = \ln 2$

3 (a) $3x+y=6$ **(b)** $3y=x+8$. Tangent does not meet curve again. Normal meets curve again at $(-9,-\frac{1}{3})$.

4 $ty=t^3x+c-ct^4$; $Q\left(\dfrac{c}{t^3}-ct^3\right)$; $PQ=\dfrac{c}{t^3}(1+t^4)^{3/2}$; $t=1$: $PQ(\text{min})=2\sqrt{2}c$

5 $ORS=2c^2$ (constant)

6 $sty+x=c(s+t)$; $s=t$ gives $t^2y+x=2ct$, the equation of the tangent at t.

7 (a) $y=\dfrac{x}{x-1}$ (or part of $xy=1$ centred on $(1,1)$).

(b) Positive part of $xy=1$ centred on $(-1,-1)$.   **(c)** Mid-points give $y=-x$.

(d) Mid-points give $y=-2x$.   **(e)** All points of 2nd and 4th quadrants.

8 $2x^2y^2=c(x^3-y^3)$ **9** $xy=\dfrac{4c^2}{9}$;   rectangular hyperbola.

10 (a) No.   Equation difficult especially for envelopes as it is difficult to pinpoint points on the curve.

(b) Circle arc in positive quadrant: $x^2+y^2=25$

(c) Ellipse quadrant: $\dfrac{x^2}{7.5^2}+\dfrac{y^2}{2.5^2}=1$

Let the ladder slip along negative parts of the axes.

Investigation 13   **(a)** Straight line: $y = x - 1$.   **(b)** Spiral: $r = \dfrac{\theta°}{90°} + 1$.

Investigation 14   Spiral.

Investigation 15   Circle; only one circle; two circles superimposed.

Investigation 16   Straight line, $x = 1$.

Investigation 17   **(a)** $(x^2 + y^2)^3 = (x^2 - y^2)^2$ **(b)** $(x^2 + y^2)^2 = x^3 - 3xy^2$

                   (c) $(x^2 + y^2)^{5/2} = x^4 + y^4 - 6x^2y^2$

Investigation 18   **(a)** parabola   **(b)** hyperbola   **(c)** ellipse.

Exercise 15.4

1 (a) Circle centre $(0,0)$, radius 2.

   (b) Half-line from origin making angle of 3 radians $(172°)$ with the positive x-axis.

2 (a) Circle reflected in $y = x$.

   (b) A rotation of $90°$; no difference if negative r values are allowed.

   (c) One circle and two circles.

   (d) Two circles and two petals.

   (e) Two full petals and one narrow petal.

   (f) Two petals NE and SW, (four petals if negative r), and three petals.

3 (a) $\tan \theta = 4$ (full straight line)   **(b)** $r = \dfrac{4 \cos \theta}{\sin^2 \theta}$ (parabola)

   (c) $r = 4$ (circle, centre $(0, 0)$, radius 4)   **(d)** $r^2 = 9 \sec 2\theta$ (rectangular hyperbola)

   (e) $r = -4 \sin \theta$ (circle)   **(f)** $r^2 = 4 \operatorname{cosec} 2\theta$ (rectangular hyperbola)

   (g) $r = \dfrac{\sin \theta + 2 \cos \theta}{\cos^2 \theta}$ (parabola)   **(h)** $r = \dfrac{3}{1 + \sin \theta}$ (parabola)   **(i)** $r = \dfrac{\sqrt{(1 - \tan^2 \theta)}}{\cos \theta}$ (petal)

4 (a) $x^2 + y^2 = 4x$ (circle)   **(b)** $x = 2$ (straight line)   **(c)** $y^2 = 9 - 6x$ (parabola)

   (d) $x^2 + y^2 = 9$ (circle)   **(e)** $xy = 9$ (rectangular hyperbola)

   (f) $(x^2 + y^2)^2 = 4(x^2 - y^2)$ (leaf)   **(g)** $(x - 4)^2 - \dfrac{y^2}{3} = 4$ (hyperbola)

Chapter 16

Exercise 16.1

1 $S_4 = 12.335$; $S_8 = 12.485$; $T_4 = 11.983$; $T_8 = 12.359$; S_8 is the best $\Rightarrow$ $\pi \simeq 3.121$.

2 (a) 0.6956   **(b)** 0.6938   **(c)** 0.69307   **(d)** $\ln 2 = 0.69315$

3 Answers should be the same, since $\Big[\ln x\Big]_2^4 = \ln 2 = 0.69315$.

4 (a) 0.994282   **(b)** 1.00003   **(c)** 1 exactly

5 (a) 1.1703   **(b)** 1.187   **(c)** Not integrable algebraically

6 **(a)** $\dfrac{\pi}{4} = 0.7853$ **(b)** $\dfrac{\pi}{4} = 0.7853$ **(c)** $\dfrac{\pi}{4}$ **7** **(a)** 0.7828 **(b)** 0.7854 **(c)** $\dfrac{\pi}{4} = 0.7853$

8 **(a)** 0.663 03 **(b)** 0.655 89 **(c)** $\sin^{-1}(0.6) = 0.6435$

9 **(a)** 1.2532 **(b)** 1.2532 **(c)** 1.2533

Exercise 16.2

1 $x = 3$ or $x = -1.5 + 2.6i$ or $x = -1.5 - 2.6i$ **2** $x = 2i$ or $-2i$ **3** $x = 1$ (repeated) or -2

4 1, 2.732 or -0.732 **5** 0 or 1 or 2   **6** 1 (three times)

7 2; $\dfrac{1}{2} + \dfrac{\sqrt{3}}{2}i$; $\dfrac{1}{2} - \dfrac{\sqrt{3}}{2}i$ **8** $+1$; -1; $+2$; -2 **9** 1; 1.151; and 2 complex roots

10 0; 1; 0.811; and 2 complex roots   **11** $+1$; -1; $+i$; $-i$ **12** 1; $-\dfrac{1}{2} + \dfrac{\sqrt{3}}{2}i$; $-\dfrac{1}{2} - \dfrac{\sqrt{3}}{2}i$

13 1; $\cos 72° + i \sin 72°$; $\cos 144° + i \sin 144°$; $\cos 216° + i \sin 216°$; $\cos 288° + i \sin 288°$

Investigation 5   **(a)**   At the intersection of $y = x$ and $y = f(x)$.

   (b)   Diverges   **(c)**   Goes round in a square.

Investigation 6   You go directly to $x = -2$.

Exercise 16.3

1 **(a)** $0 < x < 1$ **(b)** 0.671 **(c)** 0.682 **(d)** Diverges to 0 and 1; Starting from 0 or 1, oscillates.

2 **(a)** 0.73 **(b)** 0.7371 **(c)** 0.739 21 (starting at 0.7 after 2 applications)

 (d) 0.739 (slow convergence)

3 **(a)** $1.1 < x < 1.2$ **(b)** 1.146 **(c)** 1.166 **(d)** Diverges **4** **(a)** 0

5 **(a)** $0.8 < x < 0.9$ **(b)** 0.8236 **(c)** 0.8242 **(d)** 0.8242 (converges quickly)

6 **(a)** $1.7 < x < 1.8$ **(b)** 1.7645 **(c)** 1.7632 **(d)** Diverges

7 **(a)** $0.5 < x < 0.6$ **(b)** 0.568 **(c)** 0.5671 **(d)** 0.565 (slow convergence)

8 **(a)** $-2 < x < -1$ **(b)** -1.556 **(c)** -1.618 **(d)** -1.618 (converges very slowly)
   $0.6 < x < 0.7$ 0.62   0.618   0.618 (converges very slowly)

9 -0.618; 1; 1.618; Best method is by factorizing (as with question **8**).

10 **(a)** $0.6 < x < 0.7$ **(b)** 0.6204 (1 application) **(c)** 0.618 87 **(d)** 0.617 (slow)

11 **(a)** $0.6 < x < 0.7$ **(b)** 0.6412 (1 application) **(c)** 0.6414 **(d)** 0.6412 (quick)

12 **(a)** $0.7 < x < 0.8$ **(b)** 0.765 **(c)** 0.770 **(d)** 0.7043 (slow)

Note

Method **(a)** is necessary to get a starting point for any of the other methods.

Method **(b)** is cumbersome and the algorithm (automatic method) changes, depending on the partial results.

Method **(c)**, Newton-Raphson, uses an automatic formula and usually converges (gets to the result) the quickest.

Method **(d)**, Spiral, sometimes diverges (so does not work) and often takes a long time to give an accurate result.

The Newton-Raphson method is by far the most reliable for equations which cannot be solved algebraically.

Chapter 17

Investigation 1   **1**   **(a)** $y=x$ **(b)** $y<x$ **(c)** $y>x$ **(d)** $y=x^2$ (parabola)

   2   **(a)** $x+y=3$ **(b)** $x+y<3$ **(c)** $x+y>3$

   3   **(a)** $y=x^2$ **(b)** $y>x^2$ **(c)** $y<x^2$

   4   **(a)** $x^2+y^2=25$ **(b)** $x^2+y^2<25$ **(c)** $x^2+y^2>25$

   5 $-2<x<2$ or $|x|<2$ **6** $3<x+y<6$; $y>0$ and $y<2x$

   7   **(a)** $y=|x|$ **(b)** $y>|x|$ **8**   **(a)** $y=|2x-4|$ **(b)** $y=|x+1|$

   (c) $x>5$ or $x<1$

   (d) $1<x<5$ **(e) (f)** See chapter   **(g)** $x>5$ or $x<1$

   (h) $1<x<5$ The same answers.

Investigation 2   In **(a)** and **(b)**:   in **(ii)**, the negative part of the graphs are reflected in the $x-$axis;   **(iii)** and **(iv)** are the same in both cases and each is smooth, i.e. $\sin^2 x$ is a sine curve; $\sin^2 x = \frac{1}{2}(1-\cos 2x)$.

Investigation 3   **(a)** $|x|+|y|<2$ **(b)** $|y|+4|x|<x^2+4$ for $-2<x<2$

   (c) $x^2+y^2+16>8(|x|+|y|)$ for $-4<x<4$

   (d) $|y|<\cos^2 x$ for $-\dfrac{\pi}{2}<x<\dfrac{\pi}{2}$

Exercise 17.1

1 $x<3$ **2** $x>0$ **3** $x<-15$ **4** $0<x<\frac{1}{2}$ **5** $0<x<\frac{1}{4}$

6 $0<x<1$ **7** $x<-1.5$ **8** $-3<x<3$ **9** $x<-4$ or $x>4$ **10** $x<-7$ or $x>1$

11 **(a) (i)** -7 **(ii)** 7 **(b) (i)** -1 **(ii)** 14

   (c) (i) -8 **(ii)** 8 **(d) (i)** $-5\sqrt{5}$ **(ii)** $5\sqrt{5}$

12 **(a) (i)** -5 **(ii)** 5 **(b) (i)** 3 **(ii)** 8

   (c) (i) -4 **(ii)** 4 **(d) (i)** -7.07 **(ii)** 7.07

13 **(a)** Inside circle radius 3, centre $(0,0)$.

   (b) Positive part of $y=x$.

   (c) Point $\left(3,\dfrac{\pi}{4}\right)$ i.e. $x=\dfrac{3}{\sqrt{2}}$, $y=\dfrac{3}{\sqrt{2}}$

   (d) Region between circles radii 2 and 3, centre $(0,0)$.

   (e) Sector (infinite) between positive y-axis and positive $y=x$.

   (f) y-axis.

14 **(a)** Inside parabola $y=x^2$ (above).

   (b) Inside parabola $y=2-x^2$ (underneath).

   (c) Region between parabolae $y=x^2$ and $y=2-x^2$.   Area $=2\frac{2}{3}$.

15 $\dfrac{1}{12}$ **16** **(a)** 8 **(b)** $\dfrac{32}{3}$ **(c)** $64-16\pi=13.73$ **(d)** π **17** 0.222

Exercise 17.2

1 $x<3$ or $x>4$ **2** $x<1$ or $x>6$ **3** $x<1$ or $x>12$ **4** $x\geqslant0$ **5** $x\geqslant3$

6 $x\geqslant1$ or $\dfrac{3-\sqrt{5}}{2}\leqslant x\leqslant\dfrac{3+\sqrt{5}}{2}$ **7** $x<2$ or $x>3$ **8** $-3\leqslant x\leqslant-2$

9 $x<0$ or $2-\sqrt{2}<x<2+\sqrt{2}$ **10** $-2\leqslant x\leqslant-1$ **11** $-1-\sqrt{7}\leqslant x\leqslant0$ or $\sqrt{7}-1\leqslant x\leqslant2$

12 $x<-3$ or $-1<x<1$ **13** $-1<x<3$ **14** $x\geqslant-\dfrac{1}{4}$ **15** $x<-2$ or $x>0$

16 $-2<x<-1$ or $1<x<2$ **17** $x<0$ or $x>2$ **18** $-\sqrt{3}<x<0$ or $x>\sqrt{3}$

19 All values of k.   **20** $x>4$

21 **(a)** $r>2$ **(b)** $x>2$ **and** $y>\dfrac{2x}{x-2}$ **(c)** Side $x>4\sqrt{3}$

(d) Diagonals $2x$ (longer) and $2y$ (shorter): $y>2$ and $x^2>\dfrac{4y^2}{y^2-4}$

22 **(a)** $x>6$ **(b)** $x>1$ **(c)** $x>\frac{1}{6}$

23 **(a)** $m\leqslant0$ or $m\geqslant4$ **(b)** $0\leqslant m\leqslant4$ **24** Use $(p-q)^2>0$

25 **Prove** $a^3+b^3+c^3-3abc=(a+b+c)(a^2+b^2+c^2-ab-ac-bc)>0$
Hence, $a+b+c>3(\sqrt[3]{a})\,(\sqrt[3]{b})\,(\sqrt[3]{c})$

26 **(a)** $x>0$ **(b)** $x<1$ or $x>2$ **(c)** $x<0.82$ or $x>1$

Exercise 17.3

1 $x<-4$ or $x>4$ **2** $x<-5$ or $x>3$ **3** $x<-3$ or $x>1$

4 $x=2$ or $x=-2$ **5** $x=-0.5$ or -5.5 **6** $x=2$ or $x=-3$

7 $-3<x<2$ **8** $x<-2$ or $x>2$ **9** $x\leqslant\dfrac{5}{3}$ or $x\geqslant3$

10 $3\leqslant x\leqslant7$ **11** $x\leqslant\dfrac{5}{3}$ or $x\geqslant3$ **12** $x=+1$ or -1

13 No solution   **14** $x=y=1$ or $x=-1,\ y=1$ **15** $x=0.5,\ y=1.5$ or $x=-1.5,\ y=0.5$

16 $x=\frac{1}{2}n\pi$ **17** $x=45°\pm90n°$ **18** $x=45°\pm90n°$

19 $x<-\frac{1}{5}$ or $x>7$ **20** $-\frac{5}{6}<x<\frac{5}{2}$ **21** $x\leqslant\frac{5}{12}$

22 $-\frac{9}{5}\leqslant x\leqslant-\frac{7}{3}$ **23** Draw graphs – true for all x.

Past Paper Questions

General Questions

1 $\dfrac{c-b+a}{a}$ **2** **(i)** $\sqrt{2}+(3-\sqrt{2})=3$ **(ii)** $\sqrt{2}\times\sqrt{8}=4$

3 **(a)** $\dfrac{3}{2-x}-\dfrac{x+5}{2(x^2+3)}$ **4** $xy^2=36$ **5** **(i)** $x=64,\ y=16$

Inequalities

1 $-4 < x < 0$ **2** (a) $2 < x < 2.5$ or $x > 3$ (b) $-7 < x < \frac{1}{3}$

3 $\{x : x < 1$ or $x > 3\}$ **4** $x < -\frac{1}{3}$ or $x > 2$

5 $x > -1$ or $x < -\frac{4}{3}$ **6** $k = 2$; $\alpha = 60$; $0 < x < 90$ or $330 < x < 360$

Numerical Methods

1 $\frac{\pi}{3}$; $\frac{2}{3}\left(1 + \frac{1}{\sqrt{3}}\right)$ **2** $378\,667\,\text{J}$ **3** 3.977 **4** $a = 22$; 2.22 **5** $\frac{4}{3}$; 1.25

Co-ordinate Geometry

1 $OA = 7$; $OB = 9$; $\cos AOB = \frac{59}{63}$; Area $\triangle AOB = \sqrt{122}$

2 $x + y + 4 = 0$ **4** $x^2 + 8x + 2y + 13 = 0$

5 3; 4; Greatest, 6, when $\theta = 90°$, $270°$. Least, 2, when $\theta = 0°$, $180°$, $360°$.

6 (a) $(3, 3)$ (c) $(\frac{14}{3}, \frac{1}{2})$

Functions

1 $f = 2 \sin 2x$; $g = \cos 2x$; $f \cdot g = \sin 4x$ (period $\pi/2$)

2 1 and 3; $a = 0$ or -4; $a < -4$ or $a > 0$; $x = 4.1$

3 (a) $f(x) \in \mathbb{R}$, $f(x) \neq 1$ (b) x (c) $\dfrac{x+3}{x-1}$

4 (i) even (ii) periodic (iii) none

5 (a) $f(x) = \sin x$: domain $(-\infty, \infty)$; range $(-1, 1)$.
$g(x) = \sqrt{(1 - 4x^2)}$: domain $(-\frac{1}{2}, \frac{1}{2})$; range $(0, \sin 1)$.

(b) Maximum at $(-\sqrt{2}, 10 + 8\sqrt{2})$; minimum at $(\sqrt{2}, 10 - 8\sqrt{2})$; points of inflexion at $(0, 10)$, $(1, 3)$, $(-1, 17)$.

Differentiation

1 (a) $-\dfrac{(x+1)e^{-x}}{x^2}$ (b) $\dfrac{\cos x}{1 + \sin x}$ **2** $\dfrac{C}{S}$

3 (a) $x^2(1 + 3 \ln x)$ (b) $\dfrac{-(1 + \cos x + x \sin x)}{x^2}$ **4** Min $(\frac{1}{4}, \frac{8}{3})$; Area $= 1.5 - \ln 4 \simeq 0.1138$

5 $\dfrac{x}{\sqrt{1+x^2}}$; **B**; **A** does not tend to 1 as x tends to infinity; **C** has infinite gradient at $x = 0$.

Integration

1 (a) $\frac{61}{192}$ (b) $1 - \dfrac{\pi}{4}$ (c) π (d) $\ln \frac{4}{3}$ **2** $\dfrac{x^5}{5} \ln x - \dfrac{x^5}{25} + C$ **3** $\dfrac{4\pi a^3}{3}$

4 Speed $= 6\sin 2t$

$PQ = 4$

$A = \dfrac{3\pi}{2}$

5 First estimate $I = 0.879$.   Second estimate $I = 0.877$.

Trigonometry

1 3π **2** $120,\ 300$ **3** $5\cos(2t - 53)$

4 **(i)** $\theta = (2n+1)\dfrac{\pi}{3}$ or $(2n+\tfrac{1}{6})\pi$ or $(2n+\tfrac{5}{6})\pi$ **(ii)** $61.0°$; $73.8°$; $225.2°$; $10.8\,\text{m}$

5 $30°$; $150°$; $210°$; $330°$ **6** $\alpha = 6,\ \beta = 4,\ \gamma = 2$

Series

1 $1 + \dfrac{x}{2} + \dfrac{3x^2}{8} + \dfrac{5x^3}{16}$ **2** $\dfrac{2}{27}$ **3** 487 **4** $x + \dfrac{x^3}{3}$

7 **(i)** $\text{Sum} = \dfrac{n(n+3)}{4(n+1)(n+2)}$ **(ii)** $x > -\tfrac{1}{2}$; $\text{Sum to infinity} = \dfrac{(x+1)^2}{x^2}$

Differential Equations

1 $y = 4x^2$ **2** $\dfrac{2}{x+2} - \dfrac{1}{x+1}$; $y = -1 + \dfrac{(x+2)^2}{3(x+1)}$

3 $y^2 + 1 = \dfrac{cx}{x+1}$; $y^2 = \dfrac{-1}{x+1}$; $x = -1,\ y = 0$

4 **(ii)** 3 litres per minute; $\dfrac{dx}{dt} = 3 - \dfrac{3x}{50}$ **(iii)** $x\,(25\ \text{minutes}) = 38.8$; $\lim x = 50$

6 $\theta = 1000 - A\,e^{-kt}$

Permutations and Combinations

1 $350;\quad 260$ **2** 840 **(i)** 360 **(ii)** 20 **(iii)** 372

3 **(i)** 114 and 72 **(ii)** $2.009\,975$ **4** $18\,900$ **5** 110

6 **(a)** 720 **(b)** 720 **(c)** 15 **(d)** 90 **(e)** 15

Complex Numbers

1 **(a)** $5\sqrt{5}$; $-\tfrac{1}{2}$ **(b)** $\dfrac{1}{\sqrt{5}}$; $-\dfrac{1}{2}$ **2** $-\dfrac{5\pi}{6}$ **3** $x = 1,\ y = -\tfrac{1}{2}$ **4** $z = 1 - 2i$ $w = 1 - i$

5 **(ii)** $1 + 3i,\ -3$; 30 **6** $1 - 2j$; $(x^2 - 2x + 5)(x + 2)$

7 **(a)** $a = 1,\ b = 2$ **(b)** $\sqrt{3}$; $\dfrac{2\pi}{3}$ **(c)** $\left(x + \dfrac{1}{3}\right)^2 + \left(y - \dfrac{4}{3}\right)^2 = \dfrac{8}{9}$; $\pm\dfrac{2\sqrt{2}}{3} - \dfrac{1}{3} + \dfrac{4}{3}i$

Probability

1 (a) $\dfrac{11}{36}$ (b) $\dfrac{5}{9}$ (c) $\dfrac{3}{4}$ **2** (i) (a) $\dfrac{2}{9}$ (b) $\dfrac{5}{9}$ (c) $\dfrac{43}{45}$ (ii) $\dfrac{1}{4}$ **3** One

4 (a) 0 (b) $\dfrac{1}{18}$ (c) $\dfrac{73}{1296}$ (d) $\dfrac{73}{648}$ (e) $\dfrac{37}{324}$ (f) $\dfrac{857}{1296}$

5 (i) (a) $\dfrac{1}{2}$ (b) $\dfrac{5}{12}$ (ii) (a) $\dfrac{11}{15}$ (b) $\dfrac{1}{20}$ (c) $\dfrac{13}{20}$ **6** (a) (ii) $\dfrac{3}{4}$; $\dfrac{3}{10}$

Impulse and Impact

1 $\dfrac{mu^2}{3}$ **2** 0.5 **3** $\dfrac{g}{5}$; $1.12\,\text{m s}^{-1}$; $3.36\,\text{N s}$

4 (i) (a) u, $2u$ (b) $3mu$ (ii) $25\,\text{m s}^{-1}$

5 After first collision, $v(A) = \dfrac{u}{5}$; $v(B) = \dfrac{6u}{5}$. After second collision, $v(B) = \dfrac{2u}{5}$; $v(C) = \dfrac{8u}{5}$.

Centre of Mass

1 $3\mathbf{i} - 3\mathbf{j} - \mathbf{k}$; $2\mathbf{i} - \mathbf{j}$ **2** $\bar{x} = \frac{20}{7}$, $\bar{y} = \frac{5}{2}$; $41°$

3 (a) $4.4a$ (b) $3.6a$; $T_A = 0.4W$; $T_D = 0.6W$

4 $\dfrac{5h}{16}$ **5** $\dfrac{4(a^2 + ab + b^2)}{a + b}$ from O

Work, Energy and Power

1 $960\,\text{N}$ **2** $\frac{2}{3}\text{m}$; $\frac{2}{3}\text{J}$ **3** $500\,\text{N}$; $15\,000\,\text{W}$

4 (a) $1.5\,\text{m s}^{-2}$ (b) $1050\,\text{N}$; $P = 4850\,\text{N}$; $72.75\,\text{kW}$; $110\,\text{m}$

5 $25\,\text{m s}^{-1}$; $20.625\,\text{kW}$; $1250\,\text{N}$ **6** $\dfrac{mV}{M}$

Relative Velocity

1 8.4 minutes; $4.24\,\text{km}$ **2** 25 knots from $037°$; $143°$; 12 knots

3 $\mathbf{v}_B = 5\mathbf{i} + 12\mathbf{j}$; $\mathbf{v}_C = -4\mathbf{i}$; $\mathbf{r}_B = 5t\,\mathbf{i} + 12t\,\mathbf{j}$ $\mathbf{r}_C = (80 - 4t)\mathbf{i} + 240\mathbf{j}$; $16\,\text{s}$; $80\,\text{m}$

4 $u_1 = 1$; $u_2 = 4\sqrt{3}$; $\sqrt{3}\mathbf{i} + 3\mathbf{j}$ **5** $\frac{66}{13}\,\text{km}$; $\frac{448}{13}$ minutes

Vectors

1 7 **2** $\mathbf{v} = 2\mathbf{a} - 1.5\mathbf{b}$; $-\sqrt{\dfrac{5}{2}}$ **3** $\mathbf{r} = \mathbf{i} - \mathbf{k} + s(\mathbf{i} + \mathbf{j} + 4\mathbf{k}) + t(\mathbf{i} + \mathbf{j} + \mathbf{k})$

4 (a) $\mathbf{r} = 3a\mathbf{i} - 2a\mathbf{k} + t(-2\mathbf{i} + 2\mathbf{j} - 2\mathbf{k})$ (b) $\dfrac{x - 3a}{-2} = \dfrac{y}{2} = \dfrac{z + 2a}{3}$ (c) $\dfrac{12}{\sqrt{221}}$

5 $(-3, 5, 0)$; $\mathbf{r} = \begin{pmatrix} -3 \\ 5 \\ 0 \end{pmatrix} + s\begin{pmatrix} 1 \\ -1 \\ -1 \end{pmatrix} + t\begin{pmatrix} 2 \\ 1 \\ -1 \end{pmatrix}$

6 $BR/RM = 8$ **7** $2\mathbf{i} + \mathbf{k}$; $\mathbf{r} = a\mathbf{i} + a\mathbf{j} + 2a\mathbf{k} + \lambda(2\mathbf{i} + \mathbf{k})$

Projectiles

2 10 m s^{-1}; $36°\,52'$; 2.8 s; 9.6 m **3** 1.5 s **4** $38°\,43'$; $51°\,17'$

5 $\tan \beta = \tan \alpha - \dfrac{gt}{2u \cos \alpha}$; $\tan \theta = \tan \alpha - \dfrac{gt}{u \cos \alpha}$; 30 m

Circular Motion

2 $\dfrac{1}{2\pi}\sqrt{\dfrac{g}{l}}$ **3** $\dfrac{1}{\pi}\sqrt{\dfrac{g}{2}}$ revs sec$^{-1} \simeq 0.705$ revs sec^{-1} **4** $(\sqrt{3}-1)\,mg$

Motion of a particle

1 $\mathbf{v} = (t \cos t + \sin t)\,\mathbf{i} + (-t \sin t + \cos t)\,\mathbf{j}$; $\mathbf{a} = (-t \sin t + 2 \cos t)\,\mathbf{i} + (-t \cos t - 2 \sin t)\,\mathbf{j}$

2 (c) (i) 2 m s^{-2} towards the left   **(ii)** 20 m **(iii)** 12 m

3 (a) $\dfrac{9g}{25}$; $\dfrac{48mg}{25}$ **(b)** $\dfrac{7g}{25}$; $\dfrac{54mg}{25}$

4 $k \sin \omega t$; $\dfrac{k}{\omega}(1 - \cos \omega t) + u$; $-\dfrac{k}{\omega^2}\sin \omega t + \left(u + \dfrac{k}{\omega}\right)t$

5 (i) (a) $\dfrac{1}{6}$ **(b)** 800 m **(ii)** $\dfrac{u^2(3 + 2u)}{6k}$

6 $1.6 \times 10^5 \text{ km}$ **7** $t_2 = \dfrac{u(1 - e^{-0.5})}{b}$; $\dfrac{u^2(1 - 2e^{-1})}{b}$ **8** $\tan^{-1}\left(\frac{7}{24}\right)$ **9** $t = 1.5 \text{ s}$

10 (i) 20 m **(ii)** $\dfrac{\pi}{4}$

Statics

1 $45°$; $\dfrac{W}{2}$;   halfway;   all the way   **2** $\frac{1}{3}mg\sqrt{7}$ **3** $\mu \leqslant 1$

4 $T = \dfrac{(M + m)g}{\sqrt{7}}$; $x = \dfrac{(8m - M)a}{3m}$

5 (i) $\dfrac{5mg}{6}$ **(ii)** $T(BC) = \dfrac{3W}{2 \sin \phi}$; $T(AB) = \dfrac{5W}{2 \sin \phi}$

Coplanar Forces

1 $2 \text{ N} \parallel$ to BA, 2m from C;   Force 2 N along AB;   Couple of magnitude $4\sqrt{3} \text{ N m}$, sense A to C to B.

2 0.3 N m clockwise   **3 (ii)** 14.32 N **4** $\dfrac{W}{\sqrt{3}}$; $\dfrac{W}{\sqrt{3}}$ through the centre; $2W$

Index